AF389086

Flood Risk Assessments

Flood Risk Assessments:
Applications and Uncertainties

Editors

Andrés Díez-Herrero
Julio Garrote

MDPI • Basel • Beijing • Wuhan • Barcelona • Belgrade • Manchester • Tokyo • Cluj • Tianjin

Editors
Andres Diez-Herrero
Geological Survey of Spain (IGME)
Spain

Julio Garrote Complutense
University of Madrid
Spain

Editorial Office
MDPI
St. Alban-Anlage 66
4052 Basel, Switzerland

This is a reprint of articles from the Special Issue published online in the open access journal *Water* (ISSN 2073-4441) (available at: https://www.mdpi.com/journal/water/special_issues/flood_risk_applications_uncertainties).

For citation purposes, cite each article independently as indicated on the article page online and as indicated below:

LastName, A.A.; LastName, B.B.; LastName, C.C. Article Title. *Journal Name* **Year**, *Article Number*, Page Range.

ISBN 978-3-03936-938-6 (Hbk)
ISBN 978-3-03936-939-3 (PDF)

Cover image courtesy of Andrés Díez-Herrero.
Flooding the Old Segovia Mint (Central Spain) in 2014.

Contents

About the Editors

Andrés Díez-Herrero (Ph.D. Geology, MSc Hydrology) has been a scientific researcher in the Geological Hazards Division of the Geological Survey of Spain since 2005. He is a former lecturer and professor of Environmental Geology and Water Resources in the European University of Madrid, the SEK University of Segovia and the University of Castilla–La Mancha; apart from this, he has been a lecturer in the Overseas Studies Program of Stanford University and different master's programs; since 2009, he has been an associated researcher in IMDEA–Water. His research themes focus on flood hazard, risk analysis and mitigation, using geological and geomorphological methodologies as palaeohydrology.

Julio Garrote (Dr.) completed his geological degree (1999) and doctoral thesis (2013) at the Complutense University of Madrid. At first, his research topic was focused on river and terrain morphometry, evolving to river dynamics and floods. Actually, his main research focused on flood risk analysis and hydrodynamic modelling; he also collaborated with other research related to river dynamics and coastal erosion. He has worked for different Spanish Research Centers, such as Geological Survey of Spain or Hydraulic Research Center, as well as for different universities, such as University Alfonso X El Sabio, European University of Madrid, or Complutense University of Madrid.

Editorial

Flood Risk Assessments: Applications and Uncertainties

Andrés Díez-Herrero [1] and Julio Garrote [2,*]

[1] Geological Hazards Division, Geological Survey of Spain (IGME), 28003 Madrid, Spain; andres.diez@igme.es
[2] Department of Geodynamics, Stratigraphy and Paleontology, Complutense University of Madrid, 28040 Madrid, Spain
* Correspondence: juliog@ucm.es; Tel.: +34-91-394-4850

Received: 20 July 2020; Accepted: 22 July 2020; Published: 24 July 2020

Abstract: The present Special Issue brought together recent research findings in Flood Risk Assessments (FRA) and contains contributions on advanced techniques and real cases where FRA have been carried out. The 16 research contributions highlight various processes and related topics where FRA have been applied and the main benefits and improved knowledge derived from them, as well as their replicability in other study sites. The published papers can be classified into three major categories. (a) First, there are those papers focused on improving the methods and results of FRA over different scenarios of both flooding types (river flooding or flash flooding) and flooding areas (urban, non-urban, small mountain communities). (b) Second, there are studies that investigate the application of FRA to diverse topics such as "land urban planning" or "vulnerable infrastructure management (dams, power plants)". (c) Finally, there is a third group of papers which are focused on the assessment of the sources of uncertainties in FRA, with the aim of improving the results and making it more consistent with the real world.

Keywords: flood risk assessment; flood damage uncertainty; flood hazard; flood vulnerability

---

1. Introduction

Flood risk assessment (hereafter abbreviated as FRA) is a scientific/technical subdiscipline or a set of techniques regarding the quantitative analysis and evaluation of flood risk. But subdiscipline includes a notable terminological confusion that is reflected in the titles and contents of the studies, and it conditions the uncertainties and their application to various fields of management. Thus, the first purpose of this editorial introduction should be to clarify terms such as risk, risk analysis and assessment, uncertainty, and the different types of floods that can be evaluated and analyzed.

There are several definitions of 'Risk', from the academic/technical to the popular or common concepts and usages. One of the most accepted technical definitions of 'Risk' is that proposed by the United Nations International Strategy for Disaster Reduction [1]: the combination of the probability of an event and its negative consequences. This definition closely follows the definition of the International Organization for Standardization/International Electrotechnical Commission (ISO/IEC) Guide 73 [2]: "effect of uncertainty on objectives"; this standard definition was subsequently reviewed and confirmed in 2016, and therefore remains current. Many other interesting qualitative definitions have been made by the Society for Risk Analysis (SRA; [3]): (a) Risk is the possibility of an unfortunate occurrence; (b) Risk is the potential for realization of unwanted, negative consequences of an event; (c) Risk is exposure to a proposition (e.g., the occurrence of a loss) which is uncertain; (d) Risk is the consequences of the activity and associated uncertainties; (e) Risk is uncertainty about the severity of the consequences of an activity with respect to something that humans value; (f) Risk is the occurrence of some specified consequence of the activity and associated uncertainties.

Once again following the SRA glossary [3], 'Risk analysis' is the systematic process of comprehending the nature of risk and expressing risk with the available knowledge; it is often understood in a broader way. Other standards [2] define risk analysis as the process of comprehending the nature of risk and determining the level of same.

On the other hand, according to the Cambridge Advanced Learner's Dictionary, 'assessment' is "the process of considering all the information about a situation or a person and making a judgement". Meanwhile, 'analysis' is "the process of studying or examining something in an organized way to learn more about it, or a particular study of something". Following UN terminology, [1] 'Risk assessment' is a methodology of determining the nature and extent of risk by analyzing potential hazards and evaluating existing conditions of vulnerability that together could potentially harm exposed people, property, services, livelihoods and the environment on which they depend.

For the SRA, 'risk analysis' is defined to include risk assessment, characterization, communication, management, and risk-related policy, in the context of risks of concern to individuals, to public and private sector organizations, and to society at a local, regional, national or global level [3]. Otherwise, according to the ISO [2], risk assessment is the overall process of risk identification, analysis and evaluation. So the main difference between 'assessment' and 'analysis' is that the former implies evaluation and quantification, and not only a qualitative decomposition of a complex element into its components or factors. In summary, risk assessment is a part of risk analysis, comprising the quantitative evaluation of a risk.

Most of the SRA definitions of 'Risk' include the term 'uncertainty' (see aforementioned definitions). This is because most of the processes of risk assessment include uncertainties in computing the components (hazard, exposure, vulnerability), factors (frequency and magnitude) and parameters (e.g., water depth, velocity, energy, sediment load). So uncertainty is an essential item in both risk assessment and analysis.

Finally, the term 'flood' includes several different natural and artificial processes and phenomena, from the inundation of emerged areas by water to coastal flooding; and comprises both large slow river floods and flash floods. There are several classifications of flood types, using different criteria [4]: (i) flood origin (natural vs. anthropic; induced vs. aggravated); (ii) floodable area (inland vs. coast); (iii) floodable media (river vs. sea/lake); (iv) flood dynamics (deluge vs. flash flood), etc. So the correct term for applying risk assessment to floods should be enunciated in the plural: risk assessment of floods.

As a consequence of this terminological complexity, the scientific/technical literature on flood risk assessment and/or analysis is diverse and sometimes confuses its objectives and scope. However, within this multiplicity of approaches, three large sets of typologies of FRA studies can be found: general FRA, including simple flood hazard analysis and integrated analysis; the different fields of application of FRA, from land use and urban planning to emergency management; and the estimation of uncertainties in FRA, their propagation along the analysis process, and reduction mechanisms. These sets will form the organization of the papers and the editorial sections in this special issue.

2. General Studies of Flood Risk Assessment and Flood Hazard Analysis

Although studies of FRA have existed for decades, it can be considered an emerging discipline, because scientific/technical production has increased by near two orders of magnitude in the last quarter century, from 13 to more than 850 publications/year (Figure 1).

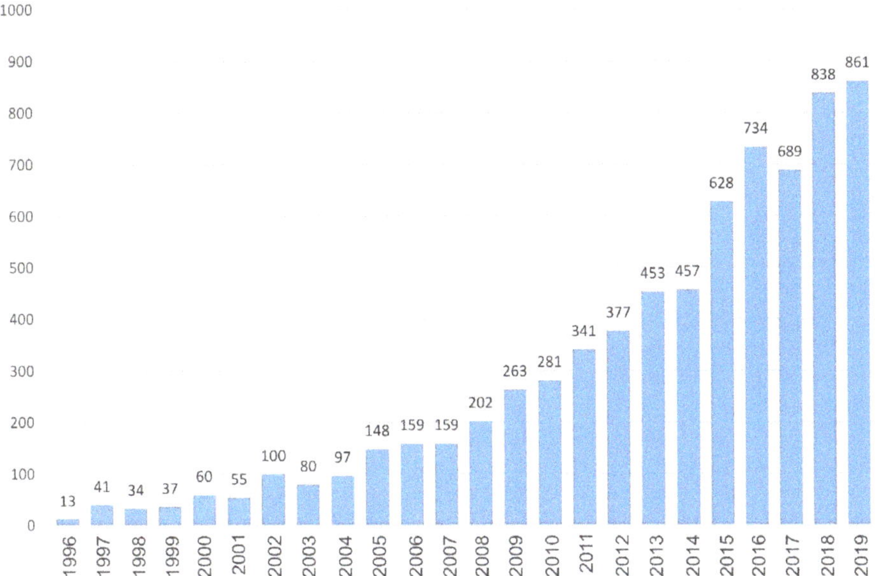

Figure 1. Bar chart of the time evolution of scientific-technical publications (records) on Flood Risk Assessments (FRA) in the Web of Science (WoS) database (search: "flood" AND "risk" AND "assessment"), from 1996 to 2019 [5].

General FRA includes both classical flood hazard analysis (FHA) and integrated true flood risk analysis (FRAn). The former, in turn, includes both flood frequency analysis (FFA: exceedance probability, return period) and magnitude estimates (e.g., discharge, water depth, flow velocity and power, transported solid load, etc.). The second also needs to analyze the exposed elements and their vulnerability (social, economic, environmental, etc.), and the integration of different factors, variables and parameters [6]. But both FHA and FRAn need to use quantitative tools, such as statistics, modeling, or empirical equations, to be true FRAs, because simple qualitative approaches cannot be considered assessments in the strict sense.

This is how the first five papers of this special issue address the topic very differently, but share the idea of the quantification of the hazard or flood risk or its factors (hazard, vulnerability, etc.).

The first addresses one of the fundamental aspects of flood hazard: flood susceptibility or 'floodability', the spatial dimension of flood hazard (flood-prone areas). The authors (Tao Fang et al., all of them from Wuhan University, China) use an advanced statistical approach to flood susceptibility, using a Poisson regression based on eigenvector spatial filtering (ESFPS) for evaluating flood risk (actually, flood susceptibility) in the middle reaches of the Yangtze River in China [7]. They employ a regression analysis to model the relationship between the frequency of alarming flood events observed by hydrological stations and hazard-causing factors from 2005 to 2012. To that end, they have selected and integrated eight factors into a Geographical Information System (GIS) environment for the identification of flood-prone basins, including: elevation (ELE); slope (SLO), elevation standard deviation (ESD); river density (DEN); distance to mainstream (DIST); Normalized Difference Vegetation Index (NDVI); mean annual rainfall (RAIN); mean annual maximum of three-day accumulated precipitation (ACC); and frequency of extreme rainfall (EXE). The results lead us to conclude that elevation, frequency of extreme rainfall, distance to rivers, NDVI and mean annual accumulation of three-day accumulated precipitation are important factors resulting in high flood risk; and ACC and NDVI are the two main factors. This means that the proposed ESFPS model can be used to map flood risks throughout basins and new recent data can function as an early warning system and help identify where hydrological stations should be built and strengthened for monitoring the water level.

An important issue in FRA, especially for certain types of floods and multi-hazard approaches, is the inventory of hazardous elements which can potentially damage vulnerable infrastructures. For example, Mei Liu et al. [8] have carried out a glacial lake inventory for the lake outburst flood (GLOF)/debris flow hazard assessment after the Gorkha earthquake in the Bhote Koshi Basin (BKB, China). This glacial lake outburst hazard assessment included two steps, glacial lake outburst potential assessment and flow magnitude assessment. The glacial lake inventory was established, and six unreported GLOF events were identified, with geomorphic outburst evidence from GaoFen-1 satellite images and Google Earth. Four GLOF hazard classes were derived according to glacial lake outburst potential and a flow magnitude assessment matrix, in which 11 glacial lakes were identified to have very high hazard and 24 to have high hazard. They found that the GLOF hazard in BKB increased after the earthquake due to landslide deposits, which increased by 216.03×106 m^3, which provides abundant deposits for outburst floods to evolve into debris flows. They suggest that, in regional GLOF hazard assessment, small glacial lakes should not be overlooked for landslide deposit entrainment along a flood route that would increase peak discharge, especially in earthquake-affected areas where large numbers of landslides were triggered.

FRA also serves to check how land use changes can modify flood hazard in terms of frequency and magnitude of extreme events and their associated damages. So Furdada et al. [9] have studied the site of a ski resort at the headwaters that led to the surpassing of a geomorphological threshold, with important consequences during flood events. They considered the interaction of the various controlling factors, including bedrock geology, geomorphological evolution, derived soils and colluvial deposits, rainfall patterns, and the hydrological response of the catchment to flood events. For example, the hydrological results of precipitation-runoff models show a major increase in discharge, reaching up to 26% for T10 and coinciding with the largest increase in the ski slopes' surface in recent years. Consideration of all the interactions leads them to conclude that a geomorphological threshold has been overcome recently (2006–2008), producing a shift in the torrential dynamics of the basin that turned into very dense, highly erosive sediment-laden flows. Present conditions and projected climate change suggest that current dynamics are likely to continue despite the efforts to stabilize the most active torrent stretches. Closely monitored interventions in the headwaters should be carried out, particularly focusing on re-vegetation.

Most FRAs usually forget to analyze one of the most important components of integrated flood risk: vulnerability. When vulnerability is analyzed, only economic vulnerability is usually included [6], and the social or environmental modalities of vulnerability are forgotten. Accordingly, studies such as the one carried out by Ferrer et al. [10], focusing on social flood vulnerability assessment in the municipality of Ponferrada (Spain), are interesting and necessary. They propose a methodology for the analysis of social vulnerability to floods based on the integration and weighting of a range of exposure and resistance (coping capacity) indicators. This needs selection and characteristics of each proposed indicator and an integration procedure based on a large-scale analytic hierarchy process (AHP). Detailed mapping of the social vulnerability index is generated at the urban land-parcel scale. It concluded that almost 77% of the affected population of Ponferrada (34,941 inhabitants) will not be able to reach a safe place, because a time to evacuate of at least 15 min is required (timespan for the water wave to reach the town). The majority of data used for the calculation of the indicators comes from open public data sources, which allows the replicability of the method in any area where the same data are available.

Finally, one of the most widely used approaches of FRAs is the quantitative estimation of future flood damages both from a social and an economic point of view. Regarding this integrated potential damage assessment, this special issue includes the paper written by Martínez-Gomariz et al., entitled "Socio-Economic Potential Impacts Due to Urban Pluvial Floods in Badalona (Spain) in a Context of Climate Change" [11]. The assessment has a twofold target: people's safety, based on both pedestrian and vehicle stability, and impacts on the economic sector in terms of direct damages to properties and vehicles; and indirect damages due to business interruption; all these are related to different return periods (i.e., 2, 10, 100, and 500 years). The main novelties of this paper are the integration of a detailed 1D/2D urban drainage model with multiple risk criteria; and the assessment of flood damages for projected future rainfall conditions (climate change scenarios, such as RCP 8.5) which enabled a comparison to be made with the current conditions, thereby estimating their potential increment. The total Expected Annual Damage (EAD) for Badalona city, considering an estimate of both direct and indirect damages, includes the breakdown of estimated damages for current (1.48 M€) and future rainfall conditions (1.93 M€), a predicted increase due to climate change of 30% in terms of EAD.

3. A Wide Range of Applications for Flood Risk Assessment

Apart from the studies dealing with general Flood Risk Assessment (FRA), a second group of manuscripts are focused on specific applications of FRA (Figure 2), from a global to a regional and local scale, and from levee structure resistance to cultural heritage risk analysis or local flood monitoring from remote sensing data. FRA application studies have not been scarce since the first years of the last century, so the volume of manuscripts range from approximately 200 in the first decade of the twenty-first century to more than a thousand manuscripts per year in the past five years. It is clear from Figure 2 that the most popular FRA application is related to planning tasks, and how resilience analyses have significantly increased in number in recent years.

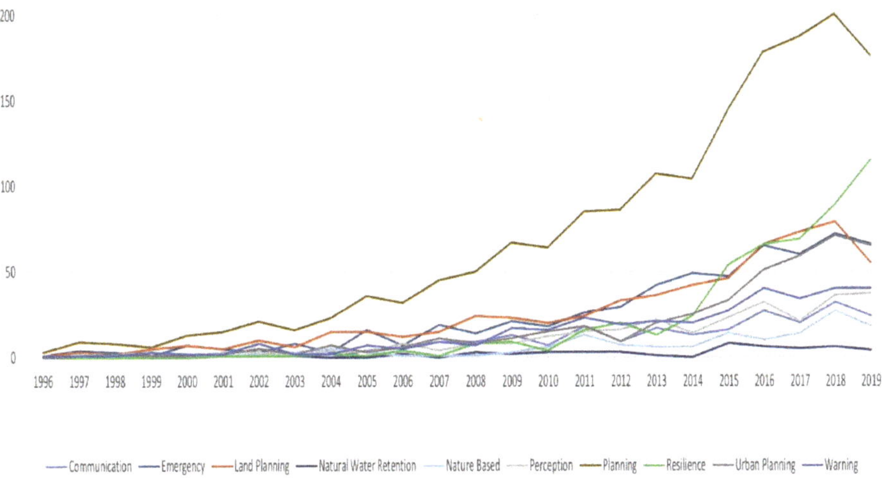

Figure 2. Temporal evolution of scientific/technical publications (records) on the main fields of application of FRA in the WOS database (search: "flood" AND "risk" AND "assessment" BY (fields of application)), from 1996 to 2019 [5].

The most globally focused research is that of Özer et al. [12], who have collected information and developed a global database on levees as fluvial defense, the International Levee Performance Database (ILPD), which aims to become a valuable knowledge platform in the field of levee safety. The ILPD comprises information on levee characteristics, failure mechanisms, geotechnical investigations and breach processes for more than 1500 cases. From all these data, the authors have extracted key conclusions, as external erosion is identified as the most frequent process for levee failures. The manuscript shows that initial failure mechanisms have an influence on breach characteristics and that failures due to instability and internal erosion are less frequent but lead to larger breaches.

Moreover, a relationship between the return period and the expected breach density during a flood event is identified, so flood events with a return period higher than 200 years produce a significant increment in breach density. Furthermore, external erosion is more likely to occur for high return periods. All these data and conclusions, in combination with hydraulic/geotechnical models, could eventually be used to complement risk assessments and to design more robust and resilient flood defenses, with less probability of catastrophic breaching.

Following with the effects of flooding on structures, Pérez-Morales et al. [13] focused their assessment on transport network vulnerability to floods. From their analysis, an optimum location of Emergency Management Centers could be derived, combined with the probable worsening of these relationships due to the effects of climate change or global warming. Given such circumstances, analysis of the vulnerability of transportation networks is a technique whose results highlight deficiencies and serve as support for future decisions concerning the transformation of the network or the installation of new emergency centers. Two main results could be highlighted from the assessment: (a) a map of present network vulnerability informs us of critical network sections that should be reinforced or protected to facilitate evacuation of the population and the movement of emergency teams in case of catastrophe; (b) when these adaptation measures are not to be carried out, solutions like finding the optimum location for a new Emergency Management Center based on reduced travel time to reach the rescue points could be the best to fill that gap. The results obtained could be used in urban planning tasks to improve the resilience of urban areas against an increase in flood episodes caused by climate change.

Garrote et al. [14] focused their application of FRA on another type of structure, which cannot usually be classified into tangible or intangible elements. These elements, such as cultural heritage (archaeological sites, historical buildings, artworks, etc.) are often located somewhere between both categories. This means they can be assigned an economic value (market value), while at the same time they represent the record of the history of human civilization, whose value is intangible. The application of FRA to cultural heritage is based on the implementation of three proposed matrices (flood hazard, flood vulnerability, and potential flood risk) in a GIS environment, which allows the estimation of the flood risk level presented by the cultural heritage elements and their classification. The flood hazard matrix uses variables such as flow depth, return period (years) and flood type (river flood or flash flood). The flood vulnerability matrix is constructed on three groups of variables (materials vulnerability, structural vulnerability, and specific content vulnerability). The qualitative characterization of the potential flood risk for the heritage site and its spatial distribution, both direct results of the methodological framework proposed here, can and should serve as a key tool for cultural heritage managers at a city, region, or state administrative level.

From the various FRA applications to different types of structures, Carreño Conde and De Mata Muñoz [15] show the applicability of Sentinel-1 Synthetic Aperture Radar (SAR) images for flooding area monitoring on a flood event scale. The authors show from their analysis that the best results from the final flood mapping for the study area were obtained using VH (Vertical-Horizontal) polarization configuration and Lee filtering 7×7 window sizes. Using pre- and post-event images with the Red-Green-Blue (RGB) method they obtained highlights of the non-permanent water bodies. The authors calibrate results in two ways: (a) comparing the flood maps obtained with the flood areas digitalized from vertical aerial photographs planned by the Hydrological Planning of the Ebro Hydrographic Confederation; and (b) comparing the results with maps of different return periods for flooded areas provided by the National Flood Zone Mapping System in Spain. The consensus between these different data sources confirmed and validated the use of Sentinel-1 SAR images to obtain flood maps with the required accuracy, and point out that they can be used as a new source of input, calibration, and validation for hydrological models to improve the accuracy of flood risk maps.

Related to the previous manuscript, Menéndez Pidal et al. [16] study and prove the effectiveness of real time integrated data provision from the Automatic Hydrological Information System (SAIH) of the Tagus River (including snow accumulation/melting processes and the forecast of inflows to the reservoir to help in the establishment of safeguards and preventive water flow releases). As an application study site of their FRA, the methodology described is shown in a real case at Rosarito and El Burguillo Reservoirs. In these two reservoirs, located in the upper mountainous part of the basin, the authors show the improvement in inflow and outflow management derived from their methodological approach which translates into an improvement in risk management against flood hazards downstream of the location of both reservoirs.

Using another specific data source, i.e., tree-ring evidence, Franco-Ramos et al. [17] carried out an assessment of lahar activity at the Pico de Orizaba Volcano (Mexico) using the debris-flow module of the rapid mass movement simulation tool RAMMS on a highly resolved digital terrain model obtained with an unmanned aerial vehicle. They use the scars found in 19 *Pinus hartwegii* trees (that served as paleo-stage indicators, PSI) to calibrate a model reconstructing the 2012 event magnitude. Using this combined assessment and calibration of RAMMS, the authors obtain a peak discharge of 78 $\text{m}^3 \text{ s}^{-1}$ for the 2012 lahar event which was likely to have been triggered by torrential rainfall during hurricane "Ernesto". Results also show that the deviation between the modeled lahar stage (depth) and the height of PSI in trees was up to ±0.43 m. These deviations are not evenly distributed spatially, as the authors show, and they are in some way related to the formation energy of landforms. The largest deviations are located on cut banks and/or in straight and exposed positions, and moderate and smaller deviations were typically found in intermediate and low-energy landforms, such as alluvial terraces. The research is another example [18,19] of the usefulness of tree-ring evidence as a data source for hydrodynamic modeling calibration and improvement of FRA.

Finally, the research of Špitalar et al. [20] entitled "Analysis of Flood Fatalities–Slovenian Illustration" constitutes the sixth and last manuscript included in this second group of research papers focused on applications of FRA. In Slovenia, they assessed the influence of different factors external to the hazard of flooding on the gravity and extent of the impact in a historical analysis (1926–2014), which includes up to 10 flooding events (from a complete record of 21 severe floods) that induced 74 casualties. Their results highlight the increasing trend in recent years of victims in cars in rural areas. From these results, an adult or elderly male is the most vulnerable to flooding effects. Focusing on demographic aspects (age and gender) of fatalities and an analysis of the circumstances of loss of life, different groups were established based on the type of flood fatality. From all these groups, victims related to the collapse of both buildings and structural mitigation measures are the most common. Evaluations of results have also shown an increasing need for improvements in communication strategies between fire fighters and the Administration of the Republic of Slovenia for civil protection and disaster relief. The assessment gives an overview of demographic, temporal and circumstantial aspects of victims, at the same time as it improves knowledge about mitigation and preventative measures that can be taken in the future and about the remaining challenges regarding how to reduce social vulnerability to flooding.

4. Uncertainties in Flood Risk Assessment

The third group of research is focused on uncertainty analysis concerning FRA (Figure 3). This last group is the one composed of a smaller number of three manuscripts. Uncertainty analyses of FRA have become more recurrent in the past decade, when the number of manuscripts has increased from a couple of dozen to more than a hundred in the last five years. This increase points towards a major idea: it is not only now necessary to perform FRA studies, but it has also become necessary to analyze their quality and the degree of trust that we can place in them, all this from the point of view of the applicability of these studies to the real world. It is clear that the volume of uncertainty analyses is lower than other topics related to FRA, and this is an area of study that still requires further scientific attention.

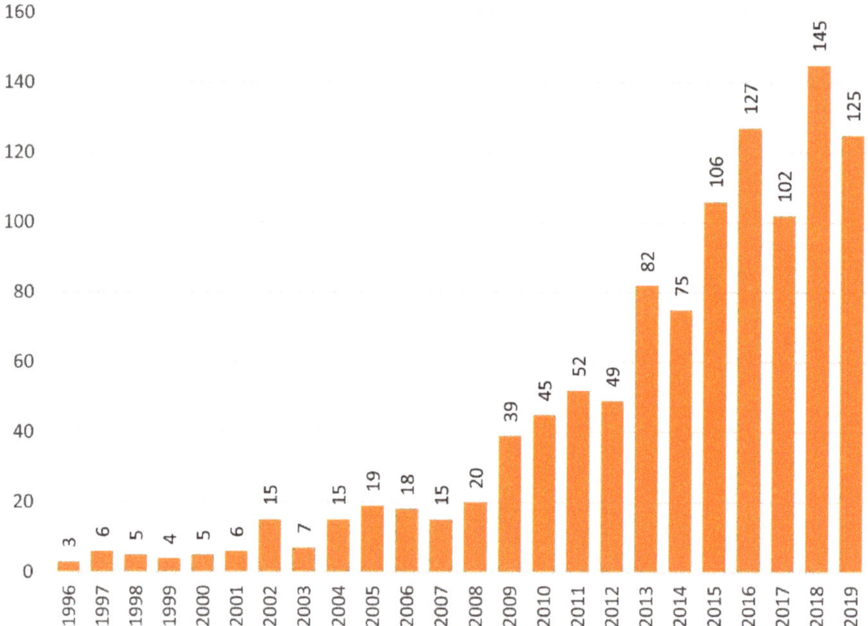

Figure 3. Bar chart of the time evolution of scientific/technical publications (records) on uncertainties in FRA in the WOS database (search: "flood" AND "risk" AND "assessment" AND "uncertainty*"), from 1996 to 2019 [5].

In the first, Garrote et al. [21] make a type of uncertainty assessment about results of meso- or macro-scale FRA (developed by combining Fuzzy DELPHI, MCA, and GIS techniques) from the use of phone calls to Emergency Services (112). In recent years, the use of social network data for flood hazards, risk analyses and calibration processes are increasing, although use of data from social networks also has its drawbacks, such as the biases associated with age and gender, and their spatial distribution. As the authors show, some of these biases can be resolved by calls made to Emergency Services, although other problems are still latent. For example, a bias does exist in the relationship between the size of the population and the number of calls made to Emergency Services. The use of data from calls made to Emergency Services and local regression models such as LOESS (locally estimated scatterplot smoothing) allowed the authors to establish the correct relationship between categorized potential risk levels (from usual FRA) and the inferred potential risk (obtained from data from Emergency Services). Therefore, based on the number of calls made to Emergency Services, we can categorize those municipalities that should be the subject of a more detailed study and those whose classification should be revised in future updates.

The research of Lastrada et al. [22] makes another kind of uncertainty assessment related to future projections of FRA under climate change scenarios (global warming). The main objective of the paper is to present a methodology to estimate climate change effect on flood hazard and flood risk using different climate models regarding a scenario of comparatively high greenhouse gas emissions (RCP 8.5). From daily temperature and precipitation data (from 2007 to 2070), and comparing results in relative terms, flow rate and flood risk variation due to climate change are estimated. In the specific case of the study site, the Meteorological Research Institute (MRI)- Coupled Global Climate Model (CGCM3) model (obtained from Spanish Meteorological Agency website for RCP 8.5) shows that climate change will cause a significant increase of potential affected inhabitants and economic damage due to flood risk. Due to the large uncertainty in climate projections that influences the estimation of future flood risk, the crucial action for authors in the case of adaptation to floods is to carry out a detailed analysis to choose the best climate model and use results in relative terms between short term (first stage) and long term (second stage) of the projection period. This evaluation enables the definition of mitigation actions in terms of cost benefit analysis and prioritization of those that should be included in Flood Risk Management Plans.

The last work, by Garrote and Bernal [23], investigated the uncertainty in FRA regarding the consideration or not of the topographic (Z value) difference between the main floor of houses and the surrounding land. To carry out this analysis, up to a total of 28 magnitude-damage functions (with different characteristics and application scales) were selected on which the effect of over-elevation and under-elevation of the main floor of the houses was simulated (at intervals of 20 cm, between −0.6 and +1 m). Each topographic configuration (over- or under-elevation) causes overestimation or underestimation of flood damage. Possible errors in economic quantification of flood damages are maximized for low flow depth values and became constant when flow depth rose to values higher than 1.5–2 m. Out of the theoretical assessment of the effects on magnitude-damage models, the authors use a real scenario where the consideration of this variable (as opposed to its non-consideration) causes an average variation in the damage estimation of around 30% (22.5% of reduction in EAD when over-elevation of main floor is considered in FRA). So, from these results, the analyzed variable can be considered as (1) another main source of uncertainty in the correct estimation of flood damage, and (2) an essential variable to take into account in a flood damage analysis for the correct estimation of loss.

Finally, a review manuscript by Guest Editors of this Special Issue, Díez-Herrero and Garrote [24] entitled "Flood Risk Analysis and Assessment, Applications and Uncertainties: A Bibliometric Review" completes this volume. The authors, through a bibliometric analysis of the most popular and complete scientific databases (Web of Science (WoS)), show us the range of flood risk analysis and assessment over time and around the world, comprising all the scientific and technical literature included in WoS. This approach may be the only way to understand the temporal and spatial evolution of this sub-discipline which interprets the emerging fields of applications and the main sources of uncertainty, overcoming limitations (in the opinion of the authors) of most other thematic reviews, which show a snapshot or a partial point of view of FRAs, focused on only a small number of selected methods and approaches. For the authors, there are four themes within FRA research that mark future trends: global change, coastal floods, economic evaluation, and psycho-social analysis of floods.

5. Conclusions

In conclusion, this special issue contains 16 studies with important results, covering several aspects of Flood Risk Assessment, from those focused on improving the methods and results of FRA over different scenarios of both flooding types (river flooding or flash flooding) and flooding areas (urban, non-urban, small mountain communities), to studies focused on the analysis of uncertainties in FRA results. While it is true that most studies focus on examples of the main FRA applications, three studies focus on the assessment of different structures (from levees to cultural heritage elements).

Author Contributions: A.D.-H. and J.G. have both conceived the idea and wrote the manuscript. All authors have read and agreed to the published version of the manuscript.

Funding: The guest editorial works have been funded by the research project DRAINAGE, CGL2017-83546-C3-R (MINEICO/AEI/FEDER, UE); and it is specifically part of the outreach work packages and dissemination tasks carried out under the coverage of the sub-project DRAINAGE-3-R (CGL2017-83546-C3-3-R; MINEICO/AEI/FEDER, UE).

Acknowledgments: Guest editors thank all the authors of the published papers, manuscripts reviewers and assistant editors for the great support to the realization of the present special issue.

Conflicts of Interest: The authors declare no conflict of interest.

References

1. UNISDR. *UNISDR Terminology on Disaster Risk Reduction*; United Nations International Strategy for Disaster Reduction (UNISDR), United Nations (UN): Geneva, Switzerland, 2009; p. 35. Available online: https://www.unisdr.org/publications (accessed on 17 January 2020).
2. ISO. *ISO GUIDE 73:2009, Risk Management-Vocabulary*; International Organization for Standardization (ISO): Vernier, Switzerland, 2009; p. 15.
3. Aven, T. *Society for Risk Analysis Glossary*; Committee on Foundations of Risk Analysis, Society for Risk Analysis (SRA): McLean, VA, USA, 2015; p. 15.
4. Díez-Herrero, A.; Lain-Huerta, L.; Llorente-Isidro, M.A. *Handbook on Flood Hazard Mapping Methodologies*, 1st ed.; Series Geological Hazards /Geotechnics, No. 2; Geological Survey of Spain (IGME): Madrid, Spain, 2009; p. 190.
5. Web of Science Database (WOS). All Databases; Clarivate Analytics. Available online: https://apps.webofknowledge.com/ (accessed on 17 March 2020).
6. Merz, B.; Kreibich, H.; Schwarze, R.; Thieken, A. Assessment of economic flood damage: Review article. *Nat. Hazards Earth Syst. Sci.* **2010**, *10*, 1697–1724. [CrossRef]
7. Fang, T.; Chen, Y.; Tan, H.; Cao, J.; Liao, J.; Huang, L. Flood risk evaluation in the middle reaches of the Yangtze River based on eigenvector spatial filtering poisson regression. *Water* **2019**, *11*, 1969. [CrossRef]
8. Liu, M.; Chen, N.; Zhang, Y.; Deng, M. Glacial lake inventory and lake outburst flood/debris flow hazard assessment after the Gorkha earthquake in the Bhote Koshi Basin. *Water* **2020**, *12*, 464. [CrossRef]
9. Furdada, G.; Victoriano, A.; Díez-Herrero, A.; Génova, M.; Guinau, M.; De las Heras, A.; Palau, R.M.; Hürlimann, M.; Khazaradze, G.; Casas, J.M.; et al. Flood consequences of land-use changes at a ski resort: Overcoming a geomorphological threshold (Portainé, Eastern Pyrenees, Iberian Peninsula). *Water* **2020**, *12*, 368. [CrossRef]
10. Tascón-González, L.; Ferrer-Julià, M.; Ruiz, M.; García-Meléndez, E. Social vulnerability assessment for flood risk analysis. *Water* **2020**, *12*, 558. [CrossRef]
11. Martínez-Gomariz, E.; Locatelli, L.; Guerrero, M.; Russo, B.; Martínez, M. Socio-economic potential impacts due to urban pluvial floods in Badalona (Spain) in a context of climate change. *Water* **2019**, *11*, 2658. [CrossRef]
12. Özer, I.; van Damme, M.; Jonkman, S. Towards an International Levee Performance Database (ILPD) and its use for macro-scale analysis of levee breaches and failures. *Water* **2020**, *12*, 119. [CrossRef]
13. Pérez-Morales, A.; Gomariz-Castillo, F.; Pardo-Zaragoza, P. Vulnerability of transport networks to multi-scenario flooding and optimum location of emergency management centers. *Water* **2019**, *11*, 1197. [CrossRef]
14. Garrote, J.; Díez-Herrero, A.; Escudero, C.; García, I. A framework proposal for regional-scale flood-risk assessment of cultural heritage sites and application to the Castile and León Region (Central Spain). *Water* **2020**, *12*, 329. [CrossRef]
15. Carreño Conde, F.; De Mata Muñoz, M. Flood monitoring based on the study of Sentinel-1 SAR images: The Ebro River case study. *Water* **2019**, *11*, 2454. [CrossRef]
16. Menéndez Pidal, I.; Hinojal Martín, J.A.; Mora Alonso-Muñoyerro, J.; Sanz Pérez, E. Real-time data and flood forecasting in Tagus Basin. A case study: Rosarito and El Burguillo reservoirs from 8th to 12th March, 2018. *Water* **2020**, *12*, 1004. [CrossRef]
17. Franco-Ramos, O.; Ballesteros-Cánovas, J.; Figueroa-García, J.; Vázquez-Selem, L.; Stoffel, M.; Caballero, L. Modelling the 2012 lahar in a sector of Jamapa Gorge (Pico de Orizaba Volcano, Mexico) using RAMMS and Tree-Ring evidence. *Water* **2020**, *12*, 333. [CrossRef]
18. Ballesteros Cánovas, J.A.; Bodoque, J.M.; Díez-Herrero, A.; Sanchez-Silva, M.; Stoffel, M. Calibration of floodplain roughness and estimation of palaeoflood discharge based on tree-ring evidence and hydraulic modelling. *J. Hydrol.* **2011**, *403*, 103–115. [CrossRef]
19. Garrote, J.; Díez-Herrero, A.; Génova, M.; Bodoque, J.M.; Perucha, M.A.; Mayer, P.L. Improving flood maps in ungauged fluvial basins with dendrogeomorphological data. An example from the Caldera de Taburiente National Park (Canary Islands, Spain). *Geosciences* **2018**, *8*, 300. [CrossRef]
20. Špitalar, M.; Brilly, M.; Kos, D.; Žiberna, A. Analysis of flood fatalities–Slovenian illustration. *Water* **2020**, *12*, 64. [CrossRef]

21. Garrote, J.; Gutiérrez-Pérez, I.; Díez-Herrero, A. Can the quality of the potential flood risk maps be evaluated? A case study of the social risks of floods in Central Spain. *Water* **2019**, *11*, 1284. [CrossRef]
22. Lastrada, E.; Cobos, G.; Torrijo, F.J. Analysis of climate change effect on flood risk. Case study of reinosa in the Ebro river basin. *Water* **2020**, *12*, 1114. [CrossRef]
23. Garrote, J.; Bernal, N. On the influence of the main floor layout of buildings in economic flood risk Assessment: Results from Central Spain. *Water* **2020**, *12*, 670. [CrossRef]
24. Díez-Herrero, A.; Garrote, J. Flood risk analysis and assessment, applications and uncertainties: A bibliometric review. *Water* **2020**, *12*, 2050. [CrossRef]

Article

Flood Risk Evaluation in the Middle Reaches of the Yangtze River Based on Eigenvector Spatial Filtering Poisson Regression

Tao Fang, Yumin Chen *, Huangyuan Tan, Jiping Cao, Jiaxin Liao and Liheng Huang

School of Resource and Environment Science, Wuhan University, Wuhan 430079, China;
fountaintop@whu.edu.cn (T.F.); tanhuangyuan@whu.edu.cn (H.T.); caojiping@whu.edu.cn (J.C.);
2018282050169@whu.edu.cn (J.L.); 2014301110077@whu.edu.cn (L.H.)
* Correspondence: ymchen@whu.edu.cn; Tel.: +86-27-687-783-86

Received: 20 August 2019; Accepted: 17 September 2019; Published: 21 September 2019

Abstract: A Poisson regression based on eigenvector spatial filtering (ESF) is proposed to evaluate the flood risk in the middle reaches of the Yangtze River in China. Regression analysis is employed to model the relationship between the frequency of flood alarming events observed by hydrological stations and hazard-causing factors from 2005 to 2012. Eight factors, including elevation (ELE), slope (SLO), elevation standard deviation (ESD), river density (DEN), distance to mainstream (DIST), NDVI, annual mean rainfall (RAIN), mean annual maximum of three-day accumulated precipitation (ACC) and frequency of extreme rainfall (EXE) are selected and integrated into a GIS environment for the identification of flood-prone basins. ESF-based Poisson regression (ESFPS) can filter out the spatial autocorrelation. The methodology includes construction of a spatial weight matrix, testing of spatial autocorrelation, decomposition of eigenvectors, stepwise selection of eigenvectors and calculation of regression coefficients. Compared with the pseudo R squared obtained by PS (0.56), ESFPS exhibits better fitness with a value of 0.78, which increases by approximately 39.3%. ESFPS identifies six significant factors including ELE, DEN, EXE, DIST, ACC and NDVI, in which ACC and NDVI are the first two main factors. The method can provide decision support for flood risk relief and hydrologic station planning.

Keywords: spatial autocorrelation; Poisson regression; eigenvector spatial filtering method; flood risk evaluation

---

1. Introduction

Floods are the most common and destructive natural disaster around the world. The dam-breaking and waterlogging events caused by floods have caused great losses of human lives and property throughout history [1]. When the water level measured in large parts of a river becomes too high and exceeds its maximum capacity, flooding could occur. A recent study ranked floods as the greatest threat to 616 cities around the world over earthquakes and storms [2]. There are many types of floods such as river floods, flash floods, urban floods, sewer flooding and coastal flooding in China. Affected by the monsoon climate and geographical conditions, China is seriously impacted by flood disasters, especially flash floods and river floods [3]. Floods in the region has a wide range of impacts and can be sudden and strong, occur for long periods, frequent or seasonal. These obvious characteristics make floods one of the most important factors restricting China's economic and social development. Hence, management and mitigation before and after a flood disaster are necessary and significant. The evaluation of risks in flood-prone areas and the construction of hydrological stations are effective means of prevention and monitoring before floods. Areas susceptible to flood must be detected in advance so that urbanization and industrialization can be prevented in these areas. In addition, some watersheds prone to flooding should be monitored with hydrological stations in the case of dangerous water levels.

Floods in different regions at different times are often caused directly by different factors [4] while climate and human activities are common factors [5]. Against the background of climate change in recent years, extreme rainy weather brings concentrated, intense and consecutive rainfall that quickly transforms into sediment and runoff [6,7]. Many cities were recently reported to encounter the highest recorded rainfall in history because massive urbanization and unwise city planning have replaced many forests and rivers with industrialized areas [8]. Vegetation is able to contain soil and some rainfall [9]. Conversely, large scale urban expansion reduces city permeability, However, the drainage functions developed in cities cannot reach the level required to swiftly discharge runoff in time [10].

Two categories of flood models have been researched and developed which is either physically based or statistically based [11]. The Storm Water Management Model, a good representative of many hydrological models, has been widely employed to simulate the flow and flood processes [12]. Apart from detailed data and complicated calculations [13], these models can only simulate the extent and velocity of flood inundation over small-scale areas through one dimensional or two-dimensional hydraulic models and require much time and hardware to compute [14]. Both of two categories have been conducted in the many previous study based on the Yangtze River [15].

Regarding statistical models that only consider correlation and cause-and-effect relationships, qualitative or semiquantitative methods such as multicriteria analytic hierarchy process (AHP) are frequently used to model the risks of disasters such as floods or landslides [16,17]. A previous study combined the AHP with a Bayesian Network to research the flooding risk in Guangzhou [18]. These models are used widely because they have good calculation performance with simpler statistical theory. Other studies also use machine learning methods such as decision tree, SVM (support vector machine) and artificial neural network (ANN) to predict the spatial distribution of flood risk over large-scale regions with accuracies higher than 80% [19,20]. Expert weighting of the AHP is relatively subjective and the different factors in different regions at different times that result in disasters tend to show distinct relative importance, hence, experiences about the weights of factors in certain areas cannot be transfer directly to others [21]. Machine learning algorithms such as ensembled decision tree model and ANN is more complicated and just like a black box so that its network structure and weight parameters are not interpretable [22]. Moreover, these models ignore the spatial autocorrelation that exists flood disasters. Considering spatial effects in these models will probably further increase their performance.

Multivariate regression usually involves several spatial factors and non-spatial attribute data. Geographic Information System (GIS) is a powerful tool to collect, process, manage and analyze spatially referenced layers in table, vector or raster format [23], and quantization calculation can mostly be easily integrated into it. Moreover, the visualization of GIS can make risk assessment accessible and easy to understand. Risk analysis of natural disasters integrated with GIS has been applied widely in the recent decades [24].

Poisson regression has been commonly applied to estimate disasters or disease risks and perform factor analysis [25,26]. The method is very suitable for modeling the data on the accumulated count of events that take place with low probability, such as diseases rate and death rates, especially when the observations approximately follow the Poisson distribution. Moreover, simple models such as generalized linear regression including Poisson regression are more suitable for small dataset and have more interpretability than complex machine learning algorithms. Aforementioned statistical models consider a weighted score or a probability as the metric of flood risk while the study considers the count of flood alarming events in history as the metric of flood risk because the frequency is also representative of occurrence rate when different hydrological stations have the same number of observations. The new data from hydrological stations currently available, allow a much better spatial pattern analysis. Moreover, some count observations often exhibit spatial autocorrelation which exists in most geospatial processes [27], according to the First Law of Geography [28]. Failure to consider spatial effects of flood alarming events based on the nature of the basin's connectivity in the regression model will lead to model uncertainty, thus, spatial autocorrelation must be included in the regression model [29,30]. Whether spatial autocorrelation exists in the observations of geographical units can be tested using Moran's I.

Eigenvector spatial filtering method is proposed to account for spatial effects [31]. This method selects a subset of eigenvectors from the spatial weight matrix that represent the spatial distribution pattern and then adds them to the ordinary Poisson regression model as independent proxy variables [32]. The linear combination of these eigenvectors filters the spatial autocorrelation out of the observations, which enables the observations in different geographical units to be independent [33,34].

In this paper, we propose the eigenvector spatial filtering Poisson regression (ESFPS) model for the estimation of flood risk, using the frequency of flood alarming events observed by hydrological stations in the middle reaches of the Yangtze River in China. Independent hazard-causing variables include elevation (ELE), slope (SLO), elevation standard deviation (ESD), river density (DEN), distance to mainstream (DIST), normalized difference vegetation index (NDVI), annual mean rainfall, mean annual maximum of three-day accumulated precipitation (ACC) and frequency of extreme rainfall (EXE), because these factors are considered to aggravate or trigger flooding or are closely associated with flood risk in the study area. The model results will be compared with the results of Poisson regression and negative binomial regression, and the best model will be used to predict the flood risk throughout the basin and analyze the spatial distribution pattern of flood risk.

2. Study Area and Data

2.1. Study Area

The focus of this study is the middle reaches of the Yangtze River spanning 108°24′–117°26′ E and 24°33′–33°14′ N and present a typical subtropical monsoon climate. The central drainage basin of the Yangtze River mainly covers three provinces—Hubei, Hunan and Jiangxi—in the central part of China, as shown in Figure 1. The weather is relatively humid with four clear seasons, and rainfall is especially abundant and is concentrated from April to August. As shown in Figure 2, according to the rainfall records from six meteorological stations in the three provinces, most of the monthly mean precipitation in the different provinces exceeds 150 mm during this period. Moreover, there are large amounts of variegated rivers and lakes, of which Dongting Lake in Hunan and Hubei Provinces and Poyang Lake in Jiangxi Province are China's largest and second-largest freshwater lake, respectively. In addition, many hydrologic stations and dams have been constructed along the Yangtze River and its main tributaries including the Han River, the Xiang River, the Yuan River and the Gan River in the study region. As a strategically important economic area, the region plays an extremely significant role in flood monitoring and forecasting. The varied topography, subtropical monsoon climate and a complex water system make this area among the highest flood prone areas especially where several destructive dam-break floods and many waterlogging events have occurred in history. The main flooding types in the area are river flooding, flash flooding and urban waterlogging caused by the former two types.

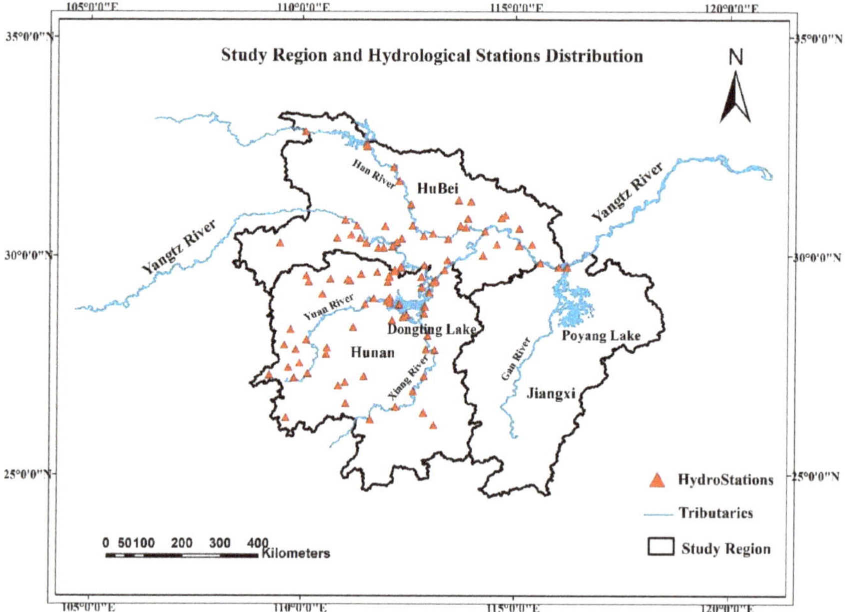

Figure 1. Study region and hydrological stations.

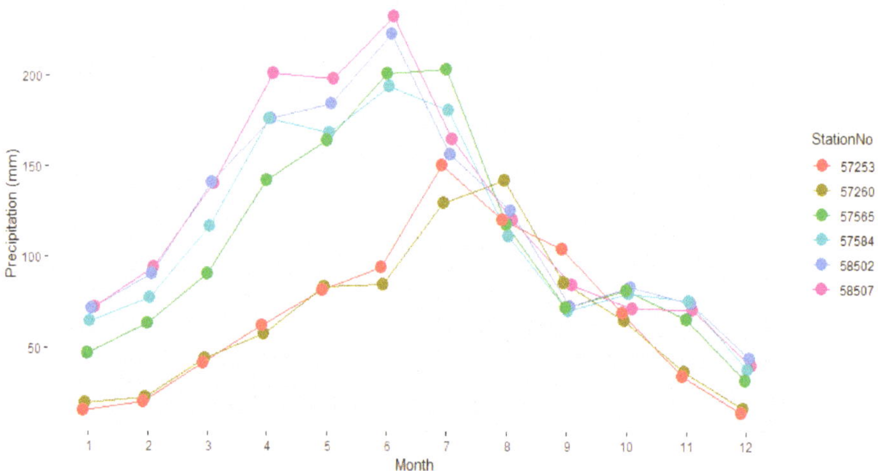

Figure 2. Monthly mean precipitation at six meteorological stations.

2.2. Data Source

In China, the major hindrance to the study of flood risk is the unavailability and scarcity of hydrological data and potential flood-related factors in digital GIS format. Geographically referenced data about hydrological stations were obtained from a list from Baidu Wenku, which contains the names, coordinates and other detailed information of all hydrologic stations around China. As Table 1 shows, daily hydrological observation data at 8:00 a.m. were acquired by means of web spider from the Hydrological Information Inquiry website built by the Hydrology Bureau from Hunan and Hubei Province. Monthly mean rainfall from 1981 to 2010 and different types of meteorological statistical data can be accessed in the National Meteorological Information Center. Topological data using SRTM DEM with 30 m resolution and MODIS NDVI raster data with 500 m resolution can be acquired from Geographical Space Cloud (http://www.gscloud.cn/). Raster data of precipitation interpolated by meteorological observations in vector format seems invalid and inaccurate due to the sparsity of meteorological stations in the study region. Hence, a nearly real-time dataset named TRMM_3B42RT short for 3-h $0.25° \times 0.25°$ merged TRMM and other satellites estimates in raster format is applied and processed to calculate statistical factors concerning precipitation during the period from 2005 to 2012 (https://precip.gsfc.nasa.gov/index.html). The vector data of the river network included in the Chinese National Basic Geographic Information System data are open to the public on the website of the National Geomatics Centre of China.

Table 1. Data sources of the study.

Data Product Name	Resolution	Data Source
GDEMDEM 30 M	30 m	http://www.gscloud.cn/
MODIS NDVI	500 m	http://www.gscloud.cn/
Hydrological Observation Data	daily	http://61.187.56.156/wap/index_sq.asp
TRMM_3B42RT	0.25 degree	https://precip.gsfc.nasa.gov/index.html
River Network	vector	http://ngcc.sbsm.gov.cn/

2.3. Derivation of Hazard Factors

The flood risk at a hydrological station is represented by the historical frequency of flood alarm events which occur when dangerous water levels higher than the historical warning level are observed. Hazard factors that affect the risk of flooding in the middle reaches of the Yangtze River mainly include topological factors, hydrological factors and trigger factors. Topological factors such as elevation, slope and elevation standard deviation come from the DEM. Hydrological factors include river density and distance to mainstream. Vegetation is also an important factor. Trigger factors are mainly derived from rainfall data because flash floods and river floods caused by concentrated heavy rainfall are dominant in the study area.

2.3.1. Frequency of Flood Alarming Events

In hydrological observations, daily reports usually contain river names, hydrological observatory names, observation time, water level, flow, rise and fall of floodwater, warning level, highest water level in history and location information. Obviously, water level that is equal or even higher than warning level at a certain time indicates a flood alarming event, which indicates an excessively high level of water and requires the attention of people to prevent flooding from occurring. In the study area, approximately more than one hundred hydrological stations have been built and functioned normally to monitor water regimens. Every province in the middle reaches of the Yangtze River has set up an open website to display and query hydrological observation data in real time. The study employs the web crawler technology to gather hydrological observation report data at 8:00 a.m. every day from 2005 to 2012. After data processing and cleaning in consideration of the validity of every observation record and functionality of every hydrological station, 104 hydrological stations and their daily report data are included for research and analysis, totaling 303,80 records during the study period. Each observation record at a certain station is considered a flood alarming event if the water level is equal to or greater than the warning level. Then, we count the frequency of all flood alarming events totaling 1647 observed by these hydrological stations. From the histogram of flood alarming events in Figure 3, most of hydrological stations have low occurrence rate of flood alarming events close to 0 and large parts of hydrological stations observed 0 times of flood alarming events in the study period. Few stations have high frequencies. In most cases, the larger the frequency is and the lower the number of hydrological stations is.

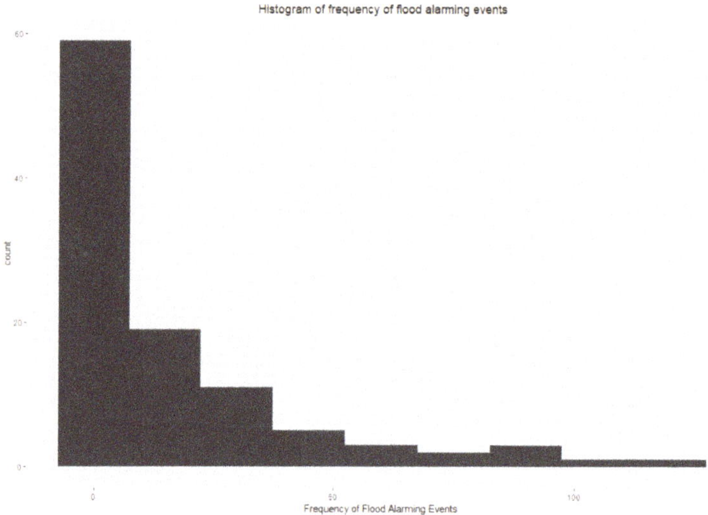

Figure 3. Histogram of the frequency of flood alarming events.

2.3.2. Topological Factors

Elevation is the most fundamental feature indicating the characteristics of the underlying surface. It is commonly acknowledged that water tends to accumulate in lower areas that are difficult to flow out from and that a hydrological station built in a relatively low area is prone to frequent flood alarming events.

Moreover, almost all of the previous studies that focused on the evaluation of flood risk considered elevation to be an important factor [17,21,35]. As is shown in Figure 4, the altitude of the study area ranges from −194 m to 5495 m and fluctuates largely. Slope can be defined as the steepness or gradient within a series of adjacent units of terrain, slope is usually measured as an angle in degrees or as a percentage and indicates the variation in elevation. In general, river currents flow faster over steep areas, reducing the hazard of flooding, while water easily accumulates on flat terrains, leading to a high and dangerous water level. Therefore, the slope of the area where hydrological stations are built must be taken into account [36]. The distribution of slope in the study area can be seen in Figure 5. Elevation standard deviation is calculated by some adjacent cells such as the neighboring cells in a 3×3 window, which represents a locally limited area. In reality, a relatively spacious areas are still susceptible to flooding because it has some local subareas with greater slope. Instead, another measurement of the elevation variation within larger ranges that needs to be included is the elevation standard deviation [18,21]. Corresponding to the resolution of the DEM and our study scale, this study uses a 5×5 neighboring window that represents an area of 150 m^2.

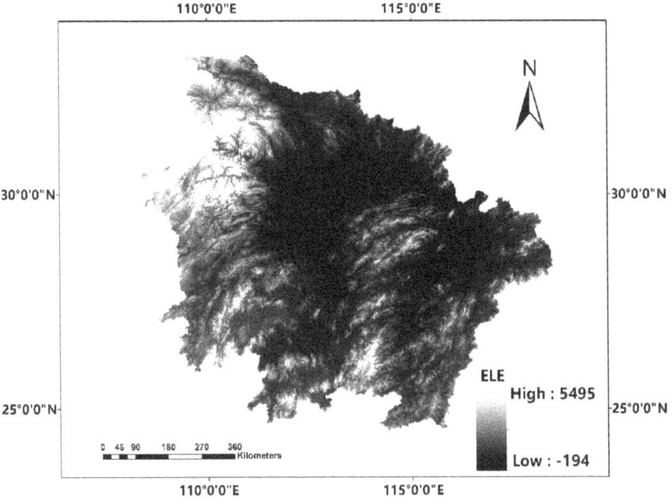

Figure 4. Elevation.

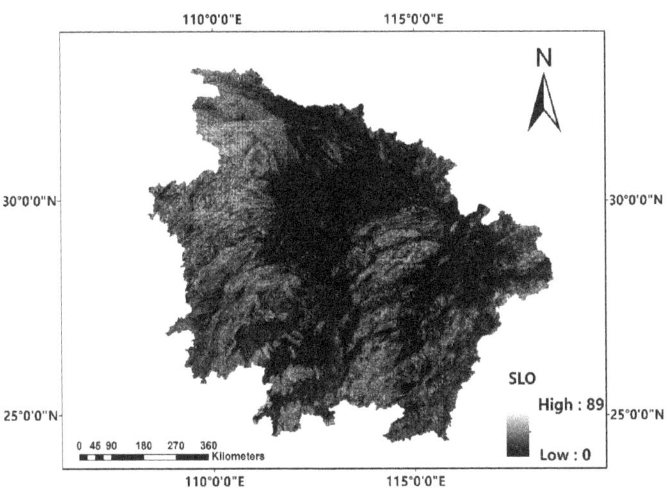

Figure 5. Slope.

The calculation method for elevation standard deviation (ESD) is shown in Equation (1):

$$\mathrm{ESD} = \sqrt{\frac{1}{n-1}\sum_{i=1}^{n}\left(\left(E_i - \overline{E}\right)\right)^2} \tag{1}$$

In the formula, E_i and $\overline{E}$ represent the cell value and mean within the window respectively and n is the number of cells in the window. Elevation standard deviation can be generated by focal statistics on DEM as seen in Figure 6.

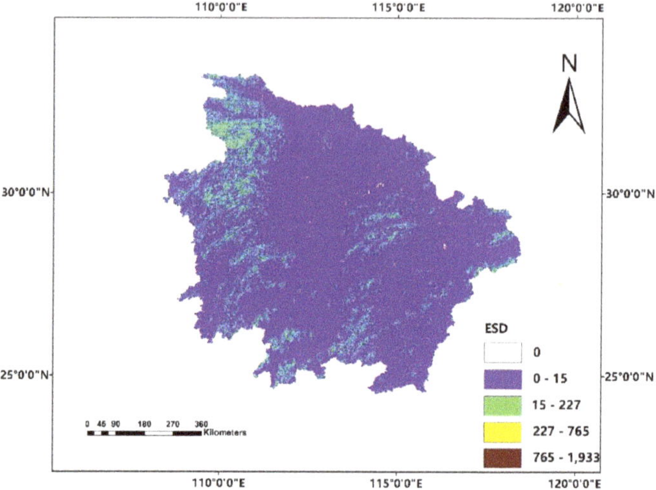

Figure 6. Elevation standard deviation.

2.3.3. Hydrological Factors

The density of a river system refers to the ratio between the summed length of the mainstream and tributaries in the basin and the area of the basin which represents the river's length within a unit of area. Climatology also reveals that the shapes of rivers have their own characteristics that allow for the discharge of surface runoff in time. River density is an important factor in controlling the occurrences of floods and river flow because these areas with high densities of rivers are prone to flooding [18]. Raster data of the river density are obtained by using the line density analysis to convert the vector data of the river network in the study region to raster data in Figure 7.

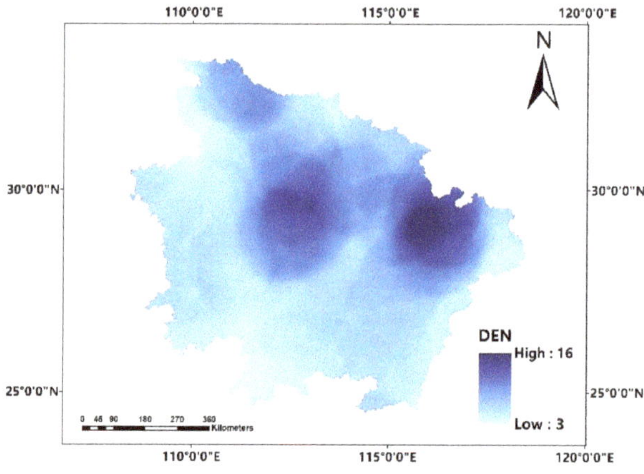

Figure 7. River Density.

Drainage proximity is measured by the distance to the mainstream or tributaries. The occurrences of flood alarming events and even flooding are related to the distribution of the drainage system [18]. A raster layer of the distance to mainstream is generated by performing Euclidean distance analysis on the mainstream network using vector data in ArcGIS10.2.1 as shown in Figure 8.

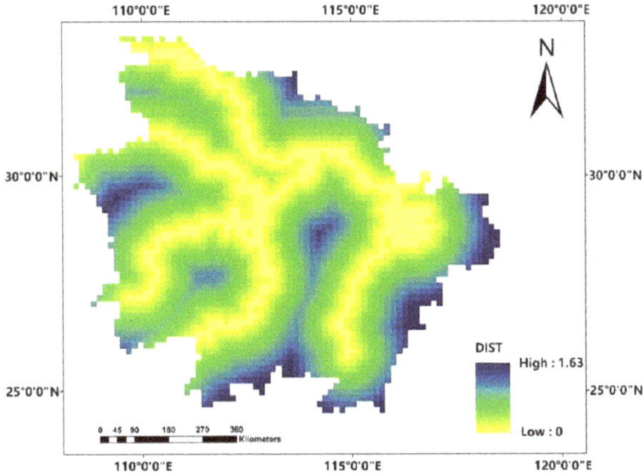

Figure 8. Extreme rainfall frequency.

2.3.4. NDVI

The effects of forest vegetation change on runoff and flooding are very significant: As forest coverage increases, the ability of river basins to intercept rainfall runoff is significantly enhanced, and the amount of rainfall converted to runoff is reduced. Meanwhile, flood peak flow decreases. Changes to vegetation cover can change soil properties and reduce soil erosion. The vegetation and the corresponding soil water absorption reduce the amount of flooding [37]. Vegetation blocks and intercepts floods and protects slopes. Therefore, NDVI, which is widely used in vegetation research of remote sensing is the best indicator factor of plant growth status and spatial distribution of vegetation density as shown in Figure 9.

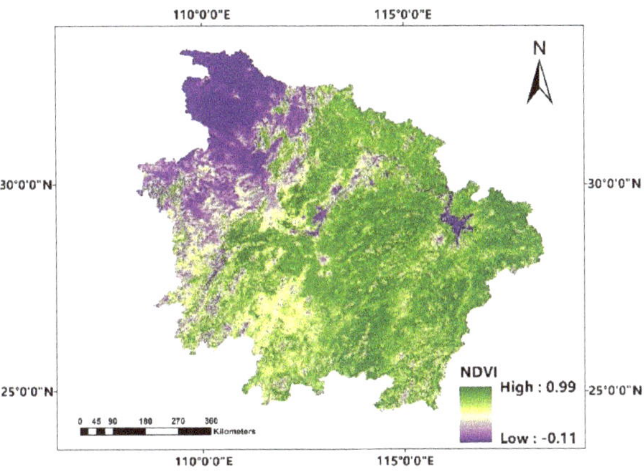

Figure 9. NDVI.

2.3.5. Trigger Factors

Statistical factors related to precipitation are derived from the processing and calculation of raster data, in which the value of every cell signifies a 3-h rainfall rate(mm/h) based on map algebra. This type of rainfall dataset has $0.25° \times 0.25°$ resolution, which geographically corresponds to 25 km $\times$ 25 km area that is similar to a county. Many state-of-the-art studies apply real-time data for more accurate simulation and analysis [38]. Considering its fine temporal resolution and better accuracy than rainfall raster data interpolated by rainfall observations of weather stations, the dataset is an ideal source for researchers to calculate trigger factors of flood. First, we accumulated six raster layers of the rainfall rate to represent daily precipitation. Second, every three consecutive raster layers were summed by using map algebra, and eventually, the raster of annual mean rainfall is produced. The detailed steps for processing the raster data are shown in Figure 10:

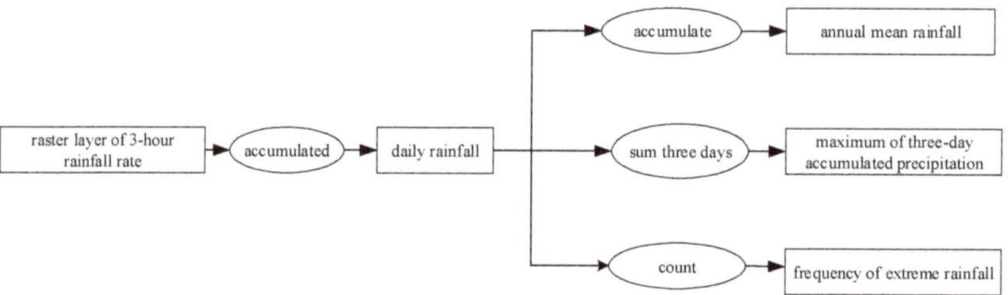

Figure 10. Flow chart of rainfall raster processing.

The raster data of the 3-h rainfall rate is accumulated ArcGIS10.2, and then, the derived raster data of daily rainfall are accumulated or counted further, and ArcPy is fit for efficient and fast batch processing.

Although annual mean rainfall is not directly related to flood alarming events and floods, it represents a meteorological pattern of different regions in the long run. Some previous studies have included similar factors that represent rainfall intensity such as annual mean rainfall or the Modified Fournier Index (MFI) [22]. The spatial distribution of annual mean rainfall is as shown in Figure 11.

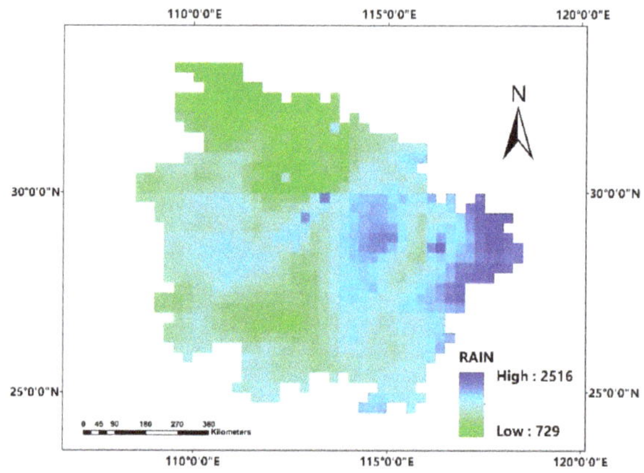

Figure 11. Annual mean rainfall.

It is well known for everyone that flood alarming events and flooding are often caused by continuous and intense rainfall in short periods. According to the resources of recorded historic flood, the maximum of three-day consecutive precipitation plays the most essential role in occurrences of dangerous water levels and floods. Therefore, the mean of the annual maximum of three-day accumulated precipitation is necessary to consider the occurrence of flood alarming events from a short-term perspective. As is seen in Figure 12, the factor ranges from 84 mm to 249 mm.

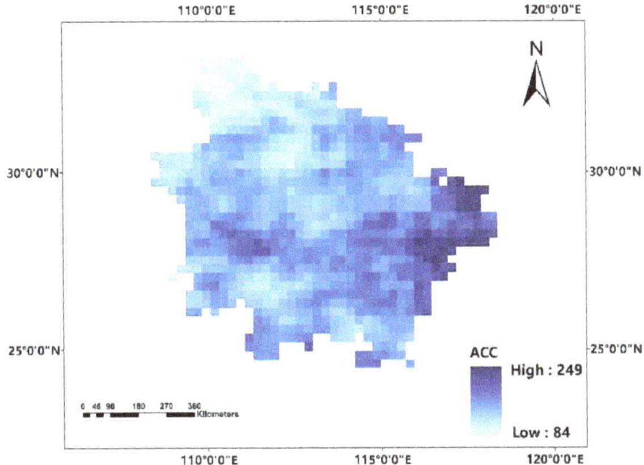

Figure 12. Mean annual maximum of three-day accumulated precipitation.

In meteorology, precipitation that is equal to or greater than 50 mm in 24 h is defined as torrential rain according to the official national standard released by China Meteorological Administration. The rise of river levels is more likely after extreme rainfall. Theoretically, more frequent extreme rainfall means a higher likelihood of flood alarming events and even flooding at worst. The count of heavy precipitation was considered by some previous studies that tried to evaluate flood risk [18]. The frequency of extreme rainfall in raster format is shown in Figure 13.

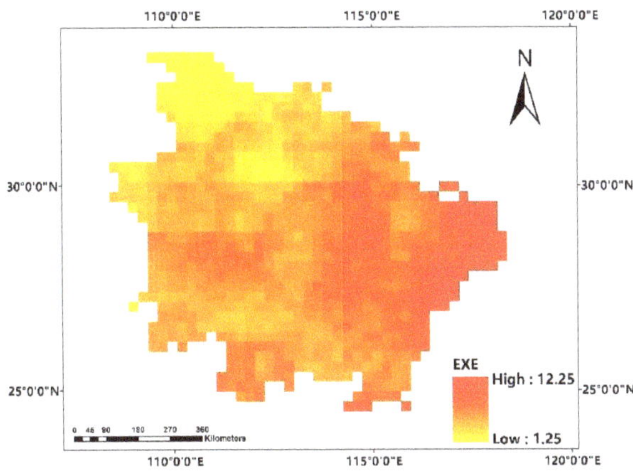

Figure 13. Frequency of extreme rainfall.

3. Methodology

The methodology of this study includes four steps: factor selection, eigenvector spatial filtering Poisson regression (ESFPS), model assessment and flood risk mapping as is shown in Figure 14. The second step is also divided into four parts: construction of a spatial weight matrix, testing of spatial autocorrelation, calculation of eigenvectors, normalization of independent variables, eigenvector selection and calculation of coefficients. The performance of ESFPS will be compared with that of the Poisson Regression (PS) and the negative binomial regression (NB). The model metrices for assessment includes fitness accuracy and generalization error. Afterwards, the best regression model is applied to compare their importance by weights of different factors that pass the significance test in a regression. Finally, this model will be used to predict the flood risk in the middle reaches of the Yangtze River.

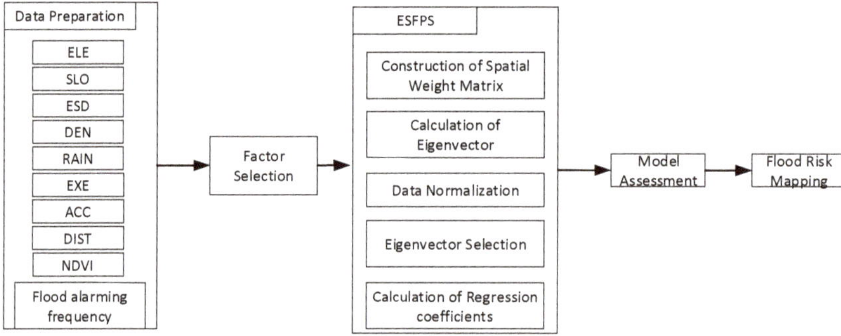

Figure 14. Flowchart of the study.

3.1. Factors Selection

Multicollinearity problems of factors are often encountered in regression problems, hence, a multicollinearity test must be performed before the regression [39]. Variance inflation factor (VIF) and tolerance are suggested as indicators of multicollinearity. To calculate the collinearity of these independent variables, every independent variable is used as the dependent variable and the remaining independent variables are used as the independent variable in the same regression equation. Tolerance and variance inflation factors can be calculated by Equations (2) and (3):

$$\text{tolerance} = 1 - R_j^2 \tag{2}$$

$$\text{VIF} = \frac{1}{tolerance} \tag{3}$$

R_j is the coefficient of determination of a regression of the j-th independent variable on other independent variables. If the tolerance is close to 0, the collinearity problem is quite serious. The larger the VIF is, the more severe the collinearity is. Generally, if the VIF is less than 5, the collinearity is not serious, and the standard for a broader point is less than 10 [40].

3.2. Eigenvector Spatial Filtering Poisson Regression

3.2.1. Construction of Spatial Weight Matrix

In this study, Thiessen polygons are applied to define the neighboring relationship between two different geographical units. In climatology, Thiessen polygons, also called Voronoi diagrams are applied to estimate the precipitation in the neighborhood area of a point-shape unit [41]. This study considers that the area covered by a river within a Thiessen polygon has similar hydrological conditions based on the basic fact that close hydrological stations on the same drainage area usually encounter high water levels and flood alarming events at the same time when flooding occurs. When a spatial weighted matrix W is constructed, the queen neighborhood is often applied to determine whether two units is neighbor each other [42]. Two polygons that share common points or edges are considered neighbors. the value in the matrix is set to 1 if the i-th unit is linked to the j-th unit, otherwise, the value is set to 0.

As is seen in Figure 15, the larger the red bubble is and the darker the color of the polygon, the more frequently flood alarming events occurred. Geographical units with high frequency tend to accumulate with each other and Thiessen polygons with low frequency are close to each other.

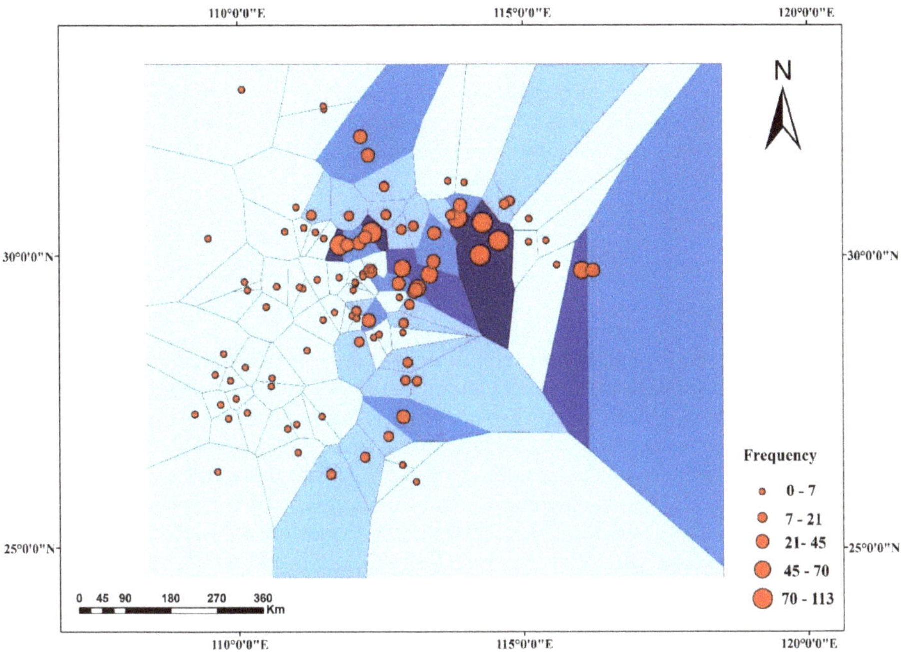

Figure 15. Graduated frequency and distribution of flood alarming events.

After performing a spatial autocorrelation test on the frequency of flood alarming events and the residuals of frequency calculated by ordinary least squares (OLS), relatively significant autocorrelation is found in Table 2: the Moran's coefficients are 0.32 and 0.23, respectively, which are higher than the expected values, −0.01 and −0.03. The R square of 0.2623 means that ordinary linear regression is not fit for modeling. A p-value that is close to zero means that significant spatial autocorrelation exists and eigenvector spatial filtering algorithms can be applied.

Table 2. Results of the spatial autocorrelation test.

	Moran's I	**Expectation**	**Variance**	**p-Value**	**R Squared**
Frequency	0.3246	−0.0097	0.0040	1.547×10^{-7}	
OLS residual	0.2255	−0.039	0.0043	2.743×10^{-5}	0.2623

3.2.2. Calculation of Eigenvectors

Before the calculation of eigenvectors, the raw spatial neighborhood matrix must be row-standardized by Equation (4):

$$C = (I - \frac{11^T}{n})W(I - \frac{11^T}{n}) \tag{4}$$

where n is the number of studied units, W is the spatial neighborhood matrix of $n \times n$, I is an identity matrix of $n \times n$, 1 is a vector of $n \times 1$ whose values are all 1, and C is called the centralized weighted matrix. The decomposition of the centralized weighted matrix can be expressed as Equation (5):

$$C = E\Lambda E^T \tag{5}$$

The decomposition of C generates eigenfunctions that contain n eigenvectors and n corresponding eigenvalues [33]. The n eigenvectors can be denoted as $E = (E_1, E_2, \cdots, E_n)$ and each of eigenvector is capable of capturing latent spatial autocorrelation at different scales [34]. Λ is an n-by-n diagonal matrix, and its diagonal elements are n eigenvalues of C, which can be denoted in descending order as $\lambda = (\lambda_1, \lambda_2, \cdots, \lambda_n)$. It should be noted that all eigenvectors are orthogonal and uncorrelated [34]. Moreover, Moran's I of every eigenvector is calculated by Equation (6):

$$\text{Moran's I}_i = \frac{n}{1^T C 1} \lambda_i \tag{6}$$

where λ_i is the i-th eigenvalue.

3.2.3. Z-Score Normalization

In a multicriterion evaluation system, the evaluation criteria usually have different dimensions and orders of magnitude due to the different natures. When the levels between the indicators differ greatly, if the analysis is performed directly with the original index values, the role of the higher-value indicators in the comprehensive analysis will be highlighted, and the effect of the lower-level indicators will be relatively weakened. Therefore, to ensure the reliability of the results and compare the weights of different factors, the original indicator data must be standardized [43]. In addition, normalization can improve the convergence speed of the model. The normalization of data involved scaling the data down to a small specific interval.

The methods of normalization include min-max normalization and z-score normalization. Z-score normalization is applied to transform nine factors before regression because the minimum and maximum of each factor are unknow in the current datasets. The Z-score transformation of the j-th factor of the i-th sample X_{ij} is as Equation (7):

$$Z_{ij} = \frac{X_{ij} - \overline{X_j}}{S_j} \tag{7}$$

where $\overline{X_j}$ and S_j are the mean and standard deviation of the j-th factor.

3.2.4. Eigenvector Selection

We choose eigenvectors whose Moran's I values are greater than or equal to a threshold as eigenvectors with larger Moran's I values indicates more spatial autocorrelation, and the threshold has been researched in some previous studies on spatial autocorrelation [32]. The parameter can also be lifted to limit the number of candidate eigenvectors and complexity of the final model and 0.25 is chosen for the threshold. In the statistical model, stepwise regression is an automatic method of selecting among independent variables under certain criterion [44]. The AIC is a comprehensive metric of goodness of fit and complexity of model and a model with lower value of AIC mean a better one. Based on the Poisson regression model, forward stepwise selection of independent variables including factors and spatial synthetic variables is chosen to conduct and if the new model has a lower AIC than the previous model and passes the significance test of residuals of spatial autocorrelation after an eigenvector is added to the model, the eigenvector will be included in the final model. If the eigenvector cannot lower the AIC of the current regression model or significant spatial autocorrelation of the residuals still exists between the predicted values output by the current model and ground truth values, the eigenvector will be omitted. When all eigenvectors have been checked, the final Poisson model is output. The detailed process is shown on the flowchart in Figure 16.

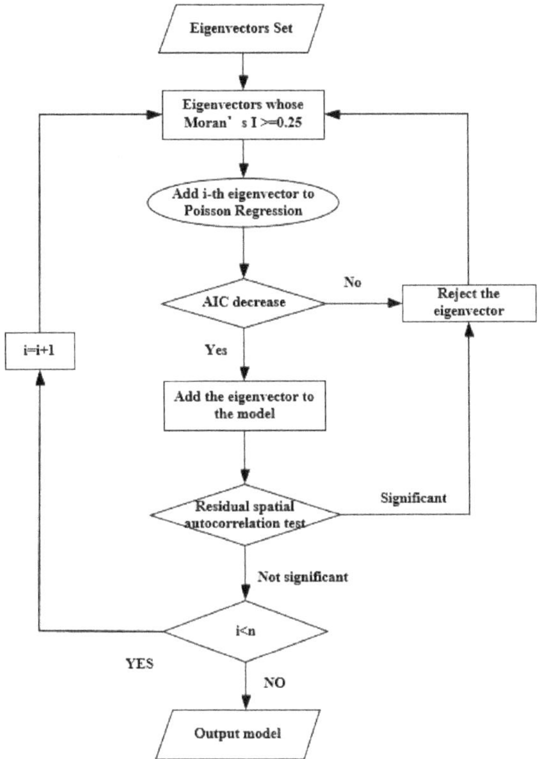

Figure 16. Flowchart of eigenvector selection.

3.2.5. Coefficient Calculation

Different from multivariate linear regression whose response variable is a continuous variable, Poisson regression assumes that the response variable, which is the frequency of flood alarming events in the study, is an observed count that follows the Poisson distribution, whose possible values of Y are non-negative integers. More importantly, this model is based on the hypothesis that a large count or frequency is rare. In the model, the Poisson incidence rate μ is regressed by its k predictors X_i ($ELE, SLO, ESD, DEN, DIST, NDVI, RAIN, ACC, EXE$) and coefficients $\beta_1, \beta_2, \ldots, \beta_9$ are estimated weight coefficients from flood alarming datasets. Poisson regression for sample i is written as Equations (8) and (9):

$$\Pr(Y_i = y_i | y_i, t_i) = \frac{e^{-\mu_i t_i}(\mu_i t_i)^{y_i}}{y_i!} \tag{8}$$

$$\mu_i = t_i \times \mu\left(X_i^T \beta\right) = t_i \times \exp(\beta_1 ELE_i + \beta_2 SLO_i + \ldots + \beta_9 EXE_i) \tag{9}$$

where t_i can be seen as an intercept of linear combinations of factors. Therefore, we can reasonably assume that frequent flood alarming events occurred with low possibility at a certain hydrological station, while most hydrological stations encountered few records of flood alarming events as illustrated in the frequency distribution of flood alarming events in Figure 3.

Similarly, the coefficient of ESFPS can be formulated as Equation (10):

$$\mu_i = t_i \times exp(\beta_1 ELE_i + \beta_2 ESD_i + \ldots + \beta_9 EXE_i + \sum_{k=1}^{l} \alpha_k E_k) \tag{10}$$

where E_k is the selected eigenvectors by the forward stepwise algorithm, and l is the number of the selected eigenvectors.

3.3. Model Assessment

The Poisson model (PS) and negative binomial regression (NB) are used to compare with ESFPS, and these models are provided with the same hazard-causing factors and data processing. Comparison of model assessment is conducted mainly by two aspects, performance of fitness and generalization error. Performance of fitness is assessed by pseudo R-squared and AIC while the metric of generalization error is the mean square error of leave-one-out cross validation.

3.3.1. Accuracy of Model Fitting

The R-squared value which is widely used in linear regression cannot be extended to Poisson regression models, and some pseudo R-squared models have been proposed to measure the interpretation of the Poisson model. In Poisson regression, the most popular pseudo R-squared measure is the function of the log-likelihoods of three models as shown in Equation (11):

$$Pseudo\ R^2 = \frac{LL_{fit} - LL_0}{LL_{max} - LL_0} \tag{11}$$

where LL_0 is the log-likelihood of the intercept-only model, LL_{fit} is the log-likelihood of the current model, and LL_{max} is the maximum log-likelihood possible.

3.3.2. AIC

The Akaike information criterion is a measure of the relative quality of statistical models for a given set of data. The AIC of one model is calculated as Equation (12):

$$AIC = 2k - 2\ln(\hat{L}) \tag{12}$$

where k is the number of parameters and $\hat{L}$ represents the maximum likelihood value for the model. Therefore, this metric can serve as the means for model comparison and selection [45]. Among several candidate models for the samples, the preferred model is the one with the minimum AIC value. In this study, AIC is not only used in the comparison of the final candidate models, but also taken as a prespecified criterion in the stepwise algorithms.

3.3.3. Leave-One-Out Cross Validation (LOOCV)

In supervised learning applications of statistical learning theory, the out-of-sample error or generalized error is a measurement of how accurately a model is capable of predicting a response value for previously unseen data [46]. It is noteworthy that stepwise regression encounters the issue of overfitting because the algorithm will search a very large number of candidate models using brute force. As for LOOCV conducted every time, only one sample is used as the test dataset and the other samples are used as training dataset. If we have k samples, we will train k times and test k times. Although this method demands frequent and complex computation, it has a high utilization rate of samples and is especially suitable for small datasets.

3.4. Flood Risk Mapping

Given hazard-causing factors in other parts of the studied region, a regression model can be used to predict the expected frequency of flood alarming events. The expected frequency obtained by Poisson regression also represents the risk of flood alarming events in the format of frequency while the probability output by the logistic regression indicates the flood risk in the format of probability [47]. After comparing the Poisson model, negative binomial model and ESFPS model, the best one will be selected to predict the expected frequency of flood alarming events throughout the central basin of the Yangtze river. In the GIS system, the linear vector data of river network is firstly transferred to point-shape vector that represents the range of the middle reaches of the Yangtze river. Secondly, these points extract the corresponding value from the raster of factors that pass the significance test. Thirdly, ESFPS transform these factors into a weighted score and output the expected frequency. Finally, point-shape vector among the whole study area is converted back into the raster format for visualization.

4. Results

4.1. Factor Selection

In multicollinearity diagnosis, the variance inflation factor and correlation coefficient matrix are calculated, and the results are plotted respectively in Table 3 and Figure 17. The correlation coefficient between slope (SLO) and elevation standard deviation (ESD) is 0. 97, close to 1 and VIF of slope and elevation standard deviance is 10.994 and 10.460 therefore one of these two factors must be excluded. The results of multicollinearity diagnosis of the remaining eight factors can be seen in Table 4 after slope is excluded and the result shows that there is hardly strong multicollinearity in these remaining eight factors.

Table 3. Variance inflation factor.

Factor	ELE	SLO	ESD	DEN	RAIN	EXE	DIST	ACC	NDVI
VIF	1.299	10.944	10.460	1.783	2.678	6.095	1.156	5.339	1.090

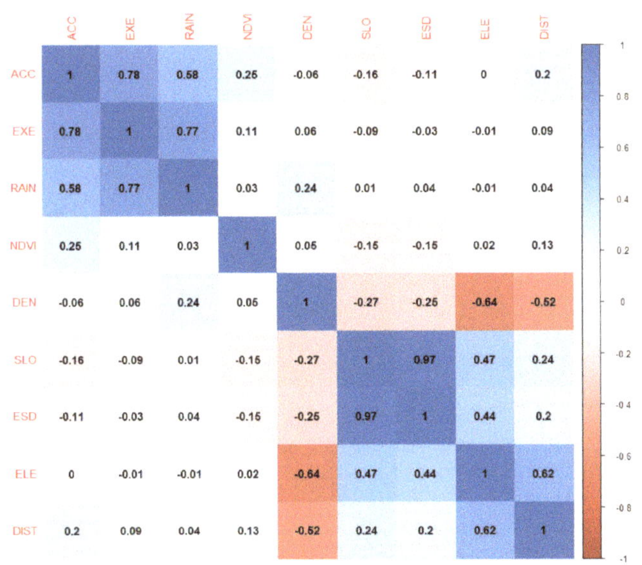

Figure 17. correlation coefficient matrix.

Table 4. Variance inflation factor after factor selection.

Factor	ELE	ESD	DEN	RAIN	EXE	DIST	ACC	NDVI
VIF	1.331	1.236	1.837	2.804	5.980	1.155	5.265	1.099

4.2. Results of Model Coefficients

The maximum threshold of the p-value that passes significant tests is set to 0.05. As shown in Table 5, the Poisson model considers seven factors including ELE, ESD, DIST, RAIN, EXE, ACC and NDVI significant while the negative binomial model only accepts ELE, DIST, ACC as significant factors. ESF-based Poisson regression model considers six factors including ELE, DEN, DIST, EXE, ACC and NDVI significant. Among the eight factors, according to their estimated coefficients and significance in Table 5, three models all shows that elevation, elevation standard deviation, distance to mainstream and NDVI are negatively correlated with the frequency of flood alarming events while the frequency of extreme rainfall is positively related to it. Moreover, annual mean rainfall cannot pass the significance test of NB and ESFPS while river density cannot pass the test of PS and NB, and ACC, DIST and ELE are all significant in three models.

Table 5. Results of the three regressions.

Model	PS		NB		ESFPS	
Factors	Coefficient	p-Value	Coefficient	p-Value	Coefficient	p-Value
ELE	−1.58272	0 ***	−0.9996	0.001683 **	−0.0066	0 ***
ESD	−0.25210	0 ***	−0.19299	0.368332	−0.0769	0.28752
DEN	−0.06189	0.11095	0.07266	0.749230	0.0630	0.000915 **
DIST	−0.56152	0 ***	−0.79702	0.000618 ***	−0.0187	0 ***
RAIN	−0.11253	0.00136 **	−0.03248	0.897488	0.0006	0.5612
EXE	0.59830	0 ***	0.58400	0.053373	1.0334	0 ***
ACC	2.876	0 ***	−0.70197	0.014845 *	5.6894	0 ***
NDVI	−0.2313	0 ***	−0.18582	0.252906	−2.3345	0 ***
Intercept	1.7188	0 ***	1.8193	0 ***	1.4003	0 ***

In Table 5, the signs of significance ***, **, * and · represent that the p-value is less than the threshold of 0, 0.001 and 0.05, respectively.

Regression equation of the PS model is written as Equation (13):

$$\mu_i = \exp(-1.58272 \times ELE_i + (-0.2521) \times ESD_i + (-0.56152) \times DIST_i + (-0.11253) \times RAIN_i \\ +0.5983 \times EXE_i + 2.876 \times ACC_i + (-0.0312) \times NDVI_i + 1.7188) \tag{13}$$

Regression equation of the NB model is written as Equation (14):

$$\mu_i = \exp((-0.996) \times ELE_i + (-0.79702) \times DIST_i + (-0.70197) \times ACC_i + 1.8193) \tag{14}$$

Regression equation of the ESFPS model is written as Equation (15):

$$\mu_i = \exp((-0.0066) \times ELE_i + 0.063 \times DEN_i + (-0.0187) \times DIST_i \\ +1.033 \times EXE_i + 5.6894 \times ACC_i + (-2.335) \times NDVI_i + 4.193 \times EV1 \\ +8.377 \times EV2 + (-4.048) \times EV7 + 1.4003) \tag{15}$$

In the equations of the ESFPS model, $EV1$, $EV2$ and $EV7$ are the selected eigenvectors as seen in Table 6.

Table 6. Eigenvectors selected by ESFPS.

Eigenvectors	Coefficient	p-Value
EV1	4.1934	0.039604 *
EV2	8.3770	0.000133 **
EV7	−4.0476	0.002899 ***

The signs of significance **, * and * represent that the p-value is less than the threshold of 0, 0.001 and 0.05, respectively.

4.3. Model Assessment

The Poisson regression model, the negative binomial regression model and the eigenvector spatial filtering Poisson regression fit the samples with the performance as shown in Table 7:

Table 7. Performances of three models.

Model	AIC	Pseudo R Squared	LOOCV	Moran's I (Expectation)
PS	1979.4	0.56	817.9884	−0.01929 (−0.0097)
NB	944.86	0.47	608.1647	0.05737 (−0.0097)
ESFPS	1040.3	0.78	430.4137	0.01958 (−0.0097)

Moran's I for the residuals of the predicted frequency of flood alarming events are all reduced to be closer to the expected value (−0.0097), and the results shows that no spatial autocorrelation exists in the residuals. While the pseudo R squared of PS (0.56) is greater than that of NB (0.47), the AIC of PS, 1979.4 is greater than that of NB, 944.86. ESFPS has the best fitness with the Pseudo R Squared, 0.78, which obviously outperforms PS and NB. Generalized error of ESFPS, 430.4137, calculated by LOOCV is less than that of NB, 608.1647 and both of them are less than generalized error of PS, 817.9884. Although the AIC of ESFPS 1040.3 is 10.1% greater than that of NB, 944.86, ESFPS is much better than PS and NB because ESFPS has better performance of fitness accuracy and generalized accuracy. ESFPS improves pseudo R squared from 0.56 to 0.78, by 39.3% and reduces generalized error from 817.9884 to 430.4137.

4.4. Spatial Pattern of Flood Risk

After the ESFPS model with better accuracy is obtained, the expected frequency throughout the basin can be calculated by the ESFPS model because the frequency of flood alarming events represents the risk of flood occurrences at a hydrological station. More hydrological stations should be planned and built in the place under high risk while obsolete stations have to be removed or repaired. As the river network in the format of raster shows in Figure 18, the darker the color is, the higher the flood risk of the river is. In the upper branch of the Hanjiang River, Poyang Lake District, Zuzhang River and Dongting Lake District, the red pixels indicate high flood risk and yellow pixels indicate an ordinary hazard level. Hydrology departments should pay attention to these sub-basins and strengthen their hydrological infrastructure. In general, sub-basins with high river density and frequent extreme consecutive rainfalls are susceptible to flood alarming events.

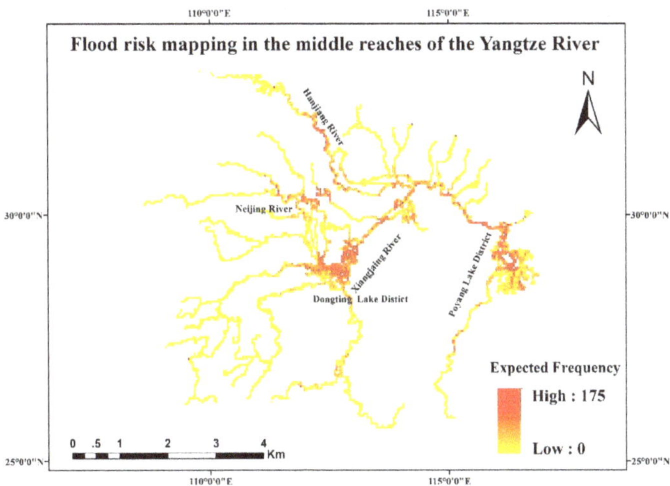

Figure 18. Predicted frequency of flood alarming risk.

5. Discussion

5.1. Improvement in the Accuracy of the Flood Risk EVALUATION Model

This study models the relationship between the frequency of flood alarming events observed on hydrological stations and nine factors using the Poisson regression model. Inclusion of spatial autocorrelation in the regression model increases the accuracy of fitting and prediction. The pseudo R squared is improved by 39.3% from 0.56 to 0.78. Regarding the generalization performance of these models, LOOCV of the ESFPS model is better than that of the Poisson model and negative binomial model. Moran's I of residuals for the ESFPS model is near to its expected value. In general, ESF-based Poisson regression is better than other counterparts.

5.2. Determination of Significant Factors

Both the PS model and the ESFPS model consider elevation, frequency of extreme rainfall, distance to mainstreams, NDVI and mean annual maximum of three-day accumulated precipitation as significant factors. Among the eight factors, according to their estimated coefficients and significance, elevation, elevation standard deviation, distance to the mainstream and NDVI are negatively correlated with the frequency of flood alarming events while frequency extreme rainfall, elevation and river density is positively related with it. Extreme rainfall events represented by frequency of extreme rainfall and mean annual maximum of three-day accumulated precipitation together play a significant role in flood events. These results are consistent with many previous researches and human experiences. Six factors that pass significant test in the ESFPS model are ranked according to the absolute value of weight coefficients in Table 8. Comparing their weights of these factors, accumulation and NDVI are the two main impact factors, with the weights of (+5.6894) and (−2.3345). Hence, short-term heavy rainfall is a leading trigger for the flood in the study area and vegetation has an important resistance to flood.

Table 8. Rank of factors.

Factor	Weight	Rank
ACC	5.6894	1
NDVI	−2.3345	2
EXE	1.0334	3
DIST	−0.0187	4
DEN	0.063	5
ELE	−0.0066	6

5.3. Limitations and Future Enhancements

The complexity of the ESFPS model is greater than that of the PS model because it requires more time to select eigenvectors. In addition, The study only focuses on the factors on the earth surface and trigger factors, and the type of soil and lithology is often considered as important factors in the previous study [48]. Flood is also influenced by how much water can be stored in the soil from previous flooding and local rainfall. Different types of soil and lithology have different water retention capacities. The study initially oversees these two factors because it tries to map flood risk and the flood-alarming events of the hydrological stations on the water instead of the whole regions spanned by the middle Yangtze River. Social and economic factors such as land use and constructions of dikes and drainage facilities should have been included in the model because the capacity of discharging runoff swiftly are represented by these factors. Subsequent research should also consider social factors such as the type of land use [49] and investment in hydrological infrastructures such as dikes and pumping stations [21]. From the perspective of data, hydrological data that ranges from 2005 to 2012 lack timeliness and the flood risk map output by the model can function as an early warning system if new recent data in larger size are used. Moreover, it is more reasonable for the conceptualization of neighborhood relationship for the spatial weight matrix to consider the topology of river network because different segments belonging to the same river possibly share similar hydrological observations and flood alarming events could be observed by hydrological stations in downstream segments and upstream segments at the same time. Hence, soil type, density of dikes and pumping stations and new neighborhood relationship will be considered in the further research.

6. Conclusions

The study proposes a Poisson regression based on eigenvector spatial filtering to model the flood risk at hydrological stations using the frequency of flood alarming events. The frequency of flood alarming events and eight factors including elevation, elevation standard deviation, river density, annual mean rainfall, mean annual maximum of three-day accumulated precipitation, frequency of extreme rainfall and NDVI are used to train the regression model. The PS, NB and ESFPS models reach the conclusion that elevation, frequency of extreme rainfall, distance to rivers, NDVI and mean annual accumulation of three-day accumulated precipitation are important factors leading to high flood risk. In the study area, heavy consecutive rainfall is the chief culprit of the flood and vegetation can reduce flood risk efficiently. Moreover, Inclusion of autocorrelation of the frequency of flood alarming events into the model robustly improved the accuracy of fitness and prediction. Stepwise regression using the AIC can efficiently select the eigenvectors that represent spatial distribution pattern of flood alarming frequency. Eigenvectors with high eigenvalues and Moran's I are accepted in the final ESFPS model. The ESFPS model can be used to map flood risks throughout basins according to the expected value of frequency that the model output after the data related to the eight factors are input into the model. In the study, flood risk mapping output by ESFPS and new recent data can function as an early warning system and help identify where hydrological stations should be built and strengthened for monitoring of water level.

Author Contributions: conceptualization, Y.C.; methodology, Y.C. and T.F.; software, J.L. and L.H.; validation, H.T.; formal analysis, T.F.; investigation, J.C.; resources, T.F.; data curation, H.T.; writing—original draft preparation, T.F.; writing—review and editing, Y.C.

Funding: This research was funded by Ministry of Science and Technology of the People's Republic of China, grant number [2017YFB0503704] and the National Nature Science Foundation of China, grant number [41671380]. And the APC was funded by [2017YFB0503704].

Acknowledgments: This work was supported by National Key R&D Program of China: [grant number 2017YFB0503704] and the National Nature Science Foundation of China [grant number 41671380].

Conflicts of Interest: The authors declare no conflict of interest.

References

1. Petrucci, O.; Aceto, L.; Bianchi, C.; Bigot, V.; Brázdil, R.; Pereira, S.; Kahraman, A.; Kılıç, Ö.; Kotroni, V.; Llasat, M.C.; et al. Flood fatalities in Europe, 1980–2018: Variability, features, and lessons to learn. *Water* **2019**, *11*, 1682. [CrossRef]
2. Schelske, O.; Sundermann, L.; Hausmann, P. *Mind the Risk—A global Ranking of Cities Under Threat from Natural Disasters*; Swiss Reinsurance Company Ltd.: Zurich, Switzerland, 2013.
3. Zhang, L.; Geng, J.; Fan, C. The comprehensive analysis of flood disasters losses in china from 2000 to 2010. *IOP Conf. Ser. Mater. Sci. Eng.* **2018**, *466*, 012023. [CrossRef]
4. Snedaker, S.; Rima, C. Chapter 4—Risk assessment. In *Business Continuity and Disaster Recovery Planning for It Professionals*, 2nd ed.; Snedaker, S., Rima, C., Eds.; Syngress: Boston, MA, USA, 2014; pp. 151–224.
5. He, Y.; Pappenberger, F.; Manful, D.; Cloke, H.; Bates, P.; Wetterhall, F.; Parkes, B. 5.16—Flood inundation dynamics and socioeconomic vulnerability under environmental change. *Clim. Vulnerability* **2013**, 241–255. [CrossRef]
6. Pei, F.; Wu, C.; Qu, A.; Xia, Y.; Wang, K.; Zhou, Y. Changes in extreme precipitation: A case study in the middle and lower reaches of the yangtze river in China. *Water* **2017**, *9*, 943. [CrossRef]

7. Hsieh, S.-H.; Liu, L.-W.; Chung, W.-G.; Wang, Y.-M. Sensitivity analysis on the rising relation between short-term rainfall and groundwater table adjacent to an artificial recharge lake. *Water* **2019**, *11*, 1704. [CrossRef]

8. Hashizume, M. 1.10—precipitation and flood hazards: Health effects, risks, and impacts. *Clim. Vulnerability* **2013**, 115–124. [CrossRef]

9. Legesse, D.; Vallet-Coulomb, C.; Gasse, F. Hydrological response of a catchment to climate and land use changes in tropical africa: Case study south central ethiopia. *J. Hydrol.* **2003**, *275*, 67–85. [CrossRef]

10. Chen, Y.R.; Yeh, C.-H.; Yu, B. Integrated application of the analytic hierarchy process and the geographic information system for flood risk assessment and flood plain management in Taiwan. *Nat. Hazards* **2011**, *59*, 1261–1276. [CrossRef]

11. Chau, K.W.; Wu, C.L.; Li, Y.S. Comparison of several flood forecasting models in yangtze river. *J. Hydrol. Eng.* **2005**, *10*, 485–491. [CrossRef]

12. Bisht, D.S.; Chatterjee, C.; Kalakoti, S.; Upadhyay, P.; Sahoo, M.; Panda, A. Modeling urban floods and drainage using swmm and mike urban: A case study. *Nat. Hazards* **2016**, *84*, 749–776. [CrossRef]

13. Sharma, S.K.; Kwak, Y.J.; Kumar, R.; Sarma, B. Analysis of hydrological sensitivity for flood risk assessment. *ISPRS Int. J. Geo-Inf.* **2018**, *7*, 51. [CrossRef]

14. Neelz, S.N.; Pender, G. *Benchmarking of 2d Hydraulic Modelling Packages*; Environment Agency: Bristol, UK, 2010.

15. Lu, C.; Zhou, J.; He, Z.; Yuan, S. Evaluating typical flood risks in yangtze river economic belt: Application of a flood risk mapping framework. *Nat. Hazards* **2018**, *94*, 1187–1210. [CrossRef]

16. Malczewski, J. A gis-based approach to multiple criteria group decision-making. *Int. J. Geogr. Inf. Syst.* **1996**, *10*, 955–971. [CrossRef]

17. Wang, Y.; Li, Z.; Tang, Z.; Zeng, G. A gis-based spatial multi-criteria approach for flood risk assessment in the dongting lake region, Hunan, central China. *Water Resour. Manag.* **2011**, *25*, 3465–3484. [CrossRef]

18. Chen, Y.; Liu, R.; Barrett, D.; Gao, L.; Zhou, M.; Renzullo, L.; Emelyanova, I. A spatial assessment framework for evaluating flood risk under extreme climates. *Sci. Total Environ.* **2015**, *538*, 512–523. [CrossRef] [PubMed]

19. Tehrany, M.S.; Pradhan, B.; Mansor, S.; Ahmad, N. Flood susceptibility assessment using gis-based support vector machine model with different kernel types. *Catena* **2015**, *125*, 91–101. [CrossRef]

20. Xiong, J.; Li, J.; Cheng, W.; Wang, N.; Guo, L. A gis-based support vector machine model for flash flood vulnerability assessment and mapping in China. *ISPRS Int. J. Geo-Inf.* **2019**, *8*, 297. [CrossRef]

21. Xiao, Y.; Yi, S.; Tang, Z. Integrated flood hazard assessment based on spatial ordered weighted averaging method considering spatial heterogeneity of risk preference. *Sci. Total Environ.* **2017**, *599–600*, 1034–1046. [CrossRef]

22. Kourgialas, N.N.; Karatzas, G.P. A national scale flood hazard mapping methodology: The case of Greece—Protection and adaptation policy approaches. *Sci. Total Environ.* **2017**, *601–602*, 441–452. [CrossRef]

23. Leggett, D.J.; Jones, A. The application of gis for flood defence in the anglian region: Developing for the future. *Int. J. Geogr. Inf. Syst.* **1996**, *10*, 103–116. [CrossRef]

24. Dawod, G.M.; Mirza, M.N.; Al-Ghamdi, K.A. Gis-based estimation of flood hazard impacts on road network in Makkah city, Saudi Arabia. *Environ. Earth Sci.* **2012**, *67*, 2205–2215. [CrossRef]

25. Mandallaz, D.; Ye, R. Prediction of forest fires with poisson models. *Can. J. For. Res.* **1997**, *27*, 1685–1694. [CrossRef]

26. Wahiduzzaman, M.; Yeasmin, A. Statistical forecasting of tropical cyclone landfall activities over the north Indian ocean rim countries. *Atmos. Res.* **2019**, *227*, 89–100. [CrossRef]

27. Betts, M.G.; Diamond, A.W.; Forbes, G.J.; Villard, M.A.; Gunn, J.S. The importance of spatial autocorrelation, extent and resolution in predicting forest bird occurrence. *Ecol. Model.* **2006**, *191*, 197–224. [CrossRef]

28. Tobler, W. On the first law of geography: A reply. *Ann. Assoc. Am. Geogr.* **2004**, *94*, 304–310. [CrossRef]

29. Getis, A. *Spatial Filtering in a Regression Framework: Examples Using Data on Urban Crime, Regional Inequality, and Government Expenditures*; Springer-Verlag: Berlin, German, 2010.

30. Getis, A.; Griffith, D.A. Comparative spatial filtering in regression analysis. *Geogr. Anal.* **2002**, *34*, 130–140. [CrossRef]

31. Murakami, D.; Griffith, D.A. Random effects specifications in eigenvector spatial filtering: A simulation study. *J. Geogr. Syst.* **2015**, *17*, 1–21. [CrossRef]

32. Chun, Y.; Griffith, D.A.; Lee, M.; Sinha, P. Eigenvector selection with stepwise regression techniques to construct eigenvector spatial filters. *J. Geogr. Syst.* **2016**, *18*, 67–85. [CrossRef]

33. Griffith, D.A. A linear regression solution to the spatial autocorrelation problem. *J. Geogr. Syst.* **2000**, *2*, 141–156. [CrossRef]

34. Griffith, D.; Chun, Y. *Spatial Autocorrelation and Spatial Filtering*; Springer-Verlag: Berlin, German, 2003.

35. Kourgialas, N.N. A flood risk decision making approach for mediterranean tree crops using gis; climate change effects and flood-tolerant species. *Environ. Sci. Policy* **2016**, *63*, 132–142. [CrossRef]

36. Wu, Y.; Zhong, P.A.; Zhang, Y.; Xu, B.; Ma, B.; Yan, K. Integrated flood risk assessment and zonation method: A case study in huaihe river basin, China. *Nat. Hazards* **2015**, *78*, 635–651. [CrossRef]

37. Rawat, P.K. Impacts of climate change and hydrological hazards on monsoon crop patterns in the lesser himalaya: A watershed based study. *Int. J. Disaster Risk Sci.* **2012**, *3*, 98–112. [CrossRef]

38. Jiang, S.; Ren, L.; Hong, Y.; Yang, X.; Ma, M.; Zhang, Y.; Yuan, F. Improvement of multi-satellite real-time precipitation products for ensemble streamflow simulation in a middle latitude basin in south China. *Water Resour. Manag.* **2014**, *28*, 2259–2278. [CrossRef]

39. Farrar, D.E.; Glauber, R.R. Multicollinearity in regression analysis: The problem revisited. *Rev. Econ. Stat.* **1967**, *49*, 92–107. [CrossRef]

40. Kutner, M.H.; Nachtsheim, C.J.; Neter, J. *Applied Linear Regression Models*; McGraw-Hill Irwin: New York, NY, USA, 2004.

41. Zhu, Q.A.; Zhang, W.C.; Zhao, D.Z. Topography-based spatial daily precipitation interpolation by means of prism and thiessen polygon analysis. *Sci. Geogr. Sin.* **2005**, *25*, 233–238.

42. Zhang, J.; Li, B.; Chen, Y.; Chen, M.; Fang, T.; Liu, Y. Eigenvector spatial filtering regression modeling of ground $PM_{2.5}$ concentrations using remotely sensed data. *Int. J. Environ. Res. Public Health* **2018**, *15*, 1228. [CrossRef]

43. Grus, J. *Data Science from Scratch: First Principles with Python*; O'Reilly Media: New York, NY, USA, 2015.

44. Hocking, R.R. A biometrics invited paper. The analysis and selection of variables in linear regression. *Biometrics* **1976**, *32*, 1–49. [CrossRef]

45. Akaike, H. IEEE xplore abstract—A new look at the statistical model identification. *IEEE Autom. Control Trans.* **1974**, *19*, 716–723. [CrossRef]

46. Kohavi, R. A study of Cross-Validation and Bootstrap for Accuracy Estimation and Model Selection. In Proceedings of the International Joint Conference on Artificial Intelligence, Montreal, QC, Canada, 20–25 August 1995.

47. Youssef, A.M.; Pradhan, B.; Sefry, S.A. Flash flood susceptibility assessment in Jeddah city (Kingdom of Saudi Arabia) using bivariate and multivariate statistical models. *Environ. Earth Sci.* **2016**, *75*, 12. [CrossRef]

48. Tehrany, M.S.; Pradhan, B.; Jebur, M.N. Spatial prediction of flood susceptible areas using rule based decision tree (dt) and a novel ensemble bivariate and multivariate statistical models in gis. *J. Hydrol.* **2013**, *504*, 69–79. [CrossRef]

49. Kazakis, N.; Kougias, I.; Patsialis, T. Assessment of flood hazard areas at a regional scale using an index-based approach and analytical hierarchy process: Application in rhodope–evros region, Greece. *Sci. Total Environ.* **2015**, *538*, 555–563. [CrossRef] [PubMed]

Article

Glacial Lake Inventory and Lake Outburst Flood/Debris Flow Hazard Assessment after the Gorkha Earthquake in the Bhote Koshi Basin

Mei Liu [1,2], Ningsheng Chen [1,*], Yong Zhang [1,2] and Mingfeng Deng [1]

[1] Key Laboratory of Mountain Hazards and Surface Process, Institute of Mountain Hazards and Environment, Chinese Academy of Sciences, Chengdu 610041, China; lmeimei90@163.com (M.L.); zhangyongcas@163.com (Y.Z.); dmf@imde.ac.cn (M.D.)

[2] University of Chinese Academy of Sciences, Beijing 100049, China

* Correspondence: chennsh@imde.ac.cn; Tel.: +86-13808171963

Received: 31 December 2019; Accepted: 5 February 2020; Published: 10 February 2020

Abstract: Glacial lake outburst floods (GLOF) evolve into debris flows by erosion and sediment entrainment while propagating down a valley, which highly increases peak discharge and volume and causes destructive damage downstream. This study focuses on GLOF hazard assessment in the Bhote Koshi Basin (BKB), where was highly developed glacial lakes and was intensely affected by the Gorkha earthquake. A new 2016 glacial lake inventory was established, and six unreported GLOF events were identified with geomorphic outburst evidence from GaoFen-1 satellite images and Google Earth. A new method was proposed to assess GLOF hazard, in which large numbers of landslides triggered by earthquake were considered to enter into outburst floods enlarge the discharge and volume of debris flow in the downstream. Four GLOF hazard classes were derived according to glacial lake outburst potential and a flow magnitude assessment matrix, in which 11 glacial lakes were identified to have very high hazard and 24 to have high hazard. The GLOF hazard in BKB increased after the earthquake due to landslide deposits, which increased by 216.03×10^6 m^3, and provides abundant deposits for outburst floods to evolve into debris flows. We suggest that in regional GLOF hazard assessment, small glacial lakes should not be overlooked for landslide deposit entrainment along a flood route that would increase the peak discharge, especially in earthquake-affected areas where large numbers of landslides were triggered.

Keywords: glacial lake outburst flood (GLOF); debris flow; Bhote Koshi; landslides; Gorkha earthquake; hazard assessment

1. Introduction

Many studies have demonstrated that most glaciers are retreating because of global warming and that the meltwater makes an important contribution to the development of glacial lakes in the Himalayas [1–5]. The sudden emptying of these lakes due to dam overflow and moraine or ice dam failure releases large volumes of water and sediment in destructive events called glacial lake outburst floods (GLOF) [6,7]. In the Himalaya region, at least 62 GLOF have been reported, which caused catastrophic destruction and fatalities in downstream regions [8–12]. The peak flood discharge can easily attain tens of thousands of m^3/s and travel more than 100 km away [8,13]. Given their high magnitude discharge and long runout distance characteristics, the GLOF impact is sometimes transboundary, especially in the Himalayas. More than 10 GLOF events originated in Tibet, and the catastrophic floods killed hundreds of people and destroyed much infrastructure downstream, causing enormous damage in Nepal and India [14,15]. As a result, GLOF hazard assessment is receiving increased attention from researchers and governments.

In previous GLOF hazard assessment studies, only glacial lakes with an area $>10^5$ m^2 or volume $>10^6$ m^3 were considered to be risky of outburst [15,16]. However, some small outbursts occurred in the high mountain regions, but are often ignored due to the limited scale of the events or difficult access [17]. Veh [18] detected 10 previously unreported GLOF from Landsat time series in a study area covering only 10% of the Hindu Kush Himalayan region. In addition, glacial lake outburst floods are highly unsteady flows characterized by pronounced changes as they propagate down to the valley [13]. The outburst flood can change from a normal flood to a hyperconcentrated flow or debris flow [19,20], and the volumes and peak discharges can increase several to ten times owing to erosion, slides from lateral slopes, sediment entrainment and bulking process along the flow path [21,22]. As an example, in Norway, a glacial lake outburst flood developed into a debris flow due to erosion and blockage, and the volume increased nearly ten times from 25,000 to 240,000 m^3 [23]. Sediment can be entrained by scouring unconsolidated deposits of the channel bed, or eroding landslide and collapse of lateral slopes [24,25]. In the seismic belts, large numbers of weak structures and broken rocks are developed along the active fracture zone, and the soils become looser after an earthquake [26]. Studies showed that large earthquakes, such as the Chi-chi earthquake and Wenchuan earthquake, trigger many collapses and landslides, resulting in an increase in loose deposits [27,28]. Although rare reported GLOF events in the Himalaya are directly triggered by earthquakes [29], the loose deposits and landslides induced by earthquakes may affect the magnitude and impact of GLOF. Therefore, it is necessary to build a GLOF hazard assessment model, considering small glacial lakes and the scenario of glacial lake outburst debris flows after earthquakes, especially in areas where many collapses and landslides have developed along the channels.

The Bhote Koshi Basin across China and Nepal, is a highspot area of glacial lakes and GLOF events (Wang and Jiao, 2015). Four glacial lakes have experienced six GLOF events since 1935 (Figure 1). Taraco Lake failed on 28 August 1935, and the GLOF damaged more than 10 hm^2 of wheat fields (Lv, 1999); The Jialongco GLOF occurred on 23 May and 29 June 2002, which caused 7.5 million yuanin economic losses in Nyalam County (Chen et al., 2007). The Cinrenmaco Lake experienced two GLOF events, first in 1964 and second on 10 July 1981. The GLOF in 1981 had the most destructive effects, in which more than 200 people were killed in Nepal, and the total losses were estimated at approximately three million dollars [14,30]. The latest GLOF event occurred on the night of 5 July 2016, which was caused by Gongbatongshaco (GBTSC) Lake in the Zhangzangbo Valley. GBTSC is a small moraine-dammed lake, with a surface area of 1.7×10^4 m^2 and it was almost empty after the outburst. Although it only released 1.1×10^5 m^3 of water, the peak discharge reached 2400 m^3/s at Khukundol, 30 km downstream from the lake, due to severe erosion and sediment entrainment [31]. The GLOF caused severe damage downstream of Bhote Koshi, damaging 77 houses, 3 bridges and the Araniko Highway, and destroying the intake dam of the Upper Bhote Koshi Hydropower Project in Nepal (Figure 1). The 2016 GLOF damage sits within the area affected by the Gorkha earthquake (magnitude M 7.8 in 2015), where extensive landslides and rockfalls were triggered on the slopes, and some landslide deposits even blocked the river [32,33]. Therefore, the small glacial lake GBTSC GLOF caused a serious disaster, which caused us to reconsider the small-lake induced GLOF hazard after the earthquake.

The aims of this study are: (1) to establish a detailed glacial lake inventory of BKB after the Gorkha earthquake, based on high resolution remote sensing satellite images; (2) to evaluate the GLOF hazard of BKB considering the scenario that outburst floods evolve into debris flows due to erosion and entrainment of loose solids.

Figure 1. Photographs of Gongbatongshaco (GBTSC) Lake and the glacial lake outburst floods (GLOF) damage caused downstream: (**a**) GBTSC Lake after outburst, (**b**) the breach and debris fan in front of the lake, (**c**) landslides and river bank collapse triggered by GLOF near Friendship Bridge and (**d**) the destroyed dam of Upper Bhote Koshi Hydroelectric Project in Nepal.

2. Study Area

The study site is located in the central Himalayas and covers latitudes 27°37′–28°31′ N and longitudes 85°40′–86°20′ E with an area of 3406 km^2 (Figure 2). Bhote Koshi, which is also called Poiqu in China, is a transboundary river with a length of 143 km. It originates in the Bangbulei Mountains in northern Nyalam County, China, flows into Nepal, and at last feeds into the Ganges River. The Araniko Highway, built along the Bhote Koshi Valley, is a key trade and transport route between China and Nepal. Approximately 200,000 people live in the watershed, among which only 2.7% of them live in China, and the infrastructure in this region is particularly vulnerable [34].

The Bhote Koshi Basin stretches across the Higher Himalaya and Lower Himalaya, and the South Tibetan Detachment System (STDS) and the Main Central Thrust (MCT) pass through it. The basin is strongly affected by seismic activity. According to the statistics of the United States Geological Survey (USGS) earthquake records (http://earthquake.usgs.gov/earthquakes), there were 213 earthquakes (magnitude larger than M 4.5) in the area of 150 km^2 around BKB from 1983 to 2016, including a M 8.3 earthquake, three earthquakes larger than M 7.0 and 79 earthquakes equal or larger than M 5.0. The latest large earthquake, M 7.8, on 25 April 2015 and its largest aftershock (M 7.3) on 12 May 2015, produced severe impact in the study area. The epicenter of the major aftershock was only 19 km southeast of Kodari. The lower part from Zhangzangbo Valley to Dolalghat, which is approximately half of the region, was located in seismic intensity zones VIII and VII, and the upper part was in the VI zone according to the seismic intensity of the Gorkha earthquake provided by the USGS National Earthquake Information Center (Figure 2).

The elevation ranges from the highest peak of Mt. Shishapangma at 8012 m to the lowest point of 591 m in Dolalghat, Nepal. Given the large relief, the landforms are different from north to south. In the north and central parts of the basin are alpine regions and gorges, while the valley becomes broader in the south. The climate also varies considerably from south to north. The Himalayan southern slope region of the basin is affected by the Indian monsoon and experiences high precipitation levels. Meanwhile, due to blockage by the Himalayan range, the warm, moist air from the Indian monsoon can hardly reach the northern part of the basin. According to monthly data obtained from the Nyalam meteorological station (3810 m a.s.l.) and the Zhangmu meteorological station (2250 m a.s.l.), the mean annual temperature ranges from 3.8 °C to 10.1 °C, and the mean annual precipitation ranges from 643.4 mm in the north to 2820.6 mm in the south.

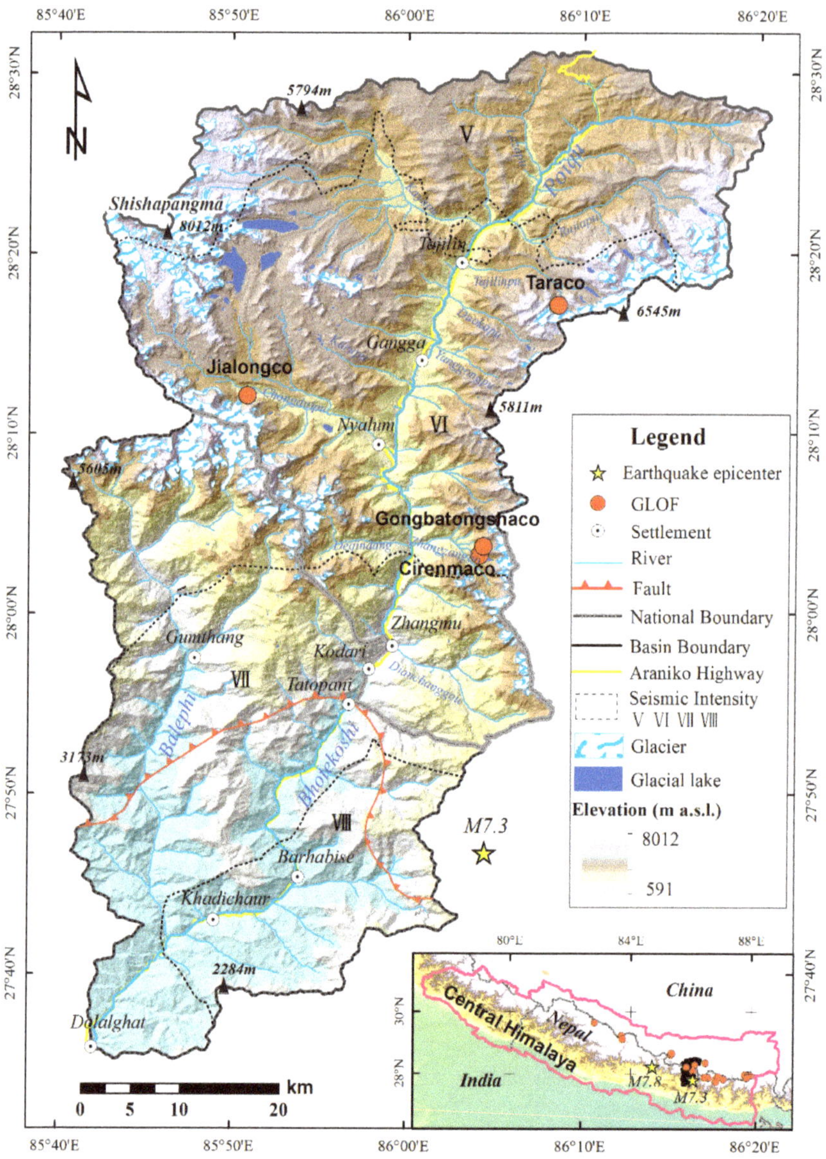

Figure 2. Study area and GLOF events in Bhote Koshi Basin.

3. Methods

3.1. Glacial Lake and Landslide Inventory Mapping

Glacial lake and landslide identification were based on GaoFen-1 (GF-1) satellite images. Twenty-two GF-1 images obtained from the China Center for Resources Satellite Data and Application (http://www.cresda.com/CN/) were Level-1A products, with cloud coverage less than 20% (Table 1). Seven images were used to map glacial lakes and fifteen images were used to map landslides pre- and post-earthquake. Geometric correction and image sharpening were conducted in ENVI 5.2 before mapping in ArcGIS 10.2. The resolution of pan sharpened images was 2 m. This high-quality imagery available allowed us to recognize glacial lakes as small as 0.01 km^2. Manual visual interpretation was used to identify glacial lakes and landslides.

Table 1. GaoFen-1 images used in this study.

Data Usage	Sensor	Product ID	Date	Cloud (%)
	PMS1	2056413	20 December 2016	13
	PMS1	2056415	20 December 2016	4
	PMS1	2056414	20 December 2016	3
Glacial lake mapping	PMS1	1929663	1 November 2016	13
	PMS1	1929664	1 November 2016	16
	PMS1	1929665	1 November 2016	6
	PMS2	1524197	14 April 2016	15
	PMS2	1242505	13 December 2015	11
	PMS2	1242506	13 December 2015	1
	PMS2	820531	22 May 2015	1
Landslide mapping post-earthquake	PMS1	827786	23 May 2015	3
	PMS2	1062062	26 May 2015	14
	PMS2	1062061	26 September 2015	0
	PMS2	1242505	13 December 2015	11
	PMS1	1251892	17 December 2015	3
	PMS2	751296	11 April 2015	0
	PMS2	598009	19 January 2015	12
	PMS2	507470	9 December 2014	1
Landslide mapping pre-earthquake	PMS3	507469	9 December 2014	19
	PMS1	646048	22 September 2014	16
	PMS1	232717	22 May 2014	2
	PMS1	142225	30 December 2013	9

All glacial lakes were verified and modified against Google Earth to see if there are some misinterpretations of the results due to the effect of terrain shadow. The characteristics and surrounding information of all lakes (larger than 0.01 km^2) were measured or estimated, aided by Topography Mission digital elevation model (SRTM DEM) (30 m) and Google Earth. These data compose a complete inventory and provide a basis for identifying dangerous glacial lakes. The inventory of the database consisted of 17 parameters, and some important attributes are explained as follows:

(a) Name: some glacial lakes were annotated according to the topographical map of 1978.

(b) Longitude and latitude: the central location of a glacial lake was calculated automatically in ArcGIS based on WGS84 coordinates.

(c) Elevation (m a.s.l.): the central elevation of a glacial lake was derived from the DEM.

(d) Dam type: moraine dam, ice dam and bedrock dam, which was specified based on remote sensing images and the topography map (1:100,000; produced in 1978).

(e) Area (km^2): the glacial lake surface area was calculated automatically in ArcGIS 10.2, based on UTM projection zone 48 on a WGS84 ellipsoid.

(f) Dam width (m): these values were estimated using Google Earth.

(g) Volume (m^3): each glacial lake's volume was estimated using Equation (1), which was established between lake areas and volumes of lake water based on data from 33 Himalayan glacial lakes measured in the field [34],

$$V_{\text{gl}} = 0.0578 A_{gl}^{1.4683} \tag{1}$$

where, A_{gl} is glacial lake area.

(h) Estimated freeboard values (1, or 0): the height of the freeboard is difficult to measure by remote sensing but is a crucial parameter that influences dam failure. Here, we estimated whether the height was larger than only a few meters (the value was 1) or indeed close to zero (the value was 0) [35], so it is a semiquantitative parameter.

(i) Potential triggering impacts: whether the mass movement around a glacial lake can enter into the lake, such as rockfalls (R), landslides (L), ice and glacier avalanches (IGA), debris flows (DF) or flood from a lake situated upstream (ULF). If there is no mass movement, the value was null. This was identified based on Google Earth and the slope maps derived from the DEM, so it is also a semiquantitative parameter.

(j) Distance to mother glacier (m): the distance between the back edge of a glacial lake to the mother glacier. If they are in contact, the value was 0; if there is no glacier around the lake, the value was set to null.

(k) Distance to the nearest settlement (m): the drainage distance from the glacial lake dam to the nearest major settlement was measured using ArcGIS 10.2.

(l) Drainage gradient (°): the average drainage gradient was estimated by a DEM-derived drainage map.

In this study, the term landslide refers to mass movement of a slope, including rockfalls, slope failure and soil slides. Most landslides can be easily identified by visual inspection for vegetation loss or deposits. If there is no vegetation in some areas, the morphology needs careful attention. Landslides are classified as pre-earthquake and post-earthquake landslides. The landslide volume (V) was estimated using a power-law landslide area–volume empirical formula (Equation (2)):

$$V_s = \alpha A_s^{\gamma} \tag{2}$$

where A_s is landslide area, α and γ are empirically calibrated scaling parameters derived from mixed soil and bedrock landslides in the Himalayas; α is 0.257, and γ is 1.36 [36].

3.2. Glacial Lake Outburst Hazard Assessment

Glacial lake outburst hazard assessment includes two steps, glacial lake outburst potential assessment and flow magnitude assessment. First, a qualitative method was used to identify glacial lake outburst potential; then, the outburst flow characteristics were determined, flood or debris flow according to loose matter along the flow path and the channel gradient, and then the magnitude at the nearest settlement was calculated. Finally, the GLOF hazard was derived by the glacial lake outburst probability and flow magnitude based on a matrix diagram, which has been widely used in flood, landslide and rock fall hazard assessments [7,37]. GLOF hazards in BKB were divided into four classes: "Very High", "High", "Medium" and "Low". The process of GLOF hazard assessment is summarized in Figure 3.

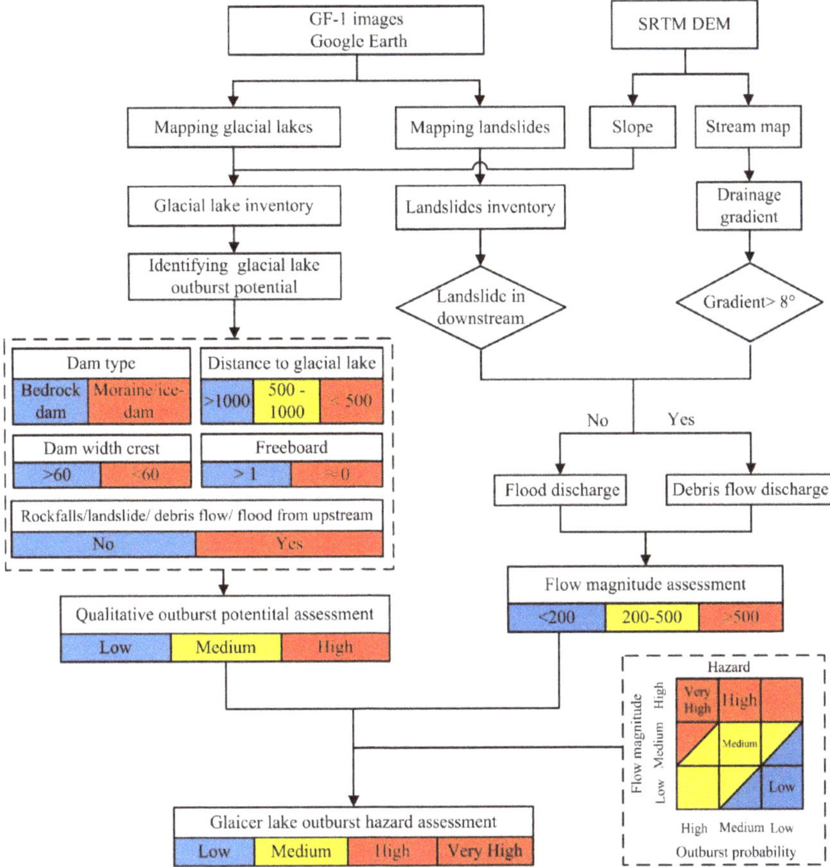

Figure 3. Flow chart illustrating the process used for creating the glacial lake inventory of Bhote Koshi Basin in 2016, classifying glacial lake outburst potential and outburst risk assessment.

3.2.1. Glacial Lake Outburst Potential Assessment

Many criteria and schemes, derived from GLOF experiences all around the world, have been proposed to assess and identify potentially dangerous glacial lakes based on GLOF [38–42]. Here, a qualitative assessment method is proposed to identify the potentially dangerous glacial lakes from three criteria. The first one is potential triggering impacts, such as rockfalls, landslides, snow and ice avalanches, debris flows and flood from a lake situated upstream [43]. Such mass movement entering the lakes trigger displacement waves that subsequently overtop and erode the dams is the most common cause of dam failure in the Himalaya [8,17,44–46]. Steep glacier surfaces that are in contact with or close to a lake are prone to ice avalanches [47–49]. In addition, steep topography is also likely to cause rockfalls and landslides, and as a glacier retreats, much glacial debris remains, which may start a debris flow under heavy rainfall or intense glacier melting [8,38,50]. The second one is dam stability. Studies show most of the GLOF events in the Himalayas are caused by moraine- or ice dam failures, and the bedrock dams are the more stable with low outburst probability [8,51]. The dam width crest is an indicator for the susceptibility of a dam to fail [52]. The third one is freeboard, which is considered a crucial parameter that influences whether a potential impact wave overtops the dam [7,49].

Five key indicators were selected to identify glacial lake outburst potential according to the three criteria. These parameters of each glacial lake were easily obtained from the 2016 glacial lake inventory database. The critical values for assessment are given in Figure 3. The key indicator was defined with qualitative probabilities high, medium and low, and considered independently. The overall potential is not the mean of the individual indicators. A high-potential outburst glacial lake must satisfy three criteria that are high, a low-potential lake has two or three low criteria and no high criteria and the rest are medium-potential lakes. Finally, three potential degrees were classified as high, medium and low.

3.2.2. Flow Magnitude Assessment

The flow magnitude is highly dependent on the peak discharge at the breach and the channel condition [53]. The peak discharge depends on the lake volume and the breach geometry [38]. For a rapid hazard assessment, complex breach processes and flow behavior are beyond the scope. In this paper, the worst breach scenario was assumed, i.e., a full breach that empties the glacial lake water completely. The maximum discharge (Q_p) was estimated using the empirical formula (Equation (3)):

$$Q_p = 2V_{gl}/t \tag{3}$$

where V_{gl} is the glacial lake volume, and t is the drainage duration in seconds, which is assumed to be 1000 s [54].

The outburst flood peak discharge increases due to erosion and entrained sediments. Thus, we first needed to judge whether an outburst flood would develop into a debris flow. The average channel gradient and unconsolidated deposits along the channel are key factors that affect whether an outburst flood evolves into a debris flow [13,55]. Erosion is found to occur where the channel gradients exceed 8° [38] and abundant unconsolidated deposits are distributed in the channel and on the slopes [56]. Channel gradients were calculated based on drainage maps derived from the DEM. The unconsolidated deposits include moraine deposits, fluvial and glaciofluvial sediments and landslide deposits. The maximum eroded sediment volume per unit channel length varies from ten to hundreds of cubic meters due to local and regional differences in geology, topography and hydrology of torrent catchments [38,54]. Therefore, it was hard to set a value certain of sediment depth or volume eroded by flood in different channels or basins. A rough assessment was used to estimate the flow magnitude to the nearest settlement. Flood peak discharge was estimated using an empirical equation (Equation (4)) [57,58]:

$$Q_{pl} = \frac{W}{\frac{W}{Q_p} + \frac{L}{VK}} \tag{4}$$

where Q_p is the flood peak discharge m³/s; W is the capacity of the lake, m³; Q_p is the peak discharge at the breach, m³/s; L is the distance from the glacial lake dam, m; and VK is an empirical coefficient equal to 3.13 for rivers on plains, 7.15 for mountain rivers and 4.76 for rivers flowing through terrain with intermediate relief [59], which here we set the value as 7.15.

For an outburst debris flow, the water source is the outburst flood. Therefore, the peak discharge of the debris flow consists of outburst flood discharge and soil particle flow. Blocking was not considered here, so the debris flow peak discharge (Q_{df}) can be calculated [60]:

$$Q_{df} = (1 + \varphi)Q_{pl} \tag{5}$$

where φ is the increase coefficient of debris flow peak discharge, which can be calculated by:

$$\varphi = (\gamma_s - \gamma_w)/(\gamma_s - \gamma_c) \tag{6}$$

where γ_s is the specific gravity of the solid material, g/cm³, and usually determined as 2.65 g/cm³; γ_w is the unit weight of water, $\gamma_w = 1$ g/cm³; γ_c is the unit weight of the debris flow, g/cm³. Studies show glacial lake outburst debris flow in Tibet is usually diluted flow [52], and the density is 1.3–1.8 g/cm^3. For the convenience of calculation, here we set the average value of γ_c as the density of GBTSC outburst debris flow, 1.55 g/cm^3.

According to Chinese debris flow prevention and control standards (DZT-0220-2006), a peak flow discharge of more than 200 m^3/s is defined as a large hazard. However, the scale of a glacial outburst flood/debris flow is usually larger than that of a rainfall-triggered debris flow [16]. Therefore, in this paper, three flow magnitude classes were established: flow discharge <200 m^3/s (low), 200–500 m^3/s (medium) and >500 m^3/s (high). Finally, the GLOF hazard was derived by the glacial lake outburst probability and flow magnitude based on a matrix diagram, which has been widely used in flood, landslide and rock falls hazard assessments [7,37]. GLOF hazards in BKB are divided into four classes: "Very High", "High", "Medium" and "Low".

4. Results

4.1. Glacial Lake Inventory

A total of 122 glacial lakes larger than 0.01 km^2 with an area of 20.38 km^2 were identified based on the GF-1 images from 2016 (see Supplementary Materials). According to the dam type, 84 moraine-dammed lakes with a total area of 16.87 km^2 accounts for the largest number and area of all lakes. These moraine-dammed lakes are mainly distributed at 5100–5400 m a.s.l. There are 25 bedrock-dammed lakes that account for 15.3% of the area of all lakes. The average area of a bedrock-dammed lake is 0.12 km^2, and are mainly distributed at 4100–4700 m a.s.l. The ice-dammed lakes are the least and smallest, occupying 10.7% and 1.9% of the total number and area. The ice-dammed lakes consist of tiny and small lakes, with a mean area of 0.03 km^2, mainly distributed at 5000–5200 m a.s.l. (Figure 4a).

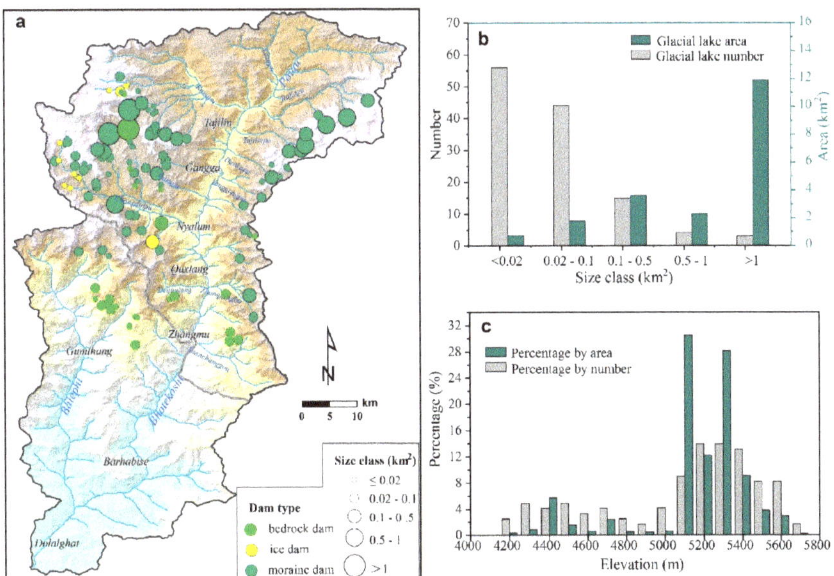

Figure 4. Characteristics of glacial lakes in 2016. (**a**) Glacial lakes distribution of different dam type and size class, (**b**) the number and area variation in size classes and (**c**) the percentage of number and area at different elevations.

As the area of glacial lakes vary greatly, from 0.01 to 5.29 km^2, we classify them into five size classes: tiny (A $\leq$ 0.02 km^2), small (0.02 < A $\leq$ 0.1 km^2), medium (0.1 < A $\leq$ 0.5 km^2), large (0.5 < A $\leq$ 1 km^2) and giant (A > 1 km^2). The percentage of numbers and areas for each size class are shown in Figure 4b.

The main size class of a glacial lake is tiny, accounting for 45.9% (n = 56) of the total number, and small glacial lakes account for approximately 36.1% (n = 44). The total area of tiny and small lakes is 12.4%. Four glacial lakes are large, and three are giant lakes, occupying 69.7% of the total area. Tha mean value of lake area is 0.17 km^2, and the largest glacial lake is Galongco, with a surface area of 5.29 km^2.

The glacial lakes are distributed at elevations ranging from 4100 to 5750 m a.s.l. and are separated into different elevation classes every 100 m (Figure 4c). Most glacial lakes are located at elevations of 5000–5600 m a.s.l., accounting for 66.4% and 86.6% of the total number and area, respectively. Approximately 27.9% of glacial lakes are located below 5000 m a.s.l. and are evenly distributed in each elevation class with an average 3.5% by number. Approximately 13.9% of glacial lakes are distributed from 5200–5300 m a.s.l. and account for 28.1% of the area of all lakes. It is noticeable that the largest percentage by area is distributed at 5000–5100, which accounts for 30.55%.

4.2. Glacial Lake Outburst Flood Hazard

The glacial lake outburst potential assessment results show that 19 glacial lakes have high outburst potential, in which all of these lakes are moraine-dammed and ice/glacier avalanche is the main potential triggering impact; 51 are medium risk, in which two are ice-dammed and nine are bedrock-dammed, and 42 lakes with an area less than 0.1 km^2; 52 glacial lakes are low, in which 11 are ice-dammed and 16 are bedrock-dammed (Figure 5a). It is noticeable that 11 out of 19 high outburst potential lakes have an area less than 0.1 km^2, and the one bedrock-dammed lake, Gongco, with an area of 2.9 km^2, has low outburst potential.

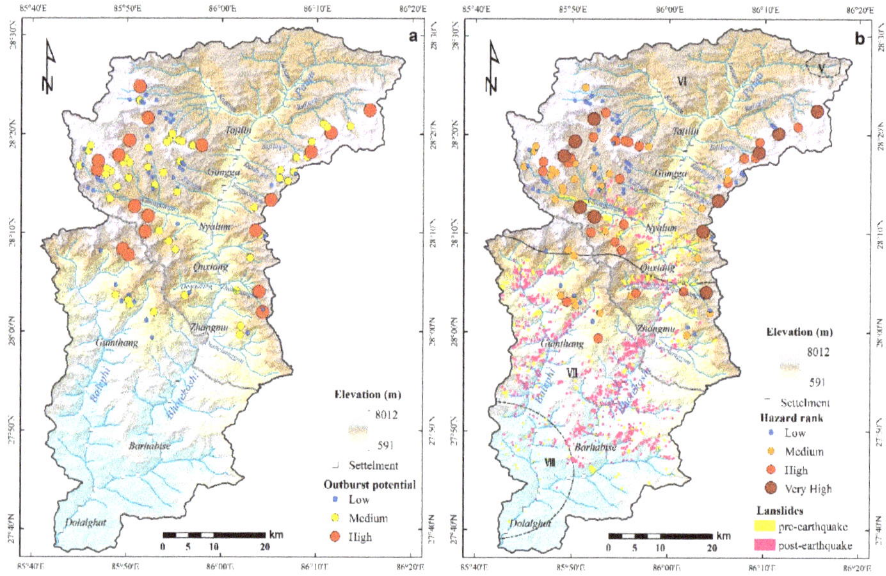

Figure 5. (**a**) Glacial lake outburst potential and (**b**) glacial lake outburst flow/debris flow hazard assessed in this study, considering landslide deposit distribution.

In this study, 1670 landslides with a total area of 18.70 km^2 were identified, in which 1183, with an area of 12.18 km^2, were triggered by the Gorkha earthquake (Figure 5b). These post-earthquake landslides vary in size ranging from 230 m^2 to 254,474 m^2. Most of the landslides are distributed in the middle and southern parts of the basin, and a large number of Gorkha earthquake-triggered landslides are concentrated in the VII region of seismic intensity. A lot of landslides were distributed in the sub-basins such as Gumthang, Deqingdang, Chongduipu, Zhangzangbo and Dianchangchanggou. Large landslides reach channels, and some small landslides are mostly located on steep slopes disconnected from a river channel. The other 487 landslides occurred before the Gorkha earthquake and have an area of 6.51 km^2. The largest mapped landslide that occurred before the earthquake is 0.81 km^2. The total landslide deposit volume is estimated at 91.67 × 10^6 m^3 before the earthquake, and the volume increased to 216.03 × 10^6 m^3 after the earthquake. Considering landslide distribution, 73 glacial lake outburst floods are highly prone to debris flow, which will increase the magnitude.

According to the glacial lake outburst potential and flow magnitude, GLOF hazard assessment results are shown in Table 2 and Figure 5. Eleven glacial lakes are identified with very high hazard, among which seven could evolve into debris flows. Twenty-four glacial lakes are high hazard, among which 11 could evolve into debris flows. Thirty-two glacial lakes are identified with medium hazard; the other 55 glacial lakes are considered to have low to no hazard to downstream areas. Four very high-hazard glacial lakes are located in the Chongduipu gully, which presents a large threat to Nyalam County, especially the giant glacial lake Galongcuo that could generate a peak flow discharge of about 224,449 m^3/s. Both Jialongcuo and Cirenmacuo have burst out twice before, and are also identified as very high hazard due to high freeboards and hanging glaciers behind the lakes. Eight glacial lakes with low-outburst probability but high-magnitude flow are considered to have high hazard. Among these lakes, Gongcuo and Darecuo are bedrock-dammed lakes and have no potential triggering impacts around the lakes, so they are considered to have a low probability of outburst. However, because of their large volumes, the outburst flows are assumed to be high. 63% (n = 22) of the very high and high glacial lakes' areas are larger than 0.1 km^2, and their peak flow discharges were larger than 1000 m^3/s. Ten small glacial lakes (area <0.1 km^2) identified as high hazard. Three small glacial lakes, No. 16 (area 0.09 km^2), No. 18 (area 0.05 km^2) and No. 81 (Nongjue, area 0.07 km^2), are considered very high hazard for they may cause peak debris flow discharges of 1118 m^3/s, 597 m^3/s and 994 m^3/s at the nearest settlement, respectively.

Table 2. Very high and high hazard glacial lakes.

Id	Name	Longitude (°)	Latitude (°)	Elevation (m)	Dam Type	Area (km^2)	Q_{pl} (m^3)	Q_{df} (m^3)	Probablity of Outburst	Flow Magnitude	Hazard
1	Qiezelaco	86.26	28.37	5532	moraine	0.26	1967		High	High	Very High
3	Cawuqudenco	86.19	28.34	5423	moraine	0.55	6666		High	High	Very High
7	Paquco	86.16	28.30	5307	moraine	0.58	7950		High	High	Very High
16		86.09	28.22	5178	moraine	0.09	726	1814	High	High	Very High
18		86.06	28.17	5194	moraine	0.05	239	597	High	High	Very High
22	Cirenmaco	86.07	28.07	4633	moraine	0.34	5087	12,717	High	High	Very High
34	Gangxico	85.87	28.36	5212	moraine	4.52	172,879		High	High	Very High
61	Galongco	85.84	28.32	5077	moraine	5.29	145,746	364,365	High	High	Very High
62		85.82	28.30	5093	moraine	0.27	1832	4580	High	High	Very High
80	Jialongco	85.85	28.21	4380	moraine	0.63	8336	20,840	High	High	Very High
81	Nongjue	85.87	28.19	4628	moraine	0.07	398	994	High	High	Very High
2	Youmojiaco	86.23	28.35	5337	moraine	0.55	4881		Medium	High	High
6	Gangpuco	86.16	28.32	5543	moraine	0.22	2355		Medium	High	High
8	Southhu	86.15	28.30	5343	moraine	0.17	1227		Medium	High	High
9	Taracuo	86.13	28.29	5257	moraine	0.23	2186		Medium	High	High
10	Tuzhuocuo	86.10	28.25	5201	moraine	0.15	1309	3272	Low	High	High
23		86.03	28.07	4486	bedrock	0.03	257	642	Medium	High	High
32	Yinreco	85.89	28.37	5245	moraine	0.28	2878		Low	High	High
40	Mabiya	85.91	28.32	5384	moraine	0.14	931		Medium	High	High
42		85.92	28.32	5345	moraine	0.08	504		Medium	High	High
43	Mulaco	85.93	28.32	5306	moraine	0.11	760		Medium	High	High
44	Xiahu	85.95	28.31	5232	moraine	0.31	3352		Medium	High	High
51	Cuonongjue	85.92	28.26	5095	moraine	0.23	2353		Low	High	High
63		85.83	28.29	5013	moraine	0.26	1863	4658	Medium	High	High
64		85.83	28.29	5050	moraine	0.06	204	511	Medium	High	High
70		85.78	28.29	5418	moraine	0.05	130	324	High	Medium	High
72		85.78	28.27	5309	moraine	0.07	184	459	High	Medium	High
83		85.87	28.17	4712	moraine	0.04	125	312	High	Medium	High
84	Daroco	85.92	28.18	4366	bedrock	0.48	10,966	27,414	Low	High	High
85		85.91	28.15	4486	ice	0.20	2468		Medium	High	High
86		85.92	28.14	4871	moraine	0.09	597		Medium	High	High
88		85.94	28.07	4524	bedrock	0.06	391	977	Low	High	High
89	Bhairab Kunda	85.88	27.99	4102	bedrock	0.06	304	760	Low	High	High
102		85.83	28.05	4250	bedrock	0.07	210	524	Low	High	High
103	Gongco	85.87	28.33	5113	bedrock	2.09	28936		Low	High	High

Note: Q_{pl} is the flood peak discharge at the nearest settlement and Q_{df} is the debris flow peak discharge at the nearest settlement.

5. Discussion

5.1. Glacial Lake Inventory

The new 2016 glacial lake inventory indicates that BKB is highly developed glacial lakes. Glacial lake inventory studies have also been conducted in other regions along the Himalayas [61–64]. Glacial lake studies in the Himalayas show that the greatest numbers and areas of glacial lakes are distributed in the central Himalaya [2,7,9,10,62,65]. To compare the glacial lakes and GLOF of BKB with other regions throughout the central Himalayas, the Gyirong River Basin (GRB), which is next to BKB with a similar area was selected and the glacial lake density (glacial lake number/basin area) and lake area per basin area (total glacial lake area/basin area) were calculated (Table 3). The results show that the glacial lake density of BKB is four times that of the central Himalayas and the lake area per basin area is four times that of GRB. The basin area of GRB is larger than BKB, and the glacial lakes density is similar, while the lake per basin area varies greatly. This is due to more large and giant lakes in BKB. According to the statistics, the largest lake is less than 0.5 km^2 in GRB, while there are seven lakes larger than 0.5 km^2 with the largest being 5.29 km^2 in BKB. Studies show the glacial lake expansion rate reaching 0.26 km^2/year in Poiqu [66], while the rate of GRB is 0.09 km^2/year [63]. Glacial lake expansion is the result of glacier retreating response to climate change. That means BKB is more sensitive to climate change than GRB.

Table 3. Comparison of glacial lake and glacial lake outburst flood among Bhote Koshi Basin, Gyirong River Basin and Central Himalaya.

Region	Basin Area (km^2)	Number of Glacial Lakes	Glacial Lake Area (km^2)	Glacial Lake Density	Lake per Basin Area
Bhote Koshi Basin	3406	122	20.38	0.04	0.0060
Gyirong River Basin [63]	4640	148	7.12	0.03	0.0015
Central Himalaya [2]	280,000	1943	203.7	0.01	0.0007

The analysis of multitemporal and high-resolution remote sensing images during the compilation of the glacial lake inventories provided a good opportunity to identify previously unreported GLOF events [67,68]. Six unreported glacial lake outburst events were found when we mapped glacial lakes from GF-1 and Google Earth. These glacial lakes have retained typical outburst geomorphic and sedimentological features, such as V-shaped breaches, debris fans and subsequent devastated channels (Figure 6). All of them were moraine-dammed lakes, and their surface areas are 0.01–0.11 km^2. Two glacial lakes (Figure 6a,b) are located in Keyapu Valley and the other four (Figure 6c–f) are in Chongduipu Valley. All glacial lakes except No. 31 are fed by glaciers, and the distances to the glaciers are less than 500 m. The V-shaped breach and debris fan of glacial lake No. 86 is the largest, and its mother glacier is thick and hangs behind the lake. The surface area of the glacial lake is 0.09 km^2 and the freeboard is much more than one. The rest of the outburst events were small scale and seemed to cause no downstream damage since no erosion was observed in downstream channels. The outburst flood formed a deposition fan at the intersection with the main channel, such as glacial lake No. 31, where the outburst deposition blocked the channel and formed a small lake. Some vegetation has covered the debris fan (glacial lake No. 86) and deposition fan (glacial lake No. 81). It shows that the glacial lake outburst occurred a long time ago. However, the outburst time (year) cannot be determined because of the lack of long-term and high-quality (low cloud cover or high-resolution) data. We traced theses lakes on Google Earth images, and it shows that outburst signs have existed since 1984. As we documented in the literature, we found there was a GLOF event in 1955 in Nyalam, but the record did not mention which lake burst out [30]. Four glacial lakes (Figure 6c–f) are located in the Chongduipu Basin upstream of Nyalam County. As we cannot be sure of the exact outburst time, the outburst magnitudes can be estimated only through debris fans. Glacial lake No. 86 had the largest outburst magnitude with a debris fan area of approximately 252,808 m^2, and the gully downstream is highly eroded. This lake may have caused damage in downstream areas. The other three glacial lake deposits are found at the intersections with the main channel (Chongduipu), which means their outburst floods did not propagate downstream. Therefore, we conclude that the GLOF event in 1955 was caused by glacial lake No. 86.

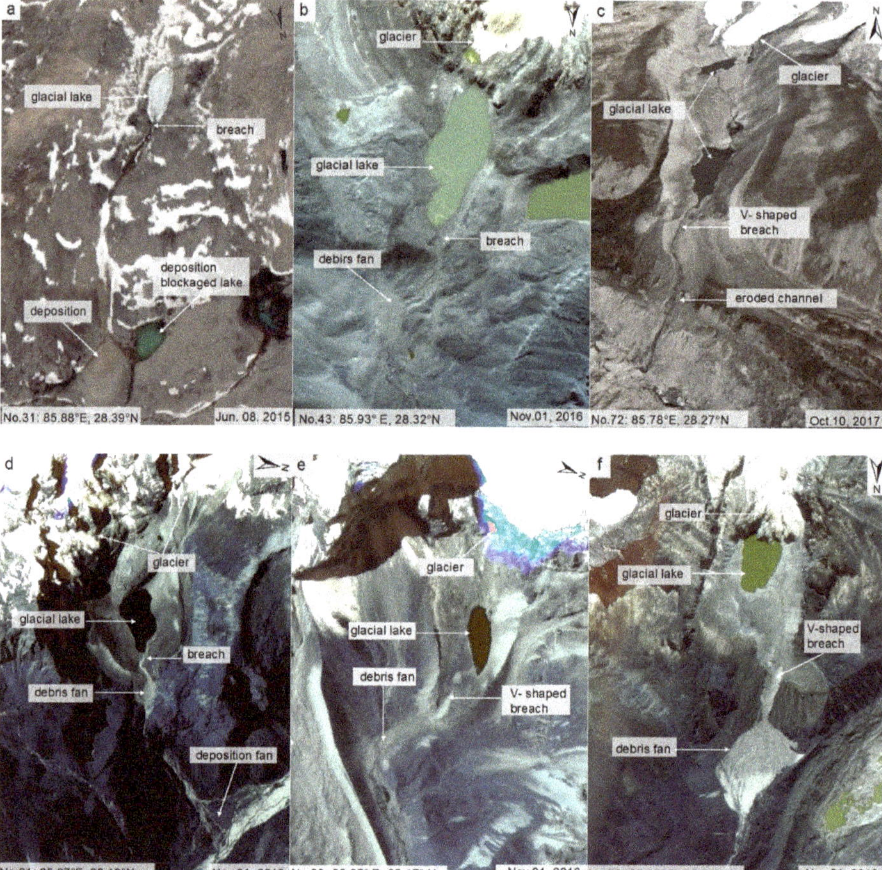

Figure 6. Unreported glacial lake outburst flood events were identified based on Google Earth images (**a**,**c**) and GaoFen-1 (GF-1) images (**b**,**d**–**f**) in the Bhote Koshi Basin.

The six undocumented outburst events found show that high-resolution remote sensing images make it possible to trace minor GLOF events that were unreported because of difficult access or few people living in high mountain regions. It also proves that the GLOF frequency is high in BKB. Other unknown and unpublished GLOF have also been found in Bhutan, Nepal and other parts of the Himalayas, based on long-time-series remote sensing data that show glacial lake changes (disappearance or abrupt shrinkage) and typical topographic features, such as exposed debris fans and sediment tails in downstream river channels [9,17,18]. A database of past GLOF events as complete as possible is essential for robust and reliable GLOF hazard assessment [69]. The gradually improving GLOF inventory helps us better understand the mechanism of GLOF and to do hazard assessment.

5.2. Glacial Lake Outburst Hazard

A GLOF is a complex process, and the hazard magnitude is determined by the outburst water volume and flood routing [8,22,38]. The outburst water volume is mainly related to the glacial lake volume [54,70]. Since the depth of a glacial lake is hard to acquire by remote sensing data and investigation, the surface area of the lake becomes an important indicator for assessing lake volume and hazard. In previous GLOF hazard assessment studies, glacial lakes smaller than 0.1 km^2 were assumed not to pose a hazard potential relevant to downstream locations [47]. Khanal et al. [34] identified 10 critical lakes in the BKB, and all of them are larger than 0.2 km^2. Indeed, small glacial lake outbursts can cause damage to downstream locations. According to the inventory of historical glacial lake outburst floods in the Himalayas, small GLOF, such as Zanaco, Geiqu and Choradari Lakes with areas smaller than 0.1 km^2, damaged downstream roads and villages [9]. On one hand, a small outburst can create a much larger outburst from another lake located downstream, for examples, the reach of GLOF from lakes Artesoncocha and Chacrucocha [43,67]. On the other hand, a small glacial lake outburst flood can transform into debris flows due to downstream sediment entrainment. For example, a small ice-dammed lake outburst in 2009 (area of 34,000 m^2) at Keara in the Andes caused damage 10 km downstream [41].

During the flood routing, landslides on the slopes enter the flood, transforming the flood into debris flows and greatly increasing the discharge, volume and impact. Ignoring the earthquake-induced landslides would underestimate the basin's GLOF hazard. If we do not consider landslides triggered by earthquakes transforming the glacial lake outburst floods into debris flows in the BKB, only nine glacial lakes are identified as having very high hazard, 16 are at high hazard, 12 are at medium hazard and 85 are at low hazard. The hazards rank of two very high hazard small glacial lakes, eight high hazard lakes and 20 medium hazard lakes, accounting for 24.6% of the total lakes, would be decreased. It leads to the GLOF hazard of BKB greatly underestimated.

In this study, the GLOF caused by GBTSC on 25 July 2016 is a good example. GBTSC is located in the Zhangzangbo Valley on the right bank of the Poiqu River, and the average gradient is 182‰ (Figure 7b). This lake was tiny; the surface area before the outburst was 0.01 km^2. After the outburst, the lake was almost empty as shown in Figure 7b. The width of the breach was 27 m, and the depth was 9 m. The peak discharge was 618 m^3/s, and increased to 4019 m^3/s at the section of the Zhangzangbo Valley mouth (approximately 7.2 km from the breach), according to the investigation and assessment report written by the Institute of Mountain Hazards and Environment (http://www.imde.ac.cn). The discharge increased by almost eight times because the outburst flood changed to a debris flow. The Zhangzangbo was in the seismic intensity zone VII and intensely impacted by the Gorkha earthquake. Many landslides were triggered along the river, and some landslide deposits blocked the river (Figure 7c,d). The loose mass volume increased to 8.74 × 10^6 m^3 in Zhangzangbo according to the landslide distribution. These landslide deposits provided rich masses for the debris flow. Once the GLOF occurred, these deposits were easily eroded and entrained, leading the flood to change to a debris flow and amplifying the discharge. Tens of thousands of landslides were triggered by the M 7.8 (Gorkha) and M 7.3 (Dolakha) earthquakes [71,72]. It will take some time to transport these landslide deposits, which accumulated on the slope or in the channel. Thus, in the region affected by strong earthquakes, we must strengthen the monitoring of high-hazard glacial lakes and pay special attention to glacial lake outburst debris flows after an earthquake.

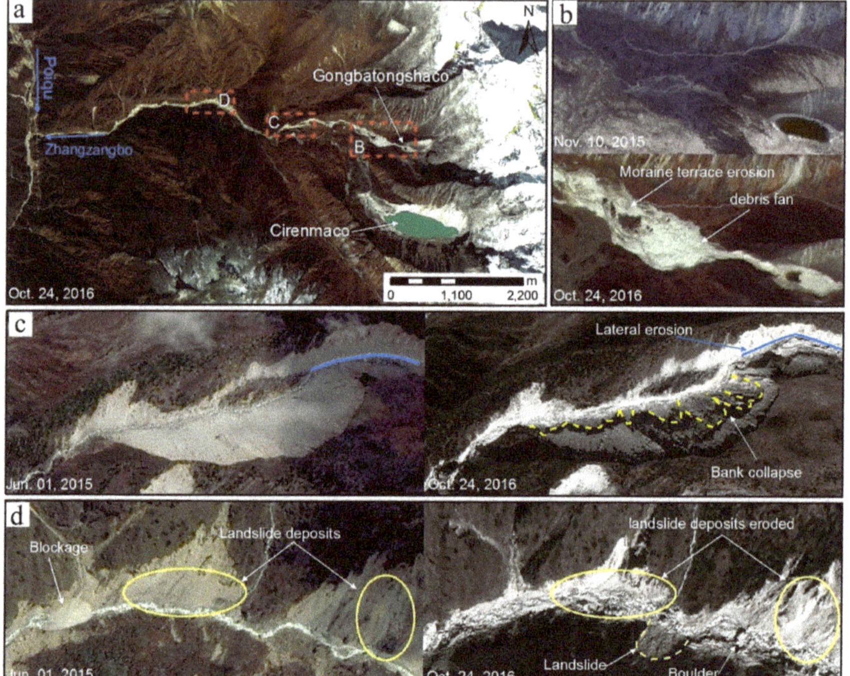

Figure 7. Comparison of channel changes before and after GBTSC GLOF along the Zhangzangbo Valley. (**a**) The flow path of GBTSC GLOF; (**b**) GF-1 images showing that the flood left a large debris fan in the front of the lake, and eroded the moraine terrace; (**c**) lateral erosion and bank collapse in the moraine terrace and the width of channel increased; (**d**) landslide deposits distributed along the channel and blocked the channel before bursting, while GLOF eroded landslide deposits and triggered bank slump and landslide afterward, causing an increase in channel width. (Note: the 1 June 2015 images of c and d are from Google Earth, others are GF-1 images).

6. Conclusions

In this study a new detailed 2016 inventory of glacial lakes in the BKB was established and six unreported GLOF that occurred before 1984 have been detected with geomorphic outburst evidence based on high-resolution remote sensing images. The BKB is one of the most hotspot small river basins for glacial lakes and GLOF in the central Himalayas. High-resolution remote sensing images are useful for detecting unreported GLOF events in high mountainous regions and sparsely populated regions, which is conducive to improving the GLOF inventory and better assessing GLOF hazard. A rough but more comprehensive method was proposed to assess GLOF hazard, which considers the probability for a flood to develop into a debris flow in the downstream, where large numbers of landslides triggered by earthquake are distributed. The GLOF hazard in BKB increases due to landslide deposits volume, which increased approximately 124.36×10^6 m^3 after the Gorkha earthquake, and 11 glacial lakes are identified as very high hazard, nine are high hazard, 32 are medium hazard and 55 are low hazard. However, about 24.6% of the all lakes' hazards would be underestimated without earthquake-induced landslides, in which most of them are small glacial lakes. Therefore, for regional GLOF hazard assessment, small glacial lakes should not be overlooked for landslide deposit entrainment along a flood route and flood eroding channel bed would increase the peak discharge, especially in earthquake affected areas where large numbers of landslides were triggered. We suggest that more attention should be paid to the very high and high-hazard glacial lakes and to improving the engineering security standard for defending against flood hazards downstream of BKB.

Supplementary Materials: The following are available online at http://www.mdpi.com/2073-4441/12/2/464/s1. Table S1: glacial lake inventory of Bhote Koshi.

Author Contributions: N.C. and M.L. conceived the original ideas and drafted the original manuscript. Y.Z. and M.D. carried out field investigation and data collection. N.C. and M.L. revised the original manuscript. All authors have read and agreed to the published version of the manuscript.

Funding: This research was funded by National Natural Science Foundation of China (Grant NOs. 41671112 and 41861134008) and the 135 Strategic Program of the IMHE, CAS (Grant NO. SDS-135-1705).

Conflicts of Interest: The authors declare there no conflict of interest.

References

1. Bolch, T.; Stoffel, M. The State and Fate of Himalayan Glaciers. *Science* **2012**, *336*, 310–314. [CrossRef] [PubMed]
2. Nie, Y.; Sheng, Y.; Liu, Q.; Liu, L.; Liu, S.; Zhang, Y.; Song, C. A regional-scale assessment of Himalayan glacial lake changes using satellite observations from 1990 to 2015. *Remote Sens. Environ.* **2017**, *189*, 1–13. [CrossRef]
3. Kang, S.; Xu, Y.; You, Q.; Flügel, W.-A.; Pepin, N.; Yao, T. Review of climate and cryospheric change in the Tibetan Plateau. *Environ. Res. Lett.* **2010**, *5*, 015101. [CrossRef]
4. Song, C.; Sheng, Y.; Wang, J.; Ke, L.; Madson, A.; Nie, Y. Heterogeneous glacial lake changes and links of lake expansions to the rapid thinning of adjacent glacier termini in the Himalayas. *Geomorphology* **2016**, *280*. [CrossRef]
5. Harrison, S.; Kargel, J.S.; Huggel, C.; Reynolds, J.M.; Vilímek, V. Climate change and the global pattern of moraine-dammed glacial lake outburst floods. *Cryosphere* **2018**, *12*, 1195–1209. [CrossRef]
6. Cenderelli, D.A.; Wohl, E.E. Peak discharge estimates of glacial-lake outburst floods and "normal" climatic floods in the Mount Everest region, Nepal. *Geomo* **2001**, *40*, 57–90. [CrossRef]
7. Worni, R.; Huggel, C.; Stoffel, M. Glacial lakes in the Indian Himalayas—From an area-wide glacial lake inventory to on-site and modeling based risk assessment of critical glacial lakes. *Sci. Total Environ.* **2013**, *468*, S71–S84. [CrossRef]
8. Richardson, S.D.; Reynolds, J.M. An overview of glacial hazards in the Himalayas. *Quat. Int.* **2000**, *65–66*, 31–47. [CrossRef]
9. Nie, Y.; Liu, Q.; Wang, J.; Zhang, Y.; Sheng, Y.; Liu, S. An inventory of historical glacial lake outburst floods in the Himalayas based on remote sensing observations and geomorphological analysis. *Geomorphology* **2018**, *308*, 91–106. [CrossRef]
10. Yamada, T.; Sharma, C. Glacier lakes and outburst floods in the Nepal Himalaya. *IAHS Publ. Publ. Int. Assoc. Hydrol. Sci.* **1993**, *218*, 319–330.
11. Kattelmann, R. Glacial lake outburst floods in the Nepal Himalaya: A manageable hazard? *Nat. Hazards* **2003**, *28*, 145–154. [CrossRef]
12. Falátková, K. Temporal analysis of GLOFs in high-mountain regions of Asia and assessment of their causes. *Acta Univ. Carol. Geogr. Univ. Karlov.* **2016**, *51*, 145–154. [CrossRef]
13. Worni, R.; Huggel, C.; Clague, J.J.; Schaub, Y.; Stoffel, M. Coupling glacial lake impact, dam breach, and flood processes: A modeling perspective. *Geomo* **2014**, *224*, 161–176. [CrossRef]
14. ICIMOD. *Glacial Lakes and Glacial Lake Outburst Floods in Nepal*; International Centre for Integrated Mountain Development: Kathmandu, Nepal, 2011.
15. Khanal, N.R.; Mool, P.K.; Shrestha, A.B.; Rasul, G.; Ghimire, P.K.; Shrestha, R.B.; Joshi, S.P. A comprehensive approach and methods for glacial lake outburst flood risk assessment, with examples from Nepal and the transboundary area. *Int. J. Water Resour. Dev.* **2015**, *31*, 219–237. [CrossRef]
16. Cui, P.; Dang, C.; Cheng, Z.L.; Scott, K.M. Debris Flows Resulting From Glacial-Lake Outburst Floods in Tibet, China. *Phys. Geogr.* **2010**, *31*, 508–527. [CrossRef]
17. Komori, J.; Koike, T.; Yamanokuchi, T.; Tshering, P. Glacial Lake Outburst Events in the Bhutan Himalayas. *Glob. Environ. Res.* **2012**, *16*, 59–70. [CrossRef]

18. Veh, G.; Korup, O.; Roessner, S.; Walz, A. Detecting Himalayan glacial lake outburst floods from Landsat time series. *Remote Sens. Environ.* **2018**, *207*, 84–97. [CrossRef]

19. Kershaw, J.A.; Clague, J.J.; Evans, S.G. Geomorphic and sedimentological signature of a two-phase outburst flood from moraine-dammed Queen Bess Lake, British Columbia, Canada. *Earth Surf. Processes Landf.* **2010**, *30*, 1–25. [CrossRef]

20. Worni, R. Challenges of modeling current very large lahars at Nevado del Huila Volcano, Colombia. *Bull. Volcanol.* **2012**, *74*, 309–324. [CrossRef]

21. Manville, V. Palaeohydraulic analysis of the 1953 Tangiwai lahar; New Zealand's worst volcanic disaster. *Acta Vulcanol.* **2004**, *16*, 1000–1015. [CrossRef]

22. Mergili, M.; Schneider, D.; Worni, R. Glacial Lake Outburst Floods (GLOFs): Challenges in prediction and modelling. In Proceedings of the 5th International Conference on Debris-Flow Hazards Mitigation: Mechanics, Prediction and Assessment, Rome, Italy, 14 June 2011; pp. 973–982.

23. Breien, H.; Blasio, F.V.D.; Elverhøi, A.; Høeg, K. Erosion and morphology of a debris flow caused by a glacial lake outburst flood, Western Norway. *Landslides* **2008**, *5*, 271–280. [CrossRef]

24. Iverson, R.M. Elementary theory of bed-sediment entrainment by debris flows and avalanches. *J. Geophys. Res. Earth Surf.* **2012**, *117*. [CrossRef]

25. Lugon, R.; Stoffel, M. Rock-glacier dynamics and magnitude–frequency relations of debris flows in a high-elevation watershed: Ritigraben, Swiss Alps. *Glob. Planet. Chang.* **2010**, *73*, 202–210. [CrossRef]

26. Pareek, N.; Sharma, M.L.; Arora, M.K. Impact of seismic factors on landslide susceptibility zonation: A case study in part of Indian Himalayas. *Landslides* **2010**, *7*, 191–201. [CrossRef]

27. Huang, R.; Li, W. Post-earthquake landsliding and long-term impacts in the Wenchuan earthquake area, China. *Eng. Geol.* **2014**, *182*, 111–120. [CrossRef]

28. Lin, C.W.; Shieh, C.L.; Yuan, B.D.; Shieh, Y.C.; Liu, S.H.; Lee, S.Y. Impact of Chi-Chi earthquake on the occurrence of landslides and debris flows: Example from the Chenyulan River watershed, Nantou, Taiwan. *Eng. Geol.* **2004**, *71*, 49–61. [CrossRef]

29. Osti, R.; Bhattarai, T.N.; Miyake, K. Causes of catastrophic failure of Tam Pokhari moraine dam in the Mt. Everest region. *Nat. Hazards* **2011**, *58*, 1209–1223. [CrossRef]

30. Xu, D. Characteristics of debris flow caused by outburst of glacial lake in Boqu river, Xizang, China, 1981. *GeoJournal* **1988**, *17*, 569–580. [CrossRef]

31. Cook, K.L.; Andermann, C.; Gimbert, F.; Adhikari, B.R.; Hovius, N. Glacial lake outburst floods as drivers of fluvial erosion in the Himalaya. *Science* **2018**. [CrossRef]

32. Guo, C.-w.; Huang, Y.-d.; Yao, L.-k.; Alradi, H. Size and spatial distribution of landslides induced by the 2015 Gorkha earthquake in the Bhote Koshi river watershed. *J. Mt. Sci.* **2017**, *14*, 1938–1950. [CrossRef]

33. Tanoli, J.I.; Ningsheng, C.; Regmi, A.D.; Jun, L. Spatial distribution analysis and susceptibility mapping of landslides triggered before and after Mw7. 8 Gorkha earthquake along Upper Bhote Koshi, Nepal. *Arabian J. Geosci.* **2017**, *10*, 277. [CrossRef]

34. Khanal, N.R.; Hu, J.M.; Mool, P. Glacial Lake Outburst Flood Risk in the Poiqu/Bhote Koshi/Sun Koshi River Basin in the Central Himalayas. *Mt. Res. Dev.* **2015**, *35*, 351–364. [CrossRef]

35. Petrov, M.A.; Sabitov, T.Y.; Tomashevskaya, I.G.; Glazirin, G.E.; Chernomorets, S.S.; Savernyuk, E.A.; Tutubalina, O.V.; Petrakov, D.A.; Sokolov, L.S.; Dokukin, M.D. Glacial lake inventory and lake outburst potential in Uzbekistan. *Sci. Total Environ.* **2017**, *592*, 228. [CrossRef] [PubMed]

36. Larsen, I.J.; Montgomery, D.R.; Korup, O. Landslide erosion controlled by hillslope material. *Nat. Geosci.* **2010**, *3*, 247. [CrossRef]

37. Lateltin, O.; Haemmig, C.; Raetzo, H.; Bonnard, C. Landslide risk management in Switzerland. *Landslides* **2005**, *2*, 313–320. [CrossRef]

38. Huggel, C.; Haeberli, W.; Kääb, A.; Bieri, D.; Richardson, S. An assessment procedure for glacial hazards in the Swiss Alps. *CaGeJ* **2004**, *41*, 1068–1083. [CrossRef]

39. Mckillop, R.J.; Clague, J.J. Statistical, remote sensing-based approach for estimating the probability of catastrophic drainage from moraine-dammed lakes in southwestern British Columbia. *Glob. Planet. Chang.* **2007**, *56*, 153–171. [CrossRef]

40. Wang, W.; Yao, T.; Gao, Y.; Yang, X.; Kattel, D.B. A first-order method to identify potentially dangerous glacial lakes in a region of the southeastern Tibetan Plateau. *Mt. Res. Dev.* **2011**, *31*, 122–130. [CrossRef]

41. Cook, S.J.; Kougkoulos, I.; Edwards, L.A.; Dortch, J.; Hoffmann, D. Glacier change and glacial lake outburst flood risk in the Bolivian Andes. *Cryosphere Discuss.* **2016**. [CrossRef]

42. Kougkoulos, I.; Cook, S.J.; Jomelli, V.; Clarke, L.; Symeonakis, E.; Dortch, J.M.; Edwards, L.A.; Merad, M. Use of multi-criteria decision analysis to identify potentially dangerous glacial lakes. *Sci. Total Environ.* **2018**, *621*, 1453–1466. [CrossRef]

43. Emmer, A.; Cochachin, A. The causes and mechanisms of moraine-dammed lake failures in the Cordillera Blanca, North Amerian Cordillera, and Himalayas. *AUC Geogr.* **2013**, *48*, 5–15. [CrossRef]

44. Westoby, M.J.; Glasser, N.F.; Hambrey, M.J.; Brasington, J.; Reynolds, J.M.; Hassan, M.A. Reconstructing historic Glacial Lake Outburst Floods through numerical modelling and geomorphological assessment: Extreme events in the Himalaya. *Earth Surf. Processes Landf.* **2014**, *39*, 1675–1692. [CrossRef]

45. Liu, J.J.; Cheng, Z.L.; Su, P.C. The relationship between air temperature fluctuation and Glacial Lake Outburst Floods in Tibet, China. *Quat. Int.* **2013**, *321*, 78–87. [CrossRef]

46. Rounce, D.R.; Mckinney, D.C.; Lala, J.M.; Byers, A.C.; Watson, C.S. A new remote hazard and risk assessment framework for glacial lakes in the Nepal Himalaya. *Hydrol. Earth Syst. Sci.* **2016**, *20*, 1–48. [CrossRef]

47. Wang, X.; Liu, S.; Ding, Y.; Guo, W.; Jiang, Z.; Lin, J.; Han, Y. An approach for estimating the breach probabilities of moraine-dammed lakes in the Chinese Himalayas using remote-sensing data. *Nat. Hazards Earth Syst. Sci.* **2012**, *12*, 3109–3122. [CrossRef]

48. Fujita, K.; Suzuki, R.; Nuimura, T.; Sakai, A. Performance of ASTER and SRTM DEMs, and their potential for assessing glacial lakes in the Lunana region, Bhutan Himalaya. *J. Glaciol.* **2008**, *54*, 220–228. [CrossRef]

49. Clague, J.J.; Evans, S.G. A review of catastrophic drainage of moraine-dammed lakes in British Columbia. *Quat. Sci. Rev.* **2000**, *19*, 1763–1783. [CrossRef]

50. Haeberli, W.; Schaub, Y.; Huggel, C. Increasing risks related to landslides from degrading permafrost into new lakes in de-glaciating mountain ranges. *Geomorphology* **2016**, *293*. [CrossRef]

51. Das, S.; Kar, N.S.; Bandyopadhyay, S. Glacial lake outburst flood at Kedarnath, Indian Himalaya: A study using digital elevation models and satellite images. *Nat. Hazards* **2015**, *77*, 769–786. [CrossRef]

52. Lv, T.B.; Li, D. *Introduction of Debris Flow Resulted from Glacial Lakes Failed*; Sichuan University Publishing House: Chengdu, China, 1999.

53. O'Connor, J.E.; Baker, V.R. Magnitudes and implications of peak discharges from glacial Lake Missoula. *GSAMB* **1992**, *104*, 267–279. [CrossRef]

54. Huggel, C.; Kääb, A.; Haeberli, W.; Teysseire, P.; Paul, F. Remote sensing based assessment of hazards from glacier lake outbursts: A case study in the Swiss Alps. *Can. Geotech. J. CaGeJ* **2002**, *39*, 316–330. [CrossRef]

55. Stoffel, M.; Huggel, C. Effects of climate change on mass movements in mountain environments. *Prog. Phys. Geogr.* **2012**, *36*, 421–439. [CrossRef]

56. Cui, P.; Zhou, G.G.; Zhu, X.; Zhang, J. Scale amplification of natural debris flows caused by cascading landslide dam failures. *Geomorphology* **2013**, *182*, 173–189. [CrossRef]

57. Cui, P.; Zhuang, J.Q.; You, Y.; Chen, X.Q.; Scott, K.M. Landslide-dammed lake at Tangjiashan, Sichuan province, China (triggered by the Wenchuan Earthquake, May 12, 2008): Risk assessment, mitigation strategy, and lessons learned. *Environ. Earth Sci.* **2012**, *65*, 1055–1065. [CrossRef]

58. Fan, X.; Tang, C.X.; Westen, C.J.V.; Alkema, D. Simulating dam-breach flood scenarios of the Tangjiashan landslide dam induced by the Wenchuan Earthquake. *Nat. Hazards Earth Syst. Sci.* **2012**, *12*, 3031–3044. [CrossRef]

59. Li, W. *Handbook of Hydraulic Calculations*; Water Publication: Beijing, China, 1980. [CrossRef]

60. Liu, J.; You, Y.; Chen, X.; Liu, J.; Chen, X. Characteristics and hazard prediction of large-scale debris flow of Xiaojia Gully in Yingxiu Town, Sichuan Province, China. *Eng. Geol.* **2014**, *180*, 55–67. [CrossRef]

61. Wang, S.; Zhang, T. Spatial change detection of glacial lakes in the Koshi River Basin, the Central Himalayas. *Environ. Earth Sci.* **2014**, *72*, 4381–4391. [CrossRef]

62. Bajracharya, S.R. Glaciers, glacial lakes and glacial lake outburst floods in the Mount Everest region, Nepal. *Ann. Glaciol.* **2009**, *50*, 81–86. [CrossRef]

63. Jiang, S.; Nie, Y.; Liu, Q.; Wang, J.; Liu, L.; Hassan, J.; Liu, X.; Xu, X. Glacier Change, Supraglacial Debris Expansion and Glacial Lake Evolution in the Gyirong River Basin, Central Himalayas, between 1988 and 2015. *Remote Sens.* **2018**, *10*, 986. [CrossRef]

64. Gardelle, J.; Arnaud, Y.; Berthier, E. Contrasted evolution of glacial lakes along the Hindu Kush Himalaya mountain range between 1990 and 2009. *Glob. Planet. Chang.* **2011**, *75*, 47–55. [CrossRef]

65. Bhambri, R.; Bolch, T. Glacier mapping: A review with special reference to the Indian Himalayas. *Prog. Phys. Geogr.* **2009**, *33*, 672–704. [CrossRef]

66. Wang, W.; Xiang, Y.; Gao, Y.; Lu, A.; Yao, T. Rapid expansion of glacial lakes caused by climate and glacier retreat in the Central Himalayas. *Hydrol. Process.* **2015**, *29*, 859–874. [CrossRef]

67. Emmer, A. Geomorphologically effective floods from moraine-dammed lakes in the Cordillera Blanca, Peru. *Quat. Sci. Rev.* **2017**, *177*, 220–234. [CrossRef]

68. Wilson, R.; Glasser, N.F.; Reynolds, J.M.; Harrison, S.; Anacona, P.I.; Schaefer, M.; Shannon, S. Glacial lakes of the Central and Patagonian Andes. *Glob. Planet. Chang.* **2018**, *162*. [CrossRef]

69. Emmer, A.; Vilímek, V.; Huggel, C.; Klimeš, J.; Schaub, Y. Limits and challenges to compiling and developing a database of glacial lake outburst floods. *Landslides* **2016**, *13*, 1579–1584. [CrossRef]

70. Clarke, G.K.C.; Mathews, W.H.; Pack, R.T. Outburst floods from glacial Lake Missoula. *Quat. Res.* **1984**, *22*, 289–299. [CrossRef]

71. Martha, T.R.; Roy, P.; Mazumdar, R.; Govindharaj, K.B.; Kumar, K.V. Spatial characteristics of landslides triggered by the 2015 M w 7.8 (Gorkha) and M w 7.3 (Dolakha) earthquakes in Nepal. *Landslides* **2016**, 1–8. [CrossRef]

72. Kargel, J.; Leonard, G.; Shugar, D.H.; Haritashya, U.; Bevington, A.; Fielding, E.; Fujita, K.; Geertsema, M.; Miles, E.; Steiner, J. Geomorphic and geologic controls of geohazards induced by Nepal's 2015 Gorkha earthquake. *Science* **2016**, *351*, aac8353. [CrossRef]

Article

Flood Consequences of Land-Use Changes at a Ski Resort: Overcoming a Geomorphological Threshold (Portainé, Eastern Pyrenees, Iberian Peninsula)

Glòria Furdada [1,2,3,*], Ane Victoriano [1,2,3], Andrés Díez-Herrero [4], Mar Génova [5], Marta Guinau [1,2,3], Álvaro De las Heras [4], Rosa Mª Palau [6,7], Marcel Hürlimann [6], Giorgi Khazaradze [1,2,3], Josep Maria Casas [1,3], Aina Margalef [8], Jordi Pinyol [9] and Marta González [9]

[1] Department of Earth and Ocean Dynamics, Faculty of Earth Sciences, University of Barcelona, c/ Martí i Franquès s/n, 08028 Barcelona, Spain; ane3790@gmail.com (A.V.); mguinau@ub.edu (M.G.); gkhazar@ub.edu (G.K.); casas@ub.edu (J.M.C.)

[2] RISKNAT Research Group, University of Barcelona, c/ Martí i Franquès s/n, 08028 Barcelona, Spain

[3] Geomodels Research Institute, University of Barcelona, c/ Martí i Franquès s/n, 08028 Barcelona, Spain

[4] Geological Hazards Division, Geological Survey of Spain & IMDEA-Water. c/ Ríos Rosas 23, 28003 Madrid, Spain; andres.diez@igme.es (A.D.-H.); alvarodelasheras@hotmail.com (Á.D.l.H.)

[5] Department of Natural Systems and Resources, Technical University of Madrid (UPM), 28040 Madrid, Spain; mar.genova@upm.es

[6] Department of Civil and Environmental Engineering, Division of Geotechnical Engineering and Geosciences, UPC BarcelonaTECH, c/ Jordi Girona 1–3 (D2), 08034 Barcelona, Spain; rosa.m.palau@upc.edu (R.M.P.); marcel.hurlimann@upc.edu (M.H.)

[7] Center of Applied Research in Hydrometeorology, Universitat Politècnica de Catalunya, BarcelonaTech, c/ Jordi Girona 1–3 (C4), 08034 Barcelona, Spain

[8] Centre d'Estudis de la Neu I la Muntanya d'Andorra/Snow and Mountain Research Center of Andorra, Av. Rocafort 21-23, Edifici Molí 3r pis, AD600 Sant Julià de Lòria, Andorra; amargalef@iea.ad

[9] Group of the Geological Hazard Prevention Map of Catalonia, Cartographic and Geological Institute of Catalonia, Parc de Montjuïc S/N, 08038 Barcelona, Spain; Jordi.Pinyol@icgc.cat (J.P.); Marta.Gonzalez@icgc.cat (M.G.)

* Correspondence: gloria.furdada@ub.edu; Tel.: +34-93-402-1370

Received: 24 December 2019; Accepted: 25 January 2020; Published: 29 January 2020

Abstract: The sensitive mountain catchment of Portainé (Eastern Pyrenees, Iberian Peninsula) has recently experienced a significant change in its torrential dynamics due to human disturbances. The emplacement of a ski resort at the headwaters led to the surpassing of a geomorphological threshold, with important consequences during flood events. Consequently, since 2008, channel dynamics have turned into sediment-laden, highly destructive torrential flows. In order to assess this phenomenon and o acquire a holistic understanding of the catchment's behaviour, we carried out a field work-based multidisciplinary study. We considered the interaction of the various controlling factors, including bedrock geology, geomorphological evolution, derived soils and coluvial deposits, rainfall patterns, and the hydrological response of the catchment to flood events. Moreover, anthropogenic land-use changes, its consequential hydrogeomorphic effects and the role of vegetation were also taken into account. Robust sedimentological and geomorphological evidence of ancient dense debris flows show that the basin has shifted around this threshold, giving rise to two different behaviours or equilibrium conditions throughout its history: alternating periods of moderate, bedload-laden flows and periods of high sediment-laden debris flow dynamics. This shifting could have extended through the Holocene. Finally, we discuss the possible impact of climate and global change, as the projected effects suggest future soil and forest degradation; this, jointly with more intense rainfalls in these mountain environments, would exacerbate the future occurrence of dense sediment-laden flows at Portainé, but also in other nearby, similar basins.

Keywords: flood torrential dynamics; geomorphological threshold; land-use changes; ski resort; hydrological response; Pyrenees

---

1. Introduction

1.1. The Geomorphological Threshold Concept

The International Association of Geomorphologists, in its glossary compiled by Goudie [1], defines geomorphological threshold as "a threshold of landform stability that is exceeded either by intrinsic change of the landform itself, or by a progressive change of an external variable". This definition derives from the remarkable ideas of [2–5], among others, also compiled and discussed by [6]. Gregory [7] defines thresholds as stages or tipping points at which essential characteristics change dramatically. They are boundary conditions separating two distinct phases or equilibria conditions. According to [8], equilibrium typically refers to a condition with no net change and is thus very dependent on the time span being considered. Thus, the dynamic equilibrium concept assumes that, in a basin, forms and processes adjust to convey water and sediment supply while maintaining a balance with the erosional resistance and the stability characteristics of the banks [9,10]. If there is a change in the balance of forces that control the state of the system, it may become unstable, a threshold can be overpassed, and it can shift from a state of dynamic equilibrium to another state. The last book chapters from [5], dedicated to thresholds and man, emphasize how human activities can lead to overpass thresholds and modify geomorphic processes, such as accelerating erosion [11].

The hypothesis that a geomorphological threshold was exceeded because of anthropogenically induced land cover changes can be applied very well to the mountain basin of Portainé (eastern Pyrenees, Iberian Peninsula; see Section 1.2 and Figure 1). In response to an ordinary rainfall event in 2008, the torrential dynamics turned dramatically into heavily sediment laden, destructive flows (Section 1.4).

Gregory [7] considers that a holistic view acknowledges the need to understand how a land surface system operates, as well as to know the relation of its component parts. In Portainé, the dynamics reveal a complex connectivity among water-soil-rock processes. Specialised research, only focused on geomorphology, hydrology, or flow dynamics, does not capture the complexity of the processes in this mountain environment. Thus, the problem should not be considered exclusively by a single discipline, but in the wider context, by looking at relationships as a whole. Consequently, our approach is based on our previous, specifically focussed works carried out with different methods and with conclusive proofs and results [12–22] and also considers the complexity of the system as suggested by Kondolf and Piégay [23], by working with the convergence of evidence. From this scope, this paper integrates and synthesizes the results of multidisciplinary research carried out in the basin of Portainé, as only through this holistic approach can the hypothetical surpassing of the geomorphic threshold after flood events be detected and characterized.

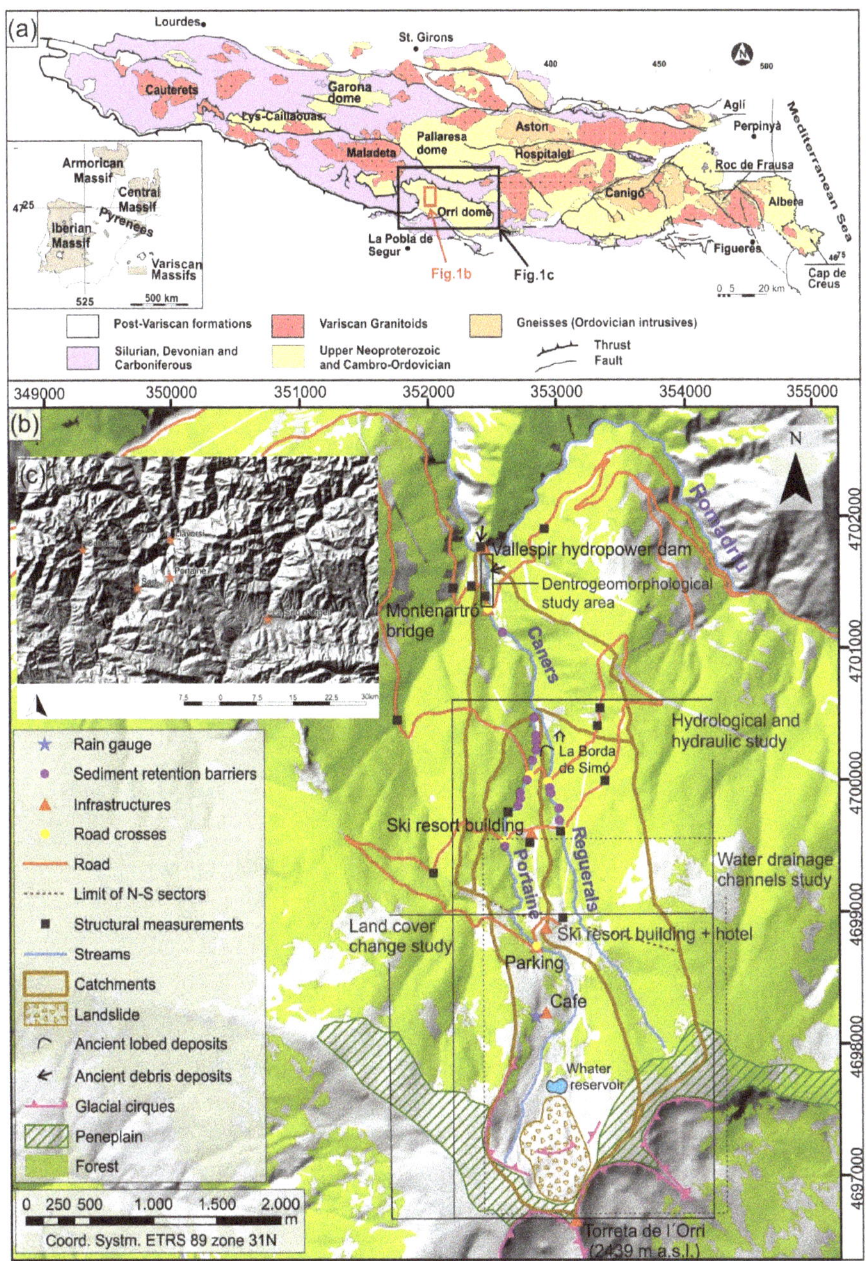

Figure 1. Study area: (**a**) Geographic and geological location of the Axial Pyrenees; Pre-Variscan materials, on which the study area is located, stand out. (**b**) Geomorphological setting and delineation of the Portainé drainage basin, including sub-areas of the study; specific location of studied sediments and anthropogenically significant elements cited throughout the article. (**c**) Locations of the rain gauges used in this study.

1.2. Study Area

The Portainé basin (Figure 1) (5.7 km^2) is located in the Pyrenees (Pallars Sobirà County, Catalonia, Spain), with elevations ranging from 2439 m a.s.l. (Torreta de l'Orri peak) to 950 m a.s.l. (Romadriu river and Vallespir hydropower dam). It includes the Portainé mountain torrent, its tributary Reguerals and, downstream, their confluence (at 1285 m a.s.l.), the lower reach named Caners (4.3, 3 and 1.5 km long; 25.2%, 31.3% and 23.1% average gradient, respectively). The ski resort of Port Ainé was set up at the headwaters and was inaugurated in 1986; its access road follows the hillslopes, repeatedly crossing the channels.

In this region, as the altitude increases, the montane forests are dominated first by *Quercus pubescens* Willd (and less frequently by *Quercus petraea* (Mattuschka) Liebl), second by *Pinus sylvestris*, which is sometimes mixed with *Abies alba* Miller and, in the subalpine belt, by *Pinus uncinata* Ramond ex DC. With respect to soil moisture conditions in ravines and rivers, *Fraxinus excelsior* L. forests and mixed deciduous forests with *Tilia platyphyllos* Scop and *Corylus avellane* L. are developed [24]. There are some bushes and herbaceous crops at medium altitudes and, from 2100 m upwards, there are pastures with patches of *Rhododendron ferrugineum* L. bushes.

The climate is Alpine Mediterranean, strongly influenced by the orography and characterized by high humidity, large variations in temperature, and irregular rainfalls (mean annual rainfall of 800 mm, mean annual temperature of 5–7 °C, [25].

1.3. Data Constraints

As rainfall is the primary trigger of torrential events, the main constraint of this study is the lack of long and continuous time series in the basin. There has been a meteorological station in Portainé (at 1985 m a.s.l.) only since 2011. Thus, for the present work, data from other stations in the region, presented in Figure 1 and Table 1, were used. There are no river gauging stations in the basin. This lack of local data determined the hydrometeorological approach (see Section 5).

Table 1. Basic information on the rain gauges and data used in this study. Data used for long series construction (see Section 5.1.1) correspond to the stations highlighted in bold characters.

Rain Gauge	Altitude (m a.s.l.)	Distance to the Portainé Basin (km)	Measuring Interval (yyyy)	
			Start	End
Portainé	1985	0	2011	-
Montenartró	1322	2.4	2010	-
Llagunes	1300	6.3	2008	-
Sort	679	6.9	2009	-
Llavorsí	850	8	1915	1999
Salòria	2451	16.2	2004	-
Capdella	1083	18.8	1932	1959
La Seu d'Urgell	849	19.22	1996	-

1.4. The 2006 and 2008 Trigger Events

Two significant torrential events occurred in 2006 and 2008. In May 2006, the Reguerals ravine obliterated the access road to the ski resort at its highest intersection. At that time, the drainage infrastructure consisted of a pipe of approximately 80 cm in diameter that was blocked by sediments, causing the passage of the dense flow over the road. This is the first documented road blockage in the area, and most likely reflects system disequilibrium around the geomorphological threshold. The night of 11 to 12 September 2008, the largest torrential event known in the Portainé basin occurred. It coincided with earthworks at the ski resort and resulted from ordinary rains (43 mm at Salòria station, see Section 5.1.3). The talus supporting the road (earth embankment) at the highest crossing of the Portainé stream became saturated and collapsed. This created an erosive breach along the ravine slopes downstream (Figure 2). Highly destructive debris flows moved along both torrents of Portainé and Reguerals, sweeping everything along its path and damaging the Vallespir dam. All intersections of the access road were damaged by the clogging of the drainage passages and the consequent deposition on the road. Since then and until 2017, nine destructive torrential avenues occurred in the basin (Table 2).

Figure 2. Helicopter view of the erosive breach created by the trigger event of 11–12 November 2008. Image property of ICGC (Institut Cartogràfic i Geològic de Catalunya).

Table 2. Historical record for the occurrence of recent torrential flows in the Portainé catchment and their effects on road intersections, indicated by their altitude (TO: total obstruction of the culvert; PO: partial obstruction of the culvert; C: talus collapse; w: water circulation) and by the downstream debris cone (E: erosion; D: deposition).

Event Date (day month year)	Portainé			Reguerals		Caners	No. of Total Road Obstructions	Debris Cone
	1965 m a.s.l.	1700 m a.s.l.	1450 m a.s.l.	1665 m a.s.l.	1465 m a.s.l.	1035 m a.s.l.		
May 2006					TO		1	
11–12 September 2008	TO, C	TO, C	TO	TO, C	TO	w	5	D
02 November 2008			TO				1	
22-23 July 2010			TO, C		TO		2	E, D
12 August 2010			TO, C		TO		2	E, D
05 August 2011		PO	TO, C	w	w		0	D
23 July 2013			TO	TO	TO		3	E, D
20 August 2014							0	
30 August 2014		PO	TO				1	
21 August 2015		PO	TO				1	E
29 August 2016	PO		PO				0	

There is barely any documented information on destructive torrential events in the Portainé basin prior to 2006; they were mostly unknown until the works presented here (see Section 6). However, since 2009 (notably, since the 2008 events), 6.6 M€ has been invested in road works and 0.5 M€ in mitigation measures, including flexible sediment retention barriers along the streams (Figure 1) [19].

2. Methodology

To address the problem, a secular time scale, useful for land use management purposes, was considered appropriate. Then, the threshold control factors can be divided into those that remain unchanged at this timescale (lithological and main macro-geomorphological characteristics), and those susceptible to change (land use, vegetation and hydrometeorology), the former representing the intrinsic control factors and the latter the extrinsic ones [20]. The catchment's hydrological response, its alteration due to the anthropogenic land-use change, and the consequent hydrogeomorphic effects must have been a consequence of the evolution, interaction and change in balance among these control factors. Thus, three levels of analysis were performed: the first being the identification and characterisation of the causes of the change, the second the characterisation of particular, local effects and the third the holistic integration of all the acquired knowledge (Table 3).

Table 3. Synthesis of the methodology used to carry out the present research, with 3 levels of analysis and considering causal control factors, local effects and global consequences in the studied basin.

1st Level of Analysis: Characterisation of Causes of Change			
Causes	**Control Factor**	**Study Method**	**Result**
Intrinsic	Lithology: Substratum	Fieldwork. Structural analysis.	Dipping of remarkable planes. Characterisation of joints and discontinuity planes.
	Geomorphology:		Characterisation of main forms, deposits, soils. Recognition of erodible deposits and soils and ancient deposits. Basin characterisation.
	Area evolution: landforms and deposits.	Fieldwork: Evidence collection. Geomorphological mapping and analysis.	
	Morphometry.	GIS morphometric analysis.	
Extrinsic	Land Use and Land Cover Changes:		
	Regional.	Historical data analysis: population and economic activities evolution. Aerial photograph series comparison.	Forest evolution: regeneration and densification.
	Headwaters.	Fieldwork: evidence collection and mapping. GIS aerial photograph series photointerpretation and analysis.	Land-use changes characterisation.
	Precipitation:		
	Regional precipitation.	Statistical trend analysis of historical rainfall series.	Regional precipitation characterisation.
	Local, basin precipitation modelling.	Elaboration of synthetical hyetographs.	Calculated precipitation return periods.
	Out of basin local precipitation data consideration.	Simulated and real data comparison.	Validation.
2nd Level of Analysis: Characterisation of Local Effects			
Different, Particular Effects	**Observed Change**	**Study method**	**Result**
Observed Effect	Changes in hydrology:	Headwater's hydrological modelling with different land cover conditions.	Validation of changes.
	Hydrogeomorphic effects: *Hydrogeomorphic processes at the headwaters.*	Fieldwork: evidence collection.	Erosion characterisation. Runoff concentration characterisation.
	Hydrogeomorphic processes in the main channels.	Record of road damages. LiDAR analysis.	Sediment balance.
	Effects in thedownstream stretch.	Fieldwork: Dendrogeomorphological sampling; geomorphological and topographical mapping; (Total station, GNNS RTK). Dendrogeomorphological analysis. GIS analysis.	Record of torrential dynamics changes downstream.
3rd Level of Analysis: Discussion and Knowledge Integration			
Holistic Comprehension of the Effects of Change			
Geomorphological Threshold Overcoming Corroboration			

3. Intrinsic Control Factors: Lithology and Geomorphology

3.1. Lithological Characterisation

Insofar as the terrain properties are a conditioning factor for the soil development and hydrological behaviour of systems, we have carried out a specific geological and structural analysis of the study area.

The Pyrenees is an Alpine intracontinental fold and thrust belt that resulted from the convergence between the Iberian and European tectonic plates from Late Cretaceous to Oligocene times [26]. In the Pyrenees, rocks ranging in age from Late Neoproterozoic to Mississippian form an elongated strip in the backbone of the chain known as the Axial Zone, which is unconformably overlain by Mesozoic and Cenozoic rocks (Figure 1). The rocks that crop out at the Portainé catchment are deformed Cambro–Ordovician metapelites from the Serdinya formation [27], also known as Seo [28] or Jujols [29]. The performed analysis consisted in measuring the structural surfaces of the bedrock at fifteen outcrops, but also in studying any other deposits found within the catchment.

The bedrock consists of slates and phyllites with a principal foliation plane, but also showing several other mesoscale surfaces. On the one hand, we recognize four structural fabrics: S0 bedding (11/019), S1 cleavage (49/203), Sr/S2 regional cleavage (40/026), and S3 subtle cleavage (53/316). These suggest several deformational episodes, with the most remarkable planes (S0 and S2) gently dipping to the north subparallel to the bedding and to the topographic surface (Figure 3; Figure S1 in the supplementary material). On the other hand, we have identified two types of fractures: tectonic jointing (two subvertical sets, parallel and perpendicular to Sr, and usually filled with hydrothermal quartz), in addition to non-tectonic fractures (opened). The latter occur much more intensely near the topographic surface or in vertical road slopes (Figure 3). Moreover, near the topographic surface the bedrock often displays considerable weathering.

Figure 3. Regional cleavage of the Cambro–Ordovician bedrock subparallel to the topographic surface. Gradual transition from bedrock (S) to colluvium (C). The black square delineates the area in the right picture.

3.2. Geomorphological Characterisation

3.2.1. Evolution of the Area and Present Features

The main geomorphological features of the area are relict remains of Neogene planation surfaces [30,31] (and references therein), small glacial cirques in the highest areas, and active fluvio-torrential and slope processes affecting the whole territory (Figure 1). The residual planation surfaces correspond to smooth mountain ridges with gentle slopes (<15°), like the one extending around the flat top of Torreta de l'Orri [28]. The incision of the main Pyrenean rivers dissecting these surfaces rejuvenated the relief [30]. On a more local scale, the Romadriu river and its tributaries, like those of the Portainé basin, originated high reliefs (500–1000 m) with narrow valleys and steep slopes (see Section 1.2). Furthermore, during the Late Miocene, the paleoclimate was warm and humid, [32], favouring intense chemical weathering.

Pleistocene glaciations affected the whole Pyrenees [33]. Around Torreta de l'Orri, four cirques —including the one forming the headwaters of Portainé basin— are the southernmost glacial features in the area. Tills and younger rock glacier moraines cannot be clearly observed as they have been eroded, reworked, and incorporated into the slope deposits by postglacial processes, and recently, by construction at the ski resort. The large areas not occupied by glaciers were affected by intense periglacial processes, which produced a large amount of debris that is widely found as thick soils overlaying the bedrock. In the headwaters, there are *grèzes litées* stratified slope deposits (Figure S2). Downwards, up to 10 m thick unconsolidated colluvium covers the hillslopes (Figure 2). Lastly, torrential deposits are accumulated along the valley bottoms and in the Portainé basin in the alluvial cone at the end of the stream, coming from the erosion of the bedrock or the colluvium. All of this large quantity of deposits and the superficially highly altered substratum are susceptible to erosion and to be entrained in the torrential floods.

3.2.2. The Basin's Propensity to Torrential Dynamics

The susceptibility of the basins to produce floods or dense flows is constrained by their morphometric characteristics. The method developed by [34] distinguishes between basins prone to torrential floods that can convey some bed load, hyperconcentrated flows or debris flows. The method is based on the relation among the following morphometric catchment parameters: (i) area, (ii) length from the cone apex to the most distal point, (iii) basin relief (vertical drop from the highest point to the cone apex), (iv) the Melton ratio (relief/$\sqrt{area}$), and (v) relief ratio (relief/length). [35] have already validated this method in Catalonia in more than 50 documented active torrential catchments, including Caners (Figure S3).

The Caners catchment has a length of 5.05 km and a Melton ratio of 0.62 km/km, which indicates its morphometric propension to generate hyperconcentrated flows.

Therefore, the basin is prone to generating, in a wide sense, dense flows, both because its morphometry and due to the availability of erodible material. This is also supported by sediment evidence: the depositional cone at the Portainé confluence with the Romadriu river (see next section). However, it is worth noting that in recent decades, before 2006, the basin produced no dense flows that mostly transported some bedload with no damaging consequences.

3.2.3. Former Evidence of Geomorphological Disequilibrium

Consistently with the torrential basin proneness, solid evidence of ancient dense flows in the basin exists in the form of characteristic sedimentary deposits, which are presented in the following subsections.

Evidence of large debris flows. Evidence of ancient, large debris flows exists at the lower end of the study area. As will be discussed in Section 7, in this case, *ancient* means that these dense flows could be generated since the middle of the 20th century and extend back to, at least, the entire Holocene. Some deposits crop out at the right margin of the Romadriu river, in front of the Caners confluence and at an altitude ranging from 5 to 10 m above the channel (Figure 1). They consist of matrix supported, disorganized and unconsolidated deposits, with heterometric sub-rounded to sub-angular clasts ranging from millimetres to decimetres, protected from erosion inside substratum cavities (Figure S4). Their facies correspond undoubtedly to debris flow deposits. They do not present enough ancient organic detritus and are highly bioturbed by insects and by the roots of the plants growing around, and so they could not be dated by ^{14}C or by OSL techniques. Nonetheless, their high topographic position indicates that they do not correspond to the recent activity of the Portainé basin.

On the other hand, in the 21 August 2015 torrential event, the sediment entrained in the Portainé stream (Table 2) was mostly retained by the flexible barriers [18]. Thus, unloading of the flow produced erosion downstream in the lower reach of Caners, with the outcropping of ancient sediments (Figure 1) that had been covered by younger torrential deposits. These sediments are shown and described in Figure 4 and correspond undoubtedly to debris flow facies. Up to three flow waves can be distinguished in some outcrops. They were preserved, as in the previously described case, in sheltered topographic irregularities of the margins and the stream channel.

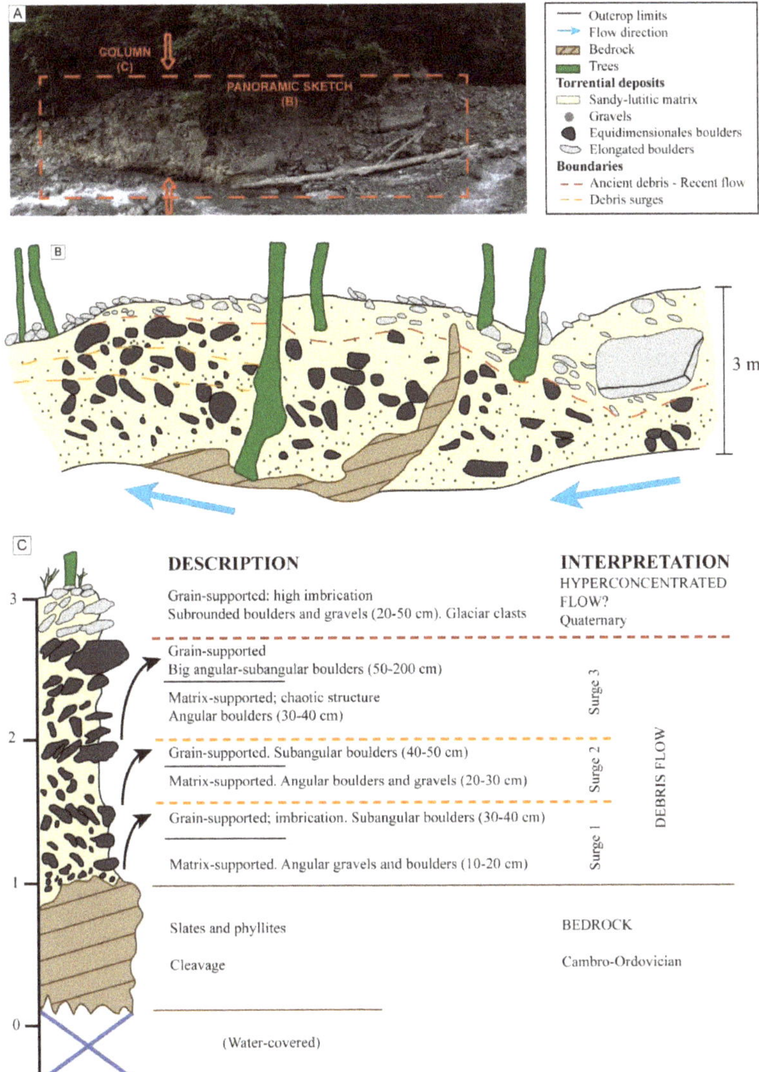

Figure 4. Study of the torrential deposits in the right margin of the lower reach of the Caners stream. (**A**) Photo of the outcrop and location of the sketch and the column. (**B**) Panoramic sketch with the interpretation of materials and torrential episodes. (**C**) Stratigraphic column with the description of the sedimentary packages identified, and the interpretation of the torrential process that generated each of them.

Evidence of dense overflows. At an altitude of approximately 1300 m a.s.l., at the "la Borda de Simó" location (Figure 1), there is solid evidence for ancient dense flows. The terrain is shaped in a clearly decametric lobed morphology (Figure S5), product of these dense flows that had overflown their streams and deposited in this gentler sloped area.

It has not been possible to date the age of the lobes to date, but some trees that grow in them have been dated. Ten *Pinus sylvestris* were sampled; their maximum age is slightly more than 71 years and their average age is a little over 61 years. Even though the precise age of these overflow deposits is unknown, it is prior to the establishment of the trees, i.e., prior to the middle of the 20th century.

4. Land-Use Changes

4.1. General Forest Regeneration: Vegetation Changes in the Lower Part of the Basin

Depopulation was a significant phenomenon throughout the twentieth century in the Pyrenees and, specifically, in the Pallars Sobirà County. Table 4 illustrates the evolution of the population of the municipality of Rialp since 1920, where Portainé belongs. Note the decrease in population, with a minimum in 1950, and a subsequent increase to the present day. Variations in population were associated with the mortality related to the civil war (1936–1939), with previous and later migration and with changes in economic activities, with the abandonment of agriculture and livestock and the birth and growth of tourism mostly since the nineties [36]. The resulting decrease in the use of the land favoured the forest regeneration. Inferring from the ages estimated in [21] (see Section 6.3), most of the trees were progressively established in the Portainé alluvial cone since the beginning of the 1950s. During previous decades, the left bank of the stream was intensely cultivated and when this stopped, present vegetation colonized the area. Summarising, the forest mainly grew and densified in the area, downward from the ski resort (Figure S6). In this regard, one inhabitant (Mr. J. Montserrat) commented about the Romadriu River: *"Not having cattle, there are more trees and the trees drink water when it rains and the floods are smaller"*. The densification of the vegetation cover suggests that torrential floods in the Portainé basin should be smaller now than in the middle of the 20th century, but what happens is just the opposite.

Table 4. Evolution of the population of the Rialp municipality since 1020 [37].

Year	1920	1930	1940	1950	1960	1970	1981	1991	2001	2011
Inhabitants	704	593	460	382	495	659	415	466	537	664

4.2. Land-Use Changes in the Headwaters

The headwater vegetal land cover has been severely disturbed since the ski resort development, as can be clearly seen in Figure 5. We conducted a detailed mapping of the land use (at several year intervals) based on vertical aerial photographs and orthophotos from 1956 to 2014 (Figure S7). Simplified land cover changes are shown in Table 5. Thus, while the forest densified in the region (see previous section), the increase in the area without vegetal cover in Portainé headwaters over time stands out, obviously leading to a decrease in infiltration and water concentration increases (see Section 5.2).

One of the most significant land cover changes is the substitution of the *Rhododendron ferrugineum* bushes' cover by the ski tracks (Figure 5, Figures S7 and S8). This bushes' cover frequently overlaps a moss and herb cover that grow in quite well-developed organic soil. During rainfall episodes, this vegetal cover intercepts and retains water and the soil facilitates infiltration. Conversely, devegetated ski runs favour surface saturation, runoff and water concentration. Fieldwork conducted in October 2014 allowed us to ascertain that, two days after a rainfall, the mosses under the bushes' cover were still wet while the ski runs were already completely dry. Thus, this land cover substitution favours a decrease in interception and infiltration and aggravates all processes related to water concentration and erosion.

Figure 5. Google earth image (2008) of the headwaters, where the Port Ainé ski resort is located. Impact on vegetation is clearly visible.

Table 5. Areas corresponding to the different types of vegetation and areas without vegetation calculated from the headwater maps (Portainé and Reguerals bsubbasins) and their increments (examples in Figure S7). Δ expressing increment.

Type of Vegetation (Area: km^2)	1956	1996–1997	2008	2014	Δ 2014–1956
Without plant cover	0.08	0.38	0.47	0.52	0.43
Meadows	0.71	0.48	0.43	0.39	−0.32
Bushes	0.36	0.15	0.12	0.12	−0.24
Open forest	0.92	0.79	0.77	0.76	−0.16
Dense forest	0.66	0.93	0.93	0.92	0.26

Besides, water drainage channels were dug obliquely to the ski slopes. This is a technique used to avoid rainwater concentration and erosion along the tracks. The water is collected and diverted to the side of the runs. In Portainé, however, this produces undesired side effects because of the channelized runoff erosion (see Section 5.2). In the ski resort, from 2005 to 2007, 0.24 km of channels was dug, 6.52 km more from 2007 to 2011, and 1.58 km more still from 2011 to 2013, giving a total of 8.34 km of such channels (Figure S8). These channels also concentrate the surficial water that is redirected to the Portainé and Reguerals streams.

5. Analysis of Rainfall and Hydrological Response

5.1. Analysis of Rainfall Behaviour

Taking into consideration the lack of rainfall and discharge data in the basin (see Section 1.3), two complementary studies were carried out. The first explores long-term rainfall data series to detect changes in rainfall patterns. The second models synthetic hyetographs. Both are presented in the following sections. Finally, the available data corresponding to the recent torrential events were compared with previous modelling.

5.1.1. Trend Analysis of Historic Rainfall Time Series

The temporal evolution of rainfall during the last hundred years was studied. The goal of this task includes detecting important changes in the precipitation patterns at the beginning of the 21st century, if they occurred. Daily precipitation was assessed, since these types of records represent the only time interval that is available for the historical records.

One of the major concerns in the analysis of rainfall trends is the creation of a long time series. As stated above, the Portainé rain gauge was installed in 2011 and only a discontinuous time series is available. Thus, data from other stations are needed to construct the time series (Figure 1; Table 1).

Fortunately, the Llavorsí rain gauge, which is located approximately 7 km north of the Portainé basin, includes data from 1917 to 1999. However, there is a large gap between 1932 and 1959 because of the Spanish Civil War. The main criterion to fill this gap and to extend the time series until the present was to take into account the data of the closest available rain gauge. Finally, data from four different rain gauges were used (Sort, Llavorsí, La Seu d'Urgell, and Capdella).

As a first step, the daily rainfall values of two gauges are compared for the time span of available data at both of them (Figure S9). The results show that the correlation between Llavorsí and Capdella is especially poor, while the other correlations are acceptable from a meteorological point of view. The disagreement between Llavorsí and Capdella, which is reflected by the much higher daily precipitation at Capdella, is caused because this rain gauge is located in another river valley (Figure 1) and at a higher altitude (Table 1). These circumstances produce higher annual precipitation at Capdella (approximately 1250 mm) than at Llavorsí (720 mm). Thus, a correction to the Capdella records is necessary to fill the gap of the Llavorsí time series.

Finally, a linear regression model between the two rain gauges is applied to estimate the daily rainfall of the final time series. The gap between 1999 and the present was directly filled by the rainfall records measured at La Seu d'Urgell and Sort.

The daily precipitation records between 1917 and 2017 for the Portainé region were used to analyse the general patterns and to detect some important changes in rainfall conditions (Figure S9). This complete time series includes two days with rainfall depths larger than 100 mm/d and seven days with values larger 80 mm/d.

From a general point of view, no change in extreme events is visible. For a detailed analysis, the occurrence of heavy rainfall events was evaluated at annual and decadal periods. The number of daily rainfall events that accumulated more than three critical values, Pcr, were calculated (Pcr equal to 30, 40, and 50 mm/d). The results do not show any significant increase in heavy rainfalls during the last hundred years. Performing a linear regression analysis, the three data series only reveal small augmentations (Figure 6b) and very low R^2-values. For Pcr = 50 mm/d the slope of the trend line is almost zero, while the one for Pcr = 30 mm/d is 0.01. Although this analysis is simple and preliminary, the results show that the higher torrential activity observed in the Portainé basin seems to be related to other factors than a recent increase in rainfall activity (discussed in previous/subsequent sections of this paper). However, a more sophisticate statistical analysis of the rainfall time series may be implemented in a future study applying tests like Mann–Kendall (see [38]).

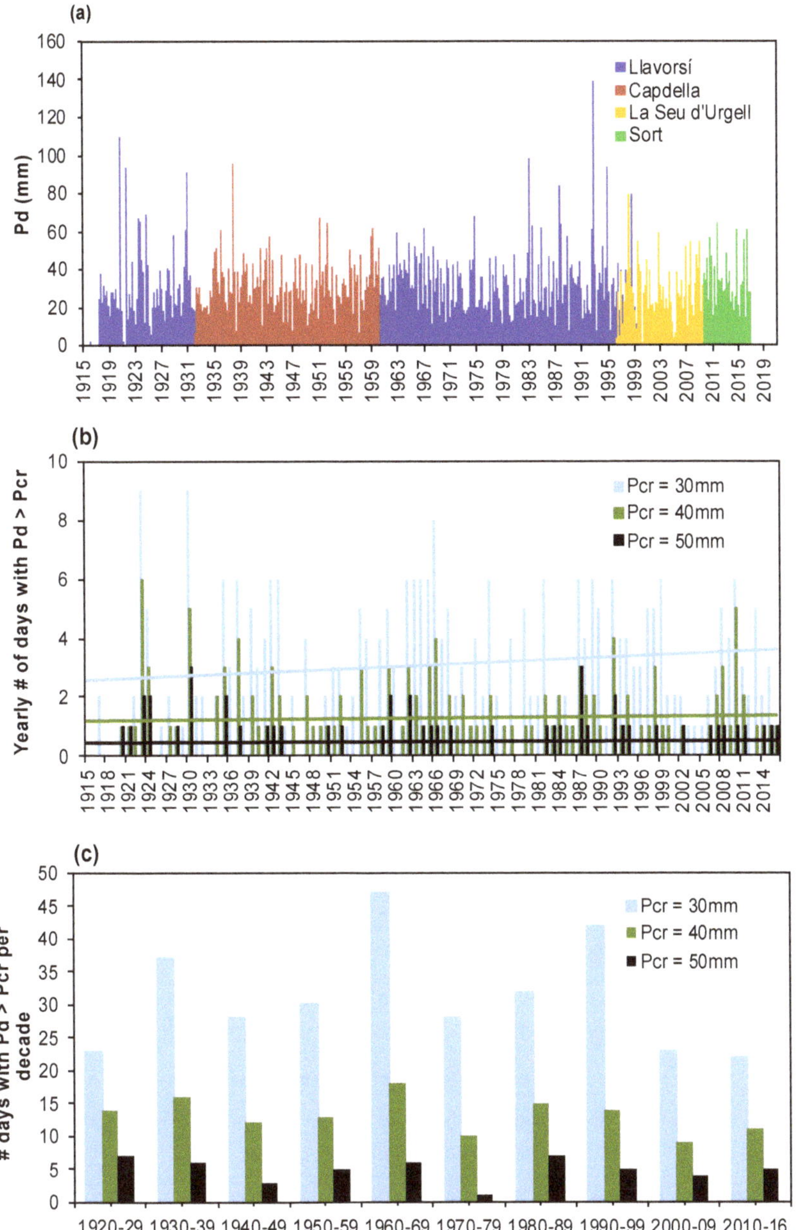

Figure 6. Analysis of rainfall trends for the Portainé area. (**a**) Complete time series of daily rainfall (P_d). (**b**) Number of heavy rainfall events exceeding critical daily rainfall (P_{cr}) per year. Linear regression models are included with the same colour as the bars. (**c**) Number of heavy rainfall events exceeding P_{cr} per decade.

5.1.2. Elaboration of Synthetic Hyetographs

To compare the hydrological response of the basin before and after the settlement of the ski resort (see Section 5.2), synthetic hyetographs were calculated [16]. As the objective was to compare two different time scenarios, a rainfall regionalisation procedure, complementary to the previous analysis (see Section 5.1.1), was considered useful enough. Based on the official publication "Maximum daily precipitation in the Peninsular Spain" [39], we used MAXIN software [40] to consistently obtain the rainfall return periods' regionalisation. The SQRT-ETmáx extreme-distribution function was used and the maximum daily rains associated with the 10, 50, 100, and 500 year return periods (T) were obtained: 85, 118, 133, and 172 mm/day, respectively.

Next, the rainfall intensities were calculated from the annual maximum daily rainfall for a time interval equivalent to the concentration time (tc) based on the formulation of [41]. For each one, a centred hyetograph was created to represent the variation of the intensity of the precipitation over 24 h (Figure 7). Other asymmetric patterns of hyetographs (event beginning biased and event end biased) were also tested but discarded.

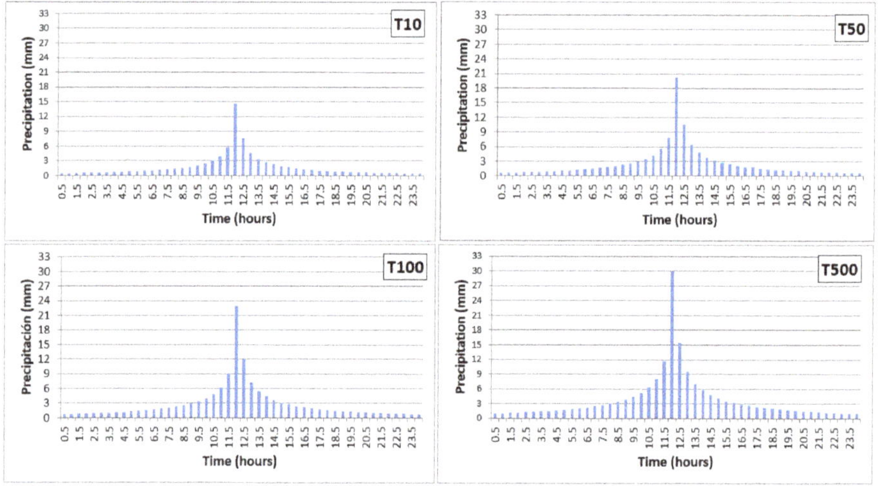

Figure 7. Synthetic hyetographs for return periods of 10, 50, 100, and 500 years.

These synthetic hyetographs obtained were used to evaluate the changes in hydrological behaviour in the headwaters of the basin, as input of rainfall–runoff models, as explained in Section 5.2.

5.1.3. Other systematic and non-systematic knowledge on precipitation

In the Pallars Sobirà county, maximum intensity and frequency rainfall events occur in spring and summer, mainly as convective storms [25]. Orography controls the generation of convective cells at the top of the drainage basins [42] which enhances precipitation, as in the Portainé summit. The torrential events in Portainé, thus, are mostly related to intense and localized convective summer rainstorms (witnesses' pers. comm.) and [25].

From 2011 to 2017, six rainfall events that produced floods were recorded at several meteorological stations at hourly intervals, including Portainé. Values vary widely, mostly due to the rainfalls' convective nature. The comparison of daily synthetic hyetographs values (T10) with the real, recorded rainfall values (Figure 8A) show that the recorded ones are smaller than the synthetic ones, which is logical because the real precipitation occurs almost annually. But when comparing the hourly rainfall intensities (Figure 8B), the real values are higher than the T10 synthetic ones, highlighting that very intense rainfalls occur almost yearly.

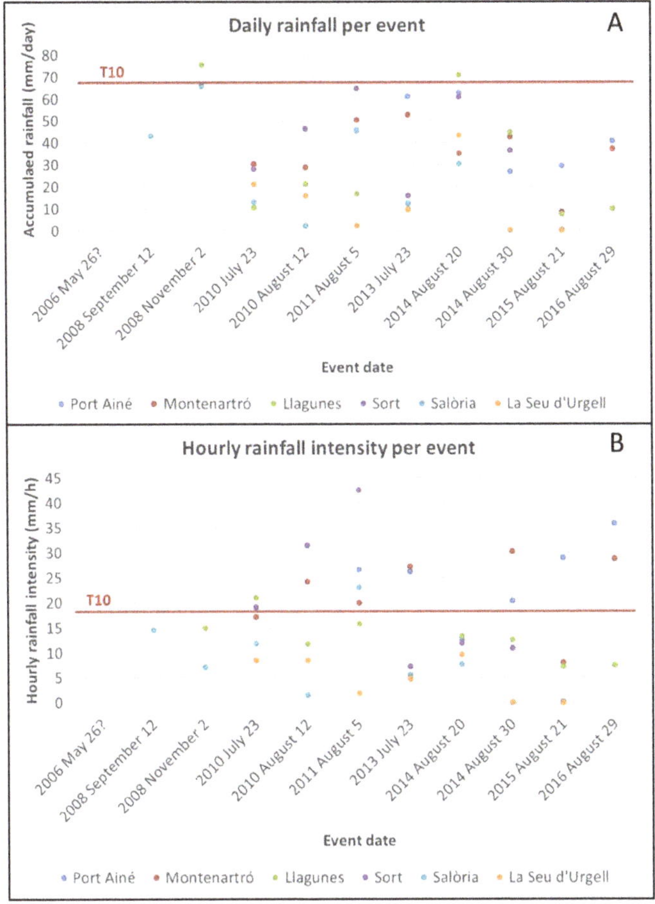

Figure 8. Comparison among the daily (**A**) and maximum hourly (**B**) rainfall intensity per event for the different meteorological stations close to and including Portainé. The calculated T10 daily rainfall value and the T10 hourly rainfall value extracted from the hyetographs, presented in Section 5.1.2, are shown with a red line.

5.2. Changes in the Hydrological Behaviour of the Basin's Headwaters

The hydrological response of the basin before and after the settlement of the ski resort was compared. The methodology used is presented broadly in [16]. To perform this comparison, hydrological methods of rainfall–runoff transformation were chosen. Two basin models were generated with GIS tools and basin model schemes were created with the HEC-HMS 4.0 software for the scenario prior to the ski resort settlement and for the current one. Three runoff models were created by using the "Curve Number" (CN) method of the U.S. Soil Conservation Service [43,44], implemented in the HEC-HMS software. The generated synthetic hyetographs (see Section 5.1.2) were used as input meteorological model. Integrating all these and by using the HEC-HMS 4.0 software, three hydrological rainfall–runoff models were built: (i) 1956, prior to the ski resort development; (ii) 1996, after completion of the ski resort; and (iii) the current scenario. To simulate the net rainfall transformation into runoff, the SCS Unit Hydrograph method was chosen; a non-existent base flow was considered since there are no permanent rivers and the torrential ravines are very dependent on the rains. For the flow wave propagation, the Muskingum method [45] was applied (the parameters x and k, needed for the method, were calculated with the NWS FLDWAV model and are shown in Table S1). Several control points, shown in Figure 9 and considered in Table 5, were used for the hydrological simulations.

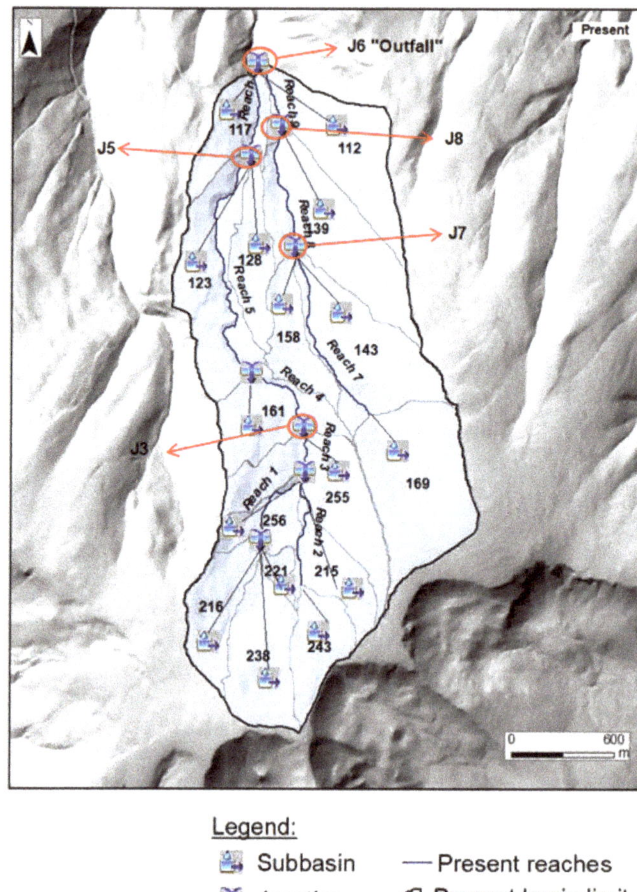

Legend:
- Subbasin — Present reaches
- Junction — Present basin limit

Figure 9. Portainé and Reguerals modelled basins and sub-basins using HEC-HMS software. The junctions or "control points" J3, J5, J6, J7, and J8 corresponding to Table 5 are highlighted.

The results (Table 6) show that there are significant differences in peak flows at the confluence of the Portainé and Reguerals torrents (J6). In general, the greater ones are detected in T10, mainly due to the soil cover influence (and therefore the CN) in response to the smaller rainfalls. Therefore, the differences decrease with increasing return periods. The land-cover alterations produced between 1956 and present are larger in Portainé than in the Reguerals headwaters. Therefore, the control point J3 is the one showing the major increase in discharge, reaching up to 26% for T10 and coinciding with the largest increase in the ski slopes' surface in recent years (Figure S10). Meanwhile, at point J8 in Regerals, a reduction of 11% is observed, which is likely due to an increase or densification in the forest cover. In general terms, the models show an increase in discharge of 15% for T10 at point J6. This increase in discharge can only be interpreted by taking into account the land-cover transformation.

Table 6. Hydrological modelling results. The different return periods considered and the consequent peak flow discharges (m^3/s) at the control points J3, J5, J6, J7 and J8 are shown. % indicates the percentage of change between the 1956 and present situations.

		Control Points (Junctions)										
		J6		J5		J3		J8		J7	
Tr	Situation	Peak Flow	%	Peak Flow	%	Peak Flow	%	Peak Flow	%	Peak Flow	%
10	Present	8.1	15	5.6	23	4.6	26	2.2	−9	1.9	−11
	1996	7.8		5.4		4.44		2.1		1.8	
	1956	6.9		4.3		3.4		2.4		2.1	
100	Present	20.5	7	12.9	14	10	15	6.7	0	5.2	7
	1996	20.3		12.8		9.9		6.3		5.1	
	1956	19.1		11.1		8.5		6.7		5.6	
500	Present	32.4	4	19.5	9	14.7	10	11.2	2	8.5	4
	1996	32.3		19.5		14.8		10.5		8.4	
	1956	31		17.7		13.2		11		8.9	

6. The Present Hydrogeomorphic Effects

6.1. The Processes in the Headwaters

Elimination of vegetation and anthropic modifications lead to changes in the headwaters' processes, increasing erosion and leading to water discharge concentration downwards.

The initial effects of rainfall water concentration over the ski slopes are rill erosion and flow sediment load increase derived from the eroded, stratified soils (see Section 3.2.1), thus generating heavily loaded flows. These entrained sediments deposit as accumulations and lobes of angular, coarse gravels when the flow energy decreases, so when the slope angle is reduced, or the flow decelerates because of obstacles. However, when the flow unloads, the cleaner, concentrated water continues moving downwards (Figure S11) and contributes to increasing discharges in the Portainé and Reguerals streams.

The drainage water channels dug on the ski runs (Section 4.2) concentrate rainfall waters efficiently. The undesired side effects are the deepening of the channels and soil degradation, new incisions, channel head new development and enhanced rill and gully erosion (deeper than 1 m) at the margins of the ski runs (Figure 10), also observed by Fidelus-Orzechowska et al. [46] and Ristić et al. [47] in the Southern Poland and East Serbia ski resorts and deeply studied and characterized by Wrónska-Wałach et al. [48] in a ski resort in Poland. In the case of Portainé, this also leads to higher water concentrations and an increase in the discharges at the lower end of the facility, coherently with the hydrological modelling (see Section 4.2).

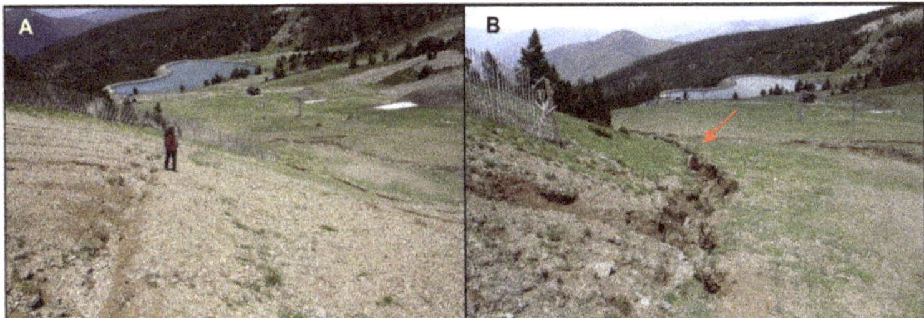

Figure 10. Pictures of the erosion on the basin headwaters due to land-use changes and enhanced by drainage water channels and concentration of runoff. (**A**) drainage water artificial channel and (**B**) gully erosion downstream; for a better appreciation of the scale, note the person in the channel highlighted with a red arrow.

6.2. Hydrogeomorphological Processes along the Main Channels

Hydrogeomorphological activity reaches its maximum intensity along the intermediate and lower reaches of the torrents, posing a severe risk to road infrastructures (e.g., road intersections). These processes have been assessed during post-event field surveys [12,13,15], but they are restricted to specific debris flow or flood events. The geomorphic change detection and quantification in this area has been systematically performed by the comparison of sequential bare-earth Digital Terrain Models (DTMs) obtained from airborne LiDAR data from 2009, 2011, and 2016 [22].

LiDAR-based DTM differencing shows a complex erosion–deposition pattern of the channels throughout time during the studied period. Erosion was the predominant dynamics along valley bottoms. The material that is eroded from the torrent bed and banks is susceptible to being transported by the flow, thus increasing its sediment load, especially during high discharge events. Other localized incisions have been identified, such as downstream and lateral scouring at sediment retention barriers, and enhanced entrenchment downstream of the road intersections. Depositional processes rather prevailed locally and mostly correspond to human-altered stretches. Indeed, these phenomena occurred due to the filling of the barriers, to the obstruction of underground drainage channels at road intersections, or to sediment being deposited in the alluvial cone during extraordinary torrential events.

During the 2009–2011 period, 22,042 m^3 and 19,204 m^3 of material were eroded and accumulated, respectively (−2838 m^3 sediment budget), whereas, in 2011–2016, even if the time period was considerably longer, the activity was less intense, and the erosion and deposition volumes were only 8308 m^3 and 8161 m^3 (−147 m^3 sediment budget). Budget segregation analyses showed the spatio-temporal distribution of those geomorphic changes (Figure 11). The Portainé torrent is the most erosive one. Caners tends to be aggradational, and Reguerals varies over time. It is noteworthy that erosion prevails in the upper sector of the catchment (headwaters). When dividing the drainage network into shorter reaches, a complex hydrogeomorphic dynamic arises as a result of the interference between the natural channel evolution and the influence of barriers. Moreover, 2008 post-event field data revealed that the most intense erosion occurred along two specific stretches of the Portainé torrent (1930–1645 m a.s.l. and 1555–1387 m a.s.l.), which became predominantly depositional after barriers were installed in 2010. These differences prove the undeniable geomorphic impact of the structural measures, that significantly alter flow hydrodynamics.

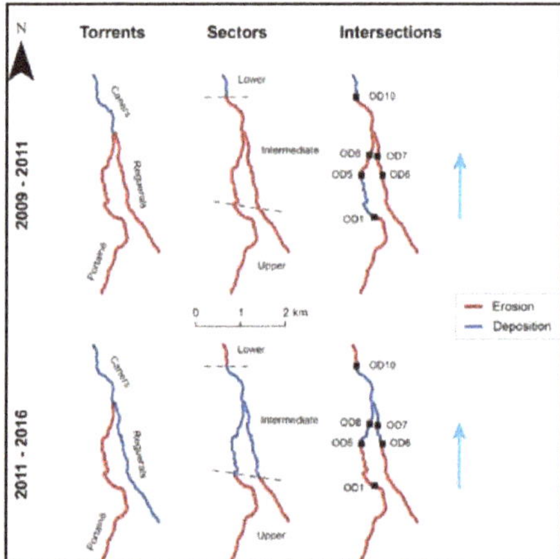

Figure 11. Sediment budget segregation by dividing the channels according to different criteria: drainage network (torrents), mean channel gradient (catchment sectors) and road intersections. Blue arrows indicate flow direction.

6.3. Record of Changes in the Torrential Dynamic in the Downstream Stretch

A dendrogeomorphological study was carried out in the cone, at the downstream stretch of Caners [21]. One hundred and sixty-six samples from 67 trees belonging to 10 different species were analysed, which are representative of the broadleaf forest in the downstream stretch (Figure 12). The main identified macroscopic indicators of flood damage were wounds, decapitations (loss of the main stem), tilting, stem burial and dead trees. Sampling mainly consisted of extracting cores using a standard Pressler increment borer; additionally, we also obtained some wedges from heavily damaged or dead trees. We dated 364 Flood Dendrogeomorphological Evidence (FDE), mainly scars and decapitations, growth changes and tree ages.

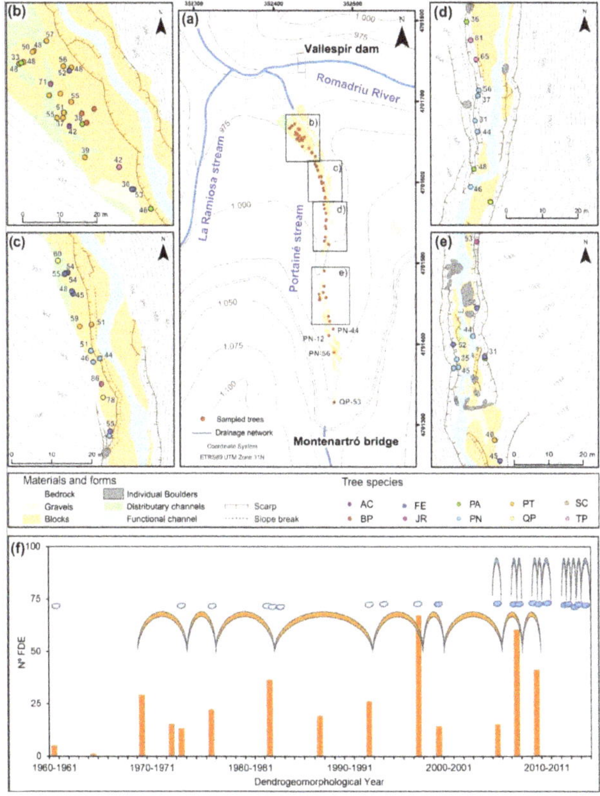

Figure 12. (**a**) Geomorphic forms at the Portainé cone and location of the 67 sampled trees. Black boxes show the stream stretches enlarged in (**b–e**) that show the main geomorphic forms and materials as well as the sampled trees, with the species indicated by colours and the estimated age by numbers. In the southern part of the study area, the scattered trees are identified with the species code and the estimated age. Species code: AC—*Acer campestre* L., BP—*Betula s.p.* L., FE—*Fraxinus excelsior* L., JR—*Juglans regia* L., PA—*Prunus avium* L., PN—*Populus nigra* L., PT—*Populus tremula* L., QP—*Quercus petraea* (Matt.) Liebl., SC—*Salix caprea* L., TP—*Tilia platyphyllos* Mill. (**f**). Number of dated FDE (columns in orange) and documented events of regional (light blue clouds) or local (dark blue clouds) rainfalls. The arcs represent the well-defined event-to-event intervals determined by the Flood Dendrogeomorphological Evidence (arcs in orange colour) and by the documented local events rainfalls since 2006 in the Portainé stream (arcs in grey colour).

We also determined ten well-defined Dendrogeomorphological Years (DYs), when the events that affected the trees were identified by the temporal overlaps of many FDEs. The patterns of FDE distribution are subject to uncertainties because of the limitations of the method, for example, the different tree-ages. However, the maximum and average ages of many tree species were similar, indicating their first establishment in the 1950s. In addition, these data were compared with the available documentary record. This multi-evidence data about temporal distribution of torrential floods and event-to-event intervals from the last 60 years demonstrate the recent change in frequency of destructive torrential flows (Figure 12f).

These dendrogeomorphological event-to-event intervals determined between 1969–1970 and 2009–1910 are shown in Figure 12f. The average recurrence interval is 4.5 years and the median, 4 years. In contrast, the documented data available on local rainfalls and debris events from 2006 to 2016 show that the average recurrence interval was almost 1 year (Figure 12f). No interval of such a high frequency occurred before, during the second half of the 20th century. Even though it is not possible to derive the type of torrential flow and its density from the FDE, the obtained results show an undeniable increase in destructive flows over the last decade. These results are in agreement with the hypothesis of surpassing a geomorphological threshold.

7. Discussion

The Portainé basin shows a very complex set of factors that characterize it and drive its dynamic behaviour. As pointed out by [49], interrelationships and scale effects may be difficult to capture in model cascade approaches. Our work presents an alternative to this model cascade approach. Next, all the factors presented above are related and compared in order to acquire a holistic understanding and to corroborate the surpassing of a geomorphological threshold in the basin.

Firstly, the lithological and geomorphological characteristics (Sections 3.1 and 3.2) define a basin prone to sediment-laden torrential flows. The folded and fractured bedrock—with major structures gently dipping subparallel to the topography, and with open fractures near the surface and major weathering—on the one hand favour its erodibility and, on the other hand the formation of deep soils and colluvial deposits. The relief evolution, together with chemical and physical weathering, produces these deposits, and so the material available to be entrained by the flows is effectively unlimited [50]. In addition, the basin morphometry also indicates its tendency to generate dense flows (see Section 3.2.2; [34]. The sediment-laden flows' activity is not continuous and uniform over time but is intermittent and episodic. There must be coinciding critical values of the hydrometeorological parameters (intensity or duration of rain, soil moisture, runoff concentration...) for them to take place. Solid evidence for the occurrence of ancient, non-dated debris flows, and the existence of, at least, a period with not very dense flows in the basin, between the middle of the 20th century and 2006 (see Sections 1.4 and 3.2.3), suggests the shifting of the torrential dynamics from highly active periods, producing very dense, sediment-laden flows, to other less active periods, with more or less bed load transport depending on the flood magnitude. Thus, the adaptive cycle [51] of this system shows that it may exist in at least two steady states, flipping from one to another as its geomorphological threshold is transgressed, as acknowledged by [7] for multiple steady state systems. In that sense, the extrinsic threshold of [50], corresponding to a certain precipitation value that triggers dense flows, can change over time depending on the state of the basin, being a lower value when steady state conditions are similar to the current ones and, thus, favour sediment-laden torrential flows.

As a first approximation, it should have been presumed that the flood magnitudes had been similar to those nowadays during the last hundred years, since rainfall does not show significant pattern changes, as seen in Section 5.1.1. It is worth noting that the values of the rainfall return periods obtained from the historical rainfall time series (Section 5.1.1) are very consistent with those obtained from synthetic hyetographs (Section 5.1.2), which makes our analyses robust. In addition, our results coincide well with those obtained by [52], who describe a period of low frequency and intensity of precipitation from 1917 to 2012, recorded in the varved sediments of Montcortés Lake (approximately

20 km southwest of Portainé). In their study, these varved sediments included clastic layers deposited because of heavy rainfalls; the dating provided by the varved layers corresponds very well to the 80–90 mm rainfall events recorded at the Capdella meteorological station, which corresponds to 2–5 years return period rainfalls. However, in Portainé, changes in discharges happen, as revealed by the hydrological modelling (Section 5.2), and they are due to land cover changes. Discharge changes due to ski resort land cover changes are evaluated quantitatively in the present work, being this pioneer in such evaluation in ski resorts. Among these changes, the anthropic land cover alteration due to the ski resort settlement is obvious: while the forest densified because of the cattle decrease during the last half of the 20th century (Section 4.1), the area without vegetal cover in the headwaters grew (Section 4.2). This caused a significant increase in discharges downstream, additionally enhanced by the drainage channels of the ski slopes (Section 4.2). This is consistent with the observations of [48], where drainage channels redirect water from the ski runs and link with the adjacent valley system, producing a new hydrological state and reactivating erosion processes downstream. When the geomorphological threshold of the basin is overcome, discharges are more effective in eroding, entraining sediment, and transporting it in the form of sediment-laden flows. The effects are similar to those observed by [47] throughout the Zubska riverbed, with the demolishing of bridges, filling of road culverts with sediment, and the destruction of the road system. It should also be recalled that the high intensity of convective, characteristic summer storms (Section 5.1.3) occur almost yearly (corresponding to a calculated return period of >T10). This indicates that the discharge increases expected for T10 (Section 5.2) can, in fact happen yearly, enhancing and aggravating the occurrence of debris and hyperconcentrated flows. As it is in hydrology, the processes and their effects are non-linear [49,53], and so it may be that this increase is even higher. Moreover, the multi-evidence dendrogeomorphological analysis, which also detected the change in flood frequency since 2006 and especially since 2008, but not before (Section 6.3), also supports the disturbance of the geomorphological equilibrium, occurred when the human induced flow concentration coincided with both the earthworks at the ski resort and intense but no extraordinary rainfalls. This type of paleohydrological evidence is especially useful in mountain basins, being a complement to, an alternative, or even the only source of data when there is no systematic and instrumental information available [54].

The erosion becomes particularly effective since the triggering events of 2006 and, especially, of 2008, which swept away the torrential margins' vegetation cover, which had stabilized the ravines until that moment, and created the erosive breach that left the fractured and weathered bedrock and colluvium deposits unprotected and therefore susceptible to erosion (Section 1.4). Thus, vegetation seems to play a key role in the stability of the basin and its dynamics in several ways and over several scales in this small basin, including its influence on the parameters that control discharges (interception, soil moisture, infiltration ...) and the stability of the torrent margins. As it was pointed out by [49,55], abrupt changes in watershed response can occur as a result of land-use change and can be particularly severe at small scale basins. Our study also corroborates this statement.

Lastly, it should be noted that the recent human intervention in the torrents, with the installation of sediment retention barriers, causes complex dynamics, where the sediment balance is altered (less sediment is exported from the basin), some very erosive stretches have turned into sedimentary ones, and vice versa. Indeed, the barriers attenuate some of the geomorphic effects. Nevertheless, it must be pointed out that sediment-laden torrential flows still occur and could continue occurring if the steady state of the basin remains. To help reverse these dynamics, the land use of the headwaters should be re-considered, and the vegetation cover re-established. This could favour a shift to other steady state characteristics of the basin, less prone to generating dense flows.

Regarding ancient dense flows, they could have been related to periods of decreased vegetal cover, especially in the headwaters. The Holocene variation of the treeline ecotone, or the zone of contact between subalpine forest and alpine meadows without arboreal vegetation, was studied through interdisciplinary palynological and charcoal study methods by Cunill et al. [56] at the upper Cardós river valley, just to the north of the Romadriu valley. Their results show well-established forests up to 2200 m a.s.l. since the beginning of the Holocene, reaching 2460 m a.s.l. from 9600 to 7600 cal BP. The first openings in the forest occur at approximately 7500 cal BP. From 5100 to 2200 cal BP, there is clear archaeological evidence of human occupation in the high mountains and there were related recurrent burns in order to maintain pastures. In the 6th century (1400 cal BP), there was a generalized intensification in human activity, with more creation and preservation of pastures and altitude cereal crops. The period of maximum deforestation occurred between the 15th and 16th centuries (500–400 cal BP), and at the end of the 19th century the forest recovered favouring a higher treeline. However, their data from charcoals show that there has been an almost uninterrupted history of fires from 10,800 cal BP until the mid-20th century, which could have been from natural on anthropic origin. Taking this into consideration, the ancient debris flows in the Portainé basin could have occurred at any time during the Holocene, related to natural or anthropic forest fires deforestation of the headwaters. Human activity since the Middle Ages and consequent clearing make the generation of these dense flows more likely. Unfortunately, it has not yet been possible to date these deposits, as commented in Section 3.2.3.

On the other hand, it is well known that mass movements are widely linked to climate change, for instance, to the melting and retreat of the glaciers (decompression of the glacial deposits and water addition) [57]. This could have been the case for the Portainé ancient, large debris flows. Furthermore, Abrantes et al. [58] recognised an early period (900–1100 years CE) characterized by intense precipitation/flooding and warm winters. Llasat et al. [59] distinguish an increase in flood events in Catalonia for the periods 1580–1620, 1760–1800, and 1830–1870 A.D. that coincide with different Western Mediterranean area basins and demonstrate a singular but common answer to climatic anomalies. These wet periods could also have favoured the generation of the Portainé ancient, undated dense flows, especially because of their coincidence with deforestation. In conclusion, while it has not been possible to precisely date the ancient Portainé debris flows deposits, their generation could have been related to anthropic or natural forest fires and deforestation, to particularly wet periods, or to a combination of both factors.

Finally, it must be considered that climate change, alteration of atmospheric composition, land-use changes, wildfires and biological invasions threaten forests in the entire Mediterranean basin. Each one of these factors or a combination of them have led to land degradation, vegetation regeneration decline, and expansion of forest diseases [60] recently. One example is the pine processionary caterpillar that is currently spreading throughout the Pyrenees [61]; it can accelerate tree mortality and favour fires and deforestation. In the Pyrenees, according to [62], the frequency and intensity of heat waves is projected to increase, and so the risk of forest fires is expected to increase as well. Furthermore, precipitation is predicted to globally decrease but the number of intense rainfall episodes will increase [63]. All these projections point to future degradation of the forest cover, the soil and, consequently, this would enhance the generation of dense sediment-laden flows in Portainé and other similar basins in the area.

The establishment of quantitative thresholds for variables that condition the triggering of slope movements is a necessary scientific exercise, and even an essential technique [64]. But when the systems are more complex (such as in this case study), combining sudden floods and dense washout flows, the establishment of thresholds must be approached from a multidisciplinary and multivariate perspective, not only in magnitude, but in time.

8. Conclusions

The Portainé basin shows a very complex set of factors, described, characterised and analysed by means of the multidisciplinary research presented here, that determines its flood dynamic behaviour. The holistic understanding of this complex watershed can only be appreciated by taking into account the interactions of the lithosphere, atmosphere, hydrosphere, biosphere, and anthroposphere. This means considering the bedrock geology, the geomorphological evolution, the derived soils and colluvial deposits, the rainfall characteristics, and the hydrological response. Its alteration due to the anthropic land-use change, the consequent hydrogeomorphic effects, flood characteristics and the role of vegetation must not be misunderstood. The consideration of all the interactions leads us to conclude that a geomorphological threshold has been overcome recently (2006–2008), producing a shift in the torrential dynamics of the basin that turned into very dense, highly erosive sediment-laden flows. Moreover, the basin has shifted around this threshold, giving rise to the two distinct behaviours or equilibrium conditions likely throughout all its Holocene history, as demonstrated by periods of moderate, bed load-laden flow dynamics and by the existence of deposits of ancient debris flows.

Present conditions and projected climate change suggest that the current dynamics are likely to continue despite the efforts to stabilise the most active torrent stretches. In order to avoid worse effects in the future, interventions in the headwaters with close surveillance should be carried out, particularly focussing on revegetation. Moreover, in future research, other nearby, similar basins should be studied in order to check whether they could suffer the same type of threshold surpass and therefore change in flood dynamics.

Supplementary Materials: The following are available online at http://www.mdpi.com/2073-4441/12/2/368/s1 (ZIP-Document). Figure S1: Regional cleavage of the Cambro-Ordovician bedrock subparallel to the topographic surface. Figure S2: *Gréze Litée* type deposit located in the headwaters of the Portainé basin. Figure S3: Diagram showing the Melton ratio versus the watershed length and the different types of torrential flows proposed by Wildford et al. (2004). Figure S4: A picture of one of the large, ancient and undated debris flow deposits located at the right margin of the Romadriu River, downstream from the Vallespir dam and in front of the confluence of the Caners stream. Figure S5: Lobed terrain morphology at the area of La Borda de Simó corresponding to overflow deposits of dense flows. Figure S6: Evolution of the vegetation downstream from the ski resort since 1945–1946, showing an increase in trees and dense forest until the present. Figure S7: Land cover maps of the Portainé basin headwaters for 1956, 1996, and present. Figure S8: Drainage channels built since 2005 and up to 2007, 2011, and 2013 in order to divert runoff and avoid erosion on the ski slopes. Figure S9: Availability of rainfall records for the four rain gauges and comparison of daily rainfall, Pd, measured at different rain gauges. Figure S10: General view of the *Rhododendron ferrugineum* cover located near the ski slopes and of a devegetated ski run. Figure S11: Results of the runoff concentration and sediment deposits at the ski resort. Table S1: Muskingum parameters values (x and k) used for each reach (Cn) of each stream (Reguerals and Portainé), and for the network situations on 1996/56 and at present. The parameters were calculated by using the NWS FLDWAV model. A constant value of x = 0.35 was assigned because the high longitudinal river slope, with null attenuation; the equation $K = 0.7*tc$ was used for k values estimation for each propagation reach (where tc is the time of concentration).

Author Contributions: Conceptualisation, G.F., A.V., and A.D.-H.; Methodology G.F., A.V., A.D.-H., M.G. (Mar Génova), M.G. (Marta Guinau), Á.D.l.H., R.M.P., M.H., G.K., J.M.C., A.M., J.P., and M.G. (Marta González); Investigation, G.F., A.V., A.D-H, M.G. (Mar Génova), M.G. (Marta Guinau), Á.D.l.H., R.M.P., M.H., G.K., J.M.C., A.M., J.P., and M.G. (Marta González); Formal Analysis, G.F., A.V., A.D.-H., M.G. (Mar Génova), Á.D.l.H., R.M.P., M.H., J.M.C., and A.M.; Writing—Original Draft Preparation, G.F., A.V., M.G. (Marta Guinau), Á.D.l.H., R.M.P., M.H., and J.M.C.; Writing—Review and Editing, G.F., A.V., A.D.-H., M.G. (Mar Génova), M.G. (Marta Guinau), and M.H. All authors have read and agreed to the published version of the manuscript.

Funding: This work was supported by the following Projects: CHARMA (MINECO, Ref.: CGL2013-40828-R), PROMONTEC (MINEICO-FEDER, Ref.: 444 CGL2017-84720-R), SMuCPhy (MINEICO, Ref.: BIA 2015-67500-R), and two Projects of MINECO (CGL2015-66335-C2-1-R and CGL2017-87631-P); by an APIF scholarship (UB) and by a FI 2018 scholarship (Secretaria d'Universitats i Recerca; Departament d'Empresa i Coneixement; Generalitat de Catalunya); and by the IGME's Program of Professional Practices.

Acknowledgments: We want to thank J. Calvet, R Juliá and M.A. Marqués (UB); C. Fañanás (Dpt. d'Agricultura, Ramaderia, Pesca i Alimentació, Generalitat de Catalunya); the FGC and ICGC personnel, the Arxiu Comarcal del Pallars Sobirà personnel; Ll. Pla (MeteoPirineu) and the hydroelectric company OPYCE (S.A).

Conflicts of Interest: The authors declare no conflict of interest. The funders had no role in the design of the study; in the collection, analyses, or interpretation of data; in the writing of the manuscript, or in the decision to publish the results.

References

1. Goudie, A. IAG Glossary Of Geomorphology (version 1.0). International Association of Geomorphologists, 2014. Available online: http://www.geomorph.org/wp-content/uploads/2015/06/GLOSSARY_OF_GEOMORPHOLOGY1.pdf (accessed on 14 February 2018).
2. Schumm, S.A. Geomorphic thresholds and complex response of drainage systems. In *Fluvial Geomorphology*; Morisawa, M., Ed.; State University of New York: Binghamton, NY, USA, 1973; Volume 13, pp. 299–310.

3. Schumm, S.A. Geomorphic Thresholds: The Concept and Its Applications. *Trans. Inst. Br. Geogr.* **1979**, *4*, 485–515. [CrossRef]

4. Brunsden, D.; Thornes, J.B. Landscape Sensitivity and Change. *Trans. Inst. Br. Geogr.* **1979**, *4*, 463–484. [CrossRef]

5. Coates, D.R.; Vitek, J.D. *Thresholds in Geomorphology*; Allen & Unwin Pty: Sydney, Australia, 1980.

6. Nanson, G.C.; Huang, Q. A philosophy of rivers: Equilibrium states, channel evolution, teleomatic change and least action principle. *Geomorphology* **2018**, *302*, 3–19. [CrossRef]

7. Gregory, K.J. *The Earth's Land Surface: Landforms and Processes in Geomorphology*; Sage Publications: London, UK, 2010.

8. Wohl, E.E. *Rivers in the Landscape.*; Wiley Blackwell: Chichester, West Sussex, UK, 2014.

9. Dust, D.W.; Wohl, E.E. Quantitative technique for assessing the geomorphic thresholds for floodplain instability and braiding in the semi-arid environment. *Nat. Hazards* **2010**, *55*, 145–160. [CrossRef]

10. Watson, C.C.; Biedenharn, D.S.; Thorne, C.R. *Stream Rehabilitation*; Cottonwood Research LLC: Fort Collins, CO, USA, 2005.

11. Wilshire, H.G. Human causes of accelerated wind erosion in California's deserts. In *Thresholds in Geomorphology. Binghamton Symposium*; George Allen & Unwin: London, UK, 1980; Volume 9, pp. 415–433.

12. ICGC. *Estudi de la torrentada de la nit del dia 11 al 12 de setembre de 2008 al barranc de Portainé (Pallars Sobirà)*; AP-046/10; Generalitat de Catalunya: Barcelona, Spain, 2010.

13. ICGC. *Nota de la visita al barranc de Portainé (Pallars Sobirà) arran del episodi de pluges dels dies 22 i 23 de juliol de 2010, AP-046/10*; Generalitat de Catalunya: Barcelona, Spain, 2010.

14. Victoriano, A.; Díez-Herrero, A.; Génova, M.; Guinau, M.; Furdada, G.; Khazaradze, G.; Calvet, J. Four-topic correlation between flood dendrogeomorphological evidence and hydraulic parameters (the Portainé stream, Iberian Peninsula). *Catena* **2018**, *162*, 216–229. [CrossRef]

15. Abancó, C.; Hürlimann, M. Estimate of the debris-flow entrainment using field and topographical data. *Nat. Hazards* **2014**, *71*, 363–383. [CrossRef]

16. De las Heras, Á. *Modificación de la respuesta hidrológica en avenidas torrenciales ante los cambios de usos del suelo en una cuenca de montaña (Portainé, Pirineo leridano)*; Universidad Politécnica de Madrid: Madrid, Spain, 2016.

17. Ortuño, M.; Guinau, M.; Calvet, J.; Furdada, G.; Bordonau, J.; Ruiz, A.; Camafort, M. Potential of airborne LiDAR data analysis to detect subtle landforms of slope failure: Portainé, Central Pyrenees. *Geomorphology* **2017**, *295*, 364–382. [CrossRef]

18. Palau, R.M.; Hürlimann, M.; Pinyol, J.; Moya, J.; Victoriano, A.; Génova, M.; Puig-Polo, C. Recent debris flows in the Portainé catchment (Eastern Pyrenees, Spain): Analysis of monitoring and field data focussing on the 2015 event. *Landslides* **2017**, *14*, 1161–1170. [CrossRef]

19. Pinyol, J.; Hürlimann, M.; Furdada, G.; Moysset, M.; Palau, R.M.; Victoriano, A.; González, M.; Moya, J.; Raïmat, C.; Fañanás, C. El barranco de Portainé (Pirineo Central): Un laboratorio in situ completo para el estudio de la actividad torrencial. In *IX Simposio Nacional sobre Taludes y Laderas Inestables*; Centre Internacional de Màtodes Numèrics en Enginyeria (CIMNE): Barcelona, Spain, 2017; Volume 2013, pp. 1165–1176.

20. Furdada, G.; Guinau, M.; Subiela, G.B.; Moraru, A. Las cuencas de Portainé y Romadriu: Susceptibilidad al cambio y factores de control antrópicos y naturales. In *Geomorfología del Antropoceno. Efectos del cambio Global sobre los procesos geomorfológicos*; García, G., Gómez-Pujol, L., Morán-Tejada, E., y Batalla, R., Eds.; Sociedad Española de Geomorfología: Palma, Spain, 2018; pp. 335–338.

21. Génova, M.; Díez-Herrero, A.; Furdada, G.; Guinau, M.; Victoriano, A. Dendrogeomorphological Evidence of Flood Frequency Changes and Human Activities (Portainé Basin, Spanish Pyrenees). *Tree-Ring Res.* **2018**, *74*, 144–161. [CrossRef]

22. Victoriano, A.; Brasington, J.; Guinau, M.; Furdada, G.; Cabré, M.; Moysset, M. Geomorphic impact and assessment of flexible barriers using multi-temporal LiDAR data: The Portainé mountain catchment (Pyrenees). *Eng. Geol.* **2018**, *237*, 168–180. [CrossRef]

23. Kondolf, G.M.; Piégay, H. *Tools in fluvial Geomorphology.*; Ohn Wiley & Sons Ltd.: Chichester, UK, 2003.

24. Carreras, J. *Flora i vegetació de Sant Joan de l'Erm i de la Vall de Santa Magdalena (Pirineus catalans)*; Col.lecció estudis; Institut d' Estudis Ilerdencs: Lleida, Spain, 1993; Volume 3, pp. 1–321.

25. Meteocat. *Atles Climátic de Catalunya 1961–1990*; Servei Meteorològic de Catalunya: Barcelona, Spain, 2008.

26. Muñoz, J.A. Evolution of a continental collision belt: ECORS-Pyrenees crustal balanced cross-section. In *Thrust Tectonics*; McClay, K.R., Ed.; Springer: Dordrecht, The Netherlands, 1992; pp. 235–246.

27. Padel, M. *Influence cadomienne dans les séries pré-sardes des Pyrénées Orientales: Approche géochimique, stratigraphique et géochronologique*; Université de Lille 1: Lille, France, 2016; Available online: http://ori-nuxeo.univ-lille1.fr/nuxeo/site/esupversions/38ae32e6-694a-470e-b34e-1b6938f8f6e8 (accessed on 28 January 2020).

28. Hartevelt, J.J.A. Geology of the upper Segre and Valira Valleys, Central Pyrenees, Andorra/Spain. *Leidse Geol. Meded.* **1970**, *45*, 167–236.

29. Laumonier, B.; Autran, A.; Barbey, P.; Cheilletz, A.; Baudin, T.; Cocherie, A.; Guerrot, C. Conséquences de l'absence de socle cadomien sur l'âge et la signification des séries pré-varisques (anté-Ordovicien supérieur) du sud de la France (Pyrénées, Montagne Noire). *Bull. la Société Géologique Fr.* **2004**, *175*, 643–655. [CrossRef]

30. Monod, B.; Regard, V.; Carcone, J.; Wyns, R.; Christophoul, F. Postorogenic planar palaeosurfaces of the central Pyrenees: Weathering and neotectonic records. *Comptes Rendus Geosci.* **2016**, *348*, 184–193. [CrossRef]

31. Ortuño, M.; Martí, A.; Martín-Closas, C.; Jiménez-Moreno, G.; Martinetto, E.; Santanach, P. Palaeoenvironments of the Late Miocene Prüedo Basin: implications for the uplift of the Central Pyrenees. *J. Geol. Soc.* **2013**, *170*, 79–92. [CrossRef]

32. Barrón, E.; Postigo-Mijarra, J.M.; Casas-Gallego, M. Late Miocene vegetation and climate of the La Cerdanya Basin (eastern Pyrenees, Spain). *Rev. Palaeobot. Palynol.* **2016**, *235*, 99–119. [CrossRef]

33. Fernandes, M.; Oliva, M.; Palma, P.; Ruiz-Fernández, J.; Lopes, L. Glacial stages and post-glacial environmental evolution in the Upper Garonne valley, Central Pyrenees. *Sci. Total. Environ.* **2017**, *584*, 1282–1299. [CrossRef]

34. Wilford, D.J.; Sakals, M.E.; Innes, J.L.; Sidle, R.C.; Bergerud, W.A. Recognition of debris flow, debris flood and flood hazard through watershed morphometrics. *Landslides* **2004**, *1*, 61–66. [CrossRef]

35. González, M.; Pinyol, J.; Camafort, M.; Vilaplana, J.; Oller, P. El mapa para la prevención de riesgos geológicos de cataluña 1:25 000: aplicación de una metodología a escala regional para la evaluación de la susceptibilidad a generar flujos torrenciales. In *VIII Simposio sobre Taludes y Laderas Inestables*; Alonso, E., Corominas, J., Hürlimann, M., Eds.; Centre Internacional de Màtodes Numèrics en Enginyeria (CIMNE): Barcelona, Spain, 2013; pp. 614–623.

36. Bosch i Pont, J.M. Evolució dels nuclis de població petits del Pallars Sobirà: Del 1850 a inicis del segle XXI. Ph.D. Thesis, Dpt. Geografia. Universitat de Barcelona, Barcelona, Spain, 2017. Available online: http://cataleg.ub.edu/record=b2244515~{|}S1*cat (accessed on 26 June 2019).

37. Instituto Nacional de Estadística. *Alteraciones de los municipios en los Censos de Población desde 1842*; Instituto Nacional de Estadística: Madrid, Spain, 2017; Available online: http://www.ine.es/intercensal/intercensal.do;jsessionid=14519F30BBA8257D11FF7EC9C3627D4C.intercensal02?search=1&cmbTipoBusq=0&textoMunicipio=rialp&btnBuscarDenom=Consultar+selecci%F3n (accessed on 26 March 2018)Last revision 14 February 2017.

38. Walega, A.; Młyński, D.; Bogdał, A.; Kowalik, T. Analysis of the Course and Frequency of High Water Stages in Selected Catchments of the Upper Vistula Basin in the South of Poland. *Water* **2016**, *8*, 394. [CrossRef]

39. Arias, J.S. *Máximas lluvias diarias en España Peninsular*; Serie monografías; Ministerio de Fomento, Dirección General de Carreteras: Madrid, Spain, 1999; p. 54.

40. de Salas, L.; Carrero, L. MAXIN: Gis application to estimate Rainfall Intensity-Duration-Frequency laws in the Spanish peninsular área. In Proceedings of the 8th European Conference on Applications of Meteorology, Madrid, Spain, 13–17 September 2007.

41. De Salas, L.; Fernandez, J.A. "In-site" regionalization to estimate an intensity-duration-frequency law: a solution to scarce spatial data in Spain. *Hydrol. Process.* **2007**, *21*, 3507–3513. [CrossRef]

42. Trapero, L.; Bech, J.; Lorente, J. Numerical modelling of heavy precipitation events over Eastern Pyrenees: Analysis of orographic effects. *Atmospheric Res.* **2013**, *123*, 368–383. [CrossRef]

43. NRCS (Natural Resources Conservation Service). *National Engineering Handbook. Part 630: Hydrology, chapter 9: Hydrologic soil-cover complexes*; US Dept. of Agriculture: Washington, DC, USA, 2004.

44. NRCS (Natural Resources Conservation Service). *National Engineering Handbook. Part 630: Hydrology, chapter 7: Hydrologic soil groups*; US Dept. of Agriculture: Washington, DC, USA, 2007.

45. Te Chow, V.; Maidment, D.R.; Mays, L.W. *Manual Hidrología Aplicada*; McGraw-Hill: Bogotá, Colombia, 1994.

46. Fidelus-Orzechowska, J.; Wrońska-Wałach, D.; Cebulski, J.; Żelazny, M. Effect of the construction of ski runs on changes in relief in a mountain catchment (Inner Carpathians, Southern Poland). *Sci. Total. Environ.* **2018**, *630*, 1298–1308. [CrossRef] [PubMed]

47. Ristić, R.; Kasanin-Grubin, M.; Radic, B.; Nikić, Z.; Vasiljevic, N. Land Degradation at the Stara Planina Ski Resort. *Environ. Manag.* **2012**, *49*, 580–592. [CrossRef] [PubMed]

48. Wrońska-Wałach, D.; Cebulski, J.; Fidelus-Orzechowska, J.; Żelazny, M.; Piątek, D. Impact of ski run construction on atypical channel head development. *Sci. Total. Environ.* **2019**, *692*, 791–805. [CrossRef] [PubMed]

49. Blöchl, G.; Ardoin-Bardin, S.; Bonell, M.; Dorninger, M.; Goodrich, D.; Gutknecht, D.; Matamoros, D.; Merz, B.; Shand, P.; Szolgay, J. At what scales do climate variability and land cover change impact on flooding and low flows? *Hydrol. Process.* **2007**, *21*, 1241–1247. [CrossRef]

50. Bovis, M.J.; Jakob, M. The role of debris supply conditions in predicting debris flow activity. *Earth Surf. Process. Landforms* **1999**, *24*, 1039–1054. [CrossRef]

51. Dearing, J. Landscape change and resilience theory: a palaeoenvironmental assessment from Yunnan, SW China. *Holocene* **2008**, *18*, 117–127. [CrossRef]

52. Corella, J.P.; Benito, G.; Rodriguez-Lloveras, X.; Brauer, A.; Valero-Garcés, B. Annually-resolved lake record of extreme hydro-meteorological events since AD 1347 in NE Iberian Peninsula. *Quat. Sci. Rev.* **2014**, *93*, 77–90. [CrossRef]

53. Ceola, S.; Botter, G.; Bertuzzo, E.; Porporato, A.; Rodriguez-Iturbe, I.; Rinaldo, A. Comparative study of ecohydrological streamflow probability distributions. *Water Resour. Res.* **2010**, *46*, W09502. [CrossRef]

54. Bodoque, J.M.; Diez-Herrero, A.; Eguibar, M.; Benito, G.; Ruiz-Villanueva, V.; Ballesteros-Cánovas, J. Challenges in paleoflood hydrology applied to risk analysis in mountainous watersheds – A review. *J. Hydrol.* **2015**, *529*, 449–467. [CrossRef]

55. Apollonio, C.; Balacco, G.; Novelli, A.; Tarantino, E.; Piccinni, A.F. Land Use Change Impact on Flooding Areas: The Case Study of Cervaro Basin (Italy). *Sustain.* **2016**, *8*, 996. [CrossRef]

56. Cunill, R.; Soriano, J.M.; Bal, M.C.; Pèlachs, A.; Rodriguez, J.M.; Pérez-Obiol, R.; Artigas, R.C. Holocene high-altitude vegetation dynamics in the Pyrenees: A pedoanthracology contribution to an interdisciplinary approach. *Quat. Int.* **2013**, *289*, 60–70. [CrossRef]

57. Lebourg, T.; Zerathe, S.; Fabre, R.; Giuliano, J.; Vidal, M. A Late Holocene deep-seated landslide in the northern French Pyrenees. *Geomorphol.* **2014**, *208*, 1–10. [CrossRef]

58. Abrantes, F.; Rodrigues, T.; Rufino, M.; Salgueiro, E.; Oliveira, D.; Gomes, S.; Oliveira, P.; Costa, A.; Mil-Homens, M.; Drago, T.; et al. The climate of the Common Era off the Iberian Peninsula. *Clim. Past* **2017**, *13*, 1901–1918. [CrossRef]

59. Llasat, M.-C.; Barriendos, M.; Barrera, A.; Rigo, T.; Barrera-Escoda, A. Floods in Catalonia (NE Spain) since the 14th century. Climatological and meteorological aspects from historical documentary sources and old instrumental records. *J. Hydrol.* **2005**, *313*, 32–47. [CrossRef]

60. Doblas-Miranda, E.; Alonso, R.; Arnan, X.; Bermejo, V.; Brotons, L.; Heras, J.D.L.; Estiarte, M.; Hódar, J.; Llorens, P.; Lloret, F.; et al. A review of the combination among global change factors in forests, shrublands and pastures of the Mediterranean Region: Beyond drought effects. *Glob. Planet. Chang.* **2017**, *148*, 42–54. [CrossRef]

61. Otsu, K.; Pla, M.; Brotons, L. Estimating the Severity of Defoliation Due to Pine Processionary Moth Using a Combination of Landsat and UAV Imagery. In *Proceedings of the IGARSS 2018—2018 IEEE International Geoscience and Remote Sensing Symposium*; Valencia, Spain, 22–27 July 2018, pp. 3979–3982. [CrossRef]

62. Füssel, H.-M. *Climate change, impacts and vulnerability in Europe 2016—An indicator-based report*; Publications Office of the European Union: Luxembourg, 2017; p. 419.

63. Pachauri, R.K.; Meyer, L.; Ypersele, J.P.V.; Brinkman, S.; Kesteren, L.V.; Leprince-Ringuet, N.; Boxmeer, F.V. Climate Change 2014 Synthesis Report The Core Writing Team Core Writing Team Technical Support Unit for the Synthesis Report. *Russ. Fed. Hoesung Lee (Republic Korea) Scott. B. Power N.H. Ravindranath.* Available online: https://www.ipcc.ch/pdf/assessment-report/ar5/syr/SYR_AR5_FINAL_full_wcover.pdf (accessed on 28 January 2020).

64. Segoni, S.; Piciullo, L.; Gariano, S.L. A review of the recent literature on rainfall thresholds for landslide occurrence. *Landslides* **2018**, *15*, 1483–1501. [CrossRef]

Article

Social Vulnerability Assessment for Flood Risk Analysis

Laura Tascón-González [1], Montserrat Ferrer-Julià [1,*], Maurici Ruiz [2] and Eduardo García-Meléndez [1]

[1] Department of Geography and Geology, Faculty of Environmental Sciences, Universidad de León, Campus de Vegazana, sn, 24072 León, Spain; lauratg88@hotmail.com (L.T.-G.); egarm@unileon.es (E.G.-M.)

[2] Department of Geography, Universitat de les Illes Balears, Edifici Guillem Colom Casasnovas, Campus de la Universitat de les Illes Balears (UIB), Carretera de Valldemossa, km 7.5, 07122 Palma de Mallorca, Spain; maurici.ruiz@uib.es

* Correspondence: mferj@unileon.es

Received: 20 December 2019; Accepted: 13 February 2020; Published: 17 February 2020

Abstract: This paper proposes a methodology for the analysis of social vulnerability to floods based on the integration and weighting of a range of exposure and resistance (coping capacity) indicators. It focuses on the selection and characteristics of each proposed indicator and the integration procedure based on the analytic hierarchy process (AHP) on a large scale. The majority of data used for the calculation of the indicators comes from open public data sources, which allows the replicability of the method in any area where the same data are available. To demonstrate the feasibility of the method, a study case is presented. The flood social vulnerability assessment focuses on the municipality of Ponferrada (Spain), a medium-sized town that has high exposure to floods due to potential breakage of the dam located upstream. A detailed mapping of the social vulnerability index is generated at the urban parcel scale, which shows an affected population of 34,941 inhabitants. The capability of working with such detailed units of analysis for an entire medium-sized town provides a valuable tool to support flood risk planning and management.

Keywords: vulnerability indicators; vulnerability cartography; flood risk; AHP; local analysis; open public data sources

1. Introduction

Floods are defined as the "temporary flooding of land that is not normally covered by water" [1]. Most river flooding is associated with the amount and distribution of precipitation in a drainage basin [2], but other reasons for flooding may include rapid snow melting in mountain ranges [3] or dam failure [4–7]. Other floods are related to coastal processes such as storm surges [8] and tsunamis [9,10]. Both river and coastal floods are mostly natural phenomena that occur with a frequency associated with a magnitude in a defined area. In general, these events are characterized by a high amount of damage, which is the result of the destructive power of water [11–13]. In addition to damage caused by the direct water impact, water's energy works as an erosive agent [14] that may undermine materials under the foundations of buildings, causing them to collapse [15]. Once the flood is over, damage may continue to increase due to electricity and water cutoffs or the spread of diseases such as cholera, leptospirosis, and typhoid fever [16–18], increasing economic and personal losses in the affected area [19]. Despite the natural origin of these floods, the main reason for damage can be identified in the human occupation of areas traditionally susceptible to flooding [20,21]. In many cases, the population is settled in flood-prone areas because no other places are available for building in the municipality. At the same time, new engineering infrastructures appear and provide a false sense of security as they do not completely prevent the land from flooding [22]. Therefore, the increased size of populations and the economic development of societies in flooding areas are the main factors in recent land-use changes and have increased society's vulnerability [23,24].

In the field of disaster risk reduction (DRR), efforts have traditionally been focused on infrastructure and technology-related measures [25]. Currently, however, there is consensus that these efforts must be accompanied by other measures related to the improvement of the resilience of people and territories in all disaster cycle phases (preparedness, response, recovery, mitigation) [26,27]. Many tasks can be applied to prevent and mitigate the effects of flooding, such as updating urban planning based on risk prevention [28], developing early warning systems [29,30] and specifically promoting people's coping capacity. Society must adapt to flooding, especially as the phenomenon becomes more common as a direct consequence of the climate change process [31,32]. This phenomenon affects the hydrological cycle, producing the following effects [33]: (i) an increase in temperature, which is reflected in an increase in evaporation and evapotranspiration and acceleration in the destruction of glaciers, (ii) drastic changes in water and snow precipitation patterns, and (iii) the ascent of sea level. Society must therefore change its way of acting, know the territory it occupies and delimit the areas most vulnerable to these phenomena. The results of these analyses will facilitate limiting the types of land use in these areas, which will minimize damage and facilitate the quick recovery of society [27].

The analysis of the components of risk is a recurring theme in the scientific literature. Various conceptual frameworks of risk have been proposed that define its main components and relationships (hazard, vulnerability, exposure, coping capacity, adaptive capacity, etc.) [25,34,35]. In all of these frameworks, there is common agreement about the two main components of flood risk: hazard (as the probability of occurrence of flood) and vulnerability (representing conditions and capacities that make a system or an individual susceptible to harm). It is relevant to highlight that the exposure is sometimes considered an independent component [36], but in this paper, exposure is included in the concept of vulnerability [15].

A review of the concept of vulnerability to floods also shows diverse conceptual frameworks that emphasize its multidimensional character [37–40], comprising physical, social, economic and environmental dimensions [41–51]. Consideration of the social dimension is one of the most common approaches in vulnerability studies. Social vulnerability includes all factors specifically related to the interactions of flood hazards with individuals, populations and communities, including the exposure of people, sociodemographic and socioeconomic profiles, employment, education, household composition, demographic structure, and the capacity of society to cope with hazards and their effects. All of these factors affect the resilience of communities to cope with flood effects, but their respective importance is not clear [25].

We perform a simplification exercise in this work and summarize the factors that affect social vulnerability into two major components: (i) the physical exposure of the individual/population/community to flooding; and (ii) resistance, which includes a set of minor components such as protection, reaction capacity and coping capacity [38].

Notably, most social vulnerability studies adopt a theoretical focus [52], although there has been an effort to perform empirical assessments in recent years [21,52–59]. These studies may present the drawbacks of using different measurement procedures, which makes it difficult to compare their results. However, the more common use of vulnerability indices as methodology better facilitates the comparison of results among different study areas [60]. Additionally, the different scale of studies is significant [56]. To date, research papers that estimate the social vulnerability of a municipality have applied a broad range of scales, with different cartographic units such as municipalities [61], municipal districts [40,62], and census blocks [57,58] or neighbourhoods [63]. Recently, new studies working with cadastral parcel data [59] have improved spatial accuracy. New public open-access databases and spatial data infrastructures provide data with high territorial precision, which opens new possibilities for more accurate analysis.

Based on the previously mentioned gaps in knowledge, the main objective of this paper is to propose a new method to assess social vulnerability to flooding by reviewing the importance of factors to map social vulnerability at a very detailed geographic scale, the cadastral parcel. This method may be applied to other areas where data are available. The proposed method is tested in the urban area of a middle-sized town, Ponferrada (Spain).

The use of indices has proved a robust method to assess social vulnerability [52,55,64–67]. Therefore, the results of the proposed method are expressed as a multicriteria index (Social Flood Vulnerability Index: FVIsocial) to show the relation between the components of social exposure (population exposure) and the resistance of people exposed to floods (protection, reaction capacity and coping capacity) for each defined geographical unit (cadastral parcel). These components are expressed as normalized quantitative indicators that are weighted based on the analytic hierarchy process (AHP). The assignment of these weights is based on work by a panel of risk experts who have conducted a pairwise comparison among indicators. The indicators are grouped into factors based on semantic meaning to simplify and improve the intelligibility of the model. The AHP process is one of the most commonly used methods when working with the integration of indicators in vulnerability studies [64,68–71].

The structure of the article is as follows. First, a description of the study area, the city of Ponferrada (Castilla y León, Spain), is presented. Ponferrada is a small town exposed to a hypothetical flood from a dam breakage. The methodology used to quantify social vulnerability is subsequently presented in a section divided into three phases: the description of the indicators used to quantify the vulnerability, the normalization to 0-1 values and the weighting of the indicators. The results map shows the application of the proposed method to the study area at the urban parcel scale. The article continues with a discussion of the advantages and limitations of the proposed methodology and ends with the main conclusions.

2. Study Site

The proposed method is applied in the urban area of the Ponferrada nucleus, a city that belongs to the municipality of the same name, which is located in the province of León (northwest of the Iberian Peninsula) (Figure 1). The municipality of Ponferrada has a total of 67,367 inhabitants distributed in 37 population nuclei. Its capital, Ponferrada, which is located on the banks of the River Sil and its affluent, the Boeza River, is the seventh-largest city of the Castilla y León region by population, with 41,858 inhabitants. It is the centre of the most important services of Bierzo County, including the district hospital, university, and regional authority. According to the population census [72], the municipality is distributed into 6 districts or demarcations. Each district is subdivided into different census sections. In total, the city is subdivided into 45 census sections, each of which averages approximately 1500 persons within the range of 625–2780 inhabitants.

The study area has been affected several times by floods. In the collected historical records [73], a total of 26 floods have been recorded in the area of Ponferrada, including not only the Sil River that runs through the city but also other streams in the area, such as Valdueza and Boeza. The flood hazard was reduced in 1960 with the construction of the Bárcena and Fuente del Azufre dams in the River Sil, located upstream of the town (Figure 1). Nevertheless, Ponferrada's river section is considered a potential flood section due to the possible downstream effect of the reservoir in a normal exploitation situation [73]. This scenario usually occurs when a large storm is expected and dam gates must be opened to be able to store the incoming water, which may lead to flood areas downstream. As a result, the floods in this area have three main triggers: (i) heavy rainfall, (ii) the exploitation of the dam, and (iii) the breakage of the dam. This research focuses on the last trigger, proposing a method to estimate social vulnerability and applying it to Ponferrada in the estimated flooded area that would be generated in the worst-case scenario in which both the Bárcena and Fuente del Azufre dams break [74] (Figure 2).

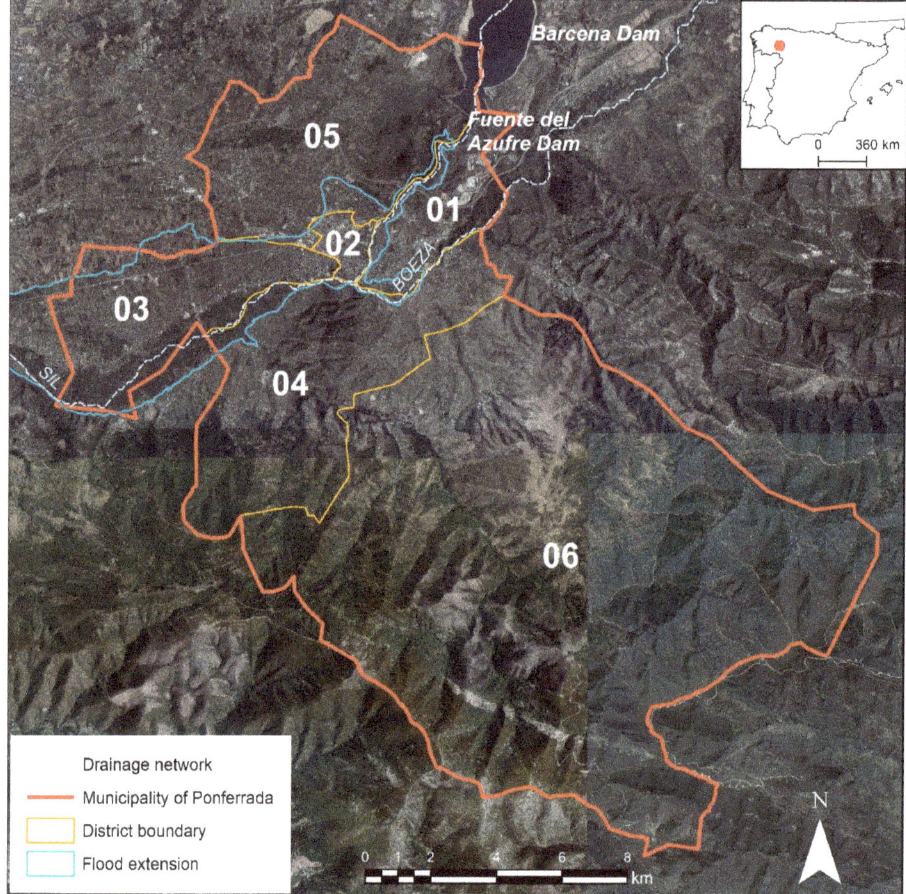

Figure 1. Location of the study area.

Of the 45 census sections, the study focuses on only 37. The eight rejected census sections coincide with sections outside of the perimeter of the flood area or those inside the perimeter but in which no urban parcels are affected. This perimeter is estimated based on break-dam modelling of the dam [73]. From these 37 sections, the assignation of data is extrapolated from neighbouring sections in the other eight census sections. The main reason for this assignment was the coincidence of two situations: on the one hand, the affected extension of the section was small; on the other hand, the affected urban parcels were closer and had social characteristics more similar to those from the neighbouring section than to those from their own section.

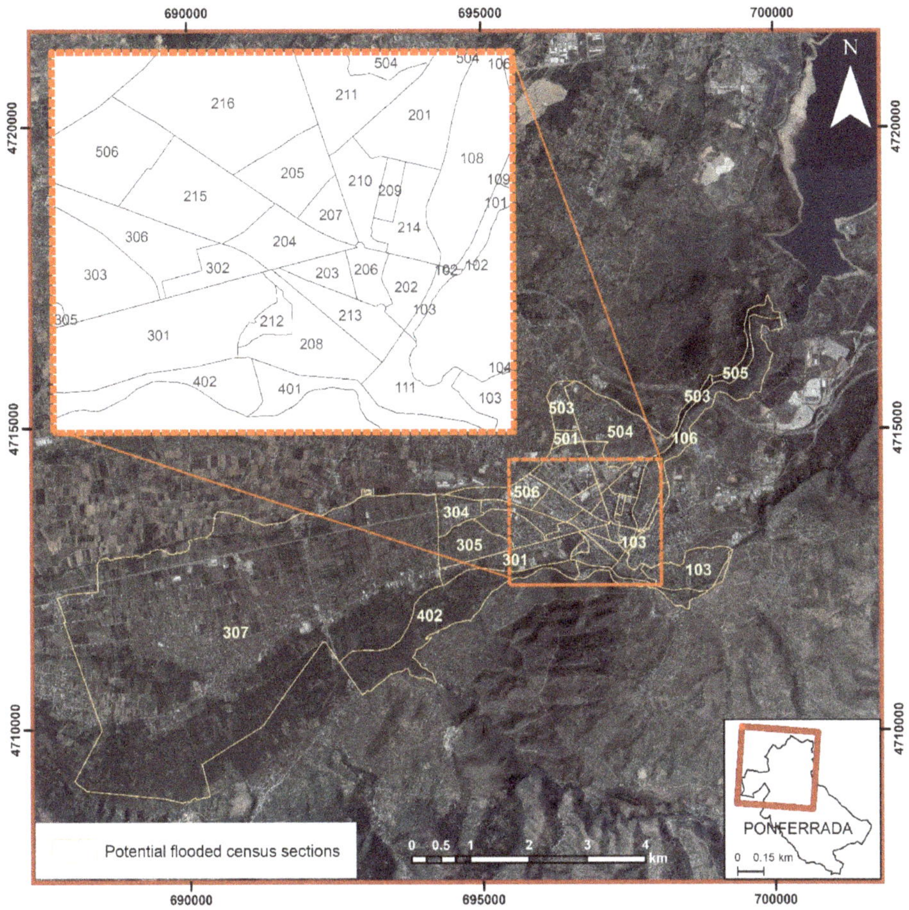

Figure 2. Census sections potentially affected by the water sheet.

3. Methodology and Data Sources

Vulnerability indices (VI) are numerical expressions that provide a statistical measure of the changes in value that vulnerability may experience. These indices are based on indicators. An indicator is a variable that reflects the state of a situation or any of its characteristics in a place at a particular time. It is expressed quantitatively as statistical information that synthesizes the data provided by the variables intended to be analysed. In the present research, all the indicators are grouped semantically into factors selected as the most representative of all the components that conform to the vulnerability. The relationships among the different elements that conform to the proposed vulnerability index are summarized in Figure 3.

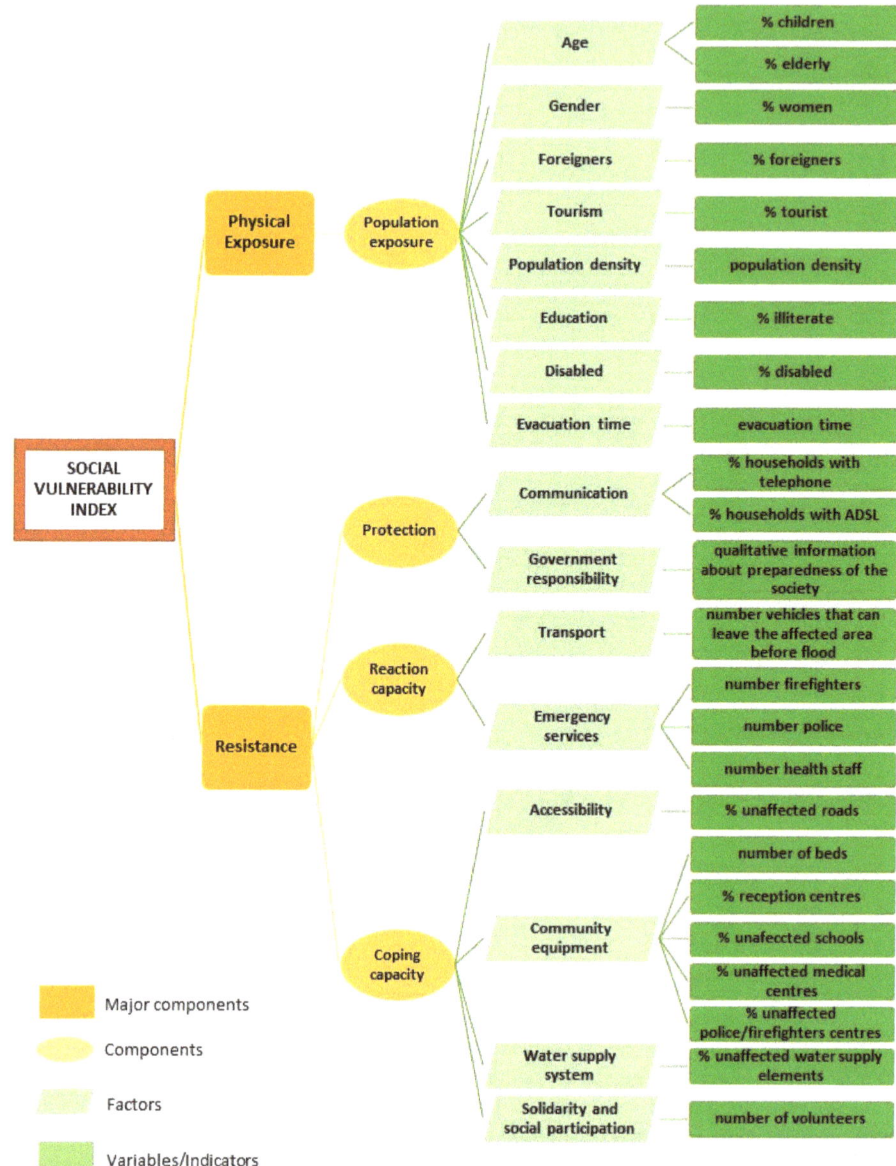

Figure 3. Relationships among the elements of the proposed vulnerability index.

Vulnerability indices are represented by equations that relate the records (weighted or not) of all the indicators (e.g., percentage of the elderly, percentage of women) with the components (e.g., exposure degree, resilience) of vulnerability. From these records, fractions are used to obtain the value of the vulnerability index, placing the indicators that increase vulnerability in the numerator and those that reduce vulnerability in the denominator [75]. In this way, the acquired results are dimensionless and allow comparison of vulnerability indices for similar indicators and for different scales in studies of different years or places.

Following these guidelines, the proposed social flood vulnerability index (FVIsocial) is measured on a scale of 0–1 and is determined by two major components: (i) the exposure degree of social indicators; and (ii) the resistance [76], as defined by Equation (1):

$$FVIsocial = (E/R)/10 \tag{1}$$

where E is the degree of social exposure (Equation (2)) and R refers to social resistance (Equation (3)).

$$E = \sum_{i=0}^{m} (v_i * w_i) \tag{2}$$

$$R = \sum_{i=0}^{m} (v_i * w_i) \tag{3}$$

where v_i is the value of each indicator (included on a scale of 0–1); w_i is the weight of each indicator (included on a scale of 0–1); and m is the number of indicators.

From Equation (1), it can be deduced that the exposure-resistance ratio will oscillate between the minimum value of 0 when the exposure degree is 0, independent of the value of the resistance, and the maximum value of 10 when the exposure degree is the maximum (1) and the resistance is the minimum (0, 1). This fraction is divided by 10 in Equation (1) to achieve a social vulnerability index with a range of 0–1. The problem of working with Equation (1) would be if resistance was 0; however, since the number of people connected via social media continues to increase [77], it is expected that every place will not reach this minimum value of 0.

Following the definition of the social vulnerability concept, the methodology developed in this research is divided into the following steps:

- Description and calculation of indicators. One added value of the proposed methodology is that the majority of these data are open access from governmental websites, although some were collected through personal enquiries.
- Indicator normalization. This process allows us to obtain dimensionless results to compare social vulnerability indices for different scales in various studies.
- Indicator weight calculation. The calculation of the weights is based on an expert panel survey following the analytical hierarchical process (AHP) methodology [78]. This method allows the evaluation of the importance of each of the indicators.

All of these phases are described in the following sections together with the data sources.

3.1. Data Sources

Most of the data required to estimate FVIsocial indicators were obtained from the open data repository of the Spanish National Statistical Database [72]. First, this database includes general social data, such as the number of inhabitants, ages, and foreign visitors, which were recorded in the last ten-yearly Spanish census in 2011. Second, this database includes spatial data at different geographic aggregation levels (municipal, districts and census sections), all of which are required in the present research. It is relevant to note that the Dirección General del Catastro in Spain has an open platform that enables the downloading of all the geographic and alphanumeric information of urban parcels [79]. To relate both attributes and geographical data, we propose the use of each indicator in its available spatial resolution and, subsequently, their disaggregation at the highest spatial resolution using geographical information system (GIS) tools. Therefore, in the last step, all the indicators should be linked to the cadastral parcel layer.

Another determining layer was the potential flood area, the perimeter of the water flood, which helped to delimit the real affected census sections and thus the study area. This map is the result of cascading dam failure modelling with a peak discharge of 2093 m^3/s according to the Boss Dambrk model [73]. It simulates one-dimensional hydrodynamic flood routing and predicts dam-break flood wave formation. Therefore, the flood area map is accompanied by a wave travel time study.

The rest of the data layers needed to estimate the indicators are as follows:

- Building doors: This is a point map containing the entrances to all the buildings, which is used to calculate the evacuation time and the population density. In general, there is only one point per urban parcel, but large buildings may show more entrances.
- Street map: This shows the communication lines inside the urban area and is an indispensable map to identify evacuation time issues; it indicates the evacuation paths.
- Road map: The access of goods during and after the emergency period is necessary to facilitate the recovery of society. This road map helps to estimate the number of access points affected by floods.
- Digital elevation model: The purpose of this map is to generate the slopes of the terrain, which are used as impedance to calculate the difficulties of evacuating people (the steeper the path, the slower the evacuation will be).

An added value of the proposed methodology is that all the information used is freely available through the Spanish government website platforms, so there is no cost to obtain updated data. The only exception is the potential flood area map of the study zone. However, many flood maps are also freely distributed.

The last data source exploited in this study comprises the different enquiries posed to (i) the population to evaluate the information on emergency response and/or (ii) public administration, consisting of the police chief, firefighter chief, dam manager chief, council tourist expert, council engineer and different NGO heads. In the first case, the idea was to randomly survey the population to evaluate how much information they have about specific protective measures in case of an emergency. In the second case, enquiries were used to obtain the following indicators:

- the level of government responsibility, whether there is an emergency plan, events to inform the population about what to do in the case of an emergency, etc.,
- percent of tourism population to obtain the internal council's data about the size of the tourism population in the town,
- the number of volunteers, to understand the size of the volunteer group that may help in the recovery period in the studied society,
- the number of emergency personnel (police, firefighters, medical workers), to identify the specialized personnel available to confront an emergency,
- the location of water supply elements to evaluate the percentage of water supply elements that will be affected by the flood and that may make the recovery process difficult.

After performing these enquiries and determining that there had been no event to inform the population, we did not implement any surveys among the population. However, if any such event is performed, it is necessary to survey the population to evaluate their level of knowledge about the procedures to follow in case of a dam failure.

3.2. Indicator Selection and Calculation

To determine the social flood vulnerability index, two tasks were carried out. First, the definitions of vulnerability factors were determined to compare different authors' approaches e.g. References [41, 44,45,75,80–87]. Second, the selection of significant indicators for each factor was performed. If there was scientific consensus in the bibliographic review, the determination of the social vulnerability factor was accomplished by means of a single indicator; when there were difficulties in assigning a unique indicator to define a factor, the method works with several indicators. The state-of-the-art processes indicate that there are uniform criteria when using the following indicators: percentage of elderly, percentage of children, percentage of women, percentage of individuals with disabilities, percentage of foreigners, number of inhabitants per km^2 and percentage of illiterate population [40,80,83,86]. This is because the intersection of the circumstances of age (elderly and children), gender (women), minority status, disability status and education (without basic education) have the greatest influence on the social burdens of natural hazards since they constitute more vulnerable groups of a population that more frequently suffer major social impacts when they are exposed to a certain risk. In the case of social resistance indicators, there is also coincidence in the number of health staff per 1000 inhabitants [80,84,85,87], the number of beds per 1000 inhabitants [82,87] and the percentage of reception centres (e.g., surface area in m^2 of sports and cultural equipment) [83,87]. Based on this information, the authors propose the number of other civil protection staff per 1000 inhabitants, considering that in some countries, there is no equivalence between health and security services. The other social vulnerability indicators selected in the present study do not show international consensus, and their use varies depending on the study area and the data availability. For instance, the number of vehicles that can leave the affected area according to the type and number of accesses (roads) is supported by one study [84], while [59] suggests that the use of vehicles is dangerous and increases damage because vehicles may be swept away with shallow waters. Although this is a true scenario, driving cars may be beneficial if there is a warning period that is long enough to allow evacuation in an orderly manner. The present study selected at least one indicator for each of the most frequently mentioned vulnerability factors in the bibliographic review, considering the ease of retrieving data and their representativeness. For example, to determine the solidarity of society, the percentage of electoral votes [83,84] cannot be applied in nondemocratic countries or in countries where the population does not trust politicians. Instead, the number of volunteers may better adjust to the solidarity factor, although it may be more difficult to collect this type of data.

As a result, there are two types of indicators according to the value that they provide: quantitative and qualitative indicators. Tables 1 and 2 show the proposed vulnerability factors for exposure and resistance and the corresponding indicator deployments. As will be explained, all the indicators were merged to define the degree of exposure and the resistance of each geographical unit based on different weight values.

Table 1. Description of the proposed factors and indicators used to analyse the degree of exposure of social vulnerability for flood events.

Factors	Exposure Indicators	Normalized Values	Spatial Scale
	QUANTITATIVE INDICATORS		
Age population	% children = *(population under 16 years/total population)* × *100* % elderly = *(population over 65 years/total population)* × *100*		Census section
Gender population	% women = *(population of women/total population)* × *100*	0%—0 100%—1	Census section
Foreign population	% foreigners = *(foreign population/total population)* × *100*		Municipal level
Tourist population	% tourist = *(tourist population (national and foreign)/total population)* × *100*		Municipal level
Population density	Population density =*Total population/surface*	Min = 0 Max = 500	Parcel level
Evacuation time	Evacuation time	Min = 0 Max = arrival time of the water sheet	Parcel level
Education	% illiterate population = *(population without studies/total population)* × *100*	0%—0 100%—1	Census section
Disabled population	% disabled = *(population with disabilities/total population)* × *100*	0%—0 100%—1	District level

Table 2. Description of the proposed social vulnerability factors and indicators of resistance to flood events.

Factors	Resistance Indicators	Normalized Values	Spatial Scale
	Quantitative Indicators		
Communications	% of households with a telephone = *(households with telephone lines/total number of cores) × 100* % of households with ADSL = *(households with Internet access/total number of households) × 100*	0%—0 100%—1	Census section
Emergency services	Number of firefighters per 1000 inhabitants = *(firefighter number/total population) × 1000*	Min = 0 Max = 1	County level
	Number of local police per 1000 inhabitants = *(local police number/total population) × 1000*	Min = 0 Max = 1.5	Municipal level
	Number of national police per 1000 inhabitants = *(n° of national police/total population) × 1000*	Min = 0 Max = 2	County level
	Number of health staff per 1000 inhabitants = *(n° of doctors and nurses/total population) × 1000*	Min = 0 Max = 2.3	Municipal level
Transport	Number of vehicles that can leave the affected area according to the type and number of accesses (roads)	Min = 0 Ma x= 3015	Water sheet level
Accessibility	% of accesses not affected by the water sheet= *(n° of unaffected accesses/total) × 100*	0%—0 100%—1	Water sheet level
Community equipment	Number of beds per 1000 inhabitants = *(n° of beds/total population) × 1000*	Min = 0 Max = 3.1	County level
	% of reception centres (surface area in m^2) (sports, cultural equipment, etc.)	0%—0 100%—1	Municipal level
	% of unaffected schools = *(n° of schools not affected/total) × 100* % of unaffected medical centres = *(n° of unaffected medical centres/total) × 100* % of unaffected police centres and firefighters = *(n° of unaffected centres/total) × 100*	0%—0 100%—1	Municipal level
Water supply system	% of unaffected water supply elements = *(n° of water supply elements not affected/total) × 100*	0%—0 100%—1	Municipal level
Solidarity and social participation	Number of volunteers	Min = 0 Max = 1.5	Municipal level
	QUALITATIVE INDICATORS		
Government responsibility	Historical record of disasters (Yes/No) Alert system (Yes/No) Emergency plan (Yes/No) Emergency drill (Yes/No) Evacuation routes (Yes/No) Courses or brochures (Yes/No)	Yes—1 No—0	Municipal level

The spatial distributions of all these indicators were mostly determined by a GIS reclassification of an administrative boundary map. Only the population density and evacuation time required more detailed spatial analyses.

3.2.1. Population Density

The areas of higher population density represent areas of higher exposure to a hazard and areas where evacuation may be more difficult, so they are very vulnerable. Although this is a relevant indicator, the personal protection data law remains the main impediment when estimating its value. For this reason, population data are only available at the census section level in open sources. To spatially disaggregate these data, it is necessary to estimate the number of dwellings in each urban parcel. This calculation is possible through the attribute database linked to the cadastral parcel layer, where it is possible to estimate the amount of properties targeted to residential use based on a unique reference cadastral identifier. The population registered in a census section is distributed homogeneously among all dwellings belonging to the same census section.

The final step to calculate the population density is to distribute the total population of each building among the number of building doors. As can be inferred, this process is impossible to perform without GIS capabilities.

3.2.2. Evacuation Time

The evacuation time is an indicator that measures the relationship between citizens' velocity of displacement and the time of arrival of the flood water. Until now, this indicator has only been used for emergency management, and very few studies have included it to define vulnerability [88]. Its importance is that it establishes a very clear threshold between those areas where there is a possibility of social recovery and those in which such recovery is impossible. The latter correspond to areas in which the evacuation time is greater than the time of arrival of the water wave.

To estimate evacuation time, calculations may be performed assuming that people evacuate on foot or by vehicle. In the case of Ponferrada, there are very few escape routes, which produces a funnel effect. For this reason, this paper focuses only on the evacuation time of pedestrians. Some studies have identified walking and driving in flood waters as the main danger for people during floods [89]. Consequently, Ref. [59] suggests moving upstairs in high buildings. This option was not considered in the present study, because without an emergency plan or information advice, people's main reaction will be to run away from the river.

The calculation of the evacuation time indicator starts with the generation of a slope map. According to this resulting topography, the time that a person requires to walk a distance is estimated. As an example, on flat terrain, it was considered that the average speed was 3.8 km/h [54]. This is a value slightly more conservative than previous estimates [90,91], but it is considered to better fit with an old population such as Ponferrada. Next, the resultant slope map is reclassified based on the inverse velocity (minutes/metres). This new map together with the street and the flood water sheet maps are the inputs to generate a cost-distance map that allows the estimation of the evacuation time for each urban parcel. Other methods that can be used to estimate the evacuation time are linked to flood evacuation simulators [92].

3.3. Indicator Normalization

In this work, normalization is used to compare and combine indicators of different natures. For this purpose, the final normalized value must be between 0 and 1. Two different methods of normalization follow. The first is applied to indicators whose values are percentages. In this case, the normalization process is immediate and involves dividing the value by 100. The second method is based on linear transformation functions [93] (Equation (4)), although this method does not limit the use of other types of functions if new indicators are incorporated in the future.

$$y = ((x - \text{Min}))/((\text{Max} - \text{Min})) \tag{4}$$

where x is the real value of the indicator, Min is the minimum established for the indicator, and Max is the maximum established for the indicator.

All transformation functions used the value 0 as a minimum value. To define the maximum value of the transformation function, we searched for generic values endorsed mostly by institutions of recognized prestige and international scope. Therefore, this value is different for each indicator, as described below and shown in Tables 2 and 3:

- Population density: The maximum value corresponds to the maximum number of people who can occupy a residential building based on the number of evacuation exits. For the example of the Ponferrada municipality, the Spanish government sets 500 as the maximum value for buildings with a single outlet, corresponding in this case with the portal of the building [94].
- Evacuation time: The minimum and maximum values are functions of the time required to reach the perimeter of the water sheet and the time that the population has available from the start of the alert until the water reaches the city. These values may vary if an emergency plan is implemented in the municipality that includes an estimated time of alert. In this indicator, the normalization values are not continuous but take a value of 0 when the evacuation times are less than the arrival times of the water sheet or a value of 1 when the time required is greater than the available time before reaching a maximum degree of exposure.
- Number of firefighters per inhabitant: The maximum value is 1 firefighter per 1000 inhabitants, which is a value defined from the recommendations of different institutions [95].
- Number of local police per 1000 inhabitants: The maximum value is established according to the local regulations of each study area. For instance, in the study case of Ponferrada, the value corresponds to 1.5 local police per 1000 inhabitants.
- Number of national police per 1000 inhabitants: In this case, the maximum value is 2, and it is determined according to the national recommendations of the study area.
- Number of health staff per 1000 inhabitants: The maximum value for the application of the transformation function is 2.3. This value is set [96] as the minimum recommended number of health personnel necessary to cover the needs of primary health care.
- Number of vehicles that can leave the affected area according to the type and number of accesses: The maximum value of this transformation function is calculated by the multiplication of the peak hour intensity value (i.e., the maximum number of vehicles per hour in a road section) by the number of available accesses that allow adequate evacuation. If the number of vehicles in the study area is higher than the previous product, the resistance will be minimal (0). In our case study, the estimation of the peak hour intensity value is derived from the average daily index. According to regional regulations, the peak hour intensity value is 12% of the achieved average daily index [97].
- Number of beds per 1000 inhabitants: The maximum value of the transformation function of this indicator corresponds to the average number of beds in each country since there is currently no optimum level of beds recommended for adequate care of the population. The reason for this lack is that these values vary according to the structure of the health system of each country [98]. For example, in Spain in 2011 and therefore in Ponferrada, the average was established as 3.1 beds per 1000 inhabitants according to data from Spanish regulations [99].
- Number of volunteers: Currently, no recommended ratio of volunteers has been established by any international organization, possibly due to the variety of the social mass that constitutes the different volunteer organizations and the lack of training standards in the field of volunteering that unifies this sector. Given this situation, this study proposes that the maximum value is 1.5, which is the same value established for the local police, who would be responsible for the organization and the direction of volunteers in the case of a disaster.
- Qualitative indicators that measure the known information according to the population affected by the natural phenomenon plus the security and warning measures: These indicators involve a questionnaire mostly aimed at analysing the perception and resistance of a society with closed answers ("yes or no") (Table 2).

Table 3. Weights of indicators that indicate the exposure degree of social vulnerability.

Indicators	Weight (w_i)
Population density	0.264
Evacuation time	0.264
% disabled	0.121
% children	0.105
% elderly	0.104
% foreign tourism	0.042
% foreigners	0.032
% national tourism	0.024
% illiterate population	0.025
% women	0.018

In the case of qualitative indicators, the proposed normalization method considers that the questions to which the answer is *yes* are equivalent to a resistance of 1, while the questions with a response of *no* equal a resistance of 0. The results from each of the questions were added and divided by the total number of answered questions.

3.4. Estimation of Indicator Weight Values (w_i)

Our hypothesis is that not all indicators will have the same importance in relation to social vulnerability, so we must assign different weights to the indicators and deploy a hierarchy. In this study, the assignment of weights to the indicators was performed by applying a qualitative method comparing pairs, known as the AHP [68–71]. This method is based on the responses of an expert panel, which, in our case, involved 15 experts including geologists, engineers, geographers, and civil workers. The panel participants answered a survey for the assessment of the weight of each indicator. Each expert independently compared all the resistance indicators and eight of the ten indicators of exposure. To assure the consistency of their answers, the consistency ratio, CR [78], was estimated. The 15 surveys showed a CR lower than 0.1, which indicates a high level of consistency. According to the results of these 15 surveys, different weights were assigned to each indicator (Tables 3 and 4). It is important to highlight that the surveys were sent together with a list of the indicators, which the experts were asked to order by importance. This information allowed for the elimination of inconsistencies.

Table 4. Weights of indicators that indicate the resistance of social vulnerability.

Indicators	Weight (w_i)
Qualitative indicators: Historical record of disasters, alert system, emergency plan, emergency drill, evacuation routes and courses or brochures	0.186
Number of firefighters, local and national police per 1000 inhabitants	0.133
% of unaffected police centres and firefighters	0.098
Number of volunteers	0.084
Number of health staff per 1000 inhabitants	0.078
% of accesses not affected by the water sheet	0.077
% of unaffected water supply elements	0.068
% of unaffected medical centres	0.058
% of reception centres (surface area (m^2) of sports centres, cultural centres, etc.)	0.049
% of schools not affected	0.048
Number of beds per 1000 inhabitants	0.044
% of households with telephone	0.035
% of households with ADSL	0.024
Number of vehicles that can leave the affected area according to the type and number of accesses (roads)	0.019

The two exposure indicators not analysed by the panel were the population density and evacuation time. For both of these indicators, the maximum weight value was assigned. There are two main reasons for this high score. First, if there is no population (density) in the affected area, the condition of social vulnerability will be minimal. In this case, it is considered that the value does not have to be equal to 0 since in certain unbuilt parcels (e.g., parks), a population may exist. Second, if the evacuation time exceeds the time of arrival of the water sheet (plus the alert time, if this information is available), the population, regardless of age, sex, nationality, education or physical capacity, will not have sufficient time to reach a safe area, so the rest of the indicators will be of secondary importance.

4. Results

The application of the proposed methodology for the assessment of the social vulnerability to floods in the study zone of Ponferrada (Spain) shows that the social vulnerability for 3330 parcels of the 6441 analysed parcels is very high, reaching the maximum value of 1 (Figure 4). This means that approximately 77% of the inhabitants affected by the water sheet will experience heavy damage from the dam break.

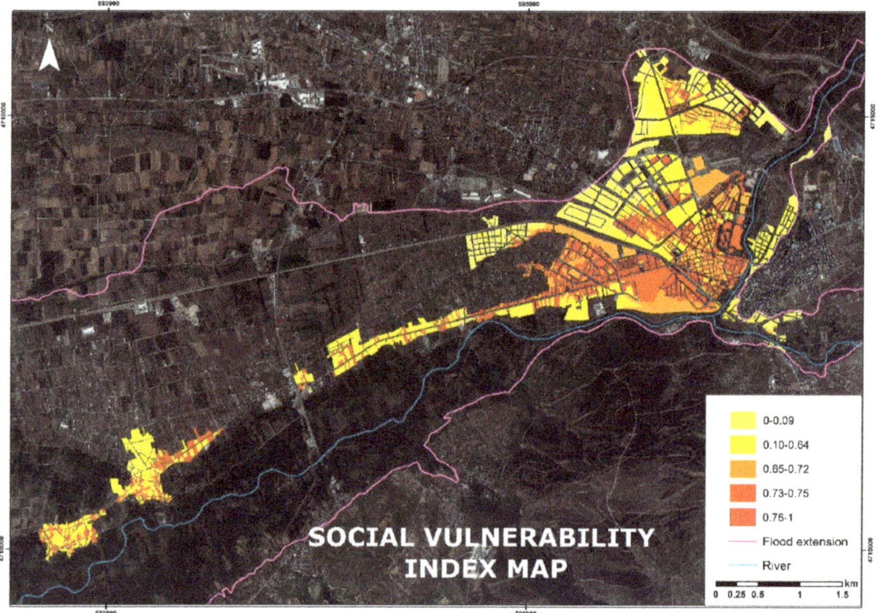

Figure 4. Map of the social vulnerability in Ponferrada due to a potential dam break. Break values of the legend correspond to the quantile distribution to better visualize the spatial distribution of index values.

In the studied case, the highest values are derived from the maximum exposure values assigned to the evacuation time indicator. In these parcels, the analysis shows that the time needed to reach an area outside the flood area is larger than the time needed by the water wave to reach the parcels (Figure 5). In addition, the value of the resistance cannot decrease this vulnerability as the resistance indicators do not have enough value to decrease the high population exposure due to the short time available to evacuate. In the rest of the parcels for which the evacuation time is low enough to reach a safe place before the water wave arrives, the social vulnerability values are very low, at less than 0.074.

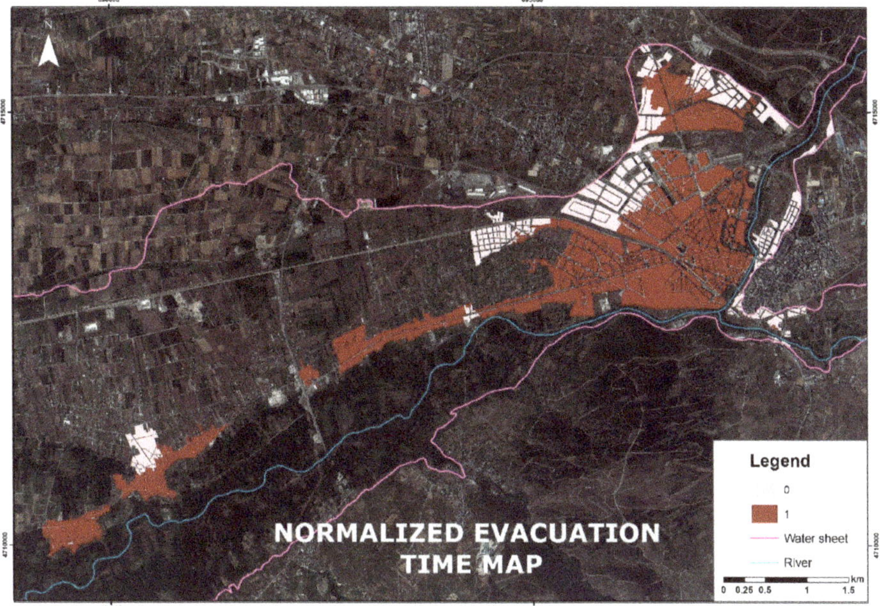

Figure 5. Normalized evacuation time map in Ponferrada.

The two other indicators that increase the vulnerability of Ponferrada are the age of the people and the population density. The impact on the vulnerability of the first indicator is due to the predominance of the elderly population compared to the number of young people in the studied society (Table 5). Therefore, the elderly indicator shows high values due to the high percentage that this ageing sector represents compared to the overall population.

Concerning the characteristics of the population, the census sectors that show the highest vulnerability values according to age data (children and elders) are in six sectors: sector 11 in district 1, sector 14 in district 2, sectors 2, 3 and 7 in district 3 and sector 4 in district 5. Except for sectors 3 and 7 of district 3, the rest of the sectors are located along the river. The distribution of women is more or less equal everywhere, and the only census sector that highlights this indicator is number 14 in district 2, which coincides with a large percentage of elderly persons. The next indicator, the foreign population, shows very low values except in sections 2 and 12 of district 2 and section 5 in district 3. All of these sections are on the riverside. With regard to the illiterate population, this indicator shows high values in section 11 in district 1, sections 1, 9 and 11 in district 2 and sections 4 and 5 in district 3. Again, all of these sections are on the riverside. Finally, the indicators related to the municipal data concerning tourism and the district data on people with disabilities present very low values. Once normalized, both indicators show values lower than 0.1.

Similar to the evacuation time, the population density indicator is analysed for the cadastral parcels (Figure 6). Upon normalizing the values, there are only two parcels in which the indicator reaches the highest possible value of 1. The rest of the parcels show very low values, at less than 0.25.

Table 5. Indicators related to age, gender, foreigners and education in Ponferrada. The highest values appear in bold font.

Census Section	% Children	% Elders	% Women	% Foreigners	% Illiterate Population
DISTRICT 1					
1	14.2	16.6	51.6	4.2	15.1
3	9.5	18	50.9	3.2	14.5
8	18.8	18.5	49.1	6.6	15.9
9	19.1	12.8	49.4	0	15.7
11	14.7	25.7	55.1	0	40.4
Average	*15.26*	*18.32*	*51.22*	*2.8*	*20.32*
DISTRICT 2					
1	3.1	32.6	49.6	1.34	37.9
2	24	12.4	47.6	20.22	13.5
3	9.1	19	52.3	3.13	11.9
4	10.1	12.5	50.4	8.12	9.3
5	15.9	13.4	52	8.94	25.2
6	7.3	28.2	50.5	9.22	21.8
7	8.1	20.8	50.2	10.1	10.4
8	9.1	12	57.7	7.57	26.2
9	8	30	50.2	8.92	34.7
10	10.5	20.1	48.9	11.71	19.2
11	3.2	29.1	46.8	0	39.9
12	18.1	17.6	48.6	22.25	21.7
13	12.9	17.1	58.8	0.4	17.7
14	12	31.2	64.8	0	16.8
15	20.8	6.9	53.8	0	7.6
16	20.3	3.7	52.5	0.8	6.4
Average	*12.03*	*19.16*	*52.17*	*7.05*	*20.01*
DISTRICT 3					
1	21	11.8	49	10.8	11.8
2	11.9	30.5	58.8	2.5	31.7
3	16.2	22.7	51.6	6.5	32.7
4	9.1	20.7	50.9	0	34.1
5	8.7	10.9	50.5	25.5	41.8
6	15.6	16.9	47.5	6.5	21.3
7	15	22.8	46.4	0	30.3
Average	*13.93*	*19.47*	*50.67*	*7.40*	*29.10*
DISTRICT 5					
4	16.8	21.9	51.8	0.5	15.1
OVERALL STUDY AREA					
Average	*13.21*	*19.19*	*51.63*	*6.17*	*22.09*

In relation to resistance, the indicator related to the scarcity of information among the population decreases resistance considerably and imposes an increase of social vulnerability [75,83,85]. The main reason for this low information level is the inadequate effort of the government and local authorities to encourage training actions to reduce vulnerability. In the case of Ponferrada, these actions must focus not only on the breakage of the dam but also on other flood phenomena caused by the overflowing of rivers, which have much higher frequency. This type of action would cause a state of alert in the population that should not be confused with a state of alarm. The population will not reach this second state once they know how to act.

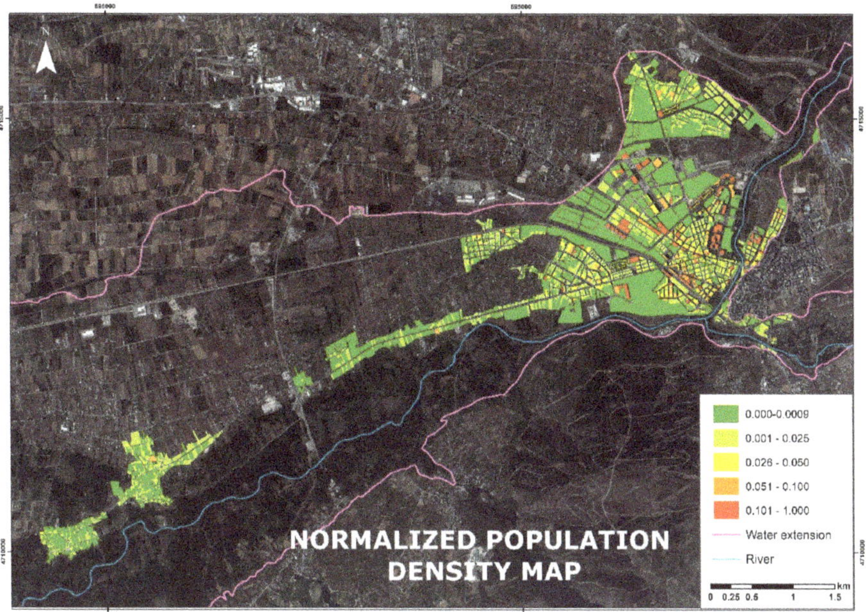

Figure 6. Normalized population density indicator in Ponferrada. Break values of the legend are adjusted to better visualize the spatial distribution of index values.

On the other hand, the indicators that increase resistance, and therefore reduce social vulnerability, are those related to communications, including those with reference to telephone and Internet access, indicators of emergency and health personnel and indicators concerning community equipment related to the number of beds. Ponferrada presents high values of these indicators since it is an area with a certain level of development and good standards of living due to the solidity of the institutions and the guarantee of services and basic rights (Tables 6 and 7).

Table 6. Values of indicators of health, security and community equipment in Ponferrada.

Indicators	Values
Number of health staff per 1000 inhabitants	1
Number of beds per 1000 inhabitants	0.788
Number of firefighters, local and national police per 1000 inhabitants	0.614
Number of volunteers	0.485
Reception centres (surface area (m^2) of sports centres, cultural centres, etc.)	0.809
% of unaffected medical centres	0.5
% of unaffected water supply elements	0.485
% of schools not affected	0.406
% of unaffected police centres and firefighters	0
Information on the population	0
% of accesses not affected by the water sheet	0.8
Number of vehicles that can leave the affected area according to the type and number of accesses (roads)	0

Table 7. Values of communication indicators calculated at the census section level but presented in this table at the district level because their values are quite homogeneous in the study area.

Indicators	District Level	
	Values	Number
% of households with a telephone	0.943	1
	0.945	2
	1	3
	1	5
% of households with ADSL	0.621	1
	0.616	2
	0.485	3
	0.744	5

5. Discussion

It is necessary to promote a culture of prevention that favours the study of the different types of vulnerability in a certain area to save millions of euros, avoid the loss of accumulated wealth and, above all, save human lives in the face of disasters [40,64,100]. The present research focuses on the social aspect and proposes a method for quantifying it based on the integration of exposure level and resistance components. With this purpose, the developed method is based on vulnerability indices as they allow dimensionless results to be obtained at the social level that can be objectively compared with vulnerability in different areas, during different periods of time and at different scales.

According to the results, the indicator that shows greater relevance concerning the increase of a society's vulnerability is the evacuation time, which is a novel indicator in this type of study; very few studies have included it to define vulnerability [88]. Its importance is that it establishes a very clear threshold between areas where there is a possibility of social recovery and those in which such recovery is truly difficult. The latter correspond to areas in which the evacuation time is greater than the time of arrival of the water sheet, so they represent locations where there is not sufficient time to reach a comfort zone. Evacuation times may increase considerably depending on two factors: the time of day and the characteristics of the population. If the disaster occurs at night, when most people are sleeping, the population's reaction time will be longer, and they will need more time to reach a safe place. At the same time, the displacement velocity will vary according to whether the evacuated population includes old people, children and individuals with disabilities, for instance. The main way to reduce these indicator values would be to increase the alarm time and developing early warning systems. Specifically, the alarm time refers to warnings carried in emergency messages through loudspeakers, lights or sounds [101]. If this time is added to the time of the water wave arrival, a person has more time to assimilate the alarm information and initiate movement for a correct evacuation. The resistance of society thus increases, and social vulnerability decreases. At the time this study was performed, Ponferrada did not have an alarm system or emergency plan to broadcast emergency messages if a dam failure occurred. Since the time for the water wave to reach the town is approximately 15 min, if messages are spread when the dam breaks, there is not enough time to save all the people in the town. This is why the values of the evacuation time indicator are so high. Nevertheless, the latest studies suggest remaining at home [89] or converting the upper floor of buildings at meeting points [59]. However, these alternatives will only be safer if they are included in emergency planning and citizens know before the event how to react.

Another indicator relevant to vulnerability studies is the population density indicator [84,86]. As expected, it is a basic indicator since if there is no population, the social vulnerability will be minimal. This is the main reason for the high weight value with respect to the rest of the indicators. In contrast to other studies, in the case of Ponferrada, the importance of this indicator lies in its weight and not in the overpopulation condition or high number of people living on each unit of the surface [75,82,84,86]. Although the calculation of its real value does not present difficulties, its normalization is complex and requires working at the building portal level to establish the evacuation capacity of people in buildings. For this reason, GIS is the only tool that can be applied in this process.

Ponferrada is an old society, which implies high vulnerability, but it has a low number of immigrants, foreigners and people with disabilities. This means that indicators related to these three groups will show lower values than in towns where there is an important economic development that functions as an immigrant call [80] or where there is an important tourism inversion [83].

The normalization of each of the indicators allows dimensionless values [75] to be obtained to enable the comparison of results not only in different periods but also in different study areas. However, not all normalization processes allow this multi-area analysis [64,102]. To achieve this goal, normalization must avoid the use of the maximum value obtained in the specific study area and work with generic concepts that can be assigned as minimum and maximum values. These generic values may be ratios, recommendations or already established standards at both national and international levels without forgetting the context of the study area. In the case of non-existent indicators, this generic value may be associated with the generic value of a similar indicator. For instance, in the case of Ponferrada, the maximum value applied to normalize the indicator of the number of volunteers was the same maximum value established for the local police. Nevertheless, this process is not always easy. The need to search for a generic value for each of the indicators in the transformation function is the step that presents the greatest difficulty when normalizing indicators.

Concerning the weights, the AHP method has shown its fitness for some of the indicators. This method has several characteristics, including the incorporation of both qualitative and quantitative indicators, the hierarchical decomposition of the problem, the organization and structure of the criteria, the mathematical sustenance of the method and the analysis of inconsistencies of the judgements issued, that contribute to the quality of the drawup of the decision process [103]. Nevertheless, this method also has some drawbacks. In this study, two problems were found in its application. The first is that the interviewed experts had some difficulties in establishing judgements about the weights of the indicators because of the large number of indicators, which hindered comparative analyses between them. For this reason, certain inconsistencies were found in the values of the weights assigned by the interviewed experts. To avoid these inconsistencies, it is recommended that in addition to the weight values, the experts establish a hierarchy of the indicators that define the exposure degree of social vulnerability and another hierarchy for the indicators that define resistance to social vulnerability. In this way, if inconsistencies are detected, possible failures in the paired comparison of the indicators can be solved, as has been done in this work. In turn, to facilitate the generation of matrices containing the weights of the indicators, a document with a brief definition of each of the indicators to be analysed should be added to the surveys to achieve homogeneity in the language and identical interpretations. The second problem that the AHP method presents is its rigor; it is not flexible in the face of future changes. This means that new indicators cannot be incorporated to define different vulnerabilities. The surveys should be repeated, incorporating the new indicators to assign the new weights.

Finally, the distribution of social vulnerability data by means of a map has been shown to be a useful tool for delimiting the areas most vulnerable to floods and to allow their diagnosis. For instance, the evacuation time map together with a map of the characteristics of buildings may help to define meeting points in high buildings spread within the flood water sheet to save people who otherwise would not reach a safe area. Therefore, working at a detailed scale proves worthwhile not only for emergency response but also for better prevention planning, improving the resilience of society. The new open-access data appear to be a perfect source to perform this type of study, although these data have limits in the scale of analysis due to the personal protection data law requirements.

6. Conclusions

Natural disasters are an increasingly popular topic at the social and economic levels since there has been a significant increase in their impact in recent decades. Therefore, it is necessary to promote a strategic and systematic prevention approach that allows the reduction of vulnerability in the face of threats, dangers and risks associated with disasters and emphasizes the need to use adequate means to increase the resilience of nations and communities in the face of disasters. The analysis of vulnerability from a social perspective has proven a valuable tool to support flood risk planning and management, when comparing the results of social vulnerability at multi-temporal and multi-spatial levels.

In the present study, the proposed method makes these comparisons possible through the use of several indicators whose normalization is based on generic values. The definition of these values has been shown to be the most difficult step of the method due to the lack of data on some of the treated indicators, so the method is open to some adjustments adapted to different study areas and types of risks. Regardless, the practicality of the proposed method has been proven by applying it in the urban area of the Ponferrada nucleus (León) in the case of a flood due to the rupture of the Bárcena dam and, in a concatenated way, to the breakage of the Fuentes del Azufre dam. The results of this application showed high social vulnerability as a consequence of the lack of time to evacuate. This indicator showed that the 15 min required for the water wave to reach the town are not enough to evacuate the inhabitants in the affected area. Almost 77% of the affected population of Ponferrada will not be able to reach a safe place.

Finally, the proposed method presents the results with high spatial variability by means of a map. This cartography allows the delimitation of the most vulnerable areas facing floods at the smallest possible scale to allow their diagnosis. This entails, among other tasks, the design of prevention, emergency and evacuation plans and decision-making with respect to the locations of new constructions.

Author Contributions: Conceptualization, E.G.-M. and M.F.-J.; methodology, L.T.-G. and M.F.-J.; data and GIS analysis, L.T.-G., M.R. and M.F.J.; AHP analysis, L.T.-G. and M.R.; writing—original draft preparation, L.T.-G. and M.F.-J.; writing—review and editing, all authors. All authors have read and agreed to the published version of the manuscript.

Funding: This research was funded by the Spanish Ministry of Science, Innovation and Universities, grant ESP2017-89045-R, by Universidad de León and by Junta de Castilla y León, grant LE169G18.

Acknowledgments: The authors would like to thank Neftalí Almarza from Confederación del Miño-Sil for his collaboration and help with all the documentation of Barcena's dam, the support of the local government and civil workers who helped us to gather all the information not available from open sources, the staff of experts who collaborated in the AHP analysis and the anonymous reviewers who improved the article with their suggestions.

Conflicts of Interest: The authors declare no conflicts of interest.

References

1. Directive 2007/60/EC of the European Parliament and of the Council of 23 October 2007 on the assessment and management of flood risks. *Off. J. Eur. Union Eur. Commun.* **2007**, *288*, 27–34.
2. Bennett, B.; Leonard, M.; Deng, Y.; Westra, S. An empirical investigation into the effect of antecedent precipitation on flood volume. *J. Hydrol.* **2018**, *567*, 435–445. [CrossRef]
3. Beniston, M.; Stoffel, M. Rain-on-snow events, floods and climate change in the Alps: Events may increase with warming up to 4 °C and decrease thereafter. *Sci. Total Environ.* **2016**, *571*, 228–236. [CrossRef] [PubMed]
4. Scholz, M. Predicting Dam Failure Risk for Sustainable Flood Retention Basins. *Sustain. Water Treat.* **2019**, 301–321. [CrossRef]
5. Prakash, M.; Rothauge, K.; Cleary, P.W. Modelling the impact of dam failure scenarios on flood inundation using SPH. *Appl. Math. Model.* **2014**, *38*, 5515–5534. [CrossRef]
6. Rico, M.; Benito, G.; Díez-Herrero, A. Floods from tailings dam failures. *J. Hazard. Mater.* **2008**, *154*, 79–87. [CrossRef]
7. Hollins, L.; Eisenberg, D.; Seager, T. Risk and resilience at the Oroville Dam. *Infrastructures* **2018**, *3*, 49. [CrossRef]
8. Ye, F.; Zhang, Y.J.; Yu, H.; Sun, W.; Moghimi, S.; Myers, E.; Liu, Z. Simulating storm surge and compound flooding events with a creek-to-ocean model: Importance of baroclinic effects. *Ocean Model.* **2020**, *145*, 101526. [CrossRef]

9.   Pignatelli, C.; Sansò, P.; Mastronuzzi, G. Evaluation of tsunami flooding using geomorphologic evidence. *Mar. Geol.* **2009**, *260*, 6–18. [CrossRef]

10.   Van Veen, B.A.D.; Vatvani, D.; Zijl, F. Tsunami flood modelling for Aceh & west Sumatra and its application for an early warning system. *Cont. Shelf Res.* **2014**, *79*, 46–53. [CrossRef]

11.   Ahmadalipour, A.; Moradkhani, H. A data-driven analysis of flash flood hazard, fatalities, and damages over the CONUS during 1996–2017. *J. Hydrol.* **2019**, *578*, 124106. [CrossRef]

12.   Rojas, R.; Feyen, L.; Watkiss, P. Climate change and river floods in the European Union: Socio-economic consequences and the costs and benefits of adaptation. *Glob. Environ. Chang.* **2013**, *23*, 1737–1751. [CrossRef]

13.   Tanoue, M.; Hirabayashi, Y.; Ikeuchi, H. Global-scale river flood vulnerability in the last 50 years. *Sci. Rep.* **2016**, *6*, 36021. [CrossRef] [PubMed]

14.   Wicherski, W.; Dethier, D.P.; Ouimet, W.B. Erosion and channel changes due to extreme flooding in the Fourmile Creek catchment, Colorado. *Geomorphology* **2017**, *294*, 87–98. [CrossRef]

15.   Oubennaceur, K.; Chokmani, K.; Nastev, M.; Lhissou, R.; El Alem, A. Flood risk mapping for direct damage to residential buildings in Quebec, Canada. *Int. J. Disaster Risk Res.* **2019**, *33*, 44–54. [CrossRef]

16.   Liu, Z.; Lao, J.; Zhang, Y.; Liu, Y.; Zhang, J.; Wang, H.; Jiang, B. Association between floods and typhoid fever in Yongzhou, China: Effects and vulnerable groups. *Environ. Res.* **2018**, *167*, 718–724. [CrossRef]

17.   Wijerathne, K.B.P.C.A.; Senevirathna, E.M.T.K. Identify the risk for leptospirosis disease during flooding periods (Special reference to Medirigiriya Divisional Secretariat Division in Polonnaruwa district). *Procedia Eng.* **2018**, *212*, 101–108. [CrossRef]

18.   Songsore, J. The Complex Interplay between Everyday Risks and Disaster Risks: The Case of the 2014 Cholera Pandemic and 2015 Flood Disaster in Accra, Ghana. *Int. J. Disaster Risk Res.* **2017**, *26*, 43–50. [CrossRef]

19.   Dewan, T.H. Societal impacts and vulnerability to floods in Bangladesh and Nepal. *Weather Clim. Extrem.* **2015**, *7*, 36–42. [CrossRef]

20.   Mustafa, A.; Bruwier, M.; Archambeau, P.; Erpicum, S.; Pirotton, M.; Dewals, B.; Teller, J. Effects of spatial planning on future flood risks in urban environments. *J. Environ. Manag.* **2018**, *225*, 193–204. [CrossRef]

21.   Nga, P.H.; Takara, K.; Cam Van, N. Integrated approach to analyze the total flood risk for agriculture: The significance of intangible damages—A case study in Central Vietnam. *Int. J. Disaster Risk Res.* **2018**, *31*, 862–872. [CrossRef]

22.   Kron, W. Changing flood risk-A re-insurer's viewpoint. In *Changes in Flood Risk in Europe*; Kundzewicz, Z.W., Ed.; CRC Press, IAHS Special Publication: Boca Raton, FL, USA, 2012; Volume 10, pp. 459–490.

23.   Hirabayashi, Y.; Mahendran, R.; Koirala, S.; Konoshima, L.; Yamazaki, D.; Watanabe, S.; Kanae, S. Global flood risk under climate change. *Nat. Clim. Chang.* **2013**, *3*, 816–821. [CrossRef]

24.   Kapucu, N. Emergency and crisis management in the United Kingdom: Disasters experienced, lessons learned, and recommendations for the future. In *Comparative Emergency Management*; McEntire, D., Ed.; FEMA: Washington, DC, USA, 2009; Chapter 4. Available online: https://training.fema.gov/hiedu/aemrc/booksdownload/compemmgmtbookproject/ (accessed on 23 January 2020).

25.   Flanagan, B.E.; Gregory, E.W.; Hallisey, E.J.; Heitgerd, J.L.; Lewis, B. A Social Vulnerability Index for Disaster Management. *J. Homel. Secur. Emerg. Manag.* **2011**, *8*. [CrossRef]

26.   Driessen, P.P.; Hegger, D.L.; Kundzewicz, Z.W.; Van Rijswick, H.F.; Crabbé, A.; Larrue, C.; Matczak, P.; Pettersson, M.; Priest, S.; Suykens, C.; et al. Governance strategies for improving flood resilience in the face of climate change. *Water* **2018**, *10*, 1595. [CrossRef]

27.   Barredo, J.I.; Engelen, G. Land use scenario modeling for flood risk mitigation. *Sustainability* **2010**, *2*, 1327–1344. [CrossRef]

28.   Muller, M. Adapting to climate change: Water management for urban resilience. *Environ. Urban.* **2007**, *19*, 99–113. [CrossRef]

29.   Durga, B.K.; Moulika, M.B.; Kumar, K.P.; Varma, L.K. Design of early warning flood detection systems. *IJETSR* **2018**, *5*, 794–799.

30.   Thielen, J.; Bartholmes, J.; Ramos, M.H.; De Roo, A. The European flood alert system ĝ€" part 1: Concept and development. *Hydrol. Earth Syst. Sci.* **2009**, *13*, 125–140. [CrossRef]

31.   Milly, P.C.D.; Wetherald, R.T.; Dunne, K.A.; Delworth, T.L. Increasing risk of great floods in a changing climate. *Nature* **2002**, *415*, 514–517. [CrossRef]

32. Oppenheimer, M.; Campos, M.; Warren, R.; Birkmann, J.; Luber, G.; O'Neill, B.; Takahashi, K. Emergent risks and key vulnerabilities. In *Climate Change 2014: Impacts, Adaptation, and Vulnerability. Part A: Global and Sectoral Aspects. Contribution of Working Group II to the Fifth Assessment Report of the Intergovernmental Panel on Climate Change*; Field, C.B., Barros, V.R., Dokken, D.J., Mach, K.J., Mastrandrea, M.D., Bilir, T.E., White, L.L., Eds.; Cambridge University Press: Cambridge, UK, 2014; pp. 1039–1099. Available online: https://www.ipcc.ch/site/assets/uploads/2018/02/WGIIAR5-Chap19_FINAL.pdf (accessed on 23 January 2020).

33. EEA-European Environment Agency. *Climate Change, Impacts and Vulnerability in Europe 2016*; EEA: Luxembourg, 2017; p. 421.

34. Richard Eiser, J.; Bostrom, A.; Burton, I.; Johnston, D.M.; McClure, J.; Paton, D.; White, M.P. Risk interpretation and action: A conceptual framework for responses to natural hazards. *Int. J. Disaster Risk Res.* **2012**, *1*, 5–16. [CrossRef]

35. Hufschmidt, G.; Crozier, M.; Glade, T. Evolution of natural risk: Research framework and perspectives. *Nat. Hazards Earth Syst. Sci.* **2005**, *5*, 375–387. [CrossRef]

36. Mishra, K.; Sinha, R. Flood risk assessment in the Kosi megafan using multi-criteria decision analysis: A hydro-geomorphic approach. *Geomorphology* **2020**, *350*, 106861. [CrossRef]

37. UNDRO. *Natural Disasters and Vulnerability Analysis*; Office of the United Nations Disaster Relief Coordinator: Geneva, Switzerland, 1980; p. 53.

38. Vargas, J. *Políticas Públicas Para la Reducción de la Vulnerabilidad Frente a los Desastres Naturalesy Socio-Naturales*; Naciones Unidas, CEPAL: Santiago, Chile, 2002; Volume 50, p. 79.

39. Kumpulainen, S. Vulnerability concepts in hazard and risk assessment. In *Natural and Technological Hazards and Risks Affecting the Spatial Development of European Regions*; Special Paper 42; Schmidt-Thome, P., Ed.; Geological Survey of Finland: Espoo, Finland, 2006; pp. 65–74.

40. Aroca-Jiménez, E.; Bodoque, J.M.; García, J.A.; Díez-Herrero, A. Construction of an integrated social vulnerability index in urban areas prone to flash flooding. *Nat. Hazards Earth Syst. Sci.* **2017**, *17*, 1541–1557. [CrossRef]

41. Dwyer, A.; Zoppou, C.; Nielsen, O.; Day, S.; Roberts, S. *Quantifying Social Vulnerability: A Methodology for Identifying Those at Risk to Natural Hazards*; Geoscience Australia Record; Geoscience Australia: Canberra, Australia, 2004; p. 92.

42. Karmakar, S.; Slobodan, P.S.; Peck, A.; Black, J. An information system for risk-vulnerability. Assessment to flood. *Int. J. Geogr. Inf. Syst.* **2010**, *2*, 129–146. [CrossRef]

43. Kappes, M.S.; Papathoma-Köhle, M.; Keiler, M. Assessing physical vulnerability for multihazards using an indicator-based methodology. *Appl. Geogr.* **2012**, *32*, 577–590. [CrossRef]

44. Birkmann, J. (Ed.) *Measuring Vulnerability to Natural Hazards: Towards Disaster Resilient Societies*; United Nations University Press: Tokyo, Japan, 2013; p. 720.

45. Dunning, C.M.; Durden, S. *Social Vulnerability Analysis: A Comparison of Tools*; U.S. Army Corps of Engineers Institute for Water Resources: Alexandria, VA, USA, 2013; p. 34.

46. Hummell, B.M.D.; Cutter, S.L.; Emrich, C.T. Social vulnerability to natural hazards in Brazil. *Int. J. Disaster Risk Sci.* **2016**, *7*, 111–122. [CrossRef]

47. Creach, A.; Chevillot-Miot, E.; Mercier, D.; Pourinet, L. Vulnerability to coastal flood hazard of residential buildings on Noirmoutier Island (France). *J. Maps* **2016**, *12*, 371–381. [CrossRef]

48. Noradika, Y.; Lee, S. Assessment of social vulnerability to natural hazards in South Korea: Case study for typhoon hazard. *Spat. Inf. Res.* **2017**, *25*, 99–116. [CrossRef]

49. Hardy, R.D.; Hauer, M. Social vulnerability projections improve sea-level rise risk assessments. *Appl. Geogr.* **2018**, *91*, 10–20. [CrossRef]

50. Aksha, S.; Juran, L.; Resler, L.M.; Zhang, Y. An analysis of social vulnerability to natural hazards in Nepal using a modified social vulnerability index. *Disaster Risk Sci.* **2019**, *10*, 103–116. [CrossRef]

51. Apotsos, A. Mapping relative social vulnerability in six mostly urban municipalities in South Africa. *Appl. Geogr.* **2019**, *105*, 86–101. [CrossRef]

52. Armaș, I.; Gavriș, A. Social vulnerability assessment using spatial multi-criteria analysis (SEVI model) and the Social Vulnerability Index (SoVI model)—A case study for Bucharest, Romania. *Nat. Hazards Earth Syst. Sci.* **2013**, *13*, 1481–1499. [CrossRef]

53. Pérez-Morales, A.; Navarro-Hervas, F.; Álvarez Rogel, Y. Propuesta metodológica para la evaluación de la vulnerabilidad social en poblaciones afectadas por el peligro de inundación: El caso de Águilas (Murcia, sureste ibérico). *Doc. D'anàlisi Geogràfica* **2016**, *62*, 133–159. [CrossRef]
54. Cano, V. Aplicación de los SIG para la generación de rutas de evacuación en caso de desastres, como ayuda para la planificación urbana: "caso costa oriental del lago de Maracaibo". *Rev. de la Fac. de Ing. UCV* **2011**, *26*, 17–26.
55. Willis, I.; Fitton, J. A review of multivariate social vulnerability methodologies: A case study of the River Parrett catchment, UK. *Nat. Hazards Earth Syst. Sci.* **2016**, *16*, 1387–1399. [CrossRef]
56. Debbage, N. Multiscalar spatial analysis of urban flood risk and environmental justice in the Charlanta megaregion, USA. *Anthropocene* **2019**, *28*. [CrossRef]
57. Cutter, S.L.; Mitchell, J.T.; Scott, M.S. Revealing the vulnerability of people and places: A case study of Georgetown County, South Carolina. *Ann. Am. Assoc. Geogr.* **2000**, *90*, 713–737. [CrossRef]
58. Prudent, N.; Houghton, A.; Luber, G. Assessing climate change and health vulnerability at the local level: Travis County, Texas. *Disasters* **2016**, *40*. [CrossRef]
59. Arrighi, C.; Pregnolato, M.; Dawson, R.J.; Castelli, F. Preparedness against mobility disruption by floods. *Sci. Total Environ.* **2019**, *654*, 1010–1022. [CrossRef]
60. Tate, E.; Cutter, S.; Berry, M. Integrated multihazard mapping. *Environ. Plan. B Plan. Des.* **2010**, *37*, 646–663. [CrossRef]
61. Bjarnadottir, S.; Li, Y.; Stewart, M. Social vulnerability index for coastal communities at risk to hurricane hazard and a changing climate. *Nat. Hazards* **2011**, *59*, 1055–1075. [CrossRef]
62. Gautam, D. Assessment of social vulnerability to natural hazards in Nepal. *Nat. Hazards Earth Syst. Sci.* **2017**, *17*, 2313–2320. [CrossRef]
63. Ebert, A.; Kerle, N.; Stein, A. Urban social vulnerability assessment with physical proxies and spatial metrics derived from air- and spaceborne imagery and GIS data. *Nat. Hazards* **2009**, *48*, 275. [CrossRef]
64. Sebald, C. Towards an integrated Flood Vulnerability Index—A Flood Vulnerability Assessment. MSc Thesis, Faculty of Geo-Information Science and Earth Observation, University of Twente, Enschede, The Netherlands, 2010.
65. Westen, C.V.; Kingma, N. Vulnerability assessment. In *Multi-Hazard Risk Assessment*; Guide Book; Westen, C.V., Alkema, D., Dam, M., Kerle, N., Kingma, N., Eds.; United Nations, University-ITC School on Disaster Geoinformation Management (UNU-ITC DGIM): Enschede, The Netherlands, 2011; pp. 255–287.
66. Papathoma-Köhle, M. Vulnerability curves vs. vulnerability indicators: Application of an indicator-based methodology for debris-flow hazards. *Nat. Hazards Earth Syst. Sci.* **2016**, *16*, 1771–1790. [CrossRef]
67. De Ruiter, M.C.; Ward, P.H.; Daniell, J.E.; AERTS, J.C.J.H. A comparison of flood and earthquake vulnerability assessment indicators. *Nat. Hazards Earth Syst. Sci.* **2017**, *17*, 1231–1251. [CrossRef]
68. Gao, J.; Nickum, J.; Pan, Y. An assessment of flood hazard vulnerability in the Dongting Lake Region of China. *Lakes Reserv. Res. Manag.* **2007**, *12*, 27–34. [CrossRef]
69. Generino, P.; Sony, E.; Proceso, L. Analytic Hierarchy Process (AHP) in spatial modeling for floodplain risk assessment. *Int. J. Mach. Learn. Comput.* **2014**, *4*, 450–457.
70. Ouma, Y.O.; Tateishi, R. Urban flood vulnerability and risk mapping using integrated multi-parametric AHP and GIS: Methodological overview and case study assessment. *Water* **2014**, *6*, 1515–1545. [CrossRef]
71. Danumah, J.H.; Odai, S.N.; Saley, B.M.; Szarzynski, J.; Thiel, M.; Kwaku, A.; Koffi, F.; Akpa, L. Flood risk assessment and mapping in Abidjan district using multi-criteria analysis (AHP) model and geoinformation techniques, (Cote d´Ivoire). *Geoenviron. Disasters* **2016**, *3*, 1–13. [CrossRef]
72. INE-Instituto Nacional de Estadística. Available online: https://www.ine.es/ (accessed on 22 April 2019).
73. CHMiño-Sil- Confederación Hidrográfica del Miño-Sil. *Evaluación Preliminar del Riesgo de Inundación (EPRI) e Identificación de las Áreas con Riesgo Potencial Significativo de Inundación (ARPSIs) del Territorio Español de la Demarcación Hidrográfica del Miño-Sil*; Ministerio de Medio Ambiente y Medio Rural y Marino: Madrid, Spain, 2011; p. 39.
74. CHMiño-Sil-Confederación Hidrográfica del Miño-Sil. *Plan de Emergencia de la Presa de Bárcena*, Unpublished material. 2012.
75. Balica, S.F. Applying the Flood Vulnerability Index as a Knowledge Base for Flood Risk Assessment. Ph.D. Thesis, Delft University of Technology and the Academic Board of the UNESCO-IHE, Institute for Water Education, Enschede, The Netherlands, 6 June 2012.

76. Tascón-González, L. Análisis metodológico para la estimación de la vulnerabilidad por inundaciones. Ejemplo de aplicación en el municipio de Ponferrada (León, España). Ph.D. Thesis, Universidad de León, León, Spain, 30 September 2017.

77. Poushter, J.; Bishop, C.; Chwe, H. *Social Media Use Continues to Rise in Developing Countries but Plateaus Across Developed Ones*; Pew Research Center: Washington, DC, USA, 2018. Available online: https://www.pewresearch.org/global/wp-content/uploads/sites/2/2018/06/Pew-Research-Center-Global-Tech-Social-Media-Use-2018.06.19.pdf (accessed on 23 January 2020).

78. Saaty, T.L. Decision making with the analytic hierarchy process. *Int. J. Serv. Sci.* **2008**, *1*, 83–98. [CrossRef]

79. Dirección General del Catastro. Available online: http://www1.sedecatastro.gob.es/ (accessed on 19 December 2019).

80. Cutter, L.S.; Boruff, B.J.; Shirley, W.L. Social vulnerability to environmental hazards. *Soc. Sci. Q.* **2003**, *84*, 242–261. [CrossRef]

81. Hearn, B.; Laska, S.; Willinger, B.; Mock, N. Gender and disasters: Theoretical Considerations. In *Katrina and the Women of New Orleans*; Willinger, B., Ed.; Tulane University: New Orleans, LA, USA, 2008; pp. 11–21.

82. Fekete, A. Validation of a social vulnerability index in context to river-floods in Germany. *Nat. Hazards Earth Syst. Sci.* **2009**, *9*, 393–403. [CrossRef]

83. Ruiz, M. Vulnerabilidad Territorial y Evaluación de Daños Postcatástrofe: Una Aproximación Desde la Geografía del Riesgo. Ph.D. Thesis, Universidad Complutense de Madrid, Madrid, Spain, 24 May 2011.

84. Holand, I.; Lujala, P.; Ketil, J. Social vulnerability assessment for Norway: A quantitative approach. *Norsk Geogr. Tidsskr. Nor. J. Geogr.* **2011**, *65*, 1–17. [CrossRef]

85. Popovici, E.-A.; Costache, A.; Dan, B.; Dogaru, D.; Sima, S. Vulnerability assessment of rural communities to floods in the western part of Romania (Banat plain). 13th International Multidisciplinary Scientific GeoConference SGEM. *J. Environ. Prot. Ecol.* **2013**, *1*, 1161–1168. [CrossRef]

86. Koks, E.E.; Jongman, B.; Husby, T.G.; Botzen, W.J.W. Combining hazard, exposure and social vulnerability to provide lessons for flood risk management. *Environ. Sci. Policy* **2014**, *47*, 42–52. [CrossRef]

87. Aroca, E.; Bodoque, J.M.; García, J.A.; Díez, A. Análisis de la vulnerabilidad social ante avenidas súbitas en zonas urbanas de Castilla y León (España). In Proceedings of the XIV Reunión Nacional de Geomorfología, Málaga, Spain, 22–25 June 2016; pp. 283–290.

88. Quezada-Hofflinger, A.; Somos-Valenzuela, M.A.; Vallejos-Romero, A. Response time to flood events using a Social Vulnerability Index (ReTSVI). *Nat. Hazards Earth Syst. Sci.* **2019**, *19*, 254–267. [CrossRef]

89. Salvati, P.; Petrucci, O.; Rossi, M.; Bianchi, C.; Pasqua, A.A.; Guzzetti, F.G. Gender, age and circumstances analysis of flood and landslide fatalities in Italy. *Sci. Total Environ.* **2018**, *610–611*, 867–879. [CrossRef]

90. Laghi, M.; Cavalletti, A.; Polo, P. *Coastal Risk Analysis of Tsunamis and Environmental Remediation*; Asian Disaster Preparedness Center: Klong Luang, Thailand, 2006; p. 98.

91. Jones, J.M.; Peter, N.G.; Natah, W.J. *The Pedestrian Evacuation Analyst—Geographic Information Systems Software for Modeling Hazard Evacuation Potential*; U.S. Geological Survey: Reston, VA, USA, 2014; p. 25.

92. Bernardini, G.; Postacchini, M.; Quagliarini, E.; Brocchini, M.; Cianca, C.; D'Orazio, M. A preliminary combined simulation tool for the risk assessment of pedestrians' flood-induced evacuation. *Environ. Model. Softw.* **2017**, *96*, 14–29. [CrossRef]

93. Garmendia, A. Valoración de impactos ambientales. In *Evaluación de Impacto Ambiental*; Garmendia, A., Salvador, A., Crespo, C., Garmendia, L., Eds.; Pearson-Prentice Hall: Upper Saddle River, NJ, USA, 2005; pp. 225–284.

94. Ministerio de Empleo y Seguridad Social. *Evaluación y Diagnóstico de las Condiciones de Evacuación en Edificios de Elevada Ocupación*, 3rd ed.; Centro Nacional de Medios de Protección: Sevilla, Spain, 2012; p. 18.

95. Hauke, A.; Georgiadou, P.; Malmelin, J.; Punakallio, A.; Päkkönen, R.; Meyer, S.; Nicolescu, G. *Emergency Services: A Literature Review on Occupational Safety and Health Risks*; European Agency for Safety and Health at Work (EU-OSHA): Luxembourg, 2011; p. 80.

96. World Health Organization. *World Health Statistics-2009*; WHO: Geneva, Switzerland, 2009; p. 149.

97. Junta de Castilla y León. *Plan Regional Sectorial de Carreteras 2008–2020*; Junta de Castilla y León: Valladolid, Spain, 2008; p. 278.

98. World Health Organization. *World Health Statistics-2017*; WHO: Geneva, Switzerland, 2017; p. 116.

99. Ministerio de Sanidad, Servicios Sociales e Igualdad. *Los Sistemas Sanitarios en los Países de la UE: Características e Indicadores de Salud 2013*; Ministerio de Sanidad, Servicios Sociales e Igualdad: Madrid, Spain, 2014; p. 99.

100. Connor, R.F.; Hiroki, K. Development of a method for assessing flood vulnerability. *Water Sci. Technol.* **2005**, *51*, 61–67. [CrossRef]

101. INSHT. *NTP 436: Cálculo Estimativo de Vías y Tiempos de Evacuación. GUÍAS de buenas prácticas*; Instituto Nacional de Seguridad e Higiene en el trabajo - Ministerio de Trabajo y Asuntos Sociales de España: Madrid, Spain, 1997; p. 6.

102. Lee, J.S.; Choi, H.I. Comparison of flood vulnerability assessments to climate change by construction frameworks for a composite indicator. *Sustainability* **2018**, *10*, 768. [CrossRef]

103. Klaus, D. Implementing the Analytic Hierarchy Process as a standard method for multi-criteria decision making in corporate enterprises—A new AHP excel template with multiple inputs. In Proceedings of the International Symposium on the Analytic Hierarchy Process, Kuala Lumpur, Malaysia, 23–26 June 2013.

 *water*

Article

Socio-Economic Potential Impacts Due to Urban Pluvial Floods in Badalona (Spain) in a Context of Climate Change

Eduardo Martínez-Gomariz [1,2,*], Luca Locatelli [3], María Guerrero [1], Beniamino Russo [3,4] and Montse Martínez [3]

[1] Cetaqua, Water Technology Centre, Environment, Society and Economics Department, 08940 Cornellà de Llobregat, Spain; maria.guerrero@cetaqua.com

[2] FLUMEN Research Institute, Universitat Politècnica de Catalunya, 08034 Barcelona, Spain

[3] AQUATEC (SUEZ Advanced Solutions), 08038 Barcelona, Spain; luca.locatelli@aquatec.es (L.L.); brusso@aquatec.es (B.R.); mmartinezp@aquatec.es (M.M.)

[4] Group of Hydraulic and Environmental Engineering, Technical College of La Almunia (EUPLA), University of Zaragoza, 50100 Zaragoza, Spain

* Correspondence: eduardo.martinez@cetaqua.com; Tel.: +34-933-124-899

Received: 14 November 2019; Accepted: 13 December 2019; Published: 17 December 2019

Abstract: Pluvial flooding in Badalona (Spain) occurs during high rainfall intensity events, which in the future could be more frequent according to the latest report from the Intergovernmental Panel on Climate Change (IPCC). In this context, the present study aims at quantifying the potential impacts of climate change for the city of Badalona. A comprehensive pluvial flood multi risk assessment has been carried out for the entire municipality. The assessment has a twofold target: People safety, based on both pedestrians' and vehicles' stability, and impacts on the economic sector in terms of direct damages on properties and vehicles, and indirect damages due to businesses interruption. Risks and damages have also been assessed for the projected future rainfall conditions which enabled the comparison with the current ones, thereby estimating their potential increment. Moreover, the obtained results should be the first step to assess the efficiency of adaptation measures. The novelty of this paper is the integration of a detailed 1D/2D urban drainage model with multiple risk criteria. Although, the proposed methodology was tested for the case study of Badalona (Spain), it can be considered generally applicable to other urban areas affected by pluvial flooding.

Keywords: pluvial floods; flood risk assessment; climate change; damages; vehicles; properties; pedestrians

---

1. Introduction

According to the latest report from the IPCC [1], heavy rainfall could be more frequent in many areas of the planet over the next years. The greater uncertainty related to precipitation with respect to other climate variables such as temperature or sea level rise, requires local studies that apply spatial and temporal downscaling techniques in order to obtain conclusive results. An extreme and moderate rainfall increment, in terms of frequency and magnitude, might aggravate the consequences of urban floods thereby causing increasing impacts on people safety and the economy of cities.

Badalona, with more than 215,000 inhabitants within its administrative limits on a land area of more than 21.2 km², is located in eastern Catalonia (Spain) and is part of the Barcelona Metropolitan Area (in Spanish AMB) (Figure 1). It is situated on the left bank of the Besòs River facing on the Mediterranean Sea, and backed by the Serra de la Marina mountain range. This city is one of the six research sites across Europe (Portugal, Spain, Cyprus, Germany, the Netherlands, and Norway) of the EU project BINGO (Bringing Inovation to Ongoing water management—a better future under climate change (2015–2019)). The project strives to provide practical knowledge and tools to end users, water managers, and decision and policymakers for a better informed decision making process under the uncertainties of climate change. The city administrative division consist in six districts and 24 subdistricts, and the average population density of the city is approximately 10,000 inhab/Km^2, although this figure can increase significantly in some districts of the old town and the city centre. The morphology of Badalona presents areas with high gradients (close to Serra de la Marina) and flat areas near the Mediterranean Sea. Moreover, the land was strongly urbanized during the last decades. These characteristics, together with the Mediterranean rainfalls characterized by high intensity and short duration, leave the city in a flood-prone situation. All these aspects put urban drainage in Badalona in the spotlight as urban flash floods happen in several critical areas with significant economic damages and high hazard conditions for pedestrian and vehicular circulation. Its drainage network is mainly a combined sewer system with 318 km of conduits, 26% of them allowing man-entry. There are nine catchments in the city, most of which discharge from the mountain to the sea. Waste water is intercepted by a main sewer placed along the coastal area that conveys it to the Besòs Waste Water Treatment Plant (WWTP) (Figure 1b).

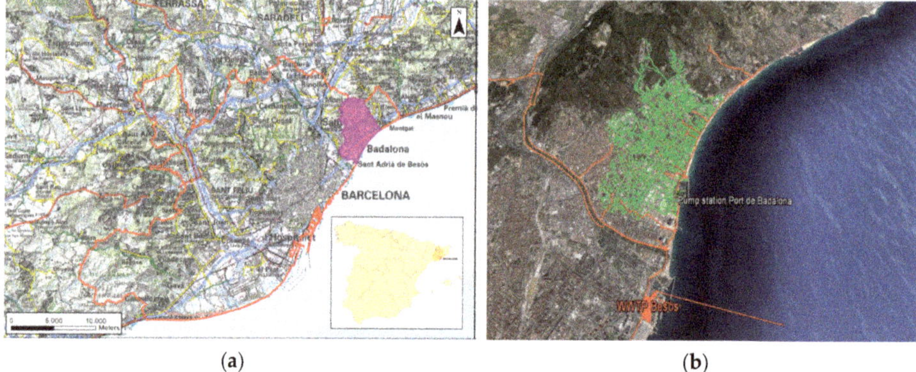

(a)   (b)

Figure 1. (a) Badalona location and Barcelona Metropolitan Area (in Spanish AMB) administrative limits (in red), and (b) Badalona drainage network. The background maps were provided by the Cartographic and Geological Institute of Catalonia (ICGC).

Therefore, according to the frequent pluvial floods in Badalona that cause sewer surcharging the need of a risk assessment arises, specially taking into account the regulations that have been established with the European Flood Directive [2]. Although in Spain all efforts to accomplish the flood risk management plans required by the European Flood Directive [2], it has to be stressed the importance of managing appropriately the ever more frequent pluvial urban floods [3].

In this research, according to a similar risk concept proposed by Turner et al. [4], a comprehensive flood risk assessment has been conducted, considering social (intangible) and economic (direct and indirect tangible) losses. The pluvial flood risk assessment carried out within CORFU project (http://www.corfu7.eu/) for the Spanish case study in the Raval District [5] has been considered as the baseline, although novel methodologies and updated criteria have been implemented.

Residents in urban areas, such as Badalona, conduct a wide variety of activities, some of which involve vehicles, during the day under different weather conditions, which makes people safety an important issue. A vehicle exposed to flooding may become buoyant and be washed away with potential injuries and fatalities. Such vehicles cause additional disruptions to traffic that is already affected by flooding, which may lead to substantial indirect economic impact, especially in urban areas. For these reasons, the main objective of the flood social risk assessment is the people's safety, focused on their stability when exposed to water flows [6] and their potential injuries due to the loss of stability of vehicles, either because there are passengers inside or pedestrians may be impacted by these massive debris.

On the other hand, in terms of economic damages, over the period 1996–2016 more than 60 flood events caused damages to properties in Badalona, summing up more than 1.8 Million Euro in 499 claims according to the data received from the Spanish insurance company Consorcio de Compensación de Seguros (CCS). The CCS is a state-owned enterprise attached to the Ministry of Economy and Enterprise that performs several functions complementing the Spanish Insurance Industry, enhancing its stability and protecting the insured. Regarding damages to vehicles, more than 160 claims have been reported in the same period (1996–2016) with a total economic damage of 478,000€ (Figure 2).

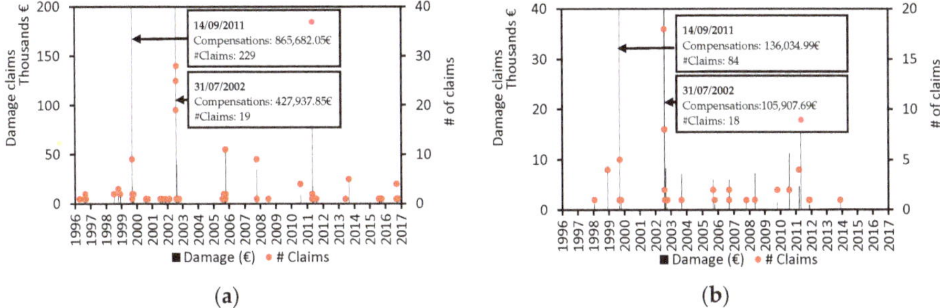

Figure 2. Claims and compensations regarding (a) properties and (b) vehicles in Badalona due to floods occurred since 1996.

The potential for cities to be damaged is notably high due to the high spatial density of people and values [7]. Studies to estimate flood damages are mainly focused on properties, using depth damages curves [7,8]. Depth damage curves can measure damage percentages or may be represented based on monetary units over surface unit, format more useful for mapping purposes. Several proposals of flood damage curves can be found worldwide, for example in:

- Australia with the Queensland Government guidelines [9].
- USA with the guidelines proposed by the U.S. Federal Emergency Management Agency [10].
- Spain (Valencia) with the criterion proposed within the EU Project CRUE [11].
- Italy with a particular development proposed by the Research Institute for Geo-hydrological Protection of Turin (Italy) for river Boesio flooding [12].
- United Kingdom with the criterion proposed in the Multicoloured Manual [13].
- Spain (Barcelona) in the framework of the EU Project CORFU [14] new damage curves focused on urban pluvial flooding were performed for the district of El Raval (Barelona).

In urban flood risk studies, direct and tangible damage assessments are common, which are related to the immediate physical contact of flood water and easily specified in monetary terms [15]. Direct and tangible damages to properties vary according to the type of property, its value, and the restitution costs to initial state [16]. Flood direct damages estimation for properties can be conducted within an urban area making the geo-referenced type of buildings and the corresponding damage curves available, and with the help of GIS tools. In addition, vehicles are susceptible of being damaged by floods, which in case of losing their stability they may collide with other urban elements or even cause injuries to residents. Therefore, the estimation of tangible and direct damages should be considered in a comprehensive damage assessment. Although, this assessment has been conducted to a lesser extend within the literature consulted, the studies found present a similar assessment that for properties by considering damage curves for vehicles. Some studies related flood damage assessment for vehicles are presented in Martínez-Gomariz et al. [17].

Furthermore, flooding also induces indirect damages that may occur—in space or time—outside the actual event [18]. The focus of this concern is on economic losses from the cascading effect of floods and business interruption that relate specifically to flooded business, which are defined as primary indirect losses [19]. The input–output method is commonly used to estimate short-term impacts [20], because it is relatively simple compared to general equilibrium models.

This paper presents a comprehensive pluvial flood risk assessment for the city of Badalona, in which a diversity of impacts has been considered. Damages assessed include intangible, by analyzing the risks for pedestrians in terms of both their own and vehicles' stability; tangible-direct damages for properties and vehicles, by proposing novel methodologies and criteria to estimate their economic damages; and tangible-indirect damages by implementing existing procedures to estimate the potential economic losses caused by the businesses interruption. Therefore, an elaborated social and economic impacts assessment has been carried out for current and future rainfall conditions, by employing as a basis a detailed 1D/2D hydrodynamic model, covering the entire drainage network of Badalona. This model is characterized in Section 2.1 followed by the description, in Section 2.2, of the rainfall conditions considered for the present and future. The methodology employed to assess the intangible (social) and tangible (economic) flood risks is presented in Section 2.3. Firstly, these risks have been assessed for the current scenario (baseline) and secondly for the future rainfall conditions (business as usual, BAU). Finally, results and conclusions regarding the risks assessment carried out for the current situation and the potential risk increase according to the expected future rainfall conditions are presented in Sections 3 and 4 respectively.

2. Materials and Methods

2.1. Hydrodynamic 1D/2D Coupled Model

A coupled 1D/2D hydrodynamic model was developed, calibrated and validated using InfoWorks ICM (www.innovyze.com). The model covers the whole municipality of Badalona with an extension of 21.7 km^2. While the 1D model reproduces the rainfall runoff processes of buildings that drain directly into the 1D drainage network, the 2D model reproduces stormwater runoff of streets, parks, and all areas that are at the terrain level and can be flooded. The 1D and the 2D model continuously interact at all 1D model nodes that are defined as "Gullies 2D". At this nodes the exchange of water is computed using experimental functions [21,22] developed at the Technical University of Catalonia (UPC) that calculate the water flow from the 2D to the 1D model and vice versa as a function of the local water level. Flow transferring took into account also inlet clogging factors estimated on the basis of field investigations [23,24]. For every node, a GIS analysis allowed to select the most appropriate UPC function based on street slope, width and type of gullies and also the number of existing gullies were associated to the closest model node. Peculiarity of the model is that part of the rainfall is directly applied to the 2D surface water model and part to the remaining 1D model. Conventional urban drainage modelling applies the rainfall directly to the 1D model [25] and then only manhole surcharge can create flooding. In Badalona, it is believed that part of urban floods come directly from areas (like streets) that have surface drainage systems with poor capacity due a lack of gullies.

The 1D model was developed in 2012 for the Drainage Management Plan (DMP) of Badalona. Within BINGO project, the model was then imported into ICM and updated to include the latest pipes and one detention tank of approximately 30,000 m^3. Overall, the 1D model includes approximately 368 km of pipes, 11,338 manholes, 11,954 sub-catchments, 62 weirs, 4 sluice gates, and 1 detention tank. The full 1D Saint-Venant equation is used to solve the sewer flow. Rainfall-runoff processes were simulated with the SWMM model (included in InfoWorks) that routes flow using a single non-linear reservoir with a routing coefficient that is a function of surface roughness, surface area, and terrain slope and catchment width. Initial losses are generally small for both impervious urban areas and green areas ($\leq$1 mm) and continuous losses for green areas are simulated using the Horton model. The extension of Badalona municipality was divided into 11,954 sub-catchments that were obtained by GIS analysis of the digital terrain model (2 m × 2 m resolution) and have areas in the range of 0.01–1 hectares in the urban areas and 1–100 hectares in the upstream rural areas. Each sub-catchment includes GIS derived information of impervious and pervious areas that are used to apply either the impervious or the pervious rainfall-runoff model. Impervious areas were not provided with continuous hydrological losses. The 2D model is based on a unstructured mesh with 199,338 cells that was created on the basis of a detailed digital terrain model (DTM) with a 2 m^2 resolution obtained by a LIDAR provided by the Cartographic and geological Institute of Catalonia (precision of 20 cm in terms of ground elevation). The size of the cells is in the range of 16–64 m^2 in the urban area and 70–400 m^2 in the upstream rural areas.

The 1D/2D model was quantitatively and qualitatively calibrated and validated by using data from 3 rainfall gauges, 14 water level sensors and flood event videos and photos. Key model performance parameters like root mean squared error, time to peak error, and absolute maximum error were computed and reported in deliverable 3.3 of BINGO project [26], where further details are provided regarding the model and its validation and calibration process.

2.2. Current and Future Rainfall Conditions Scenarios

Rainfall intensity–duration–frequency (IDF) curves, performed based on historic recorded time series data in the Fabra Observatory of Barcelona, are employed for urban hydrology studies not just in Barcelona but also across the surrounding municipalities. Therefore, these IDF curves were employed to develop the design storms, based on current rainfall conditions, as inputs for the 1D/2D coupled hydrodynamic model. Due to the proximity of Badalona to Barcelona, the IDF curves are expected to be similar in terms of both intensity and shape. Regarding future rainfall conditions, short term predictions (2015–2024) and long-term projections (2051–2100) were considered. As the short-term rainfall predictions did not indicate changes in terms of intensities, long-term projections were obtained from the project CORDEX. The EURO-CORDEX domain is a standard reference domain for Europe defined as part of the CORDEX Experiment (http://www.cordex.org/). A GCM MPI-ESM routine was employed and verified through the re-analysis of ERA-Interim. The downscaling process was carried out by employing the Regional Climate Model (RCM) COSMO-CLM.

A comparison between historic (1979–2005) and projection (2051–2100) daily rainfall data has been carried out in order to analyse variations (i.e., either increase or decrease) in extreme values from past and future rainfall intensities for different return periods. An extreme values analysis has been conducted by considering a Gumbel distribution for both samples, historic and projections, which were formed by the maximum values of 24 h rainfall data per year. As Houston et al. [27] state "given that a wet 24-h period will also contain pulses of heavier rainfall of shorter duration, it seems likely that projected uplifts in rainfall over a 24-h period will be associated with an uplift in the severity of sub-daily duration rainfall". Therefore, these intensity values have been compared from historic to future extreme values and for each considered CORDEX scenario (RCP 8.5 and 4.5). An overall increase in 24 h average intensities for RCP 8.5 was found, and a clear overall average-intensities decrease for almost all return period was observed for RCP 4.5 scenario. Hence, scenario RCP 8.5 was selected to perform the design storms for the risks assessment, as it presented the least favourable results in terms of extreme events.

The historic-future variation percentages, based on 24 h time resolution, were applied directly on the historic-real Barcelona IDF curves, based on sub-daily durations, in order to obtain the future ones and, so that future design storms can be performed based on the new IDF curves. These variations are called climate change factors and are defined as the ration between the rainfall intensity with a return period T and a duration d for a future climate scenario and the corresponding rainfall intensity in the present climate [28,29]. Thus, CORDEX results (i.e., RCP 8.4 scenario) have been employed for setting Badalona future scenario and design storms according to the climate change factors obtained (Table 1). It should be noted that there is not a linear relation between the considered scenario emissions and the amount of rainfall that will be obtained. That is to say, not necessarily a more restrictive RCP in terms of CO_2 emissions will offer higher rainfall increments. This fact was observed here for RCP 4.5, which offered an overall rainfall decrease unlike the RCP 8.5 scenario.

Table 1. Climate change factors according to RCP 8.5 scenario and related to different return periods.

Scenario/Return Period	T02	T10	T100	T500
RCP 8.5	15%	7%	2%	1%

2.3. Full Detailed Flood Risk Assessment in Urban Areas: Badalona

As introduced previously, flood damages can be classified into direct and indirect damage. The firsts occur due to the physical contact of the flood water with people, property or any other element, while the indirect damages are induced by the direct impacts and may even occur outside the flood event. Moreover, these are further classified into tangible and intangible damage, depending on whether or not these losses can be assessed in monetary values [30]. Table 2 presents a sample of flood damages classified according to the groups that were mention previously. Those considered in the present study are highlighted in bold type.

Table 2. Classification of flood damages (adapted from Velasco et al. [14]).

		Measurement	
		Tangible	Intangible
Form of damage	Direct	**Physical damage to assets:** **Infrastructure** **Contents** **Buildings** Evacuation and rescue operations Agricultural land Clean-up costs	**Fatalities and injuries** Diseases Historical and cultural losses Loss of ecological and environmental goods Inconvenience
	Indirect	Loss of industrial production Traffic disruption Emergency costs Temporary housing of evacuees **Business interruption**	Societal disruption Increased vulnerability of survivors Undermined trust in public authorities Psychological trauma

A considerable part of the literature on flood damages concerns direct tangible damage [31] while other damage types have received much less attention. However, the present study covers a greater variety of types of damages in a novel and complete social and economic pluvial flood risks assessment. Future Master Drainage Plans, in order to analyse the consequences of climate change, should consider a complete flood risk assessment like the one presented herein. It is expected that the level of security currently adopted by designers of drainage networks will be exceed more frequently, and therefore the impacts caused by floods on streets must be analysed (i.e., dual drainage concept [32]) in order to ensure a safety urban environment for pedestrians and properties.

2.3.1. Pluvial Flood Social Risk Assessment

Within the field of urban drainage and storm water management a consensus has been reached regarding the two essential hydraulic parameters (i.e., water depth and velocity) that have to be taken into account when assessing flood hazard for pedestrians. In the studies of Russo et al. [33,34] and of Martínez-Gomariz et al. [35,36], the most common flows during urban storm events, with low flow depth and high velocities, were reproduced in the laboratory through a physical model in real scale.

A sample of 26 subjects was tested in the last experimental campaign carried out by Martínez-Gomariz et al. [35,36] considering different conditions and exposure combinations (i.e., types of shoes, hands busy or free, and visibility conditions). The function $(v \cdot y) = 0.22$ m$^2 \cdot$s^{-1} was proposed as the threshold for the stability of a pedestrians exposed to a water flow produced by a pluvial flood, usually characterized by low water depths and high velocities. Based on Martínez-Gomariz et al. [35,36] results, the proposed limits for hazard delimitation in the present study are: Low hazard below the product $(v \cdot y) = 0.16$ m$^2 \cdot$s^{-1}, medium hazard for values $(v \cdot y)$ comprised between 0.16 m$^2 \cdot$s^{-1} and 0.22 m$^2 \cdot$s^{-1}, and high hazard for those above $(v \cdot y) = 0.22$ m$^2 \cdot$s^{-1}. Furthermore, it has been considered that when the water depth exceeds 0.15 m, the hazard is high, irrespective of the product of velocity and water depth. Therefore, low and medium hazard hydraulic conditions are both found below a water depth of 0.15 m. The maximum velocity that assures stability conditions for pedestrians, regardless the water depth, was established as 1.88 m$\cdot$s^{-1}. The hazard criterion for pedestrians is represented graphically in Figure 3a.

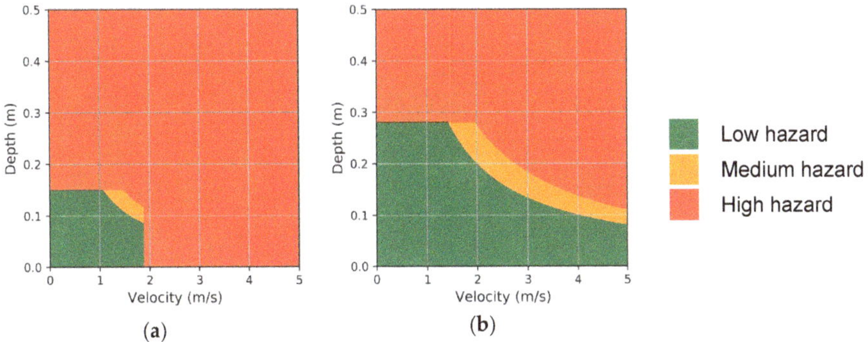

Figure 3. Hazard criteria for (**a**) pedestrians; and (**b**) vehicles.

On the other hand, the studies of Martínez-Gomariz et al. [35,37,38] present a methodology to determine the stability threshold for a vehicle, resulting from several experiments conducted to a wide variety of scaled vehicles. This methodology, which has been adopted here, enables to define a stable area in the domain flow depth-velocity for any real vehicle. According to the article published in La Vanguardia (Spanish newspaper) on the 3rd of September 2015, the three best-selling vehicles in Spain are Citroen C4, Seat Leon and Seat Ibiza. The proposed vehicle for Badalona has been the Seat Ibiza model due its lowest stability according to the methodology proposed by Martínez-Gomariz et al. [35,37], which was calculated considering a friction coefficient between tyres and road of 0.25. The Seat Ibiza model stability threshold is $(v \cdot y)_{lower} = 0.40$ m$^2 \cdot$s^{-1} and presents a buoyancy depth of 28 cm. In order to define the hazard limits, a new $(v \cdot y)_{upper}$ value is proposed based on the maximum friction coefficient ($\mu = 0.75$) proposed by Gerard [39], which yields an stability threshold of $(v \cdot y)_{upper} = 0.55$ m$^2 \cdot$s^{-1}. Thus, the proposed limits for hazard delimitation are: Low hazard below the product $(v \cdot y) = 0.40$ m$^2 \cdot$s^{-1}, medium hazard for the values $(v \cdot y)$ compressed between 0.40 m$^2 \cdot$s^{-1} and 0.55 m$^2 \cdot$s^{-1}, and high hazard beyond $(v \cdot y) = 0.55$ m$^2 \cdot$s^{-1} (Figure 3b).

In order to assess the residents' vulnerability in Badalona, a conceptual definition that considers vulnerability as the combination of exposure and susceptibilities has been adopted. The Spanish National Institute of Statistics (Instituto Nacional de Estadística, INE), provided the statistical data of current population per census districts to be employed in the assessment. By setting thresholds of the proposed statistical variables related to the citizens (i.e., indicators) for each census district, these districts can be classified according to their level of vulnerability. Two thresholds were proposed for each indicator considered, and a vulnerability score (i.e., 1 (low), 2 (medium), and 3 (high)) was granted depending on the value of the census district's indicator with respect the proposed thresholds. The indicators considered in order to define the vulnerability of each census district are: General population density, number of vulnerable infrastructure, percentage of people with critical age, and percentage of foreign people. These indicators may affect either the exposure or susceptibilities, and therefore vulnerability.

The exposure's indicators have been established as the general population density and the units of vulnerable infrastructures. The thresholds for the general population density were set using the medium density of the studied area of Badalona (10,951 inh/km^2) and the definition of the National Institute of Statistics of urban area defined as a group of minimum 10 houses in a distance less than 200 m (equivalent to 384.62 inh/km^2 in Badalona). A vulnerable infrastructure in this context is understood as a building or a specific place within the Badalona boundaries, where either a massive concentration of people can potentially occur or specific vulnerable groups of people are expected to be in the building or in their immediate vicinity. In case no vulnerable infrastructure was located within the analyzed census district, the lowest vulnerability score was given to this district (Table 3). When more than one vulnerable infrastructure is located within the district, the highest vulnerability score will be considered. If only one vulnerable infrastructure is found within the district boundaries a medium score was used for the district assessment. Two more indicators are proposed within the susceptibilities consequences' factor, namely percentage of citizens with critical age and percentage of foreign people. By defining critical age as less than 15 years old (i.e., the youngest) and over 65 (i.e., the elders), thresholds have been established as lower than 10% of the population with critical age for a low vulnerability score and higher than 33% for a high score. Other values were considered with medium score. Foreign people are not expected to know those specific most hazardous spots, which are well known by the locals. When the foreign percentage within the district is lower than 10% it will be scored with the lowest vulnerability value, and in case the percentage exceeds the 25% of the district's population it will be scored with the highest value. The previously-described indicators and thresholds are summarized in Table 3. The final vulnerability level was defined as the average score between the four indicators explained previously, which have been weighted equally. In case the average score is lower or equal than 1.5 a low vulnerability level will be established for the census district, a high level if the averaged score is higher than 2, and medium level for the rest (Table 3).

Table 3. Vulnerability criteria for pedestrians and vehicles.

Vulnerability Index/Score	Vulnerability for Pedestrians			
	Exposure		Susceptibilities	
	A General Population Density (Person/km^2)	B Units of Vulnerable Infrastructures	C % of People with Critical Age	D % of Foreign People
1 (low)	≤384.62	<1	≤25%	≤10%
2 (medium)	384.62 < X ≤ 10,951	1 ≤ X < 2	25% < X ≤ 33%	10% < X ≤ 25%
3 (high)	>10,951	≥2	>33%	>25%
Vulnerability index/score	Vulnerability for vehicles			
	Exposure		Susceptibilities	
	Vehicular flow intensity (VFI) (veh/day)		-	
1 (low)	<100		-	
2 (medium)	100 ≤ X ≤ 1000		-	
3 (high)	>1000		-	

On the other hand, in order to assess the vehicles vulnerability, three levels are proposed based on only one vulnerability indicator: The vehicular flow intensity of the different areas with road traffic. The more traffic flows the more vulnerable is a specific road. This indicator has been related to the exposure which is the only vulnerability's factor taken into account in this case. Table 3 indicates the three vulnerability levels and the thresholds considered for the vehicular flow intensity. The vulnerability maps developed for pedestrians and vehicles are presented in Figure 4.

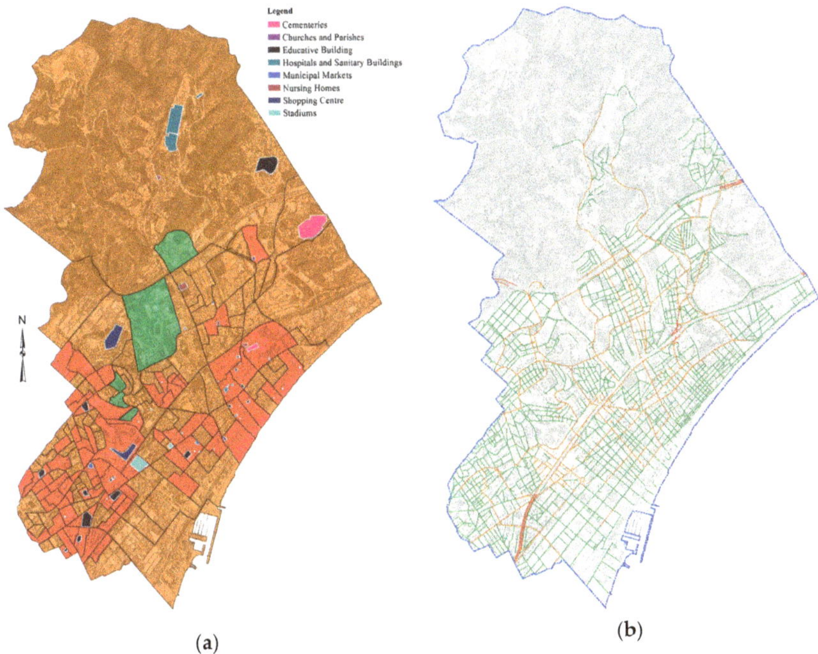

Figure 4. Vulnerability map for Badalona municipality: (**a**) Pedestrians, and (**b**) vehicles.

The selection of these indicators has been agreed with the Badalona city council technicians, who know the most regarding the vulnerabilities of the municipality. However, the indicators selected for other cities could be different because of either their specific needs or unavailability of data. Moreover, due to the data availability, limits of census districts have been the minimum area to establish a vulnerability level in this study, however it could be different in other countries.

The methodologies described previously have been applied to assess risk for pedestrians and vehicles under current and future rainfall conditions (RCP 8.5). In order to determine the risk of a specific area, hazard and vulnerability levels have been overlapped thereby establishing the risk level based on the combinations indicated in the risk matrix of Figure 5. As an example, Figure 6 illustrates the risk maps for both pedestrians and vehicles for a flood produced by a 10-year-return-period design rainfall.

		Hazard		
		Low	Medium	High
Vulnerability	**Low**	Low	Low	Medium
	Medium	Low	Medium	High
	High	Medium	High	High

Figure 5. Risk matrix for pedestrians and vehicles.

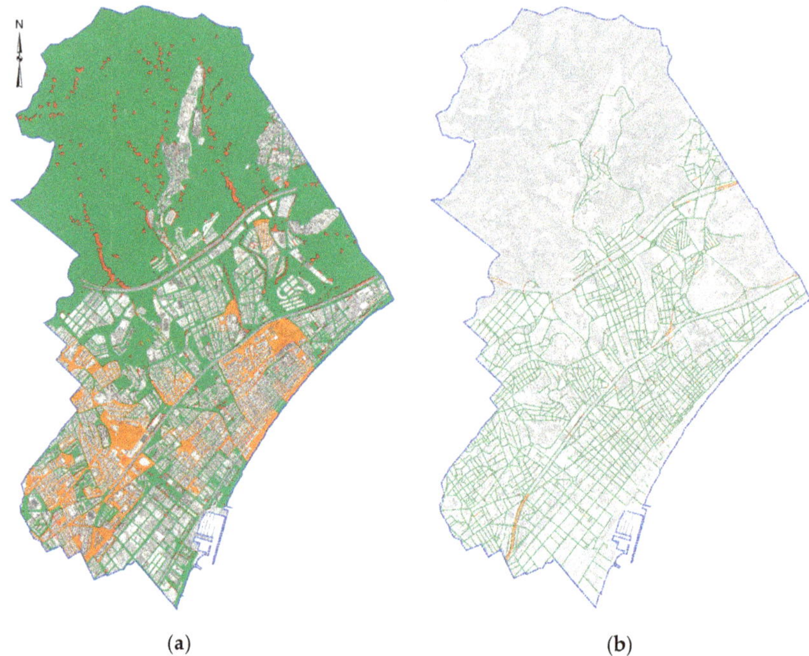

Figure 6. Risk map for a 10-year-return-period design rainfall regarding (**a**) pedestrians; and (**b**) vehicles. Current rainfall conditions.

2.3.2. Pluvial Flood Economic Risk Assessment

For the assessment of the direct damages to properties three different types of data were required: Flood maps (hazard), types of properties across the municipality (detailed land uses) and depth damage curves (vulnerability). Flood maps gather the information about hydraulic variables, namely water depths. Land use information capture the type, geometry and use of the properties in the study case area. Depth damage curves are functions that relate the type of building and its characteristics with the water depth to monetary quantify the damages produced by each flood event. Tailored depth damage curves for Badalona have been developed based on an analysis of previous flood claims in Badalona, Barcelona and some other Spanish cities. This work has been carried out with the collaboration of a flood damage surveyor [40]. Fourteen different types of properties have been taken into account to perform their corresponding damage functions as Figure 7a illustrates.

Firstly, flood maps for different design rainfalls related to 2, 10, 100, and 500 years return periods were obtained using the 1D–2D model in order to assess the water depth per each building. Secondly, the land use information was obtained by downloading a shapefile from the Web Portal of the General Directorate of the Spanish Cadastre. This georeferenced database file stores parcels information for the entire municipality, and it is freely downloadable. Therefore, the type of property, the dimension of the buildings and the situation within the case study area is known.

As a first step to assess the damages to properties, an average of the maximum simulated water depth, of those surrounding a building, was associated to each property. After this a new methodology, based on conversations with a surveyor expert in damages due to floods, is proposed to determine the water depth inside the building. The main idea in this approach is that a "sealing coefficient" (Sc) could determine the difference between the water depth outside and inside the property. It is defined as the ratio between the water depth inside and outside the building, therefore it will range from 0 (no water depth inside) to 1 (same water depth inside and outside). Accordingly, relations between water depth outside and inside the buildings have been developed, specifically for the 14 types of properties considered (Figure 7b). It was assumed that generally the residence time of the flooding is not enough for the building to leak until the water levels outside and inside become equal. Cases in which water enters directly through open doors and windows will not be considered since these are supposed to be closed because people are expected to be properly warned. However, the higher the water depth outside is, the higher the sealing coefficient will be. In other words, when the water depth outside is high enough, the water depth inside the building is expected to be the same that the outside one. Therefore, the sealing coefficient is a function of the water depth outside and also of the type of property (Figure 7b).

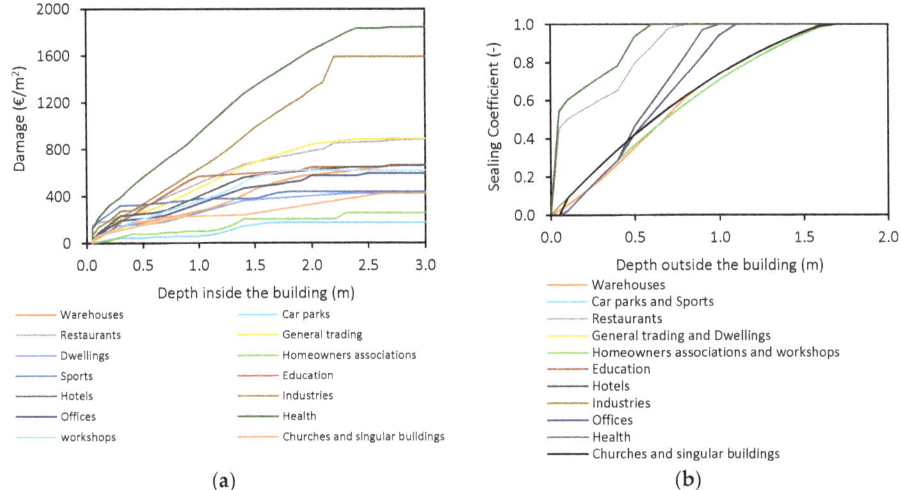

Figure 7. (a) Depth-damage and (b) sealing coefficients curves developed for Badalona case study.

In this model also a maximum of two levels of basements were considered, according to the information obtained from the Spanish cadastre. In case of the existence of basement, either one or two levels, two model parameters representing reference water depths were defined: The maximum depth and the residual depth. The former indicates that this water level is not expected to be exceed due to the steady water flow from the ground floor to the basement, and thus from the 1st basement level to the second. There may be several connections to let the water leak from one floor to the other but stairwells are supposed to be the main ones. Therefore, the maximum damage in a floor with an underneath level will be related to the maximum water depth. On the other hand, a residual water level has been defined, which determines the water volume which will be stored in this level and, hence, will not affect the underneath level.

The presence of front steps for each building across the studied urban area were taken into account also, by relating an average height of the front steps to each type of property. In this manner, the water depth in contact with the building façade will be reduced according to the front steps' height. Whereas the water depth in the ground floor will be lower than the outside one, basements act as small water storage tanks and water depths could even become higher than those present on the streets. This model was implemented in the GIS software in order to establish a water depth inside the buildings and compute the corresponding damage, based on the previously presented rules. Regarding the damage assessment for vehicles in Badalona, the methodology proposed by Martínez-Gomariz et al. [17] has been implemented for the different floods related to design storms for different return periods.

The proposed methodology developed to assess indirect damages in the context of urban floods is the input–output model, which allows to estimate the ratio of indirect damage over direct damages across the different economic sectors of the affected area. It follows the method developed within the European project PEARL (2014–2018) with the same purpose that Kowalewski [41], which is particularly relevant for the present case study, as it was specifically developed to evaluate indirect economic damage from flood events at a regional scale, although it has been downscaled to the county scale and adapted to the requirements of this case study.

Taking into account data availability and uncertainty, the input–output method offered the most effective solution to reach the indirect damage estimation objective. The data sources are the input–output table (IOT) for Catalonia (the latest published is for 2014) together with employment figures categorized by economic sector for Catalonia and Barcelonès County [42], and the historical data of claims paid by the Spanish insurance company (CCS). The IOT is a matrix that registers the use of factors of production by each economic sector in the production of final goods and services of a certain economy. From the Catalonian IOT, downscaling to county-level is based on simple rules relative to the size of employment in each sector in this county. Drawing on Leontief's production functions—cross-sectorial interdependencies of intermediate goods and final demand—the model interprets the data recorded in the table and assess the economic consequences of external shocks to the system (e.g., resulting from flood events). The final input–output account then captures the supply and demand interactions of the economic sectors downscaled to sub-regional level.

As a novelty, this version considers the IOT of Catalonia as baseline, downscaling to county level—Barcelonès—compared to the previous work, which uses national tables as baseline and downscale to the regional level. An exogenous shock is then introduced, obtained from the paid claims of CCS historical data. For Badalona 41 cases registered were included. This dataset is previously re-classified by economic sectors: from 4 types of land uses (industry, offices, retail, and vehicles) to 10 types of economic sectors used in the IOT. From the interdependencies of sectors, the model is able to simulate how damages spread across the economy, which is interpreted as indirect damages. This allows to estimate an average ratio of indirect to direct damages, in order to obtain the ratio to estimate this relationship for any direct damage quantity estimated for Badalona. Finally, the ratio is applied to the estimated direct damages per return period to obtain the total EAD under all proposed scenarios.

3. Results

This study has provided qualitative and quantitative results. The firsts are the maps allow to spot the critical areas for the city and also make possible a visual comparison of risks under current and future rainfall conditions (Figures 8 and 9).

The knowledge of the city from the Badalona city council technicians enabled the validation of the results, by identifying accordingly the hot spots of the city. The zoomed area in Figures 8 and 9 is one of the most affected by floods historically and the models responded appropriately. From Figures 8 and 9 the expected increase of damages (tangible and intangible) under a business as usual situation (i.e., do nothing under future rainfall conditions) in case of a flood caused by a 100-year return period rainfall can be visually identified.

On the other hand, the obtained quantitative results in terms of high risk areas are presented in Figure 10 together with the climate change factors associated to each return period considered. According to these outcomes an important percentage increment of high risk area for pedestrians for low return periods (39% for T2 and 42% for T10) and for vehicles (17% for T2 and 39% for T10). However, both for T2 and T10 the increment of high-risk area in terms of absolute values is not significant, particularly for vehicles. Whereas, higher return-period rainfalls (T100 and T500) would cause important increase of high risk areas and thus their expected percentage increment would be translated into relevant absolute-value increments of high risk area. Risk for pedestrians is predicted to increase up to 9% and 3% for T100 and T500 respectively. Finally, risk for vehicles would increase up to 18% and 5% for 100-year and 500-year return period respectively.

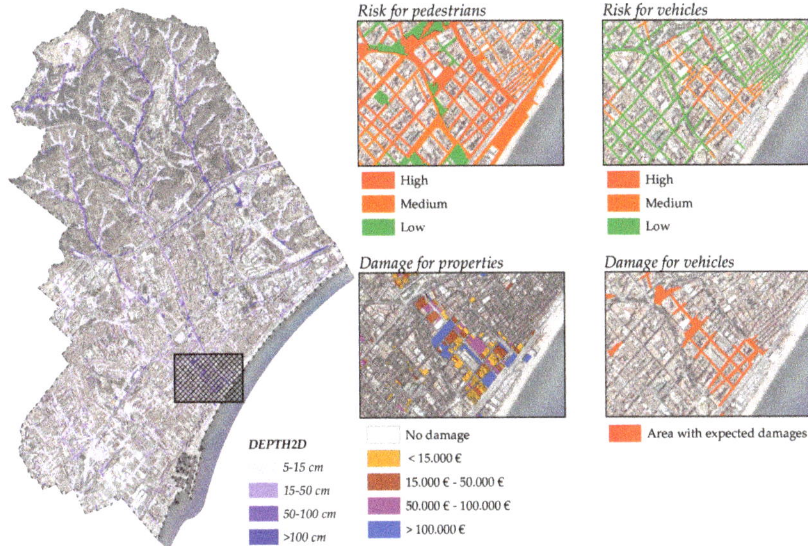

Figure 8. Flood related to a 100-year return period design storm under current rainfall conditions and the corresponding calculated damages within a zoomed area (black frame in the flood map).

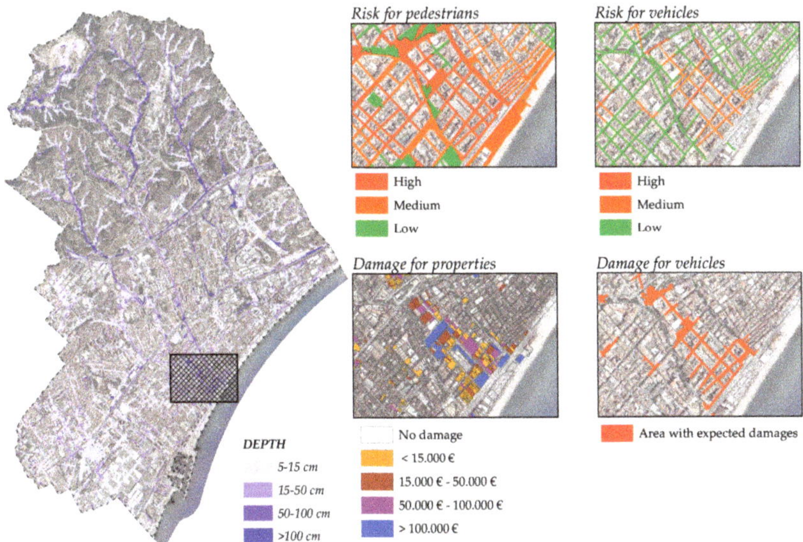

Figure 9. Flood related to a 100-year return period design storm under future rainfall conditions and the corresponding calculated damages within a zoomed area (black frame in the flood map).

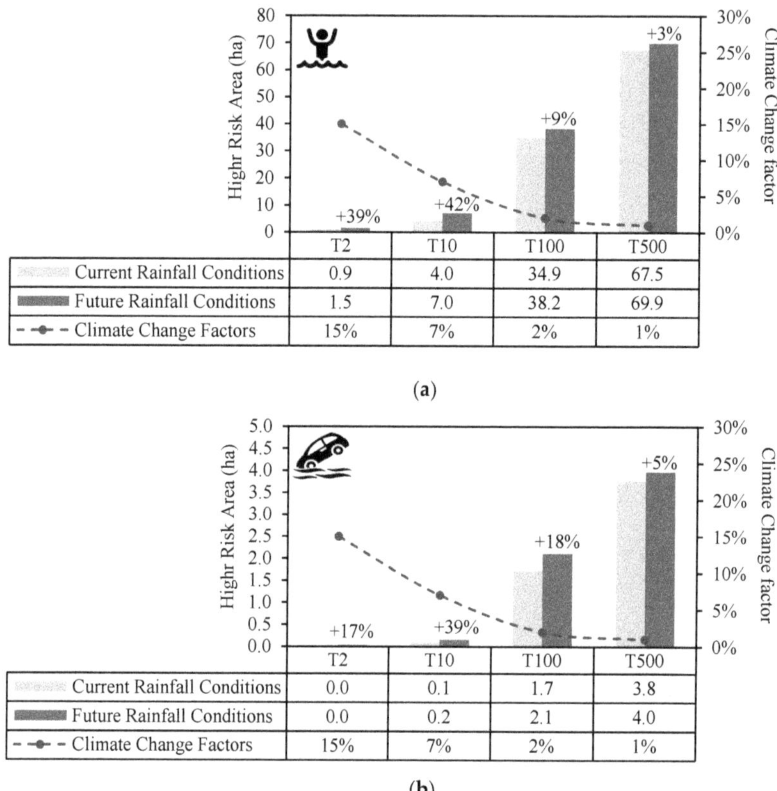

Figure 10. High risk areas for (a) pedestrians and (b) vehicles) according to current and future rainfall conditions.

The expected annual damage (EAD) is the indicator used here to measure changes in flood impacts according to expected future rainfall conditions (RCP 8.5). Once the direct and indirect damages are estimated according to the methodologies described, the EAD can be obtained. It may be calculated through the integration of the area under the curve that Figure 11 shows, result from plotting probability of occurrence of the damage and its monetary value [43].

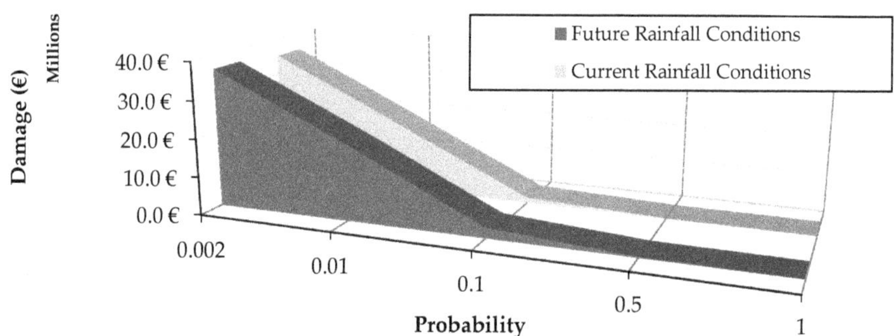

Figure 11. Damage-probability curve for current and future rainfall conditions. Both direct and indirect damages are considered.

Regarding indirect damages, the application of the Input/Output Method yields an estimation of indirect damages percentage over the expected direct damages of 32%. This ratio has been applied to each return period (i.e., T2, T10, T100, and T500) to calculate the EAD. The sectorial distribution of the direct impacts, used for the estimation of the indirect damages, yielded results in the form of a range of values. For the EAD estimation, the average value has been used. The results have been contrasted with relevant indirect damage studies, obtaining positive results. Carrera et al. [44] obtained results in the range of 0.19–0.22 for indirect damages in an Italian case study, while Hallegatte et al. [45] presented results in the range of 0.13–0.44 for a case study in the USA. The validation with real data has been proved difficult since insurance figures only show the accepted claims for businesses that have contracted the premium cover for business closure due to a weather event, which is a very limited number.

The total EAD, considering both direct and indirect damages estimation, is presented in Table 4, which includes the breakdown of estimated damages for current and future rainfall conditions. According to these results it is found a predicted increase of 30% in terms of EAD.

Table 4. Direct and indirect damages estimated for the different return periods and both current and future rainfall conditions.

Current Rainfall Conditions				
Return Period (years)	2	10	100	500
Probability	0.5	0.1	0.01	0.002
Direct damage to properties (€)	80,011.49€	1,023,592.69€	13,129,075.21€	24,440,088.96€
Direct damage to vehicles (€)	31,237.45€	130,035.63€	1,146,728.53€	2,655,441.99€
Indirect damage due to businesses interruption (€)	33,374.68€	346,088.50€	4,282,741.12€	8,128,659.29€
Damage (€)	144,623.62€	1,499,716.81€	18,558,544.86€	35,224,190.24€
Expected Annual Damage (EAD)	1,482,776.71 €			
Future rainfall conditions (RCP 8.5)				
Return Period (years)	2	10	100	500
Probability	0.5	0.1	0.01	0.002
Direct damage to properties (€)	139,454.19€	1,926,720.27€	14,289,199.05€	25,364,007.72€
Direct damage to vehicles (€)	62,300.55€	199,121.48€	1,228,946.61€	2,770,779.08€
Indirect damage due to businesses interruption (€)	60,526.42€	637,752.53€	4,655,443.70€	8,440,436.04€
Damage (€)	262,281.16€	2,763,594.28€	20,173,589.36€	36,575,222.84€
Expected annual damage (EAD)	1,929,913.89€			

4. Conclusions

As an important part of the research carried out in the BINGO project, a comprehensive flood risks and damages assessment was conducted, obtaining estimations of how projected future rainfall conditions may affect the economy and citizens' safety in the municipality of Badalona. In order to assess social and economic risks caused by floods in Badalona, firstly a detailed 1D/2D hydrodynamic model was developed and calibrated to provide estimations of the floods that could occur related to different return periods (i.e., 2, 10, 100, and 500 years). The risk definition here considers the combination of hazard and vulnerability, being the vulnerability concept formed by susceptibility and exposure. Risks considered in this study may be divided into intangible (social) and tangible (economic) damages, and also direct and indirect. Different methodologies have been proposed in this article in order to assess these risks in Badalona, with the aim of understanding how projected future rainfall would affect the city. The obtained outcomes can be useful to evaluate the efficiency of flood adaptation measures in Badalona, which can be prioritized based on risk reduction criteria with the results presented herein. Moreover, the city council, thanks to the obtained results, will be able to design new drainage elements according to future rainfall conditions to guarantee a certain level of safety conditions. Although, the city hot spots were well known by the city council technicians, the obtained results allowed them to better understand, quantify and analyze the cost of inaction. This study may be seen as an extension or enhancement of classical Master Drainage Plans, which scope traditionally ends by designing pipes and tanks for a specific security level (i.e., a design storm related to a return period). This conventional approach is far different from the concept of dual drainage which proposes to give response to what will happen on the streets once this security level is exceeded (i.e., rainfalls higher than the design storm). The criteria and methodologies presented in this article include details enough to be replicated in other cities.

Author Contributions: Hydrodynamic 1D/2D coupled model, B.R. and L.L.; Current and Future rainfall conditions scenarios, E.M.-G. and M.M.; full detailed flood risk assessment methodology, E.M.-G., M.G., M.M. and B.R.; flood risk assessment implementation and analysis of results; E.M.-G. and L.L.; writing—original draft preparation, E.M.-G.; writing—review and editing, L.L., M.G.; visualization, M.M.; supervision, B.R.; project administration, M.M.; funding acquisition, M.M.

Funding: This research was funded by EU H2020, Grant Agreement No. 641739.

Acknowledgments: The authors thank the Spanish insurance company Consorcio de Compensación de Seguros (CCS) for its important role in this research. Without its collaboration by providing claims data the damage model would have not been calibrated properly.

Conflicts of Interest: The authors declare no conflict of interest.

References

1. Intergovernmental Panel on Climate Change (IPCC). *Climate Change 2014: Synthesis Report*; IPCC: Geneva, Switzerland, 2014.
2. The European Parliament and the Council of the European Union. *Directive 2007/60/Ec of the European Parliament and of the Council of 23 October 2007 on the Assessment and Management of Flood Risks (Text with EEA Relevance)*; The European Parliament and the Council of the European Union: Brussels, Belgium, 2007.
3. Zhou, Q.; Mikkelsen, P.S.; Halsnæs, K.; Arnbjerg-Nielsen, K. Framework for economic pluvial flood risk assessment considering climate change effects and adaptation benefits. *J. Hydrol.* **2012**, *414–415*, 539–549. [CrossRef]
4. Turner, B.L.; Kasperson, R.E.; Matson, P.A.; McCarthy, J.J.; Corell, R.W.; Christensen, L.; Eckley, N.; Kasperson, J.X.; Luers, A.; Martello, M.L.; et al. A framework for vulnerability analysis in sustainability science. *Proc. Natl. Acad. Sci. USA* **2003**, *100*, 8074–8079. [CrossRef] [PubMed]
5. Velasco, M.; Russo, B.; Cabello, À.; Termes, M.; Sunyer, D.; Malgrat, P. Assessment of the effectiveness of structural and nonstructural measures to cope with global change impacts in Barcelona. *J. Flood Risk Manag.* **2018**, *11*, S55–S68. [CrossRef]
6. Hofmann, J.; Schüttrumpf, H.; Hofmann, J.; Schüttrumpf, H. Risk-Based Early Warning System for Pluvial Flash Floods: Approaches and Foundations. *Geosciences* **2019**, *9*, 127. [CrossRef]
7. Freni, G.; La Loggia, G.; Notaro, V. Uncertainty in urban flood damage assessment due to urban drainage modelling and depth-damage curve estimation. *Water Sci. Technol.* **2010**, *61*, 2979–2993. [CrossRef] [PubMed]
8. Martínez-Gomariz, E.; Guerrero-Hidalga, M.; Russo, B.; Yubero, D.; Gómez, M.; Castán, S. Desarrollo y validación de curvas de daño y estanqueidad para la estimación de daños por inundaciones en zonas urbanas españolas. In Proceedings of the VI Jornadas de Ingeniería del Agua JIA2019, Toledo, Spain, 1 October 2019; p. 4.
9. Department of Natural Resources and Mines (Queensland Government). *Guidance on the Assessment of Tangible Flood Damages*; Department of Natural Resources and Mines: Brisbane, Australia, 2002.
10. Federal Emergency Management Agency (FEMA); Department of Homeland Security. *Mitigation Division Multi-hazard Loss Estimation Methodology. Flood Model. Hazus-HM MR5 Technical Manual*; FEMA: Washington, DC, USA, 2015; p. 499.
11. Francés, F.; García-Bartual, R.; Ortiz, E.; Salazar, S.; Miralles, J.L.; Blöschl, G.; Komma, J.; Habereder, C.; Bronstert, A.; Blume, T. *Efficiency of Non-Structural Flood Mitigation Measures: "Room for the River" and "Retaining Water in the Landscape"*; EU: Brussels, Belgium, 2008.
12. Luino, F.; Cirio, C.G.; Biddoccu, M.; Agangi, A.; Giulietto, W.; Godone, F.; Nigrelli, G. Application of a model to the evaluation of flood damage. *Geoinformatica* **2009**, *13*, 339–353. [CrossRef]
13. Penning-Rowsell, E.; Viavattene, C.; Pardoe, J.; Chatterdon, J.; Parker, D.; Morris, J. *The Benefits of Flood and Coastal Risk Management: A Handbook of Assessment Techniques*; Middlesex University Press: London, UK, 2010; p. 90.

14. Velasco, M.; Cabello, À.; Russo, B. Flood damage assessment in urban areas. Application to the Raval district of Barcelona using synthetic depth damage curves. *Urban Water J.* **2015**, *13*, 426–440. [CrossRef]

15. Messner, F.; Meyer, V. *Flood Damage, Vulnerability and Risk Perception–Challenges for Flood Damage Research*; UFZ-Diskussionspapiere, No. 13/2005; Springer: Berlin, Germany, 2006.

16. Grigg, N.S.; Helweg, O.J. State-of-the-Art of Estimating Flood Damage in Urban Areas. *J. Am. Water Resour. Assoc.* **1975**, *11*, 379–390. [CrossRef]

17. Martínez-Gomariz, E.; Gómez, M.; Russo, B.; Sánchez, P.; Montes, J.-A. Methodology for the damage assessment of vehicles exposed to flooding in urban areas. *J. Flood Risk Manag.* **2019**, *12*, e12475. [CrossRef]

18. Thieken, A.; Ackermann, V.; Elmer, F.; Kreibich1, H.; Kuhlmann, B.; Kunert, U.; Maiwald, H.; Merz, B.; Müller, M.; Piroth, K.; et al. Methods for the evaluation of direct and indirect flood losses. In Proceedings of the 4th International Symposium on Flood Defence: Managing Flood Risk, Reliability and Vulnerability, Toronto, ON, Canada, 6–8 May 2008; p. 10.

19. Hammond, M.J.; Chen, A.S.; Djordjević, S.; Butler, D.; Mark, O. Urban flood impact assessment: A state-of-the-art review. *Urban Water J.* **2015**, *12*, 14–29. [CrossRef]

20. Ranger, N.; Hallegatte, S.; Bhattacharya, S.; Bachu, M.; Priya, S.; Dhore, K.; Rafique, F.; Mathur, P.; Naville, N.; Henriet, F.; et al. An assessment of the potential impact of climate change on flood risk in Mumbai. *Clim. Chang.* **2011**, *104*, 139–167. [CrossRef]

21. Gómez, M.; Russo, B. Methodology to estimate hydraulic efficiency of drain inlets. *Proc. ICE Water Manag.* **2011**, *164*, 81–90. [CrossRef]

22. Russo, B.; Sunyer, D.; Velasco, M.; Djordjević, S. Analysis of extreme flooding events through a calibrated 1D/2D coupled model: The case of Barcelona (Spain). *J. Hydroinform.* **2015**, *17*, 473–491. [CrossRef]

23. Gómez, M.; Rabasseda, G.H.; Russo, B. Experimental campaign to determine grated inlet clogging factors in an urban catchment of Barcelona. *Urban Water J.* **2013**, *10*, 50–61. [CrossRef]

24. Gómez, M.; Parés, J.; Russo, B.; Martínez-Gomariz, E. Methodology to quantify clogging coefficients for grated inlets. Application to SANT MARTI catchment (Barcelona). *J. Flood Risk Manag.* **2019**, *12*, e12479. [CrossRef]

25. Henonin, J.; Russo, B.; Mark, O.; Gourbesville, P. Real-time urban flood forecasting and modelling–a state of the art. *J. Hydroinform.* **2013**, *15*, 717–736. [CrossRef]

26. BINGO Project. *Deliverable D3.3: Calibrated Water Resources Models for Past Conditions, H2020 BINGO. Bringing Innov. to onGOing Water Management–a Better Future*; under Clim. Chang. Grant Agreem. n° 641739; EU: Brussels, Belgium, 2019.

27. Houston, D.; Werrity, A.; Bassett, D.; Geddes, A. *Pluvial (Rain-Related) Flooding in Urban Areas: The Invisible Hazard*; Joseph Rowntree Foundation: York, UK, 2011.

28. Larsen, A.N.; Gregersen, I.B.; Christensen, O.B.; Linde, J.J.; Mikkelsen, P.S. Potential future increase in extreme one-hour precipitation events over Europe due to climate change. *Water Sci. Technol.* **2009**, *60*, 2205–2216. [CrossRef]

29. Arnbjerg-Nielsen, K. Quantification of climate change effects on extreme precipitation used for high resolution hydrologic design. *Urban Water J.* **2012**, *9*, 57–65. [CrossRef]

30. Smith, K.; Ward, R. *Floods: Physical Processes and Human Impacts*; John Wiley & Sons: Chichester, UK, 1998; ISBN 0-471-95248-6.

31. Merz, B.; Kreibich, H.; Thieken, A.; Schmidtke, R. Estimation uncertainty of direct monetary flood damage to buildings. *Nat. Hazards Earth Syst. Sci.* **2004**, *4*, 153–163. [CrossRef]

32. Djordjevic, S.; Prodnovic, D.; Maksimovic, C. An approach to simulation of dual drainage. *Water Sci. Technol.* **1999**, *39*, 95–103. [CrossRef]

33. Russo, B. Design of Surface Drainage Systems According to Hazard Criteria Related to Flooding of Urban Areas. Ph.D. Thesis, Technical University of Catalonia, Barcelona, Spain, 2009.

34. Russo, B.; Gómez, M.; Macchione, F. Pedestrian hazard criteria for flooded urban areas. *Nat. Hazards* **2013**, *69*, 251–265. [CrossRef]

35. Martínez-Gomariz, E. *Inundaciones Urbanas: Criterios de Peligrosidad y Evaluación del Riesgo Para Peatones y Vehículos*, 1st ed.; Universitat Politècnica de Catalunya: Barcelona, Spain, 2016; ISBN 978-84-617-4904-1.

36. Martínez-Gomariz, E.; Gómez, M.; Russo, B. Experimental study of the stability of pedestrians exposed to urban pluvial flooding. *Nat. Hazards* **2016**, *82*, 1259–1278. [CrossRef]

37. Martínez-Gomariz, E.; Gómez, M.; Russo, B.; Djordjević, S. A new experiments-based methodology to define the stability threshold for any vehicle exposed to flooding. *Urban Water J.* **2017**, *14*, 930–939. [CrossRef]
38. Martínez-Gomariz, E.; Gómez, M.; Russo, B.; Djordjević, S. Stability criteria for flooded vehicles: A state-of-the-art review. *J. Flood Risk Manag.* **2018**, *11*, S817–S826. [CrossRef]
39. Gerard, M. *Tire-Road Friction Estimation Using Slip-Based Observers*; Lund University: Lund, Sweden, 2006.
40. Martínez-Gomariz, E.; Guerrero-Hidalga, M.; Russo, B.; Yubero, D.; Gómez, M.; Castán, S. Desarrollo y aplicación de curvas de daño y estanqueidad para la estimación del impacto económico de las inundaciones en zonas urbanas españolas. *Ing. Del Agua* **2019**, *23*, 229–245. [CrossRef]
41. Kowalewski, J. Methodology of the Input-Output Analysis. Available online: https://www.econstor.eu/handle/10419/48249 (accessed on 6 February 2019).
42. IDESCAT, S.I. of C. Population Growth. Available online: http://www.idescat.cat/emex/?id=080193&lang=en (accessed on 6 February 2019).
43. Meyer, V.; Priest, S.; Kuhlicke, C. Economic evaluation of structural and non-structural flood risk management measures: Examples from the Mulde River. *Nat. Hazards* **2011**, *62*, 301–324. [CrossRef]
44. Carrera, L.; Standardi, G.; Bosello, F.; Mysiak, J. Assessing direct and indirect economic impacts of a flood event through the integration of spatial and computable general equilibrium modelling. *Environ. Model. Softw.* **2015**, *63*, 109–122. [CrossRef]
45. Hallegatte, S. An Adaptive Regional Input-Output Model and its Application to the Assessment of the Economic Cost of Katrina. *Risk Anal.* **2008**, *28*, 779–799. [CrossRef]

Article

Towards an International Levee Performance Database (ILPD) and Its Use for Macro-Scale Analysis of Levee Breaches and Failures

Işıl Ece Özer *, Myron van Damme and Sebastiaan N. Jonkman

Department of Hydraulic Engineering, Faculty of Civil Engineering and Geosciences, Delft University of Technology, 2628 CN Delft, The Netherlands; m.vandamme@tudelft.nl (M.v.D.); s.n.jonkman@tudelft.nl (S.N.J.)
* Correspondence: i.e.ozer@tudelft.nl

Received: 2 December 2019; Accepted: 22 December 2019; Published: 30 December 2019

Abstract: Understanding levee failures can be significantly improved by analysing historical failures, experiments and performance observations. Individual efforts have been undertaken to document flood defence failures but no systematically gathered large scale, open access dataset is currently available for thorough scientific research. Here, we introduce an efficiently structured, global database, called International Levee Performance Database (ILPD), which aims to become a valuable knowledge platform in the field of levee safety. It comprises information on levee characteristics, failure mechanisms, geotechnical investigations and breach processes for more than 1500 cases (October 2019). We provide a macro-scale analysis of the available data, aiming to provide insights on levee behaviour based on historical records. We outline common failure mechanisms of which external erosion is identified as the most frequent for levees. As an example, we investigate flood events occurred in Germany (2002, 2013) and examine breach characteristics of hundreds of failures. It is found that initial failure mechanisms have an influence on breach characteristics and that failures due to instability and internal erosion are less frequent but lead to larger breaches. Moreover, a relation between the return period and the expected breach density during a flood event is identified. These insights could complement flood risk assessments.

Keywords: levee failures; database; flood risk; flood defences; levee breach

---

1. Introduction

The majority of the global population is located in flood prone coastal areas and deltas. Between 1980 and 2015, 3563 flood events have been reported in Europe and more severe floods are expected in the near future [1,2]. Flood defence failures can lead to major catastrophes in terms of loss of life and economic damage. Improved flood defence measures are required to safeguard flood prone areas from floods in order to save many lives and avoid considerable damage costs. Aiming to prevent this, flood prone countries make substantial investments in activities related to monitoring, maintenance and reinforcement of flood defences. However, our understanding of levee failure mechanisms is still limited. Despite extensive research on individual mechanisms and processes of failures, the composition of levees, their behaviour during critical conditions and the modelling of their failures still remain uncertain [3]. There is a need for better validation and calibration of models or, in other words, better insight in their uncertainty.

Since the full-scale experiments are challenging and costly and the spatial variability in subsoil typically plays an important role, historical levee failures can be used to provide insights into the real failure processes and conditions. However, most of the analyses are performed with limited data from failed levees, since most of the evidence is washed away during real failures. This highlights the need for systematically and globally collected data and a sharing platform not only to advance understanding of levee failures and their mechanisms but also to improve failure and breaching models. These insights could eventually contribute to the development of accurate techniques for innovative designs of flood defence systems and support decisions and actions affecting levee safety [4].

Several databases and data collections already exist in the field of dam safety (Table 1). Apart from general information on the available dams around the world [5,6], these databases may have a particular focus on dam breaches [7,8], landslide dam failures [9], large dam failures [10], internal erosion of dams [11], dam incidents [12] and loss of life for risk assessments [13,14]. Most of these databases include general information about the dams, such as geometry, type, reservoir capacity and in some cases information on their failures, for example, time to failure, breach information, peak discharge. Databases are also used for the analysis of accident statistics in other fields of Civil Engineering. Typical examples include global reports on natural disaster impacts and frequencies [15], remote sensing-based flood information [16] or data on flood protection standards [17].

In the field of levee safety, there is a small number of open access databases with some information available, see Table 1. Some online databases provide general information on levees at a regional [18] or at a national level without a particular focus on their failures [19,20]. Individual efforts have been undertaken to document levee failures after disasters, for example, New Orleans in 2005 [21,22], Germany in 2002 [23] but no systematically gathered large scale, open access datasets are available for thorough scientific research. In this dearth of information, some researchers have constructed data collections specifically on levee failures [24,25]. Characteristic of these mostly table-formatted databases is that only generic information on failures is provided, for example, general information on the levee geometry, material, breach information, failure mechanism. The databases identified for levees do not include any detailed geotechnical or hydraulic information that would allow more in-depth analysis of historical failures. This limits the use of these databases for detailed analysis of the failure processes, development of empirical levee breach models or validation of process-based models.

Aiming to improve the understanding of these failure mechanisms and breaching and to enhance the reliability analyses of flood defences, we set up an International Levee Performance Database (ILPD) to facilitate sharing of data on levee characteristics, failure mechanisms, geotechnical investigations and breach processes. The objectives of this study are (1) to present a comprehensive overview of the first version of the ILPD (containing more than 1500 cases) and (2) to perform a macro-scale analysis of the available data on levee failures and breaches, aiming to demonstrate the potential of the database in providing insights on levee behaviour and supporting flood risk assessments based on historical records. First, we elaborate on the purpose and structure of the ILPD and establish a terminology which is used for describing levee failures. We provide basic statistics on the currently (October 2019) available ILPD data to illustrate the overall characteristics of the database. Second, as an example for the use of a sub-dataset from the ILPD, we investigate the flood events in Germany, occurred in 2002 and 2013 and examine breach characteristics of over a hundred failures in order to outline the most common influence factors at an event-level. We also introduce two breach density parameters, which are related to the expected amount of levee breaches and their width, intended to be used for the risk assessment of levees. Finally, we highlight the operational obstacles and strategies to further improve the ILPD and discuss the use of event-level analysis for risk assessments.

Table 1. An overview of the example databases in the field of flood risk, dam and levee safety.

Database/Reference	Field of Application	Number of Cases	Failures Included	Data Type [a]	Accessibility	Active (Y/N)
Peng and Zhang [9]	Landslide dams	1044 cases	Yes	General information	Open access	No
Utah State University database [13]	Dam failures	174 cases	Yes	General information with a particular focus on loss of life data for dam safety risk assessments	Open access	No
Flood Protection Standards (FLOPROS) database [17]	Flood protection standards	179 cases for design layer 68 cases for policy layer	No	Detailed information on design and policy standards (not on the actual flood defences or their failures)	Open access	No
Association of State Dam Safety Officials (ASDSO) database [12]	Dam failures	14 detailed case studies	Yes	Detailed information such as photos, videos, general information and lesson learned, reports, failure mechanisms	Open access	No
United States Bureau of Reclamation (USBR) database [14]	Dam failures focused on loss of life	60 cases	Yes	General information with a particular focus on loss of life	Open access	Yes
National Performance of Dams Program (NPDP) database [5]	Dam	>10,000 cases but few failure cases	Yes	General information including emergency plan, population at risk, storage capacity, failure mech., consequences, lessons learned	Open access	Yes
Dam Accident Database [6]	Dam failures	900 cases	Yes	General information, including some detailed data (hydrographs, reports, photos, etc.)	Not publicly available	No
Dartmouth Flood Observatory of Large Floods [16]	Remote sensing-based flood information	4700 flooding events	Yes	General information including duration of the event, loss of life, damage, severity, effected area, magnitude of the flood, etc.	Open access	Yes
The International Disaster Database (EM–DAT) [15]	Disaster, including flooding	>10,000 cases	No	Detailed information on the disaster related data	Open access	Yes
Froehlich [7]	Dam failures	43	Yes	General information on the breach formation	Not publicly available	No
Foster et al. [10]	Large dam failures	136	Yes	General information	Not publicly available	No

Table 1. *Cont.*

Database/Reference	Field of Application	Number of Cases	Failures Included	Data Type [a]	Accessibility	Active (Y/N)
Xu [8]	Dam failures	1443 cases	Yes	Detailed information	Not publicly available	No
ERINOH database [11]	Internal erosion of dams and levees	120 failure cases	Yes	General information, also including some detailed data (hydrographs, breach info., reports, photos, maps, etc.)	Not publicly available	No
Van Baars and van Kempen [24]	Levee failures	337	Yes	General information	Not publicly available	No
Italian Levee Database (INLED) [19]	Levee information	Currently, a few cases	Yes	General information	Not publicly available	Yes
Danka [25]	Levee failures	1004 cases	Yes	General information	Open access	No
Department of Water Resources (DWR) levee database [18]	Urban and non-urban levee failures in California	215 cases	Yes	General information	Open access	No
U.S. Army Corps of Engineers database [20]	National levee information of USA	>10,000 cases (no separation for failures)	Yes	Detailed information on levee system evaluation and inspection, flood risk communication, flood plain management and risk assessment	Open access	Yes

[a] General information includes geometry, construction time, type, failure mechanisms, capacity, material, peak outflow rate, breach information and also qualitative information.

2. Towards an International Levee Performance Database (ILPD)

2.1. Purpose of the ILPD

The main purpose of the ILPD is to provide a global platform for systematically collecting and sharing data on levee performances to facilitate research. Particularly, ILPD promotes (a) learning from what went wrong during past flood events to prevent floods from occurring in the future and (b) learning about how to model failure processes and how failure mechanisms might be correlated. By facilitating this, the database can also enable the systematic validation of models and the evaluation of model accuracy. More specifically, the ILPD has been developed to provide collected datasets on:

- Actual failures during extreme catastrophic events, such as levee failures in New Orleans, LA, USA [21] and failures during the levee construction phase;
- Failures in small- and full-scale experiments, such as levee breach experiments in the Netherlands [26];
- General investigations on the performance of flood defence systems, for example, the case of New Orleans [27] or of the coastal floods in France [28];
- Detailed information on some earthen dam failures, as these show similarities with levee failures;
- Information on the consequences (e.g., damage, loss of life, flooded area, etc.) per extreme event.

2.2. The Design and the Structure

To allow for a wide range of data type and easy dissemination, the database is linked to a website from which data can be downloaded freely (leveefailures.tudelft.nl). All information stored in the ILPD is open access and sources are automatically provided along with the data upon downloading from the database.

Some extreme events, like Hurricane Katrina in 2005 or the 1953 floods in the Netherlands, caused multiple levee failures at different locations. To capture this, data is structured by events and failures, each associated to a unique ID's in the ILPD. One event can consist of several individual failures, each containing specific data. The failure cases which are clustered in the same event are either caused by the same hazard event or are part of the same experimental program. This way, data on each specific levee failure adds to the understanding of the real failure processes, whereas information on the overall consequences, such as total damage or loss of life, can be linked to events.

The ILPD has a three-level structure with increasing degree of detail. The first level (Level.1: Metadata) provides general information on individual failure cases as well as on flood events in a table-based format (exportable to .csv files), for example historical levee failures in Hungary provided by [25]. This generic qualitative and numeric data may also include information on the consequences, such as total damage, loss of life and flooded area. Level.1 data mainly consists of information on the levee (geometry, location, material, type, etc.), flow boundary conditions (return periods, max. water level, peak flow discharge, etc.), breach (width, invert level, peak discharge, etc.), soil properties, management organization, documentation (video, photo, etc.) and references. The second level (Level.2: Intermediate) includes well-documented information on historical failures, such as loading conditions and soil profiles and modelling results; or hindcasted information based on field investigations of actual failures (e.g., New Orleans in 2005 [21,22]). The detailed information aims to provide more insight into the physical processes and to facilitate rapid analysis of failures. Since every dataset contains unique information, providing detailed data in a table-based format severely limits the ease of analysis of different datasets simultaneously. Thus, a more structured format is preferred in order to support rapid analysis of data from several detailed datasets (exportable into JSON files). Lastly, the third category (Level.3: In-depth) contains detailed information on failure processes including time-dependent data on the hydrodynamic loading conditions and geotechnical information. These dataset (exportable to .pdf/.csv files) mostly consist of experimental data, such as those provided by [29].

A screenshot of the opening page of the website is provided in Figure 1. The interactive map allows for navigation to all available failure cases. Each marker on the map represents the location of a failure case. Underneath, the cases pictured on the map are listed. When the case of interest is selected, basic information of the failure is presented and the media (if present) can be viewed. After selecting one or more cases, a click on the button 'Export selection' gives a pop-up window in which the type of data to be downloaded can be chosen. On the left side of the map, several filter criteria can be set. The database can be filtered based on time span, country, case study type, defence type or failure mechanism. Moreover, more filter criteria are available using the 'Add filter criterion' feature and a combination of different filter criteria can be made as well.

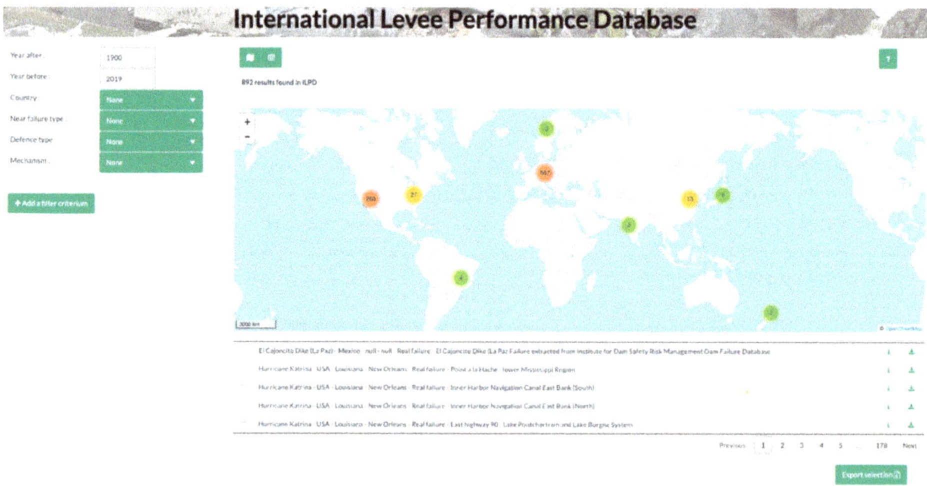

Figure 1. A screenshot of the opening page of the ILPD (leveefailures.tudelft.nl).

2.3. Categorization of Levee Failure Mechanisms

Given that various terminologies exist for levee failure mechanisms and other related parameters, a challenge in setting up an international database is to establish a list of uniform, globally acknowledged definitions. Flood defences are hydraulic structures whose primary objective is to provide flood protection along the coasts, rivers, lakes and other waterways [30]. A levee (also referred to dike or embankment) is a water retaining structure consisting of soil (fully or partly) with a sufficient elevation and strength to be able to retain the water under extreme circumstances [31]. A typical levee cross section is given in Figure 2. Earthen levees, which form a large part of the existing flood defence systems, can be constructed with (a) homogeneous soil (homogeneous earthfill), (b) several soil types (layered structure) or (c) a hard structure (levee structure combination), such as a levee with a floodwall. Examples of other types of flood defence systems are dams, dunes, storm surge barriers or temporary flood defences.

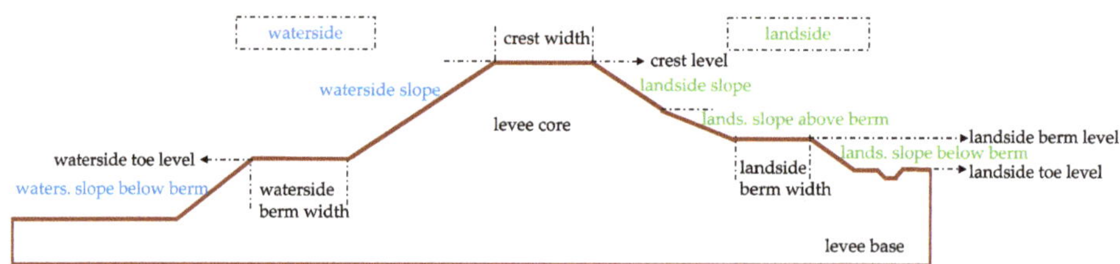

Figure 2. General levee cross section, showing the main elements of a levee included in the ILPD.

Levees can fail when their ultimate limit state is exceeded causing them no longer to fulfil their water retaining function [32]. The most commonly encountered levee failure mechanisms and their most common contributing factors are given in Figure 3 [31]. Hydraulic failures occur due to insufficient height, whereas insufficient strength leads to geotechnical failures [33]. Breaching of a levee refers to the loss of integrity or a major geometric change [32,34]. However, occurrence of a failure mechanism does not necessarily lead to a breach. For example, significant amount of overflow may cause severe floods without leading to a breach in the structure. Moreover, since the occurrence of the initial failure mechanism might trigger other mechanisms [33], a sequence of multiple failure mechanisms can be observed on the levee. For example, failures may be initiated with overtopping, followed by external erosion that causes a breach at the latter stage (Figure 3). All identified failure mechanisms are entered in the database in a chronological order.

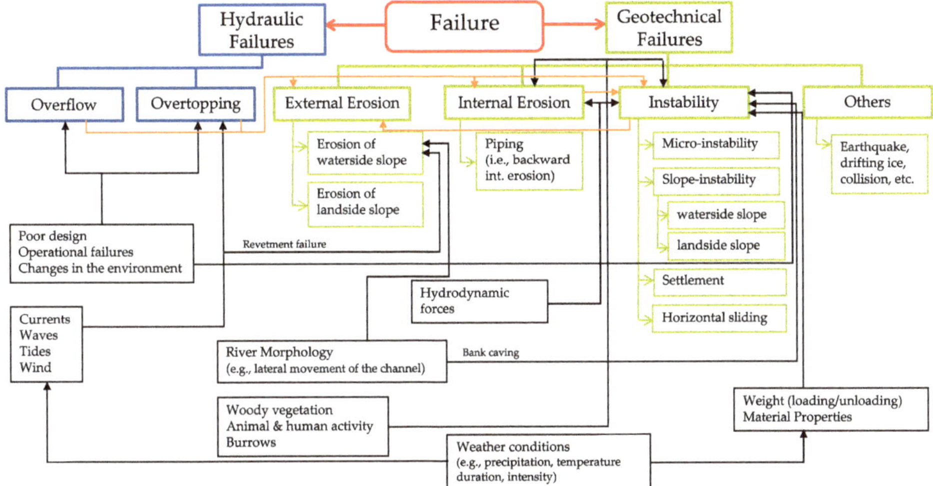

Figure 3. An overview of most relevant levee failure mechanisms and their most common contributing factors [9,31].

The most relevant levee failure mechanisms observed in the database are defined briefly here [35].

- Overtopping and Overflow—Overflow occurs when still water level is higher than the crest level of the levee. Whereas, overtopping is observed when still water level remains below the crest level but waves run-up and pass the crest level.
- External erosion—External erosion occurs when the slope of the levee is not sufficiently resistant to the hydraulic loads, that is, when the shear stress induced by flows exceeds the critical value associated with the nature of the materials of the levee [31]. Currents and waves are the main aggravating factors of external erosion which can occur on the landside or waterside slope of the levee. Overtopping/overflow of a levee can induce major damages linked to external erosion, especially on the landside slope.
- Internal erosion—Internal erosion, which refers to a generic event, is initiated by hydrodynamic forces acting on soil particles within a levee foundation which are carried downstream by seepage flow [31]. In this process, migration of material particles induced by pore pressure and flow forms channels within the foundation soils. These pipes undermine the structure of the levee and lead to failure. Internal erosion related failure mechanisms consist of—concentrated leaks, backward erosion, contact erosion and suffusion [36]. Backward erosion, known as piping, is typically most relevant for levees. It occurs if uplift, seepage, heave and piping occur respectively. Seepage also

increase the likelihood of instability because of changes to pore pressure distribution within the levee. Uplift pressure in foundation soils can generate major instability.

- Slope instability (i.e., instability)—Instability occurs when the forces (i.e., excess pore pressure) on a levee are higher than the shear resistance which is determined by the soil's shear strength. Landside slope instability occurs due to the infiltration of water into the levee body and its foundation, leading to forcing of the levee body and decreasing shear strength of the soil. Whereas, waterside slope instability occurs due to sudden drawdown of the outside water level after heavy saturation of the levee body. In this situation, the pore pressures at the base of the potential slide plane stay high, while the horizontal pressure or support from the river water is reduced.
- Micro instability—Micro instability occurs when the seepage water causes the phreatic surface to rise and reach the waterside slope of a levee. The term "micro-" is used to distinguish the stability problems related to this phenomenon from the slope instability which essentially concern the whole levee body directly.
- Settlement—Settlement is a deformation mechanism in vertical direction that can mainly lead to insufficient crest height to prevent failure mechanisms like overtopping/overflow.
- Horizontal sliding—Similar to instability of the landside slope, sliding occurs along the base of the levee body. In this case, the main driving force is the horizontal force of the water exerted on the waterside slope. This mechanism is typically an issue for levees which are made of relatively light material such as peat, where the effective stresses at the base are very low.

3. Macro-Scale Analysis of Levee Failures

3.1. General Database Statistics

A large number of failure cases are collected all around the world and their basic statistics are reported in this section to provide an overview of the characteristics of the database. Currently (October 2019), the ILPD includes 1572 failure cases of which 1538 are the failures of flood defences occurred in different time periods and 34 of them are full- and small-scale experiments collected from different countries (Figure 4a). The available information mainly concerns the failures of levees (1418 cases) and levee structure combinations (14 cases) occurred along rivers (89%), coasts (10%) and canals (1%) but also of some earthen dams (106 cases) due to their similar composition to levees. The majority of the levee failures before 1900s mostly occurred in the Carpathian Basin in Hungary, as reported by [25,37] but these only contain generic information. The other cases reported in the ILPD, occurred in the last 100 years, are mostly from Germany, The Netherlands, USA, UK and Czech Republic (Figure 4b). It is noted (and further discussed in Section 4) that the geographical and temporal distributions shown in Figure 4 only represent the data currently included in the database.

Most failure entries in the database contain general information (Level.1: 1498 cases), such as levee and breach characteristics. As an example, Figure 5 shows that the crest height of the failed levees, which is defined as the differences between the toe level and the crest level, varies mostly between 1–5 m, whereas the crest width is in the range of 2–5 m. The database also includes more detailed information on the Level.2 (59 cases) and the Level.3 (14 cases) data.

Most of the events in the database are associated with single failures. However, 38% of the failures occurred during events (65 in total) with more than one failure. Examples for the documented events with the large number of breaches are in Germany (2002) with 111 failures, Netherlands (1953) with 97 failures and Czech Republic (1997) with 33 failures.

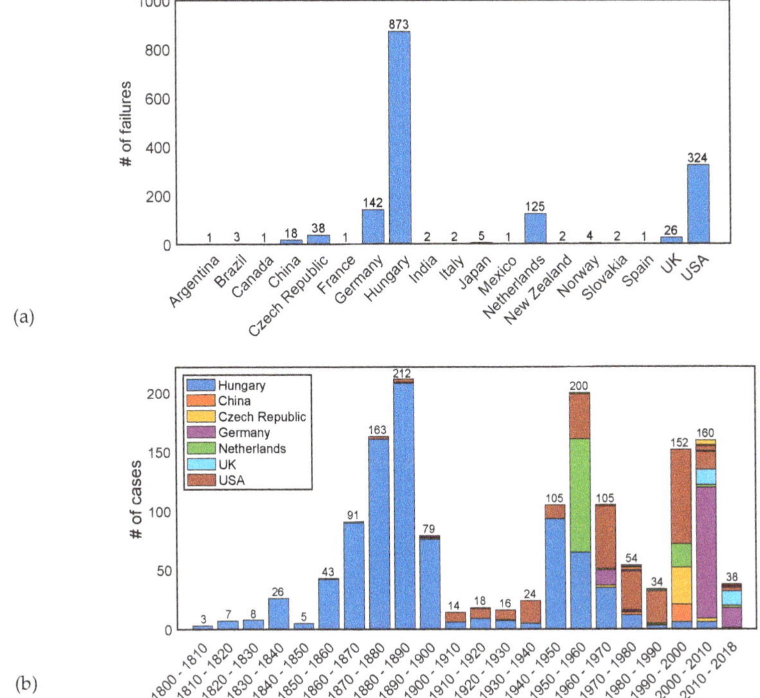

Figure 4. Statistics on the failures (**a**) per country, (**b**) per time period entered in ILPD.

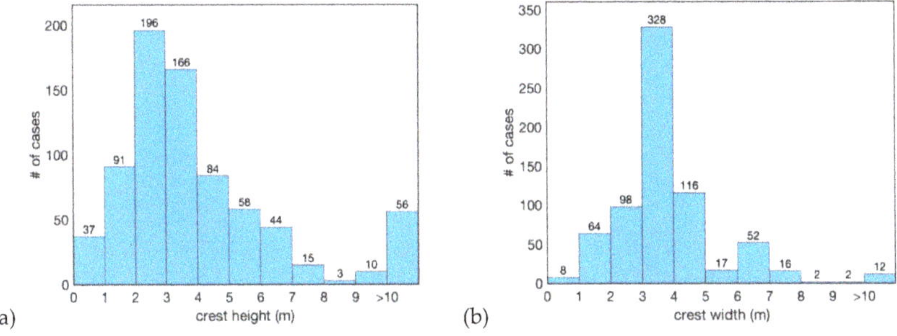

Figure 5. Statistics of the levee information (**a**) crest height, (**b**) crest width (m) per number of cases.

Failure Mechanisms

Analysis of the identified failure mechanisms within the collected cases in the database (Figure 6a) shows that more than half of the failures, of which their causes are known, occurred due to external erosion (61.5%), internal erosion (16.8%), instability (14.2%), overflow/overtopping (2.5%). Besides, the rest of the failures were due to other causes (4.7%), including man-made failures. It is noted that excluding the Hungary dataset [25], which forms the majority of the database, does not significantly affect this distribution (Figure 6a).

A comparison between observed failure mechanisms for levees and earthen dams (Figure 6b) reveals that external erosion of the slope is more likely to occur in levees. This can be explained by the fact that water level and discharge in rivers, as well as currents and waves in seas, are mostly affected by meteorological and hydrological conditions which are more uncertain than for dams. Whereas, water levels in the reservoirs behind dams are generally more controlled. Moreover, the reason for a more frequent occurrence of internal erosion in earthen dams is associated with the higher hydraulic head differences due to the larger size of dams.

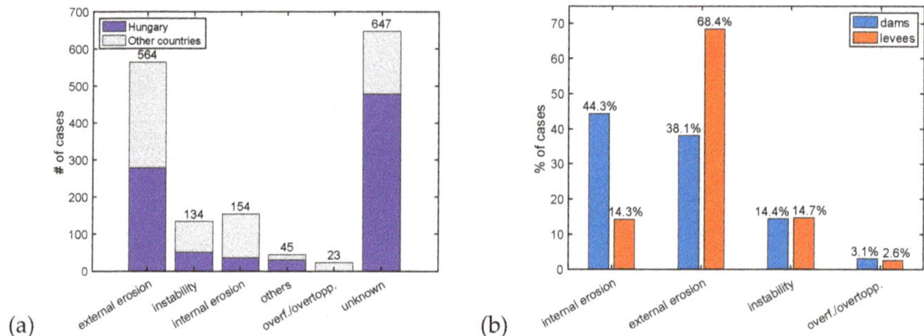

Figure 6. Statistics on (**a**) the observed failure mechanisms in ILPD (**b**) a comparison between earthen dams and levees for the most common failure mechanisms.

Levee breaching was observed for most of the failure cases that are entered in ILPD. Failures that did not lead to a breach at the later stage were mostly due to overflow/overtopping (23 cases). Information on the breach characteristics, given in Figure 7, shows that there is a considerable variation in breach depth and width values. Previous studies on breach models reveal that, under certain conditions, breach depth and width are correlated during some of the breach development stages [26,38]. However, breaching is a complex process that depends on many different factors (e.g., structure type, loading conditions, soil characteristics, etc.), thus the relation between breach depth and width may not always be explained by a simple relation model (Figure 7c).

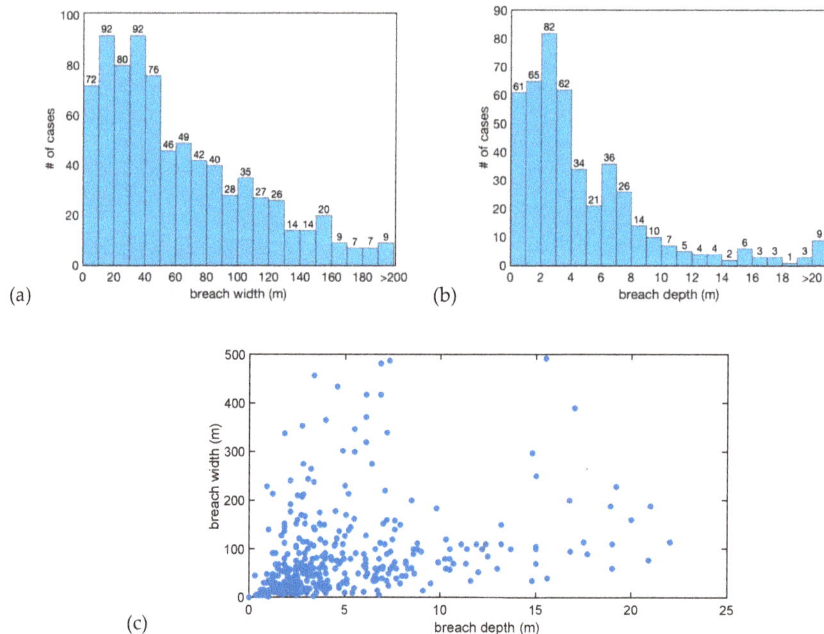

Figure 7. Statistics on (**a**) breach depth (450 cases), (**b**) breach width (785 cases) and (**c**) breach depth vs. width for (402 cases, Pearson coefficient, r = 0.164).

3.2. Investigation of the 2002 and 2013 Failures in the Elbe Region, Germany

Further analysis of data from the ILPD at the event-level can provide insights in identifying (1) typical vulnerabilities and common failure mechanisms, (2) breach characteristics and (3) density of breach occurrence. Although events and levee systems are diverse, this information can be used to inform and improve (local) flood risk assessment. As an example, sub-dataset from the ILPD, information has been analysed on the performance of river levees along the Elbe tributaries in Germany, which flooded in 2002 and 2013 (Figure 8a). A general overview of the flood events is provided first, followed by a macro-scale analysis of the failures and the associated breaches.

3.2.1. Overview of the 2002 and 2013 Flood Events

The database reports on 111 levee failures along the Mulde and the Elbe rivers observed during the flooding event of August 2002, in Saxony and Saxony-Anhalt regions, Germany, as a result of extreme meteorological conditions, followed by extreme discharges and water levels [39]. Floods occurred mainly due to overflow/overtopping and breaching of levees at many locations. Return periods of river discharges exceeded 500 years at some tributaries of the Elbe and the return period along the Elbe and Mulde varied between 100–300 years [23,40]. Considering that the design return periods for the flood defences in Germany is usually 100 years, the levees along the rivers were significantly overloaded. Incomplete flood warnings, bad maintenance of flood defence structures and a lack of awareness were recognized as the weaknesses of the flood risk management [41]. The total damage was estimated around 11.6 billion euros which is the highest amount for a damage caused by a natural hazard in Germany.

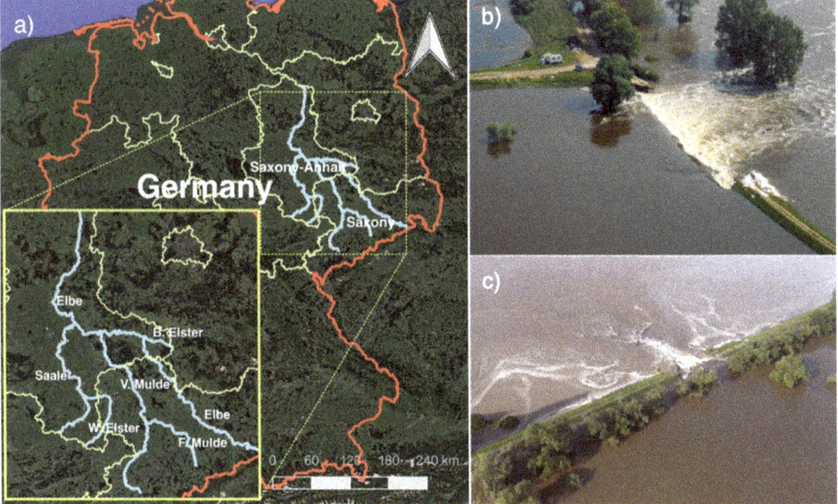

Figure 8. (**a**) Investigated rivers along the Elbe tributaries in Germany, which flooded in 2002 and 2013, (**b**) Fischbeck, 2013 (source: [42]), (**c**) Breitenhagen, 2013 (source: [43]).

Eleven years later, in June 2013, another big flood hit large parts of the same regions in Germany where multiple levee failures occurred. ILPD includes data from 17 cases of levee failures occurred in Saxony–Anhalt during this event (Figure 8a). Return periods of discharges were estimated between 50–500 years depending on the location. In this flooding event, the most seen failure mechanism was instability. Contrary to 2002, water levels in 2013 were mostly close to the crests of the levees, hence only few overflow/overtopping cases have been observed. Due to the high peak discharge, two catastrophic levee failures occurred in the central part of the Elbe River, namely Breitenhagen and Fischbeck (Figure 8b,c). The first failure near Breitenhagen occurred due to instability of the landside slope which later resulted in a 150 m wide breach and inundated an area of 80 km^2 [44]. Simultaneously, the second failure near Fischbeck initiated with large cracks followed by settlement of the landside slope, resulting in a 100 m wide breach within hours. The main failure type is recorded as instability induced by internal erosion.

Although both the 2002 and 2013 events were large scale floods with severe consequences in history, the main differences between the two were as follows. In June 2013, heavy precipitation (total 170.5 mm within 24 h) in combination with high soil moisture levels, which in nearly 40% of Germany were at the highest levels since 1962 [45,46], resulted in levee breaches and flooding mainly in the central Elbe, Mulde and Saale catchments. Instead in 2002, extraordinary precipitation (record breaking rainfall of total 312 mm within 24 h) was the main driven mechanism [41]. Although hydrological conditions and flood levels were more severe compared to the 2002 event, the cost of damage in 2013 was much lower (6–8 billion euros) and fewer levee failures occurred. This was most likely the result of more effective flood management after the 2002 event, in particular more effective disaster management and improvements in maintenance of flood defences [41].

3.2.2. Analysis of the Failures

After these events, data was collected on levee failure cases, including location, geometry, levee structure, subsoil structure, vegetation, breach geometry and failure time. According to post-investigations [23,47], different degrees of damage have been observed on the levees. Based on the change in levee cross section, the breaches occurred in 2002 and 2013 are classified into three groups (Figure 9) as (a) partial failure (10%), where the breach depth was less than the crest height; (b) total failure (26%), when the crest was completely washed away; and (c) total failure with scour (41%), when, in addition to the crest, the soil beneath the toe level was also eroded. In most of the cases, the causes of levee failures were attributed to a combination of loads and local conditions (e.g., old breaches, tree roots, poor maintenance) [23].

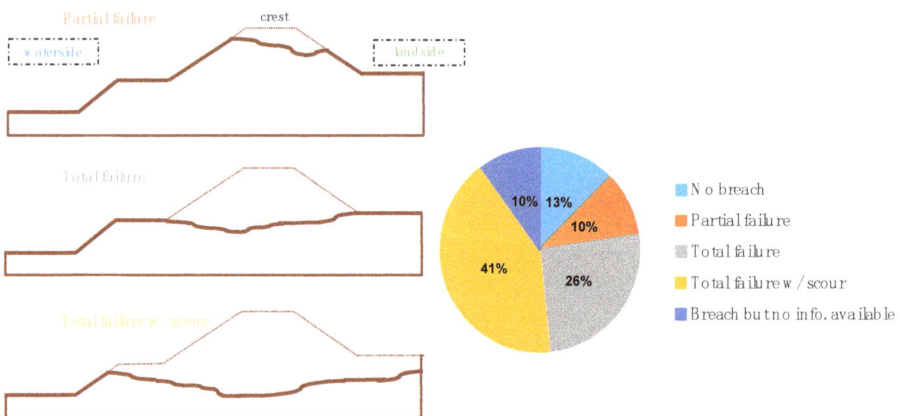

Figure 9. Damage degrees of the failed levees in Elbe region, Germany, 2002 and 2013 (adapted from [23]).

An overview of the main failure mechanisms is given in Table 2. In 2002, there were more external erosion due to the overtopping/overflow cases, whereas in 2013, internal erosion and instability of the landside slope were the most common failure mechanisms.

Table 2. Overview of the main failure mechanisms for the 2002 and 2013 events.

Main Failure Mechanisms	Number of Cases		
	2002 Event	**2013 Event**	**Total**
External Erosion	59 (53.2%)	5 (29.4%)	64 (50%)
Instability	26 (23.4%)	6 (35.3%)	32 (25%)
Internal Erosion	14 (12.6%)	5 (29.4%)	19 (14.8%)
Overflow/Overtopping	9 (8.1%)	1 (5.9%)	10 (7.8%)
Unknown	3 (2.7%)	0 (0%)	3 (2.4%)
Total	111	17	128

During the progress of these failures, multiple failure mechanisms were observed. In Figure 10, the occurrences of the different main failure mechanisms are associated with the initial mechanisms that have triggered them. Failure initiation mechanisms were mostly overtopping which was later followed by external erosion (58 cases), instability (21 cases) and internal erosion (7 cases). At some locations, overflow/overtopping (10 cases) occurred but did not lead to a levee breach. For example, four different levee breaches during the 2002 event have initiated with an instability of the landside slope which slipped away partially and the remaining part continued to be eroded gradually by overflowing water (external erosion).

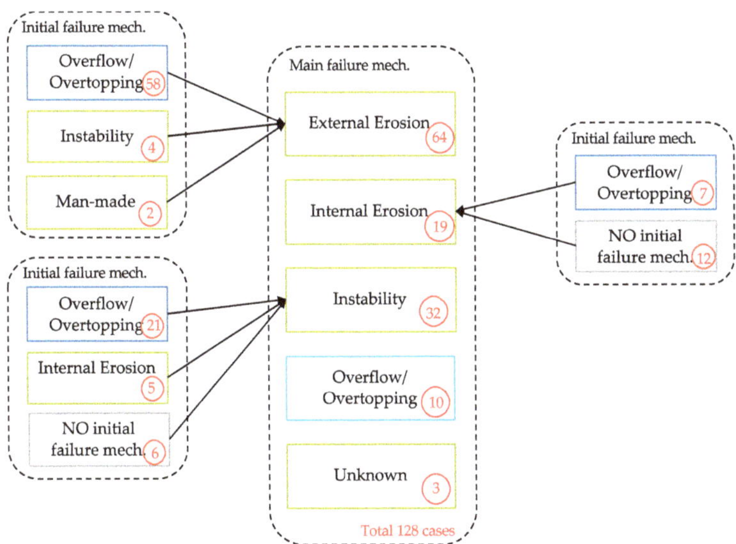

Figure 10. Initial and main failure mechanisms of the failed levees in the Elbe region, Germany, occurred in 2002 and 2013.

3.3. Levee Breach Analysis

Although various breach prediction models have been developed empirically [48–50] and physically [51–53], there is still limited insight in the characteristics of breaches during real events. In this section, it is demonstrated for the Elbe case how information from the ILPD can be used to analyse breach dimensions, including the relationship with the failure mechanisms. Another important knowledge gap concerns the number of breaches and their width (i.e., breach density) which can be expected during flood events.

3.3.1. Analysis of the Failures

Data collected on the 128 levee failures that occurred during these two flooding events have been used to analyse the breach parameters. Table 3 shows the total breach width and the average breach width per failure for each damage type, classified according to Figure 9. If the damage to the crest increases (from partial failure to total failure with scour), the average breach width per failure becomes also larger. Although width and depth of levee breaches are not necessarily linearly correlated (Figure 7c), this analysis shows that when the breach is deep, then it is more likely to be wide as well. This is also in line with some breach growth models (e.g., [26]) which predict a lateral and vertical erosion in the later stages of the breaching process.

Table 3. Breach width analysis per damage type, Germany 2002 and 2013.

Damage Type	Num. of Cases	Total Breach Width	Avg. Breach Width per Failure
Partial Failure	14	466 m	33 m
Total Failure	33	1740 m	53 m
Total Failure w/scour	52	3901 m	75 m

In order to further analyse the breach characteristics of the failures, the relation between breach parameters and failure mechanisms is assessed. First, the ratios between breach depth and crest level (i.e., relative breach depth) are compared with the breach width per main failure mechanisms (Figure 11a). It can be observed that, by normalizing the breach depth (as breach depth/crest level), the breach width is also larger for the failures with total scour. Moreover, the main failure mechanisms of most of the large breaches are observed as instability or internal erosion. Second, a comparison between breach surface (as breach depth × breach width) and geometry of the levees (i.e., cross section area) is given in Figure 11b per failure mechanisms. The levee surface could affect the breaching in different ways. A larger levee is associated with a larger hydraulic head (more forcing) but would also require more erosion during breaching (more resistance). Figure 11b shows that levees with large cross sections tend to have larger breach surfaces. Furthermore, the failures with large breaches (>500 m^2) are due to instability, with only a few failures caused by internal erosion. Variations in breach surface for similar levee geometry might be explained by the differences in hydraulic head conditions. Moreover, recent studies show that even if the hydraulic load is the same, the duration of the load is also an important parameter in development of breach surfaces [54].

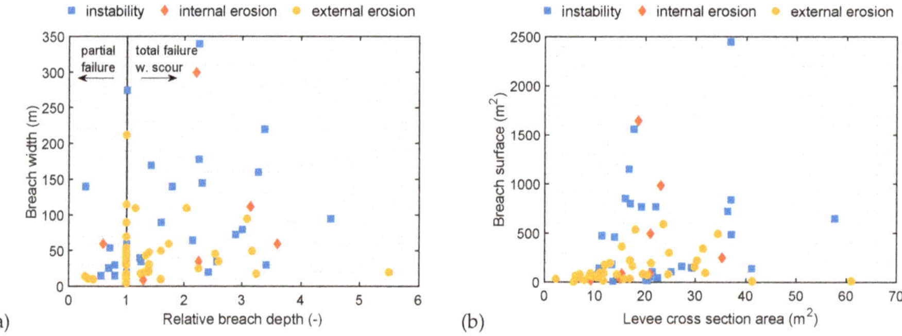

Figure 11. Relation between (**a**) relative breach depth and breach width per failure mechanism; (**b**) breach surface and geometry of the levees per failure mechanism.

The understanding of failure patterns could be further deepened by checking how the failure initiated. Thus, the following analysis of the levee failures considers both their initial and main mechanisms. In Figure 12, the total and average breach width per failure mechanisms are compared.

For instance, it can be seen that, if an external erosion case is initiated by instability, it is more likely to have a wider breach per failure than for an external erosion case that is triggered by overflow/overtopping. This can be explained as follows. When an instability occurs, most of the times a relatively large part of the crest is pushed away by internal forces with the following sequence of external erosion making the damage larger. Whereas, if the failure starts with overflow/overtopping which by time erodes the crest externally (external erosion), the levee is most likely to have smaller damages compared to the previous case. Another interesting observation is that some types of failure occur more often (e.g., external erosion triggered by overflow/overtopping) but with a relatively smaller average breach width per failure (i.e., smaller damage).

It is also noticed in Figure 12 that when an instability failure mechanism is initiated by internal erosion (e.g., piping), it is more likely to have wider breaches, whereas if it is initiated by overflow/overtopping, the size of the breach is smaller than the previous case but still larger than an instability failure without any initial mechanism. Figure 13 gives a schematization of these three cases. This trend can be explained by the fact that internal erosion, occurring directly in the subsoil, creates an extra hydraulic pressure below the crest. By time, the piping disconnects the upper part of the levee from its foundation, undermining the stability of a large section of the levee. In general, this analysis shows that the underlying failure mechanisms are of importance in the breach development.

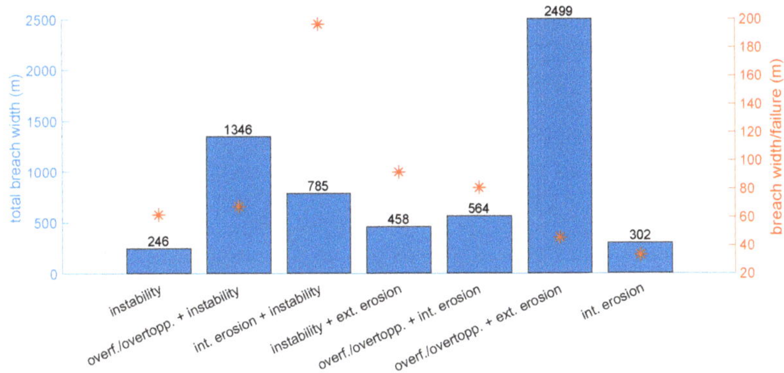

Figure 12. Breach width analysis per failure mechanism occurred during the flooding event in Germany, 2002 and 2013.

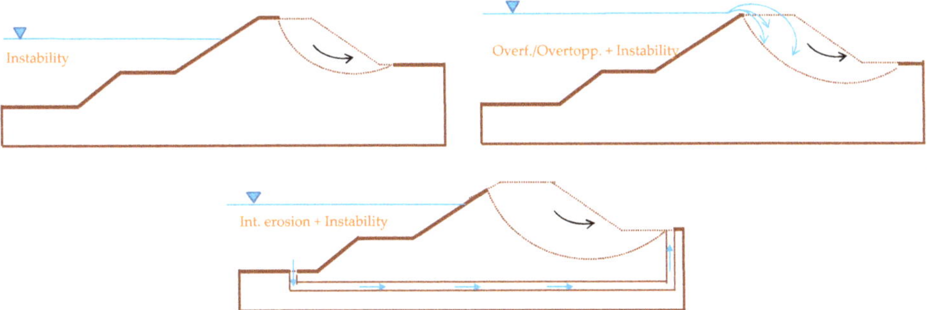

Figure 13. A schematization of an instability failure with three possible initiation mechanisms.

3.3.2. Breach Density Analysis

One of the important aspects to consider in flood risk assessments is the levee breach density, which is related to the expected amount of levee breaches and their width. Risk assessments often assume that a breach can occur at random locations and often focus on a single breach and/or consider multiple breaches less likely [55]. However, actual flood events often comprise of multiple failures and breaches. Thus, it would be beneficial to estimate the expected breach density in order to complement the risk assessment of levees. Previously, a study [56] had addressed the distance between breaches as a function of overtopping rate for the two coastal levees failed during the events of New Orleans 2005 and Denmark 1976. However, related analyses have not been performed for river flood events yet.

In order to analyse the breach density, we first define two parameters, namely *Failure intensity* (km^{-1}) and *Breach width ratio* (-), given in Equations (1) and (2) below,

$$Failure\ intensity\ =\ ^{\#\,of\,failures}/_{river\,length} \tag{1}$$

$$Breach\ width\ ratio\ =\ ^{total\,breach\,width}/_{river\,length} * 100 \tag{2}$$

where total breach width and river length are expressed in km. Failure intensity represents the occurrence of failure cases along the investigated river normalized by its length, whereas the breach width ratio corresponds to the ratio (in percentage) between the sum of all registered breach widths and the river length.

We analyse these parameters and compare with the return periods of discharge per each river for three flood events with multiple failures occurred in Germany in 2002 and 2013 (discussed previously) and also in Czech Republic in 1997, which are all stored in ILPD. The Czech event, also known in history as "The Great Flood of 1997", mainly affected the Oder and Morava basins with multiple levee failures, of which 27 have been recorded in the ILPD. For both countries the river levees were supposed to be designed for events with return periods of 100 years [57].

The calculated parameters for the 1997 and 2002 events range between 0.02 and 0.68 km^{-1} for the failure intensity and between 0.08% and 4.5% for the breach width ratio, as shown in Figure 14. For the 2013 event, the values of the calculated breach width ratio (between 0.1%–0.4%) and failure intensity (between 0.03–0.1 km^{-1}) are smaller than for the 2002 event, which can be explained by the strengthening of the levees after 2002, as explained in Section 3.2. However, although the conditions during the 2013 event was more severe, return periods found in the literature were either assessed with smaller values than for the 2002 event [58] or defined too general, that is ">100 years" [59,60]. Thus, the values for the 2013 event were not included in the rest of the analysis.

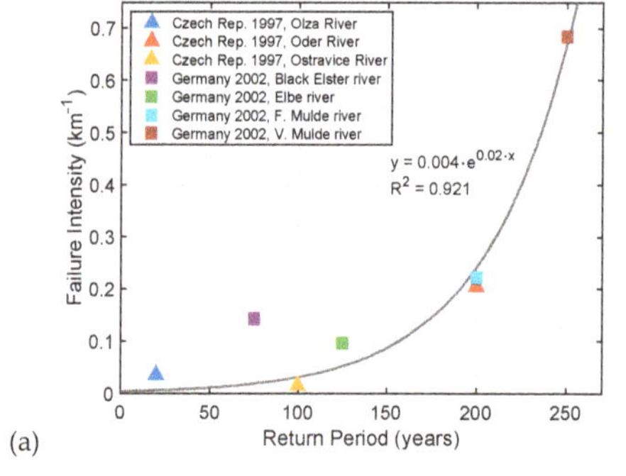

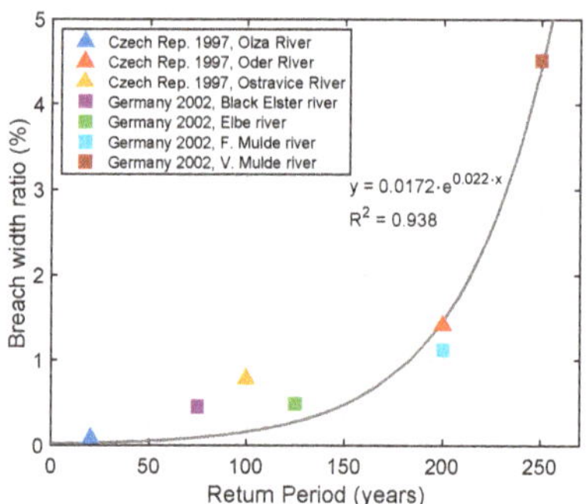

Figure 14. (**a**) Failure intensity (km^{-1}) and (**b**) Breach width ratio (-) along the rivers investigated during Germany, 2002 and Czech Republic, 1997 events.

Considering the events of Germany 2002 and Czech Republic, it is noticed that when the return period (T) increases, the failure intensity and the breach width ratio also increase. For the seven rivers considered, a non-linear regression analysis using an exponential fitting function has been performed on the available data for both breach density parameters (trust-region optimization algorithm available in the MATLAB *fit* function). The resulting functional relations and the corresponding R^2 values are given in Table 4. For instance, considering a river stretch of 50 km with a 125-year event, one would expect approximately five failures with a total width of 300 m according to these relations. Whereas, a 250-year event would lead to 22 failures with a total width of 1700 m for the same stretch. The functional relations given below are obtained from a limited amount of data. If more information regarding cases from different events will become available in the future, it can be used to validate the relations found and to refine the analysis.

Table 4. Regression functions and R^2-values for the breach density parameters and return periods.

Breach Density Parameter	Regression Function	R^2 Values
Failure intensity (km^{-1})	*Failure intensity* $= 0.004 \times e^{(0.02 \times T)}$	0.921
Breach width ratio (-)	*Breach width ratio* $= 0.0172 \times e^{(0.022 \times T)}$	0.938

For the same events investigated above, we also explore the occurrences of different failure mechanisms and the degree of damage for different return periods (Figure 15a). In general, the total number of failures increases with the change in the return period. However, as it is given the Figure, external erosion is more likely to occur for high return periods. This can be explained by the fact that a high return period is related to high water levels and river discharges, which in turn leads to overtopping/overflow followed by an external erosion. Moreover, it is shown in Figure 15b that higher return periods lead to larger degrees of damage on the levees.

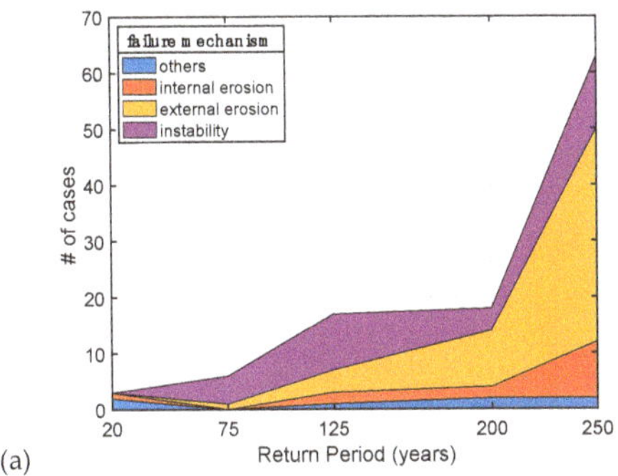

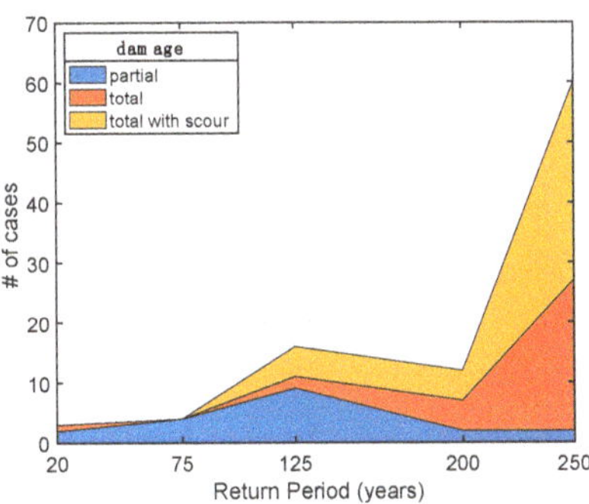

Figure 15. Distribution of (**a**) failure mechanisms and (**b**) degree of damage for different return periods (*T*) along the rivers investigated during Germany, 2002 and Czech Republic, 1997 events.

4. Discussion

4.1. ILPD

The ILPD is expected to become a global platform and scientific tool for various purposes, such as to advance the understanding of failure mechanisms and breaching of flood defences and to improve model development and validation. The first available version of the ILPD presented in this paper provides good coverage of some regions, for example, the Netherlands for the 20th century and it is already being used to support detailed studies, such as (1) geotechnical analysis of individual failures [44] and (2) detailed breach analysis [61]. In all other cases with only generic data, entries could still serve as a starting point for researchers to collect more detailed information. It is foreseen that the future expansion of the database will further offer valuable information to the scientific community as well as to the public and private sectors. More extensive datasets will give new insights into the field of flood risk and will stimulate the development and validation of more accurate techniques and modelling tools. This could eventually contribute to improving design methods of flood defences and to supporting risk assessments related to levee safety.

However, some operational obstacles must be overcome in order for the ILPD to become broadly applicable and representative. One of the main obstacles that we encountered is the issue of data sharing. It is believed that most of the detailed information is kept as an internal source since making the data public is a sensitive matter in many cases. A possible reason is that levee authorities and governments in many countries prefer not to advertise events whereby their levees failed. For instance, countries with few reported failure cases (e.g., China, Italy) in Figure 4 are not necessarily safer than the others (e.g., the Netherlands or Hungary) since many failures are known through media or registered information (e.g., EU documentation). This in fact highlights the importance of being transparent about data sharing and collaborations.

Another difficulty that arises while obtaining data is the language barrier. Across the world, detailed reports of flood events are obviously written in the official language of the country (e.g., Japanese). Especially when it concerns large quantities of data, extracting information from these reports for the ILPD becomes challenging. A way to potentially counteract this issue would be, for example, to request experts of international committees (e.g., ICOLD) to enter data related to their country into the database. However, this poses a new challenge, namely a higher demand on maintaining the database and providing support. Although there are some standard and ILPD-specific definitions for failure mechanisms and other parameters, people tend to use their own terminology, thus generating inconsistencies between cases. This is the reason why only the project members are currently allowed to enter data in the website. Thus, improving the ILPD towards a global, uniform database would require very systematic and intense data collection, also with the involvement and commitment of local stakeholders and levee managers.

4.2. Using Event-Level Analysis for Risk Assessments

Even though flood events and defence systems have specific characteristics and conditions, insights from the macro-scale analysis of the historical failures can be used to complement the (local) flood risk assessments. Firstly, the analysis of actual events could serve as calibration or reference for risk evaluations by highlighting dominant failure mechanisms and breach characteristics. For instance, as concluded in this study for the Elbe region, although some mechanisms occur more often, the damage that they cause on the levee can be less compared to the ones that has less occurrence. Likewise, it has been shown that it is important to consider not only the final failure mechanism that cause a breach but also the underlying processes that initiate the failure. These observations can be used by the local authorities as a starting point in assessing specific situations.

Another point is that current risk assessments generally tend to pay limited attention to failure scenarios with multiple breaches [55]. It is normally assumed in river levees that the occurrence of one breach reduces the expected likelihood of other breaches, as the inflow through the breach limits water levels downstream (i.e., retention effect). However, observations from the river floods studied here show that multiple breaches do occur during actual river floods, particularly when the system is overloaded by "design floods". Thus, the observed breach densities and their identified relation with the return periods could be used, for instance in combination with fragility curves, to make more informed (simplified) risk assessments. Although in this study a general overview on the breach density parameters has been given, further research on this topic is recommended. A probabilistic analysis could be included by updating the failure probabilities at the failure locations based on the local information affecting the strength of the levee, such as vegetation type, old breaches, changes in soil profile.

5. Conclusions

In this paper, we introduced the International Levee Performance Database (ILPD), aiming to create a global information-sharing platform to facilitate research on levee performances. Besides, we conducted a macro-scale analysis of the currently available data. We outline common failure mechanisms of which external erosion of the slope is identified as the most frequent for levees and internal erosion for earthen dams. As an illustrative use of an ILPD sub-set, we examined breach characteristics of over a hundred failures during the flood events occurred in Germany (2002, 2013). Based on this analysis, we identified potential linkages between initial failure mechanisms, main failure mechanisms and the eventual breach characteristics. For instance, it is concluded that initial failure mechanisms play an important role in defining breach characteristics and that failures due to instability and internal erosion are less frequent but lead to a larger breach size. Based on events with multiple failures included in the database, we also identified a relation between the return period and the expected breach density during a flood event. Such relation can be improved and validated on cases from different events that will become available in the future.

The ILPD currently contains over 1500 entries covering historical failures, experiments and other performance observations. Even though we focused on the analysis of data at an event-level, ILPD sub-datasets can be used for more detailed analyses of individual failure processes, for instance, to investigate how occurrence of failure mechanisms is related to levee and loading characteristics or to analyse breach properties in more detail. Further research can also focus on combining information on levee failures from ILPD with information on historical flood levels and floodplain development [62]. Thus, the aim of the database for the future is to provide extensive and high-quality datasets to support the development and validation of accurate methods and models for failure mechanisms and breaching of flood defences. This is currently only restricted by the limited amount of data shared in the database, which is why a joint effort of the international scientific community, private companies and governments is required to make the ILPD complete and representative. The insights provided by the analysis of the historical flood events contained in the ILPD, in combination with hydraulic/geotechnical models, could eventually be used to complement risk assessments and to design more robust and resilient flood defences, with a smaller likelihood of catastrophic breaching.

Author Contributions: Conceptualization, I.E.Ö. and S.N.J.; methodology, I.E.Ö.; formal analysis, I.E.Ö.; data curation, I.E.Ö. and M.v.D.; writing—original draft preparation, I.E.Ö.; writing—review and editing, I.E.Ö., M.v.D. and S.N.J.; visualization, I.E.Ö.; supervision, M.v.D. and S.N.J. All authors have read and agreed to the published version of the manuscript.

Funding: This research was funded by Nederlandse Organisatie voor Wetenschappelijk Onderzoek (NWO TTW), grant number 13861.

Acknowledgments: We would like to thank to Torsten Heyer and Niklas Drews for providing data on the levee failures during Germany 2002 and 2013 flooding events; to Jozsef Danka for sharing his database; and to Wim Kanning for his valuable comments. Job Kool and Marijn de Rooij are also acknowledged for their contributions on the development of the ILPD.

References

1. EEA. European Past Floods. Available online: https://www.eea.europa.eu/data-and-maps/data/european-past-floods/ (accessed on 29 October 2019).
2. Özer, I.E.; van Leijen, F.J.; Jonkman, S.N.; Hanssen, R.F. Applicability of satellite radar imaging to monitor the conditions of levees. *J. Flood Risk Manag.* **2018**. [CrossRef]
3. Schweckendiek, T.; Vrouwenvelder, A.C.W.M.; Calle, E.O.F. Updating piping reliability with field performance observations. *Struct. Saf.* **2014**, *47*, 13–23. [CrossRef]
4. Özer, I.E.; van Damme, M.; Schweckendiek, T.; Jonkman, S.N. On the importance of analyzing flood defense failures. In Proceedings of the Flood Risk 2016, 3rd European Conference on Flood Risk Management, Lyon, France, 17–21 October 2016; E3S Web of Conferences.
5. NPDP. National Performance of Dams Program. Available online: http://npdp.stanford.edu/data_library (accessed on 1 November 2019).
6. Fry, J.J.; Le Bourget du Lac, F.; Courivaud, J.R.; Blais, J.P. Dam Accident Data Base DADB–The Web Based Data Collection of ICOLD. In Proceedings of the 13th Conference on British Dam Society and the ICOLD European Club Meeting, Canterbury, UK, 22–26 June 2004.

7.   Froehlich, D.C. Embankment dam breach parameters. In Proceedings of the Hydraulic Engineering 1987 National Conference, Williamsburg, WA, USA, 3–7 August 1987. ASCE.

8.   Xu, Y. Analysis of Dam Failures and Diagnosis of Distresses for Dam Rehabilitation. Ph.D. Thesis, Hong Kong University of Science and Technology, Hong Kong, China, 2010.

9.   Peng, M.; Zhang, L.M. Breaching parameters of landslide dams. *Landslides* **2012**, *9*, 13–31. [CrossRef]

10.   Foster, M.; Fell, R.; Spannagle, M. The statistics of embankment dam failures and accidents. *Can. Geotech. J.* **2000**, *37*, 1000–1024. [CrossRef]

11.   Fry, J.J. Dam failures by erosion: Lessons from ERINOH database. In Proceedings of the 6th International Conference on Scour and Erosion (ICSE-6), Paris, France, 27–31 August 2012; Société Hydrotechnique De France.

12.   ASDSO. Lessons Learned from Dam Incidents and Failures. Available online: http://damfailures.org/ (accessed on 10 November 2019).

13.   McClelland, D.M.; Bowles, D.S. *Estimating Life Loss for Dam Safety Risk Assessment, A Review and New Approach*; Institute for Dam Safety Risk Management, Utah State University: Logan, UT, USA, 2002.

14.   USBR. *Reclamation Consequence Estimating Methodology, Dam Failure and Flood Event Case History Compilation*; Technical Report; U.S. Department of the Interior Bureau of Reclamation: Washington, DC, USA, 2015.

15.   CRED. The International Disaster Database (EM–DAT). Available online: https://www.emdat.be/ (accessed on 30 May 2019).

16.   DFO. Dartmouth Flood Observatory of Large Floods. Available online: http://floodobservatory.colorado.edu/Archives/index.html (accessed on 30 May 2019).

17.   Scussolini, P.; Aerts, J.C.; Jongman, B.; Bouwer, L.M.; Winsemius, H.C.; de Moel, H.; Ward, P.J. FLOPROS: An evolving global database of flood protection standards. *Nat. Hazards Earth Syst. Sci.* **2016**, *16*, 1049–1061. [CrossRef]

18.   DWR. DWR Levee Breach Database. Available online: http://www.dwr-lep.com (accessed on 30 May 2019).

19.   Barbetta, S.; Camici, S.; Maccioni, P.; Moramarco, T. National Levee Database: Monitoring, vulnerability assessment and management in Italy. In Proceedings of the EGU General Assembly, Vienna, Austria, 12–17 April 2015.

20.   USACE. National Levee Database. Available online: http://nld.usace.army.mil/ (accessed on 30 April 2019).

21.   Seed, R.B.; Bea, R.G.; Athanasopoulos-Zekkos, A.; Boutwell, G.P.; Bray, J.D.; Cheung, C.; Cobos-Roa, D.; Harder, L.F., Jr.; Moss, R.E.; Pestana, J.M.; et al. New Orleans and hurricane Katrina. III: The 17th street drainage canal. *J. Geotech. Geoenviron. Eng.* **2008**, *134*, 740–761.

22.   Seed, R.B.; Bea, R.G.; Athanasopoulos-Zekkos, A.; Boutwell, G.P.; Bray, J.D.; Cheung, C.; Cobos-Roa, D.; Ehrensing, L.; Harder, L.F., Jr.; Pestana, J.M.; et al. New Orleans and Hurricane Katrina. II: The central region and the lower Ninth Ward. *J. Geotech. Geoenviron. Eng.* **2008**, *134*, 718–739.

23.   Horlacher, H.B.; Heyer, T.; Carstensen, D.; Bielagk, U.; Bielitz, E.; Müller, U. Analysis of dyke breaks during the 2002 flood in Saxony/Germany. In Proceedings of the LARS 2007–Catchment and Lake Research, Arba Minch, Ethiopia, 7–11 May 2007. FWU Water Resources Publications.

24.   Van Baars, S.; van Kempen, I.M. The causes and mechanisms of historical dike failures in the Netherlands. *e-Water J.* **2009**, *2009*, 3–14.

25.   Danka, J. Dike Failure Mechanisms and Breaching Parameters. Ph.D. Thesis, Hong Kong University of Science and Technology, Hong Kong, China, 2015.

26.   Visser, P. Breach Growth in Sand-Dikes. Ph.D. Thesis, Delft University of Technology, Delft, The Netherlands, 1998.

27.   Sills, G.L.; Vroman, N.D.; Wahl, R.E.; Schwanz, N.T. Overview of New Orleans levee failures: Lessons learned and their impact on national levee design and assessment. *J. Geotech. Geoenviron.* **2008**, *134*, 556–565. [CrossRef]

28.   Vinet, F.; Lumbroso, D.; Defossez, S.; Boissier, L. A comparative analysis of the loss of life during two recent floods in France: The sea surge caused by the storm Xynthia and the flash flood in Var. *Nat. Hazards* **2012**, *61*, 1179–1201.

29.   Bisschop, F. Erosion of Sand at High Flow Velocities. Ph.D. Thesis, Delft University of Technology, Delft, The Netherlands, 2018.

30.   Jonkman, S.N.; van den Bos, J.P. *Flood Defenses Lecture Notes CIE5314*, 2nd ed.; Delft University of Technology: Delft, The Netherlands, 2017.

31. Sharp, M.; Wallis, M.; Deniaud, F.; Hersch-Burdick, R.; Tourment, R.; Matheu, E.; Seda-Sanabria, Y.; Wersching, S.; Veylon, G.; Durand, E. *The International Levee Handbook*; CIRIA: London, UK, 2013.

32. Kok, M.; Jongejan, R.; Nieuwjaar, M.; Tanczos, I. *Grondslagen voor Hoogwaterbescherming*; Ministerie van Infrastructuur en Milieu en het Expertise Netwerk Waterveiligheid: Delft, The Netherlands, 2016. (In Dutch)

33. Kanning, W. The Weakest Link: Spatial Variability in the Piping Failure Mechanism of Dikes. Ph.D. Thesis, Delft University of Technology, Delft, The Netherlands, 2012.

34. TAW. *Water Retaining Soil Structures*; Technical Report; Rijkswaterstraat—Technical Advisory Committee on Water Defenses: Utrecht, The Netherlands, 1999.

35. Van Damme, M.; Ozer, I.E.; Kool, J.J. *Guidance of the International Levee Performance Database (ILPD)*; Technical Report; Delft University of Technology: Delft, The Netherlands, 2019.

36. Bridle, R. ICOLD Bulletin 164: Internal erosion of existing dams, levees and dikes and their foundations. In Proceedings of the Bulletin 164: International Comission on Large Dams, London, UK, 15 May 2014.

37. Nagy, L. Estimating dike breach length from historical data. *Period. Polytech. Civ. Eng.* **2006**, *50*, 125–138.

38. Morris, M.W. Breaching of Earth Embankments and Dams. Ph.D. Thesis, The Open University, Milton Keynes, UK, 2011.

39. Horlacher, H.B.; Bielagk, U.; Heyer, T. Analyse der deichbrüche an der elbe und mulde während des hochwassers 2002 im bereich sachsen. *Res. Rep.* **2005**, *9*, 82. (In German)

40. Thieken, A.H.; Müller, M.; Kreibich, H.; Merz, B. Flood damage and influencing factors: New insights from the August 2002 flood in Germany. *Water res.* **2005**, *41*. [CrossRef]

41. Thieken, A.H.; Kienzler, S.; Kreibich, H.; Kuhlicke, C.; Kunz, M.; Mühr, B.; Müller, M.; Otto, A.; Petrow, T.; Pisi, S.; et al. Review of the flood risk management system in Germany after the major flood in 2013. *Ecol. Soc.* **2016**, *21*, 51.

42. Jüpner, R.; Henning, B. Deichbruch fischbeck–zwei jahre danach. *Wasser und Abfall* **2015**, *11*, 16–20.

43. Weichel, T. *Failure of the Breitenhagen Levee 2013–Drone Film*; Landesbetrieb füe Hochwasserschutz und Wasserwirtschaft Sachsen-Anhalt: Breitenhagen, Germany, 2013.

44. Kool, J.J.; Kanning, W.; Heyer, T.; Jommi, C.; Jonkman, S.N. Forensic analysis of levee failures: The Breitenhagen case. *Int. J. Geoeng. Case Hist.* **2019**, *5*. [CrossRef]

45. Für Gewässerkunde, B. *Das Juni-Hochwasser des Jahres 2013 in Deutschland*; BfG Bericht: Koblenz, Germany, 2013.

46. Jüpner, R. Coping with extremes–experiences from event management during the recent Elbe flood disaster in 2013. *J. Flood Risk Manag.* **2018**, *11*, 15–21. [CrossRef]

47. Gocht, M. *Deichbrüche und Deichüberströmungen an Elbe und Mulde im August 2002*; Technical Report; Ökonomische Beratung für Wasser und Umwelt, Water and Finance: Berlin, Germany, 2004.

48. Wahl, T.L. Uncertainty of predictiona of embankment dam breach parameters. *J. Hydraul. Eng.* **2004**, *130*, 389–397. [CrossRef]

49. Froehlich, D.C. Embankment dam breach parameters and their uncertainties. *J. Hydraul. Eng.* **2008**, *134*, 1708–1721.

50. Kakinuma, T.; Shimizu, Y. Large scale experiment and numerical modeling of a riverine levee breach. *J. Hydraul. Eng.* **2014**, *140*. [CrossRef]

51. Temple, D.M.; Hanson, G.J.; Neilsen, M.L.; Cook, K.R. Simplified breach analysis model for homogeneous embankments: Part 1, Background and model components. In Proceedings of the 25th Annual United States Society on Dams (USSD) Conference, Salt Lake City, UT, USA, 6–10 June 2005.

52. Morris, M.W. *Breach Initiation and Growth: Physical Processes (FLOODsite Report T06-08-11)*; Technical Report; HR Wallingford: Wallingford, UK, 2009.

53. Michelazzo, G. Breaching of River Levees: Analytical Flow Modelling and Experimental Hydro-Morphodynamic Investigations. Ph.D. Thesis, University of Braunschweig, Braunschweig, Germany, 2014.

54. Curran, A.; De Bruijn, K.M.; Kok, M. Influence of water level duration on dike breach triggering, focusing on system behaviour hazard analyses in lowland rivers. *Georisk Assess. Manag. Risk Eng. Syst. Geohazards* **2018**, 1–15. [CrossRef]

55. Jonkman, S.N.; Vrijling, J.K.; Kok, M. Flood risk assessment in the Netherlands: A case study for dike ring South Holland. *Risk Anal.* **2008**, *28*, 1357–1373. [CrossRef]

56. Bernitt, L.; Lynett, P. Breaching of sea dikes. In Proceedings of the 32nd Conference on Coastal Engineering (ICCE No 32), Shanghai, China, 30 June–5 July 2010.

57. Riha, J.; Pohl, R.; Escuder, I.; Martinez, F.J.; Laasonen, J.; Isomaki, E.; Tourment, R.; Maurin, J.; Mallet, T.; Deniaud, Y.; et al. *European Levees and Flood Defenses, Inventory of Characteristics, Risks and Governance (EURCOLD LFD-WG Meeting 2016)*; Technical Report; ICOLD: Lyon, France, 2016.

58. LHW. *Bericht über das Hochwasser im Juni 2013 in Sachsen-Anhalt: Entstehung, Ablauf, Management und Statistische Einordnung*; Technical Report; Landesbetrieb für Hochwasserschutz und Wasserwirtschaft Sachsen-Anhalt (LHW): Magdeburg, Germany, 2014. (In German)

59. Schröter, K.; Kunz, M.; Elmer, F.; Mühr, B.; Merz, B. What made the June 2013 flood in Germany an exceptional event? A hydro-meteorological evaluation. *Hydrol. Earth Syst. Sci.* **2015**, *19*, 309–327. [CrossRef]

60. Thieken, A.H.; Bessel, T.; Kienzler, S.; Kreibich, H.; Müller, M.; Pisi, S.; Schröter, K. The flood of June 2013 in Germany: How much do we know about its impacts. *Nat. Hazards Earth Syst. Sci.* **2016**, *16*, 1519–1540. [CrossRef]

61. Van Damme, M. An analytical approach to predicting breach width and breach hydrographs. *Nat. Hazards* **2019**. under review.

62. Kilianova, H.; Pechanec, V.; Brus, J.; Kirchner, K.; Machar, I. Analysis of the development of land use in the Morava River floodplain, with special emphasis on the landscape matrix. *Morav. Geogr. Rep.* **2017**, *25*, 46–59. [CrossRef]

Article

Vulnerability of Transport Networks to Multi-Scenario Flooding and Optimum Location of Emergency Management Centers

Alfredo Pérez-Morales [1,†], Francisco Gomariz-Castillo [1,2,*,†] and Pablo Pardo-Zaragoza [1,†]

[1]   Department of Geography, University of Murcia, 30100 Murcia, Spain; alfredop@um.es (A.P.-M.); pardozaragoza@gmail.com (P.P.-Z.)
[2]   Euro-mediterranean Water Institute (IEA), 30100 Murcia, Spain
*   Correspondence: fjgomariz@um.es; Tel.: +34-968-899-851
†   These authors contributed equally to this work.

Received: 24 April 2019; Accepted: 5 June 2019; Published: 8 June 2019

Abstract: Floods are the climatic factors that cause more significant impacts on transportation infrastructures. This circumstance could get worse, taking into account climate change effects. The literature points out different adaptation measures to minimize the possible increasing effects caused by climate change. Among them is the improvement of the vulnerability of a transport network and Emergency Management Systems. The effective management of emergencies is of vital importance to minimize the potential damage resulting from a catastrophe. Given such circumstances, analysis of the vulnerability of networks is a technique whose results highlight deficiencies and serve as support for future decisions concerning the transformation of the network or the installation of new emergency centers. The main objective of this research is to highlight the vulnerability of the road network in a variety of multi-contingency scenarios related to flooding and to identify the optimal location for a new emergency management center based on that analysis. The results obtained could be used in urban planning tasks to improve the resilience of urban areas in the face of an increase in flood episodes caused by climate change.

Keywords: flood risk assessments; vulnerability of networks; emergency management; geographic information systems; open source

---

1. Introduction

The quality of transportation planning and management is critical for the functioning of a city, and issues of urban climate change adaptation and mitigation require attention. The list of impacts is extensive, covering structural and material damages of many different types and associated disruption of transportation services for users depending on the type of transportation facility, its location relative to waterways, and the types of materials and design used [1,2].

According to [3], there are two fundamental options for risk management in the transportation sector. One is by mitigation measures in cities related to greenhouse emissions (transport system as a threat of climate). The second option is reducing risk using adaptation (climate change as a threat to the transportation system). The vulnerability of transportation assets to physical hazards, stemming from the existing spatial and economic organization of the city and its transport system, depends on a range of factors. Among those, the following can be suggested: (1) design and spatial layout of transport infrastructure; (2) basic urban form—high-density settlements with mass transit systems, or mostly motorized transportation with low densities; (3) availability of resources to keep transportation systems functioning in both disaster and non-disaster conditions. However, carrying out this type of task in previously occupied spaces is a very difficult task and, therefore, there is no alternative but to concentrate efforts on structural actions to protect the transport network or the improvement of emergency management as a measure of reducing vulnerability and increasing resistance.

Some techniques, such as the analysis of the vulnerability of transport networks [4], have acquired particular attention. According to [5], the vulnerability of a network refers to the capacity to continue operating following a disruption. In other words, the degree of susceptibility of a network to specific incidents that may lead to reduced service or accessibility levels.

In the case of floods, analysis of the vulnerability of the network status is a complex task determined by how the hazard takes place. This type of analysis mainly requires performing specific simulations that show those sections of roads that are most sensitive to being interrupted or even isolated. The number of simulations depends on the combination of floods in different channels crossed by the roads. When the obtained interruption scenarios are treated in a synthesized way, they provide a final map of the network's vulnerability to flooding. This cartography facilitates identification of the most affected sectors, and according to [6], this type of study is among those known as a *full scan approach*. Some examples can be seen in the literature: For Australia [7] and Hong Kong [8].

Once the most common disruptions are identified, strategies focus on how to solve those connection problems in order to improve accessibility in the face of any event. There are different ways to do this [9]. On the one hand, there are those that increase the network connections, so that the effects of interruption will be less severe during flooding since alternative routes can be taken. Some examples are given in [10–15]. However, such *densification* or redundancy in the network is not always possible, nor always the most appropriate. Newly built-up areas increase the demands on emergency services, leading to the need for alternative solutions. In light of the above, the possibility arises of relocating existing emergency management centers or of increasing their number in places where the distance from existing centers exacerbates their isolation. In other words, rather than increasing the number of network sections, a commonly called location-allocation analysis of emergency management centers is usually developed in order to reduce the response time in areas where help is needed [16–19]. To identify possible new sites through an objective decision-making process, a map of the vulnerability of a network to flooding is needed.

Various methodologies have been followed in the analysis of vulnerability of the transport network to several natural and human hazards; among them are: complex transport and networks theory [13,20–22], statistical confidence [23] and, perhaps the most widely used, applied GIS graph theory [5,12,24]. The present study uses a combination of GIS and graph theory. Based on the principles of graph theory, the network is understood as a set of nodes or vertices interconnected by arcs or sections [25]. Accordingly, this is useful to verify the number of sections and nodes that remain connected to a network after been flooded. It also helps to recognize the elements that are critical for improving safety. As regards location allocation methodologies based on GIS a, there is an extensive literature on this type of methodologies [26–28]. This work has as a substantial novelty, which is the consideration of vulnerability to transport networks as a determining factor for the choice of locations of new centers of emergency management.

The aim of the present study is to propose and evaluate a simple method to analyze the vulnerability of networks in the face of multiple flood scenarios and obtain the optimal location of a new (Proposed) Emergency Management Center (PEC) to minimize those losses of accessibility that can not be resolved by an increase in network density. The proposed method has been evaluated in the urban and suburban setting of Murcia, an area with severe flooding problems.

2. Description of the GIS Approach to Evaluate Vulnerability and Obtain an Optimum Location of PEC

The work process proposed in this paper (described in the following subsections) has been implemented using Open Source software. To store and analyze vectorial data, the research used Relational Database Management System (DBMS) PostgreSQL 9.5 with PostGIS 2.3 and PgRouting extensions. PostGIS [29] is an extension to PostgreSQL which adds spatial capabilities by adding spatial data types, functions, operators, and indices for spatial data handling. On the other hand, PgRouting adds network analysis functionalities to PostgreSQL / PostGIS databases. Finally, the Quantum GIS (QGIS) program and Python scripting language were used to automate the tasks involved in the processing and management of spatial information.

Lastly, the statistical analysis of the results was performed with R-CRAN [30]. The advantage of R-CRAN is that it includes a large number of advanced analysis functionalities, such as the robust version of one-way ANOVA used in this paper to identify the optimal location.

2.1. Design Workflow

Figure 1 presents a conceptual scheme that summarizes the work process. Three work blocks were established. The first block corresponds to an analysis of the vulnerability of the road network according to the accessibility of different points from existing emergency management centers. The block is subdivided into four: (A) data gathering and filtering; (B) preprocessing of the information; (C) costs matrices; (D) Origin-Destination Matrices (ODM) and cartographic results. After establishing the most difficult to reach zones, the second work block (E) is devoted to developing a method to improve the problems related with sites that show vulnerability by selecting a new site for a PEC. The third block defines the process of validation and evaluation of the results, which includes expert evaluation and the decision-making bodies. This section includes the proposal of decisions to be made and the final location of the PEC.

The main concepts described in the conceptual scheme (Figure 1) and in the rest of the subsections are:

- Network: The road network consists of a system of interconnected carriageways which are designed to carry road vehicles.
- Service area: A network service area is a region that encompasses all accessible streets (that is, streets that are reached by an emergency center first).
- PEC: New (Proposed) Emergency Management Center (PEC).
- Cost Matrix: The OD cost matrix finds and measures the least-cost paths along with the network from multiple origins to multiple destinations.
- Flood scenarios: Each of the possible combinations without repetition of the flood zones.
- Network scenarios: Each one of the simulations made in the displacements from the emergency centers to each node of the network, considering the interruptions that may occur due to the flood scenarios.

2.2. Block I: Obtain Current Centers

The work process is initialized with the introduction and filtering of the required information: (a) Network layer and (b) flood polygons. As an example, in the present study, the physical aspect -hazard- is characterized using spatial information from the National Cartographic System of Flood-prone Zones (SNCZI in its Spanish initials) [31], the most precise open source on flood-prone areas and water depth with different return periods. It is available in raster form (0.5 × 0.5 m) and was drawn up following the principles of Directive 2007/60 [32] on the evaluation and management of flood risks. The raster associated with a return period of 100 years (RP100) was chosen for this study as flood-prone area and water depth; Return period of 100 years refer to the amount of time that passes on average (100 years) between consecutive events of similar magnitude. RP100 was chosen for this study, among other reasons because it ensures the variability and representativeness of the results, especially as it is a source that facilitates reproduction in other contexts because of its spatial cover level.

In this case, the human factor is represented by the road transport network. The researchers used the Cartociudad project [33], although it is also common to use Open Street Map [34,35]. For Spain, the Cartociudad method is more precise because it draws on information provided by official sources of the National Geographic Institute (IGN), The Land Registry (DGC), National Institute of Statistics (INE) and the Post Office (all abbreviations in Spanish). This cartography of the Spanish road network breaks down the information into road types. The categories considered in the study area are summarized in Table 1.

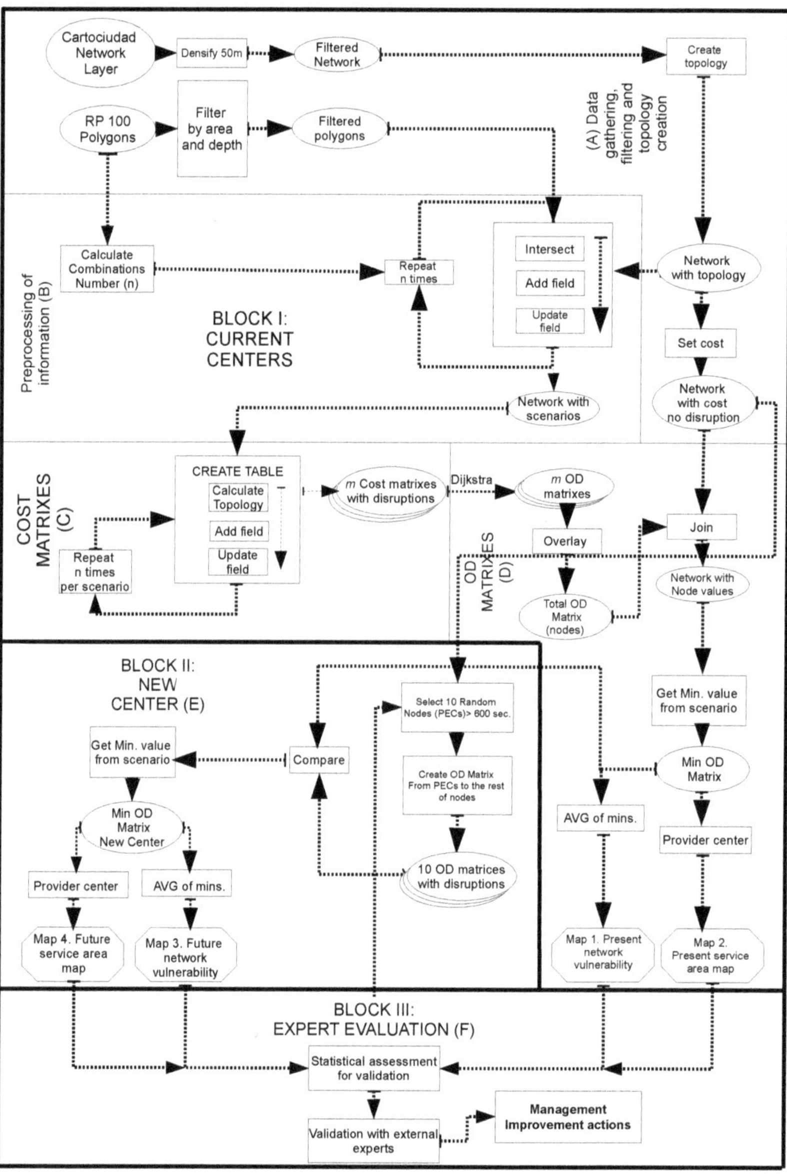

Figure 1. Conceptual scheme of the work carried out. Flow line connectors show the direction that the process flows; Rectangles show spatial instructions or actions; Ellipses show partial results, and are the input information of the next process; as a result, octagons show final maps.

Table 1. Categories of roads considered in the study area.

Road Type	Length (km)
Primary road	345.3
Secondary road	114.2
Tertiary road	147.2
Street	1097.4
TOTAL	1704.1

2.2.1. (A) Data Gathering, Filtering and Topology Creation

Once the information requirements have been identified, the first step in (A) is to filter the vector files of the road network (using PostgreSQL) and flood-prone areas. In the first case, despite the excellent quality of this type of file in areas such as those selected in the case study, two problems commonly arise:

- The first problem relates to the length of the vectors that usually configure the road layer. The lines corresponding to primary roads are usually too long, which diminishes the spatial resolution and impoverishes the cartography results. For this reason, the network must be segmented into stretches of 50 m.
- This type of network cartography usually lacks topological structure, while its generation usually involves adding spatial information to the road network, such as the relations existing between the different elements. To do this, the researchers used PgRouting as network manager. PgRouting is an extension for PostGIS which allows analyses based on topological networks using SQL. By means of the PgRouting function *pgr_createTopology* we attach topology to the network. With this process, a table of roads with the created topology is obtained and another of nodes (Network with topology in Figure 1), both used as entry parameters for the vulnerability network analysis.

Once the above is solved, the researchers calculate and assign a cost to each section. The concept of cost in this type of study implies that using the road network is not free and is calculated by reference to the cost unit adopted. Several types of cost can be distinguished (financial, political, etc.), the most common being the time needed to move from the point of origin to a destination [9]. In the present work, we use the time lost in seconds to pass road network section according to the following equation:

$$C_t = \frac{L}{v \cdot (100/360)} \tag{1}$$

where C_t is the cost in terms of time (seconds) in passing a section t, L the length of the section t in meters and v the velocity or speed in km/h.

The speed will vary with the type of vehicle and road. In this case, since emergency services are the users of the network, the researchers consider the mean speed as being ≈20% above the maximum speed permitted for each road type, and take into consideration the characteristics of the types of vehicle that usually take part in rescue missions [36]. Table 2 shows the speeds used in the calculations.

Incorporating the estimated speed in emergency situations to the road network allows us to obtain the *Network with cost non-disruptions* for emergency vehicles or those of similar characteristics.

Table 2. Estimated speeds in emergencies used in costs calculations.

Road Type	Estimated Speed
Primary road	140
Secondary road	100
Tertiary road	80
Street	60

Concerning the spatial information for flood-prone areas with a return period of 100 years (RP100 polygons in the scheme), data treatment focuses on vectorizing and simplifying based on the exclusion criterion of the depth of the water layer and area flooded. In this work, a threshold of 50 cm or more depth and a surface area greater than 5 m^2 covered by water are established as factors that prevent emergency vehicles from passing [37]. The vehicles used by emergency services can pass through a considerable depth of water. However, the most significant problem is not the exhaust pipe since most have a vertical air intake of about 3 meters, but a loss of stability, making it difficult to pass the mentioned thresholds [38]. The polygons below these two values have been excluded from the analysis since they are considered passable by vehicles.

2.2.2. (B) Preprocessing of Information

Preprocessing of the spatial data (B) combines the above-treated information, the road network, and flood-prone areas, to calculate the Cost Matrices (C) and OD Matrices (D) between the management centers and the different nodes to the network.

Regarding the calculation of flood scenarios, the parameters necessary to obtain it are the flood zones. All possible scenarios are obtained with a combination without repetition of k items from an n items set, according to the following equation:

$$C_{n,k} = \frac{n!}{k! \cdot (n-k)!} \tag{2}$$

These combinations ($C_{n,k}$) represent the number of combinations of different flood-prone areas; these are the different groups of k elements that can be formed by these n elements (flood prone areas), so that two groups differ only if they have different elements. This process has been automated by a script from phyton and QGIS to obtain flood scenarios in vector format.

2.2.3. (C) Cost Matrices

The non-disruption network scenario (*Network with cost non-disruption*) must be related to the combinations of possible floods to obtain network disruption scenarios. This procedure is usually carried out by a simple geoprocess that physically eliminates those sections of the road network that intersect with the disruption [24]. However, because the flood-prone areas usually cut the network into several sectors, the researchers have chosen a more practical solution that facilitates the multiple processing of the cost matrices for each network scenario. In those places where the water level would impede the passage of vehicles, a high cost is assigned to the node affected (>10,000 sec.). So, when the cost of using that particular route is calculated, the section would be considered invalid, and the cost of reaching the next node, preferably using an alternative route, would be calculated. If this were not done, the cost would be excessive, and it would difficult to identify poorly connected or isolated zones. This process is carried out automatically using the spatial functions of PostGIS. Table 3 includes an example to create one of the columns for the flood scenario 1.

Table 3. example to create one of the columns for the flood scenario 1.

```
—Create a column
alter table network add column cost1 numeric;
—For each section, fill this column with the normal cost
update network set cost1 = cost;
—Calculate if the network has some section cut.
—If you have it, you will proceed to put a cost of 10000.
—To know if the section is cut, it will be enough to know if a section
—intersects a polygon
—In case of intersection, the cost will be applied,
—otherwise it will be left as is:
update network
set cost1 = 10000 where id in
(select network.id from network, polygons
where st_intersects(polygons.geom, network.geom)
and polygon = 'F1');
```

2.2.4. (D) O-D Matrices

Once the costs matrices have been calculated, the ODM (D) are obtained with the distances between the different elements of the network. The main objective is to obtain a table of network nodes (destinations) and the accumulated cost (sec.) for a vehicle from every emergency management center (origin, *m*). The iterative calculation is made for each costs matrix and emergency management center using PgRouting and the shortest path algorithm, or Dijkstra's algorithm [39]. The theory underlying this algorithm is that all the routes from the point of origin to all other points of destination are explored until the shortest one is found.

The algorithm provides the *m* ODM in which the costs associated with each emergency center (time taken to reach the different network nodes) are summarized for each network scenario. The *m* tables are unified by a procedure known as spatial overlay, thus giving Total ODM (nodes) with $n \cdot m$ columns, aare joined to the layer of network nodes Network with cost non-disruptions. From this result, the minimum value for each network scenario (Min ODM) is obtained using PostGIS. This database manager let us identify which management center get to first and faster to a particular node of the network (Figure 2).

The above process of Block I terminates with map 1 *Present network vulnerability* (Figure 5) from the mean of the Min ODM for each point and network scenario thus providing the integrated vulnerability value in one table (*Mean minimum column*, Figure 2). This map will show the minimum mean access to each node (accessibility increases as the cost of reaching the destination node decreases). The vulnerability results allow map 2 to be constructed for each control center, *Present service area map* (Figure 9a). The maps represent the control center that will reach each scenario more frequently and faster and identify the center's immediate field of action and where it is preferable for a vehicle from another center to attempt to do so.

This process is similar to the previous one, but instead of selecting the minimum cost associated with the destination node, the researchers use the one associated with each emergency center.

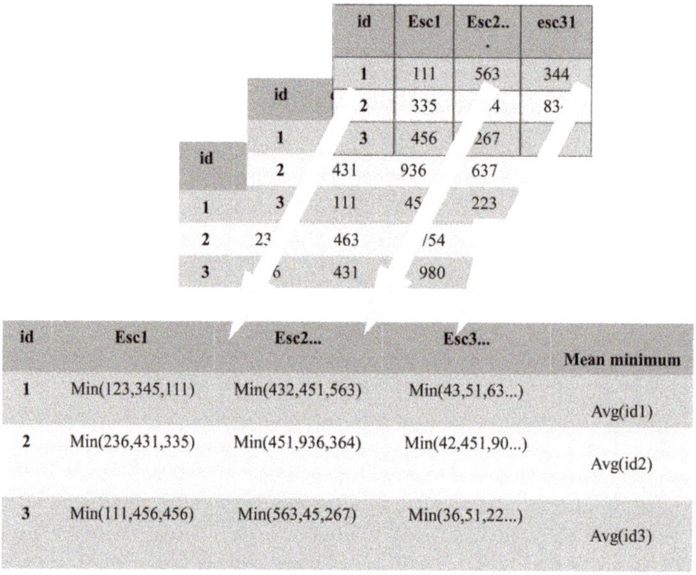

Water **2019**, 11, 1197

id	Esc1	Esc2..	esc31
1	111	563	344
2	335	4	83
3	456	267	

id	Esc1	Esc2	
1	431	936	637
2	111	45	223
3	23	463	/54
	6	431	980

id	Esc1	Esc2...	Esc3...	Mean minimum
1	Min(123,345,111)	Min(432,451,563)	Min(43,51,63...)	Avg(id1)
2	Min(236,431,335)	Min(451,936,364)	Min(42,451,90...)	Avg(id2)
3	Min(111,456,456)	Min(563,45,267)	Min(36,51,22...)	Avg(id3)

Figure 2. Process for constructing the table of means of the minimum times. Example for the case study ($n = 31$ scenarios and $m = 3$ emergency centers).

2.3. Block II: Optimal Location of New PEC

The second block aims to identify possible optimal sites (nodes in the road network) to build a PEC, which will reduce the vulnerability of the network in terms of costs (time to reach all nodes). The procedure to identify this PEC in the road network is summarized below:

1. Identification of candidate PEC. Ten possible nodes for locating the PEC are chosen for computational purposes. The points selected must satisfy the following criteria: the journey cost or response time from the existing emergency control centers to these new points in a normal situation (with no flooding) must not be more than 600 s , which is the minimum response time usually considered in the bibliography on optimal localizations [40]. This will ensure that the optimal location of PEC will not be excessively close to one of the three existing centers. It is, therefore, understood that the spatial distribution will affect (but will not determine) the new site.

2. Calculation of ODM from the selected nodes to the rest of the nodes of the network for every flood scenario. As an example, in the study case, the result gives ten matrices with the dimensions (1×31; center/network scenarios) for each possible location of PEC. The matrices depict the costs (time from each node selected to destination nodes) for each network scenario. These ten matrices, one for each new PEC, are compared with the *Min ODM* for block I. In this way, the results for the new points will have been compared with the minima obtained in block I. If the new time taken to arrive at the same destination exceeds the previous minimum, the previous time remains, and if the response time from the new center is improved, the cost of the new one is added. This procedure is very similar to that depicted in Figure 2, which, on this occasion, is repeated ten times (once for each candidate).

3. Once the previous calculations have been made, the mean access times are compared in order to identify the optimal location of the new PEC, selecting the candidate that minimizes this time. This process is based on a statistical hypothesis test (one-way ANOVA). To evaluate the statistical requirements of normality and homoscedasticity necessary in this type of parametric test, Kolmogorov–Smirnov's test for more than 50 cases [41] and Levene's test were used [42]. If these requirements are not met, a robust version of one-way ANOVA using trimmed means

and Welch correction is used [43]. When the ANOVA F-test is significant, a post hoc analysis is carried out using linear constraints for multiple comparisons, based on T3 Dunnet's method [43].

4. Once the optimum point in terms of time has been identified statistically, the means of the minimum network scenarios for each center, including the new center (PEC), are calculated. By proceeding iteratively for each node selected, the vulnerability value of the network for the existing centers concerning the candidate center is obtained. To evaluate differences in access time in the new scenario, the current network scenario and the future scenario (based on a new PEC) are compared using a robust t-test for repeated measurement based on trimmed means [43].

5. Finally, maps 3 (*Future network vulnerability*) and 4 (*Future service area*) are constructed (Figure 9a,b). Map 3 is made in a similar way to map 1, that is, with the results of the mean of the minima of the table *Mean ODM New*. Map 4 is made like map 2 that is, selecting center name instead of cost.

2.4. Block III: Expert Evaluation

Once the results are obtained, the third block defines the process of validation and evaluation of the results. This block is outside the implemented process but is essential because it involves the agencies responsible for emergency management. This section includes the decisions that must be taken to solve the problems involved in the new PEC as a function of factors that were not included in the above process (such as suitability as a function of the cadastral value, decisions related with the final management, etc.).

3. Study Area

The analysis process proposed in this study was carried out in the urban and suburban setting of Murcia, Alcantarilla, Santomera, and Beniel municipalities, in south-eastern Spain (Figure 3). It is an area featured with severe flooding problems due to the sometimes extreme behavior of the hydrographic network that drains it [44].

Flood-prone areas of the study area are characterized by a large number of wide riverbeds that, in flood situations, would interrupt the road network, and which must be taken into account when calculating possible scenarios. To simplify such calculations without losing information in the process, the riverbeds can be grouped based on topography. Five zones were considered as flood prone area scenarios (Figure 4): Segura river, Guadalentin river, North area (*Ramblas* draining the northern slopes), South area (*Ramblas* draining the southern slopes) and Northeast area (*Ramblas* draining the north-eastern slopes). As a result of Equation (2), 31 possible scenarios were obtained in the study area by combining the five possible flood zones that were summarized in the Network with scenarios.

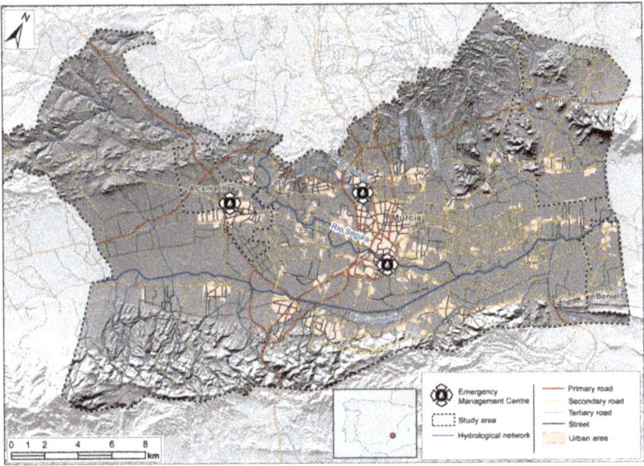

Figure 3. Study area including urban area, road network, and emergency management centers.

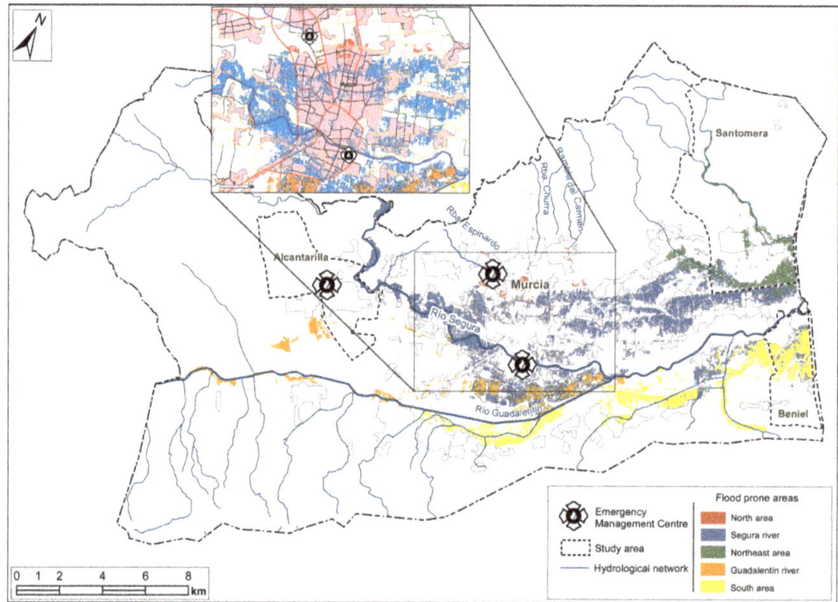

Figure 4. Flood-prone areas scenarios.

The road network is characterized by its high density and, in general, extends over the river beds that drain the area. As a result of channeling, the two main rivers that drain the area, the Guadalentin and Segura, have remained mostly unaffected by urban development and the road network crosses them using bridges. The danger that represents these two rivers comes from overflow. However, the other waterways (the *Ramblas* or ephemeral streams, which mostly follow a rectilinear path) have been occupied or, in some instances, incorporated into the road system. As a result, every time one or several of these streams flood, the problems of accessibility between the road network's points are exacerbated.

Concerning its morphometric characteristics (Table 4), the network is divided into 21319 sections of differing hierarchical importance, making a total of 1704.4 km, 82 % of which are in the municipal area of Murcia, although the highest density is in Alcantarilla, where urban growth and the increase in the number of inhabitants emphasizes the development that has taken place in an area of reduced dimensions. The ratio between road kilometers and several inhabitants is the opposite of that in Santomera.

Table 4. Morphometry of the network in the study area.

Municipality	Area (km^2)	%	Population	%	Length (km^2)	%	Density (km/km^2)	Road (Net./Pop.)
Alcantarilla	16.241	3.25	41,406	8.39	117.919	6.92	7.26	2.85
Beniel	10.092	2.02	11,057	2.24	27.041	1.59	2.68	2.45
Murcia	429.804	85.91	425,465	86.22	1469.677	86.23	3.42	3.45
Santomera	44.174	8.83	15,547	3.15	89.768	5.27	2.03	5.77
Total	500.311	100	493,475	100	1704.405	100		

It should be pointed out that the whole population of the municipalities conforming the study area is served by three main emergency response centers. (i) *Espinardo*, responsible for the northern zone of the city of Murcia,(ii) *Infante*, responsible for the southern zone on the right bank of the river Segura, and (iii) *Alcantarilla*, which serves the municipality of the same name and part of the western zone of Murcia. The equipment and manpower of each of the centers are roughly equal, although *Infante* has traditionally had a slight advantage in this respect.

4. Results

4.1. Present Road Network Vulnerability Map

The map in Figure 5 summarizes the results of the network vulnerability analysis concerning displacements (measured in seconds) for the three centers that are operational at present. Two main points are evident. First, there is a concentration of road interruptions in the north and south of the city of Murcia and the north-east of the study area, in the municipalities of Santomera and Beniel. These areas, which have the highest values (yellow, orange, and red), are frequently the most difficult to reach in terms of time, due to breaks in the road network. They are, therefore, the most sensitive, and it is here that the road network needs particular attention. Second, there are two possible ways of improving the situation.

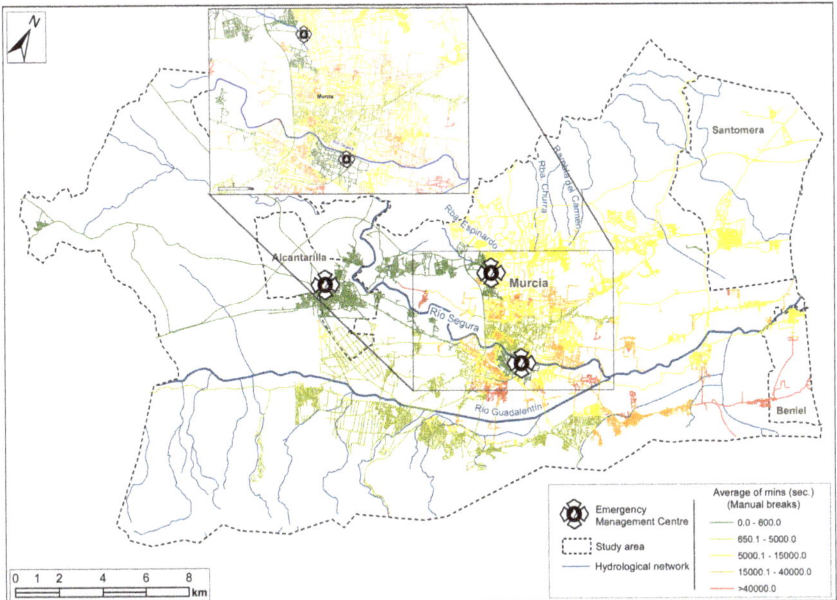

Figure 5. Map of present network vulnerability.

In the city of Murcia, structural initiatives are recommended (controlling flooding or providing alternative itineraries in the north and south). However, in the eastern part of the study area, there is no immediate response to an emergency situation since it is far from the existing emergency management centers, making it the prominent site for a new PEC.

4.2. Vulnerability of Future Network Following the Optimal Siting of a New Command Center

An analysis of the variance based on the Welch correction was carried out to identify the point that reduces the vulnerability of the network to the greatest extent. The results obtained indicated that the optimal point had a significant effect on access in time ($F(9, 51598) = 42.347$, $p < 0.0001$). Figure 6 depicts the mean access time (arithmetic and trimmed means) in increasing order by interval and homogeneous groups derived from the lincon post hoc test. The groups concentrate network scenarios whose p values infer that the mean access time is not significantly different; for example, the network scenarios of the group (a) will not differ significantly among themselves and will involve significantly higher mean access times than those included in group (b). The group representing the shortest access time is group (d), which includes network scenarios 1 ($M = 8030$, $CI = 179$) to 6 ($M = 8263$, $CI = 190$), with no significant difference between them in terms of mean access time, which can be regarded as optimal since the access time is lower than for the rest of the network scenarios. From network scenario 7 ($M = 8506$, $CI = 185$), the access times are considered significantly different from the times of the network scenarios of group d).

Having decided on the best location for the new PEC (selected node 1), the vulnerability of journeys is evaluated once again with all four centers. Figure 7 points to a statistically significant reduction in the mean access time ($T(17056) = 56.877$, $p < 0.0001$). The results represented cartographically in Figure 8a shows an improvement over those obtained for the present vulnerability map.

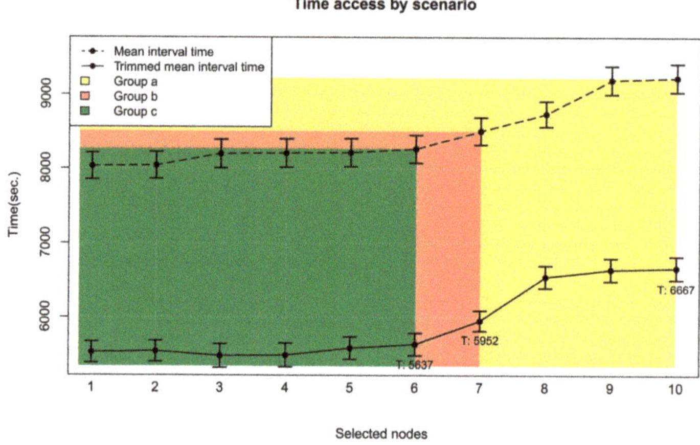

Figure 6. Mean access time (arithmetic and trimmed means) for each new PEC and scenario. Main effects, selected nodes factor (mean and 95% confidence interval). Means with different letters (rectangles) are significantly different (lincon multiple comparisons, $p < 0.05$).

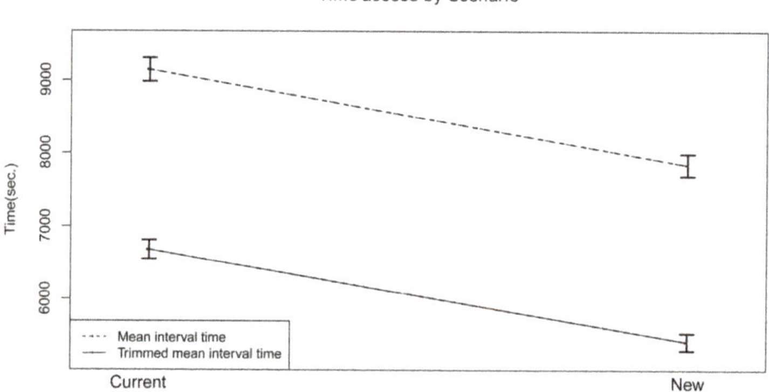

Figure 7. Main effects, network scenario factor (mean and 95 % confidence interval). Effects in terms of diminution in mean time (arithmetic and trimmed means) of present network vulnerability and future network vulnerability.

Other issues are worth mentioning, too. First, despite the reduced vulnerability, high values are still obtained for subnetworks of the hub and spoke type [45], which remain highly vulnerable since they are frequently affected and represent a loss of operational efficiency in this zone (in red in Figure 8a,b). This type of isolated areas can be observed to the south and southeast of the *Infante* station and in the network, sections corresponding to the municipality of Beniel, where high-cost values are due to the interruptions caused by two or three flood-prone zones in critical sections. The first case is of note because of the proximity in Euclidean distance to the control center, which underlines the need for this point to be secure.

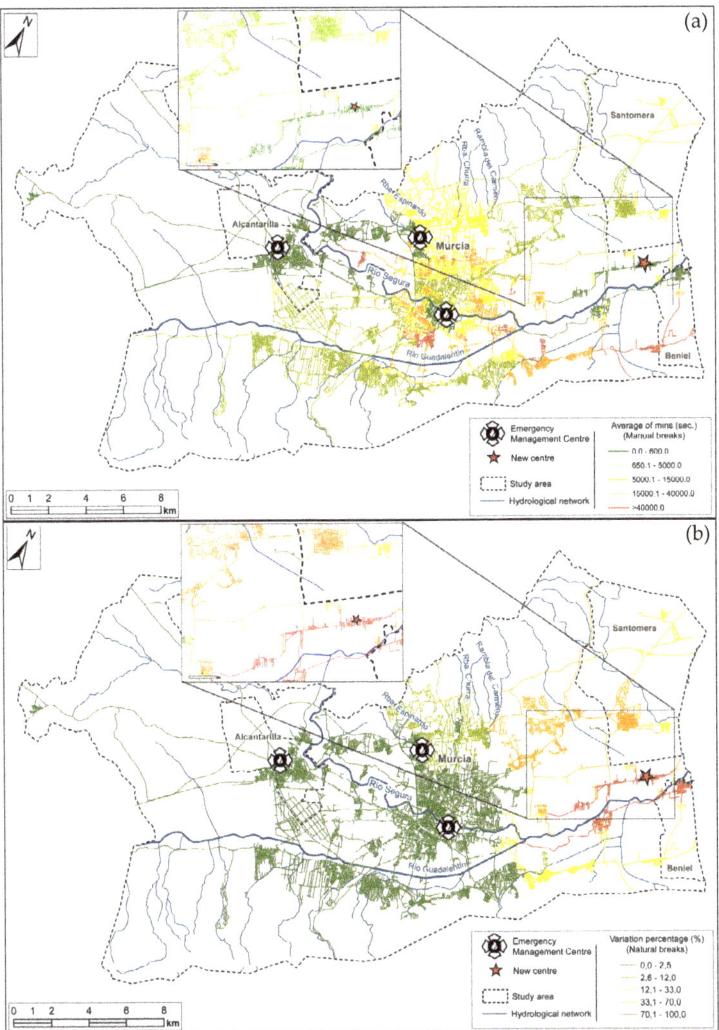

Figure 8. (**a**) Maps of future vulnerability scenarios; (**b**) Percentage of variation.

Secondly, interpretation of the map helps reveal the least and most suitable locations for the control centers. The best example of the first is represented by the Espinardo center, whose location on the road coinciding with a river flood bed substantially limits eastward journeys; in this case, the

PEC would slightly alleviate the problem, as can be seen in Figure 8b. With regard to optimal sites, the existing Alcantarilla center serves a wide zone since its radius of action is not affected by interruptions.

5. Discussion

The cartography corresponding to emergency service areas allows us to discuss the results obtained. In maps of Figure 9a,b, the researchers identify the sections of the network that can be accessed more quickly and frequently by a control center. From a comparison of both figures, it can be seen that the method followed to find the optimal location of the PEC has made it possible to equally distribute the centers spatially. On the other hand, quantitative analysis of the differences in values between both maps depicted in Table 5 suggests that the above-mentioned improvement has mainly been produced in the sections served by Espinardo center, whose service area is reduced by 45.8%. This is followed by the Infante center, with only 9.5 km difference (1.2%). For its location concerning the PEC, Alcantarilla remains unchanged. As a whole, the new center represents an accessibility gain of 16.4% concerning the current vulnerability network.

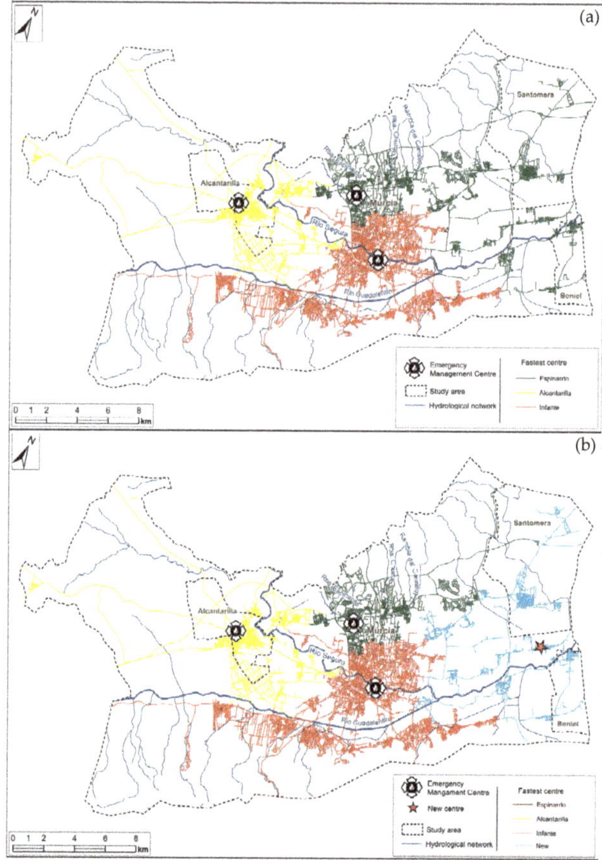

Figure 9. (**a**) Map of the service areas in the current vulnerability network and (**b**) with the new center.

Table 5. Results obtained with the lengths of the sections with the new center included, and difference concerning the present situation.

Center (Node)	Present Network Length Emergency Service Area (km)	% (100 = Total Network)	Length Emergency Service Area with New Center (km)	% (100 = Total Network)	% of Variation (100 = Present Emergency Service Area)	Difference (km) (Present-New Center)
Espinardo (1006)	591.2	34.6	320.1	18.8	−45.8	271.1
Alcantarilla (1304)	324.8	19.0	324.8	19.0	0	0
Infante (6050)	788.1	46.2	778.6	45.7	−1.2	9.5
New (9700)	—	—	280.6	16.4	—	—
Total	1704.1	100	1704.1	100	−16.4	280.6

Despite the improvements that represent the installation of a new center, there are still some road sections with substantial problems of isolation in flood situation. These results are consistent with those obtained by other works that use graph theory through SIG [5,15,34,46]. Given this circumstance and given that the problem of flooding could be increased taking into account the projections of climate change and increased exposure [47], this type of methodologies could lead to an improvement in efficiency in transport policies focusing his efforts in the reinforcement of these, especially critical sections.

This type of actions tends to focus on high-capacity roads as they are critical elements of the network due to a large amount of traffic they channel [48]. However, in the case of emergency management, it is also essential to highlight the importance of individual sections that, even with little traffic, are critical for some localities, especially when there are hardly any alternative routes because they are spaces with very low network density, which is common in peri-urban areas or mountain areas [49]. From the above, it can be inferred that vulnerability analysis of the networks provides information of interest, both in management tasks and in transportation planning.

On the other hand, there are still informational deficiencies and possible improvements that could be made: for example, with regard to the vulnerability analysis, considering the workflow of the physical state of the road or the average amount of traffic of each section to improve the calculation of the cost; in this context, with suitable information regarding roads and the volume of traffic, cost calculations can be modified simply by including the BPR (Bureau of Public Roads) formulation [50] for estimating link resistance and travel time. About the location procedure, the weaknesses of the proposed methodology could be improved through multi-criteria analysis techniques [51,52]. In this sense, the methodology proposed works with ten nodes (the available spaces with which municipal authorities in spatial planning usually deal). However, such decisions may be influenced by the needs of the population, which would carry more weight in justifying the new location rather than distances, costs, and isolation of the network's nodes. These are new challenges that open up the possibility of improving this type of study.

Once the above limitations have been overcome, the method can be used and replicated perfectly in other areas of study following the standardized procedure that summarizes the scheme of Figure 1. However, the only difficulty to carry out the above would be related to the issues of boundary conditions. In the case of this study, those limits are marked by orography, however, in other cases where the method is applied, this maybe very dificult.

6. Conclusions

According to the latest reports of the intergovernmental panel on climate change, there are still significant gaps in knowledge about the possible impacts that climate change may have on transport systems. Some relevant studies [53,54] outline some of the vital research needs and challenges. At the local scale, climate projections models that incorporate unique attributes of the urban transportation system are lacking. However, downscaled assessment is a prerequisite for a transportation planner to identify facilities and locations that are vulnerable to the impacts of climate-related events like floods.

At this sense, the proposed methodology provides two leading solutions for urban policymakers. First, a map of the present network vulnerability informs about critical network sections that should be reinforced or protected to facilitate population evacuation and the movement of emergency teams in case of catastrophe. Second, when these adaptation measures are not to be carried out, solutions, like finding the optimum location for a new PEC based in terms of reduced travel time to reach the rescue points seems, could be the best ones to solve that gap. To reduce subjectivity in that decision, the proposed method enhanced the procedure with the cartographic and statistical results of the current network's vulnerability developed in the first solution. Results of the present work show that a new emergency management center reduces present vulnerability and compensates the deficiencies in the service areas of other stations, mainly those who, due to their location, are severely restricted in their action by the presence of riverbeds, such as what occurs in Espinardo in the case of a recent study. Despite the advantage that the new station presents, there are two types of network sections whose vulnerability is not reduced. First, those in which several flood-prone areas converge. In those cases, although there is a center close to them in terms of Euclidean distance, accessibility problems are not solved since many of the sections that give access are affected. Such situations can be seen in the study area in the municipal district of Beniel and south Murcia. Second, there are also other specific examples in which the connection to the leading network depends on one connection that is frequently affected by more than one flood zone, so they are quickly isolated. One of these is to the west of Murcia. As mentioned above, to solve these problems, actions of a preventive nature, such as planning, do not fit well, and decisions are forced to go through structural actions.

Author Contributions: Conceptualization, A.P.-M., F.G.-C. and P.P.-Z.; Data curation, A.P.-M. and P.P.-Z.; Formal analysis, A.P.-M. and F.G.-C.; Funding acquisition, A.P.-M.; Investigation, A.P.-M., F.G.-C. and P.P.-Z.; Methodology, A.P.-M., F.G.-C. and P.P.-Z.; Software, A.P.-M., F.G.-C. and P.P.-Z.; Validation, P.P.-Z.; Visualization, A.P.-M. and F.G.-C.; Writing—original draft, A.P.-M., F.G.-C. and P.P.-Z.; Writing—review & editing, A.P.-M. and F.G.-C.

Funding: This research was funded in part by the Ministerio de Economía, Industria y Competitividad, Gobierno de España, grant number CGL 2016-75996-R, project *Variabilidad espacio-temporal de las inundaciones en la cuenca mediterránea española desde 1300 A.D.: procesos atmosféricos, hidrológicos e interacciones con la actividad humana*.

Conflicts of Interest: The authors declare no conflict of interest.

References

1.	Hunt, A.; Watkiss, P. *Literature Review on Climate Change Impacts on Urban City Centres: Initial Findings*; OECD: Paris, France, 2007.

2.	Board, T.R.; Council, N.R. *Potential Impacts of Climate Change on U.S. Transportation: Special Report 290*; The National Academies Press: Washington, DC, USA, 2008.

3.	Mehrotra, S.; Lefevre, B.; Zimmerman, R.; Gerçek, H.; Jacob, J.; Srinivasan, S. Climate change and urban transportation systems. In *Climate Change and Cities: First Assessment Report of the Urban Climate Change Research Network*; Rosenzweig, C., Solecki, W., Hammer, S., Mehrotra, S., Eds.; Cambridge University Press: Cambridge, UK, 2011; pp. 145–177.

4.	Berdica, K. An introduction to road vulnerability: What has been done, is done and should be done. *Transp. Policy* **2002**, *9*, 117–127. [CrossRef]

5.	Rodríguez Núñez, E.; Gutiérrez Puebla, J. Análisis de vulnerabilidad de redes de carreteras mediante indicadores de accesibilidad y SIG: Intensidad y polarización de los efectos del cierre de tramos en la red de carreteras de Mallorca. *GeoFocus. Revista Internacional de Ciencia y Tecnología de la Información Geográfica* **2012**, *12*, 374–394.

6.	Jenelius, E.; Mattsson, L.G. Developing a Methodology for Road Network Vulnerability Analysis. In Proceedings of the Nectar Cluster 1 Seminar, Molde, Norway, 12–13 May 2006, pp. 1–9.

7.	Taylor, M.A.P.; Sekhar, S.V.C.; D'Este, G.M. Application of Accessibility Based Methods for Vulnerability Analysis of Strategic Road Networks. *Netw. Spat. Econ.* **2006**, *6*, 267–291. [CrossRef]

8.	Chen, B.; Lam, W.; Sumalee, A.; Li, Q.; Zhi-Chun Li, Z.C. Vulnerability analysis for large-scale and congested road networks with demand uncertainty. *Transp. Res. Part A Policy Pract.* **2012**, *46*, 501–516. [CrossRef]

9.	Mattsson, L.G.; Jenelius, E. Vulnerability and resilience of transport systems—A discussion of recent research. *Transp. Res. Part A Policy Pract.* **2015**, *81*, 16–34. [CrossRef]

10.	Burgholzer, W.; Bauer, G.; Posset, M.; Jammernegg, W. Analysing the impact of disruptions in intermodal transport networks: A micro simulation-based model. *Decis. Support Syst.* **2013**, *54*, 1580–1586. [CrossRef]

11.	Chen, X.Z.; Lu, Q.C.; Peng, Z.R.; Ash, J.E. Analysis of Transportation Network Vulnerability Under Flooding Disasters. *Transp. Res. Rec. J. Transp. Res. Board* **2015**, *2532*, 37–44. [CrossRef]

12.	Jenelius, E.; Petersen, T.; Mattsson, L.G. Importance and exposure in road network vulnerability analysis. *Transp. Res. Part A Policy Pract.* **2006**, *40*, 537–560. [CrossRef]

13.	Jenelius, E.; Mattsson, L.G. Road network vulnerability analysis of area-covering disruptions: A grid-based approach with case study. *Transp. Res. Part A Policy Pract.* **2012**, *46*, 746–760. [CrossRef]

14.	Pregnolato, M.; Ford, A.; Robson, C.; Glenis, V.; Barr, S.; Dawson, R. Assessing urban strategies for reducing the impacts of extreme weather on infrastructure networks. *Open Sci.* **2016**, *3*. [CrossRef]

15. Sohn, J. Evaluating the significance of highway network links under the flood damage: An accessibility approach. *Transp. Res. Part A Policy Pract.* **2006**, *40*, 491–506. [CrossRef]

16. Badri, M.A.; Mortagy, A.K.; Alsayed, C.A. A multi-objective model for locating fire stations. *Eur. J. Oper. Res.* **1998**, *110*, 243–260. [CrossRef]

17. Chevalier, P.; Thomas, I.; Geraets, D.; Goetghebeur, E.; Janssens, O.; Peeters, D.; Plastria, F. Locating fire stations: An integrated approach for Belgium. *Soc. Econ. Plan. Sci.* **2012**, *46*, 173–182. [CrossRef]

18. Liu, N.; Huang, B.; Chandramouli, M. Optimal Siting of Fire Stations Using GIS and ANT Algorithm. *J. Comput. Civil Eng.* **2006**, *20*, 361–369. [CrossRef]

19. Tzeng, G.H.; Cheng, H.J.; Huang, T.D. Multi-objective optimal planning for designing relief delivery systems. *Transp. Res. Part E Logist. Transp. Rev.* **2007**, *43*, 673–686. [CrossRef]

20. Berdica, K.; Mattsson, L.G. Vulnerability: A Model-Based Case Study of the Road Network in Stockholm. In *Critical Infrastructure: Reliability and Vulnerability*; Murray, A.T.; Grubesic, T.H., Eds.; Springer: Berlin/Heidelberg, Germany, 2007; pp. 81–106.

21. Chen, A.; Yang, H.; Lo, H.K.; Tang, W.H. A capacity related reliability for transportation networks. *J. Adv. Transp.* **1999**, *33*, 183–200. [CrossRef]

22. D'Este, G.; Taylor, M.A.P. Modelling network vulnerability at the level of the national strategic transport network. *J. East. Asia Soc. Transp. Stud.* **2001**, *4*, 1–14.

23. Das, I.; Kumar, G.; Stein, A.; Bagchi, A.; Dadhwal, V.K. Stochastic landslide vulnerability modeling in space and time in a part of the northern Himalayas, India. *Environ. Monit. Assess.* **2011**, *178*, 25–37. [CrossRef]

24. Arşik, İ.; Sibel Salman, F. Modeling Earthquake Vulnerability of Highway Networks. *Electron. Notes Discret. Math.* **2013**, *41*, 319–326. [CrossRef]

25. Tinkler, K.J. *An Introduction to Graph Theoretical Methods in Geography*; Geo Abstracts Ltd.: London, UK, 1977.

26. Church, R.L. Location modelling and GIS. In *Geographical Information Systems*; Longley, P., Goodchild, M., Maguire, D., Rhind, D., Eds.; John Wiley & Sons: New York, NY, USA, 1999; pp. 293–303.

27. Fiedrich, F.; Gehbauer, F.; Rickers, U. Optimized resource allocation for emergency response after earthquake disasters. *Saf. Sci.* **2000**, *35*, 41–57. [CrossRef]

28. Hasnat, M.; Islam, R.; Hadiuzzaman, M. Emergency Response During Disastrous Situation in Densely Populated Urban Areas: A GIS Based Approach. *Geogr. Tech.* **2018**, *13*, 74–88. [CrossRef]

29. Obe, R.; Hsu, L. *PostGIS in Action*, 2nd ed.; Manning Publications: Greenwich, CT, USA, 2015.

30. R Core Team. *R: A Language and Environment for Statistical Computing*; R Foundation for Statistical Computing: Vienna, Austria, 2018.

31. MAGRAMA. *Guía Metodológica Para el Desarrollo del Sistema Nacional de Cartografía de Zonas Inundables*; Ministerio de Agricultura, Alimentación y Medio Ambiente: Madrid, Spain, 2011.

32. European Communities. *Council Directive 2007/60/EC of 23 October 2007 on the Assessment and Management of Flood Risks*; Technical Report; Office for Official Publications of the European Communities: Luxemburg, Belgium, 2007.

33. Más, S.; García, A.; González, A.; Rubio, J.; Velasco, A.; González, J.; Ruiz, C. *CartoCiudad: Una apuesta colaborativa de las Administraciones Públicas en el ámbito de los callejeros*; XI Jornadas sobre Tecnologías de la Información para la Modernización de las Administraciones Públicas: Zaragoza, Spain, 2010; pp. 6–9.

34. Bono, F.; Gutiérrez, E. A network-based analysis of the impact of structural damage on urban accessibility following a disaster: The case of the seismically damaged Port Au Prince and Carrefour urban road networks. *J. Transp. Geogr.* **2011**, *19*, 1443–1455. [CrossRef]

35. Open Street Map Contributors. 2017. Available online: https://planet.osm.org or https://www.openstreetmap.org (accessed on 18 January 2017).

36. Moriarty, K.D.; Ni, D.; Collura, J. Modeling Traffic Flow Under Emergency Evacuation Situations: Current Practice and Future Directions. In Proceedings of the 86th TRB Annual Meeting, Washington, DC, USA, 21–25 January 2007; pp. 1–13.

37. Conesa-García, C.; García-Lorenzo, R.; Pérez-Cutillas, P. Flood hazards at ford stream crossings on ephemeral channels (south-east coast of Spain). *Hydrol. Process.* **2017**, *31*, 731–749. [CrossRef]

38. Martínez-Gomariz, E.; Gómez, M.; Russo, B.; Djordjević, S. A new experiments-based methodology to define the stability threshold for any vehicle exposed to flooding. *Urban Water J.* **2017**, *14*, 930–939. [CrossRef]

39. Dijkstra, E.W. A note on two problems in connexion with graphs. *Numer. Math.* **1959**, *1*, 269–271. [CrossRef]

40. Caunhye, A.M.; Nie, X.; Pokharel, S. Optimization models in emergency logistics: A literature review. *Soc. Econ. Plan. Sci.* **2012**, *46*, 4–13. [CrossRef]
41. Royston, P. A remark on algorithm AS 181: The W-test for normality. *J. R. Stat. Soc.* **1995**, *44*, 287–321. [CrossRef]
42. Levene, H. Robust tests for equality of variances. In *Contributions to Probability and Statistics: Essays in Honor of Harold Hotelling*; Olkin, I., Hotelling, H., Eds.; Stanford University Press: Palo Alto, CA, USA, 1960; pp. 278–292.
43. Wilcox, R.R. *Introduction to Robust Estimation and Hypothesis Testing*; Academic Press: London, UK, 2012.
44. Gil-Guirado, S.; Espín-Sánchez, J.; Del Rosario Prieto, M. Can we learn from the past? Four hundred years of changes in adaptation to floods and droughts. Measuring the vulnerability in two Hispanic cities. *Clim. Chang.* **2016**, *139*, 183–200. [CrossRef]
45. O'Kelly, M.E. A geographer's analysis of hub-and-spoke networks. *J. Transp. Geogr.* **1998**, *6*, 171–186. [CrossRef]
46. Ratliff, H.; Sicilia, G.; Lubore, S. Finding the n Most Vital Links in Flow Networks. *Manag. Sci.* **1975**, *21*, 531–539. [CrossRef]
47. IPCC. *Climate Change 2014: Synthesis Report. Contribution of Working Groups I, II and III to the Fifth Assessment Report of the Intergovernmental Panel on Climate Change*; IPCC: Geneva, Switzerland, 2014; p. 169.
48. Lhomme, S.; Diab, Y.; Laganier, R.; Serre, D. Urban technical networks resilience assessment. In *Resilience and Urban Risk Management*; Serre, D., Barroca, B., Laganier, R., Eds.; CRC Press: London, UK, 2012; pp. 109–117.
49. Cova, T.; Church, R. Modelling community evacuation vulnerability using GIS. *Int. J. Geogr. Inf. Sci.* **1997**, *11*, 763–784. [CrossRef]
50. Sheffi, Y. *Urban Transportation Networks: Equilibrium Analysis With Mathematical Programming Methods*; Prentice-Hall: Upper Saddle River, NJ, USA, 1984.
51. Esmaelian, M.; Tavana, M.; Santos Arteaga, F.J.; Mohammadi, S. A multicriteria spatial decision support system for solving emergency service station location problems. *Int. J. Geogr. Inf. Sci.* **2015**, *29*, 1187–1213. [CrossRef]
52. Di Matteo, U.; Pezzimenti, P.M.; Astiaso Garcia, D. Methodological Proposal for Optimal Location of Emergency Operation Centers through Multi-Criteria Approach. *Sustainability* **2016**, *8*, 50. [CrossRef]
53. Hyman, R.C.; Potter, J.; Savonis, M.; Burkett, V.; Tump, J. Why Study Climate Change Impacts on Transportation? In *Impacts of Climate Change and Variability on Transportation Systems and Infrastructure: Gulf Coast Study, Phase I*; Savonis, M.J., Burkett, V., Potter, J., Eds.; Department of Transportation: Washington, DC, USA, 2008; pp. 1–34.
54. Neumann, J.E.; Price, J.; Chinowsky, P.; Wright, L.; Ludwig, L.; Streeter, R.; Jones, R.; Smith, J.B.; Perkins, W.; Jantarasami, L.; et al. Climate change risks to US infrastructure: Impacts on roads, bridges, coastal development, and urban drainage. *Clim. Chang.* **2015**, *131*, 97–109. [CrossRef]

Article

A Framework Proposal for Regional-Scale Flood-Risk Assessment of Cultural Heritage Sites and Application to the Castile and León Region (Central Spain)

Julio Garrote [1],*, Andrés Díez-Herrero [2], Cristina Escudero [3] and Inés García [3]

[1] Department of Geodynamics, Stratigraphy and Paleontology, Complutense University of Madrid, 28040 Madrid, Spain
[2] Geological Hazards Division, Geological Survey of Spain (IGME), 28003 Madrid, Spain; andres.diez@igme.es
[3] Risk and Emergencies Management in Cultural Heritage, Castile and León Regional Government, 47014 Valladolid, Spain; escremcr@jcyl.es (C.E.); garsanie@jcyl.es (I.G.)
* Correspondence: juliog@ucm.es; Tel.: +34-91-394-4850

Received: 26 December 2019; Accepted: 21 January 2020; Published: 23 January 2020

Abstract: Floods, at present, may constitute the natural phenomenon with the greatest impact on the deterioration of cultural heritage, which is the reason why the study of flood risk becomes essential in any attempt to manage cultural heritage (archaeological sites, historic buildings, artworks, etc.) This management of cultural heritage is complicated when it is distributed over a wide territory. This is precisely the situation in the region of Castile and León (Spain), in which 2155 cultural heritage elements are registered in the *Catalog of Cultural Heritage Sites of Castile and León*, and these are distributed along the 94,226 km^2 of this region. Given this scenario, the present study proposes a methodological framework of flood risk analysis for these cultural heritage sites and elements. This assessment is based on two main processing tools to be developed in addition: on the one hand, the creation of a GIS database in which to establish the spatial relationship between the cultural heritage elements and the flow-prone areas for different flood return periods and, on the other hand, the creation of a risk matrix in which different variables are regarded as associated both to flood hazard (return period, flow depth, and river flooding typology) and to flood vulnerability (construction typology, and construction structural relationship with the hydraulic environment). The combination of both tools has allowed us to establish each cultural heritage flood risk level, making its categorization of risk possible. Of all the cultural heritage sites considered, 18 of them are categorized under an Extreme flood risk level; and another 24 show a High potential flood risk level. Therefore, these are about 25% to 30% of all cultural heritage sites in Castile and León. This flood risk categorization, with a scientific basis of the cultural heritage sites at risk, makes it possible to define territories of high flood risk clustering; where local scale analyses for mitigation measures against flood risk are necessary.

Keywords: flood risk; cultural heritage; meso-scale; flood hazard; flood vulnerability; Castile and León; Spain

---

1. Introduction

The analysis of river flood risk is a basic aspect in any attempt to integrate river management, and it is included in the European Flood Directive (60/2007/EC). Flood risk analysis has multiple approaches based upon the type of elements at risk. These elements could be classified as tangible and intangible (thus bringing the analysis into the areas of economic or social risk, respectively) or as direct or indirect. However, elements such as cultural heritage (archaeological sites, historical buildings, artworks, etc.) are often located somewhere between both categories. This means they can be assigned an economic value (market value), while at the same time they represent the record of the history of human civilization, whose value is intangible.

This previously mentioned duality, but also mainly the irreversibility, of flood damage on cultural heritage (which cannot be reproduced once it has been destroyed) makes the prediction and assessment of the flood risks that may affect these sites a critical task for their preservation. Given this situation, the first natural disaster management strategies that included cultural heritage among its objectives began to be developed in the 90s. Among these initiatives, one could highlight the "Carta del Rischio" [1], developed by the Italian Central Institute for Restoration since 1992, or the "Noah's Ark" [2] project of the European Union, launched in 2002.

There are many processes (natural or not) that can affect cultural heritage sites. Thus, references [3,4] classify them into two major categories: unpredictable disasters and predictable deterioration. The former group would include Natural Disasters, and this subgroup, in turn, would include River Floods. These natural disasters, more frequent in recent years possibly due to global climate change, have a strong impact on cultural heritage [5,6], much more than the predictable deterioration linked to cultural heritage elements of hundreds of year in age. For this reason, in 2009 the European Union financed the project "Cultural Heritage Protection against Flooding" with the aim of developing preventive strategies for the conservation of cultural heritage.

As these cultural heritage conservation projects began to develop, the first studies on the interaction between flood risk and cultural heritage began to appear. Lanza [7] was possibly the first to analyze this relationship in detail, focusing on the historical city of Genova. Subsequently, and also on a scale of analysis that considers the effects of river floods on historic cities with a high number of cultural heritage elements, mention should be made of the work in the city of Prague [8]; in the Thai city of Ayutthaya [9,10]; in the Iranian archaeological sites of Persepolis and Naghsh-e-Rostam [11]; in New Taipei, Taiwan [12]; in the city of Seville [13]; in the archaeological site of Mogao Grottoes, located southeast of the Chinese city of Dunhuang [14]; and in the city of Florence [15]. With a different approach, works such as [16] evaluate different physical processes associated with floods and the most common damages that these processes can cause on heritage elements (e.g., Díez-Herrero [17]). Meanwhile, Herle et al. [18] analyze the effect of floods on cultural heritage not from the point of view of the effect of surface water but from that of the structural problems associated with changes in the conditions of the terrain subsurface.

Most of the studies mentioned above focus on the damage to the cultural heritage associated with a single flood event (e.g., [8–11,13]), or the possible damage level to heritage due to floods with different return periods (e.g., [12,14,15]). Among the assessments that consider the "return period" variable in their analyses, most are based on the use of hydrological models for the simulation of the transformation process "precipitation-runoff" for the estimation of peak flows. While the analysis of Arrighi et al. [15] uses the coupled hydraulic 1D-2D model HEC-RAS to determine the flooded area, flow depths and flow velocities, [11,13] both try to analyze the effects associated with extreme events, with periods of flooding recurrence over 500 years.

The present assessment tries to overcome some of the limitations found in previous works, with the aim of defining the level of potential flood risk (even qualitatively) for the cultural heritage sites of a wide region, such as the Autonomous Community of Castile and León (Central Spain). To achieve this objective, the flood hazard associated with different return periods will be analyzed, as well as the flood vulnerability of the different types of cultural heritage catalogued within the studied region. Another type of vulnerability analysis, social and economic vulnerability against flash flood risk, has recently been done for this region [19,20].

As mentioned above, some previous assessments of cultural heritage vulnerability have lags and uncertainties (subjectivity in used criteria, generalized BICs or "cultural heritage site" classifications, etc.), which do not render them invalid. Therefore, as improvements to the previously made assessments, the data considered for flood hazard assessment will be both hydrological–hydraulic as well as historical–documental and geomorphological (all of them linked to the information available in the Flood Prone Areas Cartography National System). From the point of view of flood hazard analysis, taking into account not only the return period of flooded areas but also the water depth and time lag, which improve the quantification of flood hazard. From the point of view of flood vulnerability analysis, taking into account the variety of cultural heritage elements catalogued, the approach has focused on the typology of the cultural heritage elements and its relationship with the hydraulic environment. Based on these variables, a matrix of potential flood risk has been developed for a flood risk categorization of cultural heritage sites. A GIS geodatabase helps us to integrate flood hazard and vulnerability values for each cultural heritage site. The quantification of the flood risk level in each cultural heritage site has allowed us to analyze the spatial distribution of this risk (clustering of flood risk results), as well as the ordered categorization of the patrimonial elements according to their risk level. All this will allow a scientifically based selection of single or geographically-linked groups of cultural heritage sites for the micro-scale assessment of their flood risk.

2. Materials and Methods

Castile and León is the most extensive region of Spain (and the third largest in Europe; more so than several European countries). It is located in the central-northern sector of the Iberian Peninsula and covers an area of 94,226 km^2. From the hydrological point of view, the main basin is that of the Duero River and its tributaries; although it also participates in the basins of the Tajo, Miño-Sil, and Ebro rivers. The population of Castile and León is 2.41 million inhabitants (data at 1 January 2018; with a low population density of 25.67 pop./km^2), spread out over 2248 municipalities, mainly in the cities. Regarding the cultural heritage, this region has catalogued 2155 sites of cultural interest, some of them declared World Heritage by UNESCO (BICs affected by 500-year return period flood are showed in Figure 1); being the region of Spain with the most heritage assets and one of the largest in Europe and around the World.

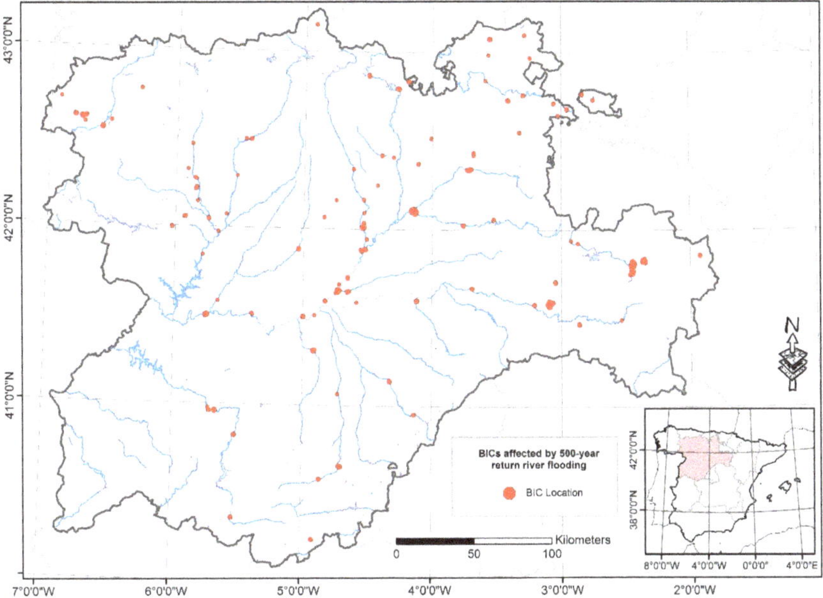

Figure 1. BICs location map for Castile and León region.

Figure 2 shows a methodological flow chart developed for the estimation of flood risk in the cultural heritage elements or sites (hereinafter referred to as BICs) analyzed. It shows the previous studies (not included in this work) that serve as data sources for this study, the methodological stages, and the techniques used and the final results.

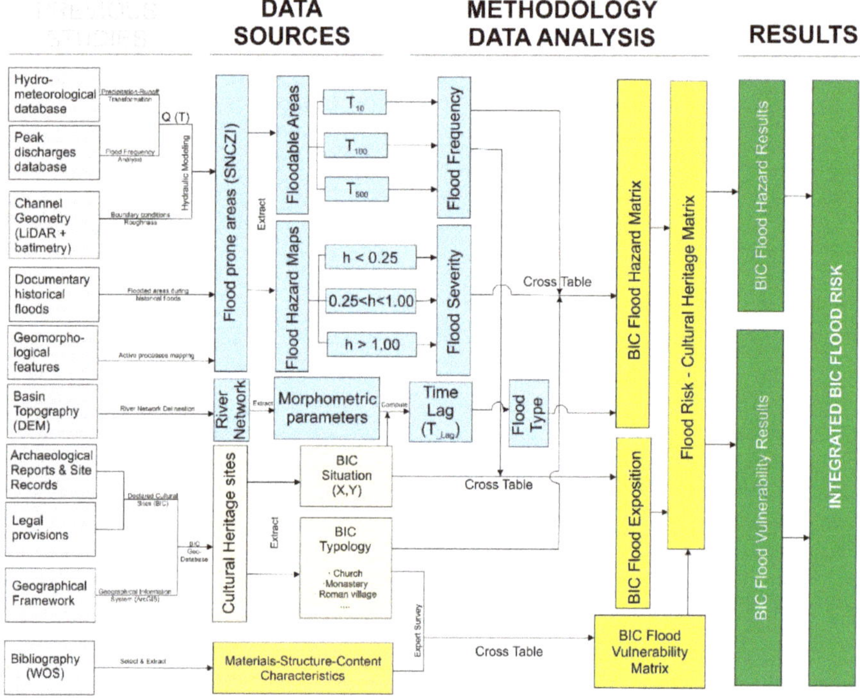

Figure 2. Methodological flow chart developed for the estimation of potential flood risk in the cultural heritage elements or sites. "T" is return period (in years), and "h" is water depth (in meters).

2.1. Flood-Prone Area Delimitation and Associated Frequency

In the present study, the delimitation of flood-prone areas for different return periods has been based on the information available in the Floodable Zones Mapping National System (hereinafter SNCZI; https://sig.mapama.gob.es/snczi/) for the territory of Castile and León. In this spatial database, four return periods of overflowing are included, associated to 10, 50, 100, and 500 years, so both ordinary floods (high frequency) and extraordinary floods (low frequency) are considered at an spatial resolution of 1×1 m.

For the delimitation of the floodable zones associated with the different return periods, three independent and complementary information sources have been used: hydrological–hydraulic data, documentary records of historical floods, and geomorphological information. The methodology used for the combination and integration of these data sources is described in [21,22].

In total, the number of floodable zones identified (areas with significant potential flood risk; hereinafter, ARPSIs) within the territory of Castile and León amounts to 405, with a total length of 3676 km. This set of river reaches gives rise to a total floodable area, for the 500-year return period, of 1585 km², only 1.7% of the extension of Castile and León.

2.2. Cultural Heritage Sites in Castile and León

The identification of the patrimonial elements considered in this study comes from the BIC Catalog of Castile and León (https://servicios.jcyl.es/pweb/portada.do). They are collected in a georeferenced database in several polygon shapefiles, with associated attribute tables, in ArcGIS shapefile format (ESRI Geosystems). From the set of patrimonial elements, selection was made of those that were located spatially in the areas for which information was available regarding the floodable areas for different return periods. Thus, of the 2155 BICs (cataloged in February 2019) located in the territory of the Castile and León region, only 145 have been considered in this study for their interference with floodable areas described in the previous section.

All the BICs considered in the present work have legislative protection due to their importance and contribution to the cultural heritage of the Castile and León region. This means that the regional government must maintain and conserve the elements in the best possible conditions and limit the possible degradation or destruction processes that could affect them.

2.3. Flood Risk—Cultural Heritage Matrix Development

The development of risk matrices for heritage buildings is an analytical methodology that has already been used previously [13,15,23]. However, the complexity or degree of detail of the matrix developed in this work significantly exceeds that of previous proposals. In the case of the works of Ortiz et al. [13,23], the matrix developed did not assess only the effect of floods, but it was a multi-risk analysis, including rainfall events, temperature, and anthropogenic factors as fires, vandalism, or tourist pressure. Additionally, the effect of the floods was considered qualitatively [13] or simply based on the distance to the river channels [23]. In reference [15], the matrix only considered two variables, the return period (hazard factor) and the potential damages (vulnerability factor), and both qualitatively.

In the present work, the flood risk analysis of the Castile and León BICs (Figure 1) uses three complementary matrices: the analysis of flood hazard, the vulnerability analysis of BICs against floods, and the estimation of flood risk associated with the BICs (Figure 2).

The hazard matrix was developed using the information and characteristics of the floodable areas of the Castile and León region stored in the spatial database of the SNCZI. Three variables associated with the flood characteristics and their possible influence on the flood hazard in the BIC locations have been used: (1) places affected by floods with a return period of 10, 100, and 500 years; (2) water depth ("h" in Figure 2) for each of these three return periods; (3) flood type (flash-flood or gradual). The identification of the flood type, as USACE [24] considered, has been based on the variable T_lag (time lag, or the time between the center of mass of precipitation hyetograph to the center of mass of runoff on the hydrograph; used for the classification of floods as flash-floods [25,26]) as a measure of the available response time against floods. This variable takes into account the size of the watershed in the estimation of its value, with different formulations to estimate the T_lag value depending on whether the river basin has an area lower or higher than 350 km^2. Water depth is one of the most used variables in the analysis of the structural stability of buildings [27] and has a strong influence regarding this stability [28]. Water depths greater than 1 m (depending on their combination with other variables, such as flow velocity and building typology) greatly compromise the structural stability of buildings. In our hazard matrix, the flow depth boundaries have been developed from those used by USACE [24], adding a higher level on the scale. The combination of these three variables defines the six levels of flood hazard (Table 1).

Table 1. Flood hazard matrix for cultural heritage sites.

FREQUENCY		SEVERITY [1]		
		Flood Type/Characteristic Times (Hours)/Basin Area (km^2)		
		Flash Flood		Normal Flood
Return Period (T, Years)	**Depth (m)**	**Small Basin** $T\_lag < 2$ h $A < 350$ km^2	**Large Basin** $T\_lag < 6$ h $A > 350$ km^2	$T\_lag > 6$ h $A > 350$ km^2
T 10	>3.6	6	6	6
	1.8–3.6	6	5	5
	0.9–1.8	5	4	4
	<0.9	4	3	2
T 100	>3.6	6	6	5
	1.8–3.6	5	5	4
	0.9–1.8	4	4	3
	<0.9	3	2	1
T 500	>3.6	6	6	5
	1.8–3.6	5	4	4
	0.9–1.8	3	3	2
	<0.9	2	1	1

[1] Flood hazard levels: 1, Low; 2, Medium-Low; 3, Medium; 4, Medium-High; 5, High; 6, Extreme. Background colors are related to different levels in all tables and figures.

For the vulnerability matrix, the Castile and León BICs have been classified into 6 main categories and 16 sub-categories (Table 2). All of them can be regarded as real estate and range from buildings of a civil nature to hydraulic infrastructures, as well as archaeological sites or defensive structures of cities and towns. The main criterion for grouping them into classes and sub-classes (unlike their classification in the BIC Catalog of Castile and León) has been their degree of general vulnerability to floods. This vulnerability has been considered dependent on the BIC building characteristics (mainly type of construction material and structure) and BIC contents (furniture, pictures, documents, etc.), and its more or less direct relationship with water conduction or defense against floods. BICs may be built using different materials and structures and contain multiple types of elements, so the BIC vulnerability level was defined as a weighted sum of the maximum vulnerability level in each category (construction material, construction structure, and content) by an expert criterion. As content is the most vulnerable element, it had the highest weight in opposition to construction material (lowest weight factor).

Table 2. Flood vulnerability matrix for cultural heritage sites.

Column groups: columns *Earth* through *Concrete* form **BIC Materials Vulnerability** (*Rubble*, *Ashlar (Small)*, *Ashlar (Great)* are sub-columns of **Stone**), summarised by **V_1—BIC Materials Vulnerability (MAX)**; columns *Foundation* through *Crypts & Cellars* form **BIC Structural Vulnerability** (*Narrow (<0.5 m)*, *Wide (>0.5 m)*, *Low (< 2 m)*, *High (>2 m)* are sub-columns of **Walls**), summarised by **V_2—BIC Structural Vulnerability (MAX)**; columns *Special Pavements* through *Documents* form **BIC Content Specific Vulnerability** (*Furniture*, *Sculpture*, *Grabados*, *Paintings*, *Documents* are sub-columns of **Artworks**), summarised by **V_3—BIC Content Vulnerability (MAX)**. The first three groups are together the **Specific Flood Vulnerability Regarding Characteristics of BICs**; the Materials and Structural groups fall under **BIC Continent Specific Vulnerability**.

Building Type	BIC Type	Earth	Adobe	Timber	Rubble	Ashlar (Small)	Ashlar (Great)	Mortar-Masonry	Bricks	Bedrock	Concrete	V_1—BIC Materials Vulnerability (MAX)	Foundation	Narrow (<0.5 m)	Wide (>0.5 m)	Low (< 2 m)	High (>2 m)	Buttres	Vaults	Crypts & Cellars	V_2—BIC Structural Vulnerability (MAX)	Special Pavements	Render & Plaster	Furniture	Sculpture	Grabados	Paintings	Documents	V_3—BIC Content Vulnerability (MAX)	Total BIC Vulnerability (0.1*V1 + 0.2*V2 + 0.7*V3)
	Proposed Value	5	4	3	2	1	1	2	2	1	1		1	3	2	2	3	1	3	4		2	3	4	3	1	6	6		
Civil	City, Village	5	4	3		1	1	2	2		1	5	1	3	2		3	1	3		3	2	3	4	3		6	6	6	5
	Palace/House			3		1	1	2	2			3	1		2		3	1	3	4	4	2	3	4	3		6	6	6	5
Industrial	Factories			3		1	1	2	2		1	3	1		2		3	1	3		3	2	3	4					4	4
Religious	Monastery	5		3		1	1	2	2			5	1		2		3	1	3	4	4	2	3	4	3		6	6	6	6
	Convent	5		3		1	1	2	2			5	1		2		3	1	3	4	4	2	3	4	3		6	6	6	6
	Church			3		1	1	2	2			3	1	3	2		3	1	3	4	4	2	3	4	3		6		6	5
Defensive	Castels	5	4	3		1	1	2	2			5	1		2		3	1	3	4	4	2	3	4	3		6	6	6	6
	Towers	5				1	1	2	2			5	1		2		3	1	3		3	2		4	3				4	4
	City Walls		4		2	1	1	2	2			4	1		2		3	1			3									3
Archaeological sites	Ancient sites (Roman and Visigotic)			2	1	1		2				2	1	3	2	2					3	2			3				3	3
Buried/Ruins	Proto-historic sites	5	4	3	2					1		5		3		2					3				3				3	3
	Pre-Historic sites									1		1				2					2				3	1			3	2
Hydraulic Infrastructures (unuseful)	Mills/Harbour					1	1	2	2		1	2	1		2	2			1		2									2
	Bridges					1	1	2	2		1	2	1		2	2			1		2									2
	Canals					1	1	2	2		1	2	1		2				1		2									2

[1] Flood vulnerability levels: 1, Low; 2, Medium-Low; 3, Medium; 4, Medium-High; 5, High; 6, Extreme.

Finally, the potential risk matrix (Table 3) assumes a combination of the previous matrices, in which the combination of the six hazard levels and the six vulnerability levels generates the defined potential risk levels. The risk level has also been classified into six categories, from a low potential risk to an extreme potential risk. Not weighted factors have been used for the combination of hazard and vulnerability matrices, but the variable vulnerability was considered as more important in flood risk level final classification, again based on expert criterion and the cited technical literature. Finally, these matrices are implemented within a GIS, as conditional operations to obtain the values of hazard, vulnerability, and potential risk associated with the Castile and León BICs.

Table 3. Flood potential risk matrix for cultural heritage sites.

Flood Risk [1]		Flood Vulnerability					
		1	2	3	4	5	6
	1	1	1	2	3	4	4
	2	1	2	3	3	4	4
Flood	3	2	2	3	3	4	5
Hazard	4	2	3	3	4	5	5
	5	3	3	4	4	5	6
	6	3	4	4	5	6	6

[1] Flood risk levels: 1, Low; 2, Medium-Low; 3, Medium; 4, Medium-High; 5, High; 6, Extreme.

With respect to the interpretation of these flood risk levels, they can be interpreted as follows:

Risk levels 1 and 2 (Low and Medium-Low) correspond to cases where the level of risk can be accepted, and no important damages are expected for those cultural heritage sites.

Risk level 3 (Medium) corresponds to a case where the flood risk for the cultural heritage is at the limit of acceptability and should be monitored regularly. Low frequency or low vulnerability makes the flood risk acceptable.

Risk level 4 (Medium-High) is linked to a case where the level of risk is just not acceptable. The frequency or the vulnerability makes the flood risk not acceptable and a further analysis at micro-scale should be carried out with the objective to get a lower flood risk level.

Risk level 5 y 6 (High and Extreme) corresponds to a case where the level of risk is unacceptable and needs further micro-scale detailed assessments to define risk mitigation measures.

3. Results of Flood Risk Assessment

3.1. Results of Flood Hazard Assessment

From the point of view of the flood hazard analysis associated with the Castile and León BICs, as has been stated in the previous section, the data sources come from the flood depth maps derived from the detailed hydraulic models developed for the SNCZI. The intersection of this GIS layer with that which contains the spatial location of BICs gave us a result in which, for the return period of 10 years, the number of affected BICs amounts to 75; for the return period of 100 years, the number of affected BICs reaches 105; and finally, for the return period of 500 years, the number of BICs is 145.

For the return period of 10 years, the flood depth for the 75 BICs presents an average value of 1.38 m (with a standard deviation of 1.35 m); and in a total of 40 BICs, the depth is greater than 1 m. In the case of the return period of 100 years, the mean flood depth for the 105 BICs has an average value of 1.54 m (with a standard deviation value of 1.40 m); there are 58 BICs with a depth greater than 1 m. Finally, in the case of the return period of 500 years, the mean flood depth for the total of 145 BICs has a value of 1.71 m (with a standard deviation value of 1.68 m), there being a total of 80 BICs in which the depth is greater than 1 m. This set of values (Figure 3) indicates the high degree of variability in the flood depth values that affect the BICs in the study area.

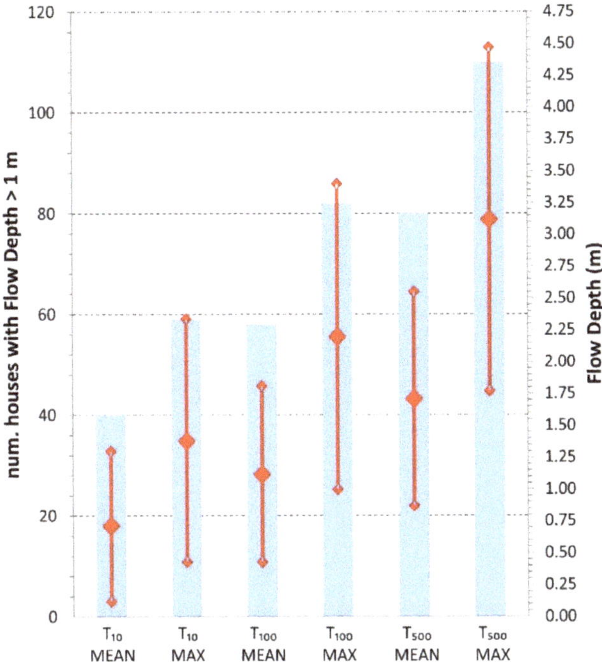

Figure 3. Flow depth values distribution (mean and 0.5 standard deviation, red line and dots) for the different return period considered. Blue bars show the number of BICs (considering average water depth and maximum water depth) where flow depth is up 1 m.

Regarding the response time of the basin, measured from the variable "Time Lag (t_lag)", it is observed that in 37 BIC locations the value of "t_lag" is less than 2 h; and in other 27 locations this variable is between 2 and 6 h. These data indicate that in 64 of the 145 BICs affected by floods, the response time to the flood is less than 6 h.

Based on the results presented above, and through an iterative conditional process, the results of the flood hazard variable linked to the Castile and León BICs are shown in Figure 4. This graph shows the existence of 41 BICs with a High or Extreme hazard (33 and 8, respectively), and 73 BICs with a hazard above Medium. Meanwhile, the BICs that show Medium-Low and Low hazard values represent a total of 53 (24 and 29, respectively). The spatial distribution of the BICs and their flood hazard is reflected in the map in Figure 5, where the BICs with high hazard values are distributed around the peripheral municipalities and on the Pisuerga River axis; while the low hazard BICs are in the central part of the Castile and León region.

3.2. Results of the Flood Vulnerability Assessment

As previously stated, from the point of view of flood vulnerability analysis, the data sources come from the BIC Catalogue of the Castile and León region.

Thus, the set of 145 BICs viewed as susceptible to being affected by floods with a return period of 10, 100, or 500 years, were classified into 16 BIC types, grouped in 6 major categories. The distribution of flood vulnerability values based on the BIC building and content show that religious BICs are mainly the most vulnerable to floods. This is due to their buildings and, in particular, to the vulnerability of their contents. Aside from religious BICs, civil and industrial BICs (plus the castles, categorized into defensive BICs) show a high vulnerability to floods. On the other hand, archeological sites and mainly hydraulic structures show the lower vulnerability levels. It is important to keep in mind that just those types of BICs do not have contents, but only building (materials and structure). And the contents of the BICs just show the highest vulnerability values.

The spatial distribution of the BICs classified according to their vulnerability is shown in Figure 6.

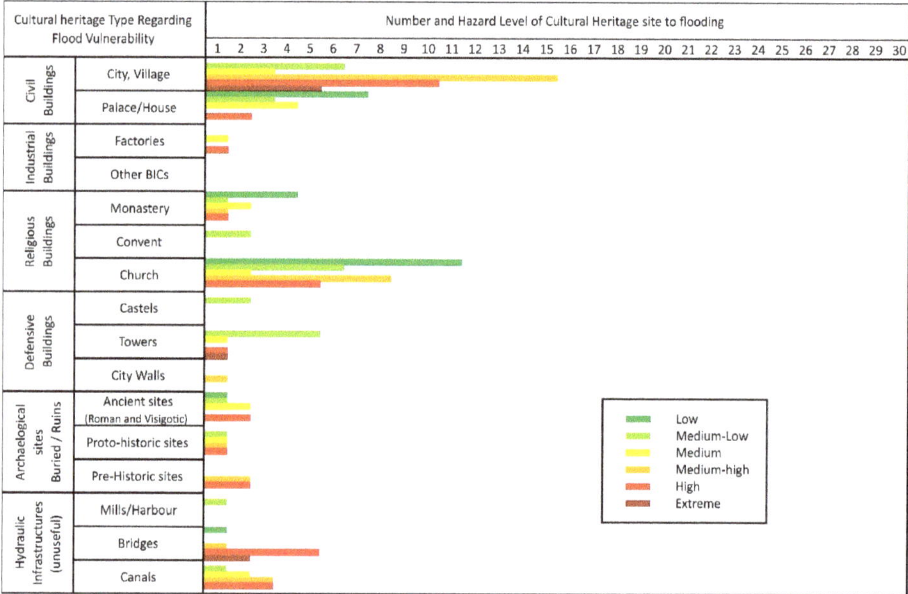

Figure 4. Distribution of flood hazard levels for the different cultural heritage types defined for Castile and León region.

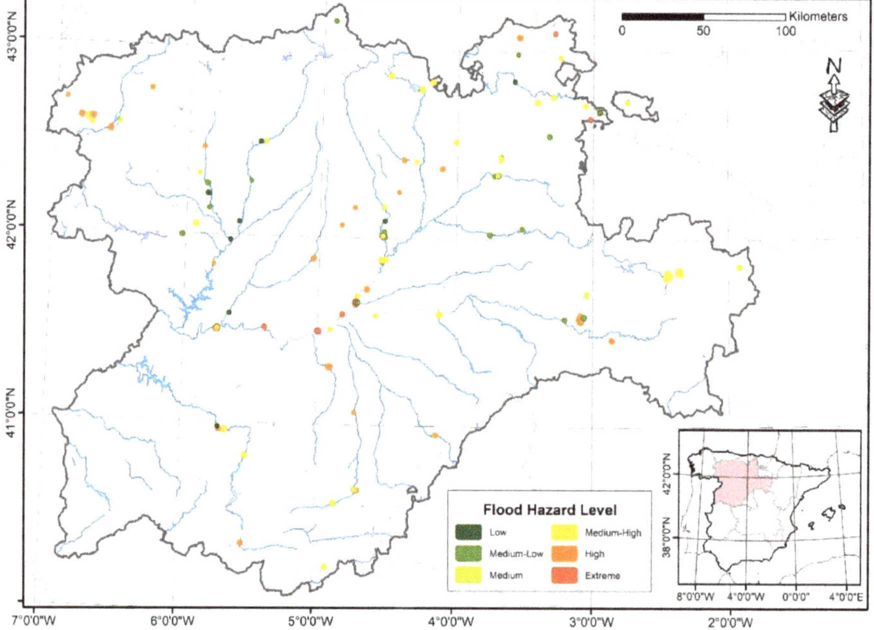

Figure 5. Spatial distribution of BICs in Castile and León region and their associated flood hazard level.

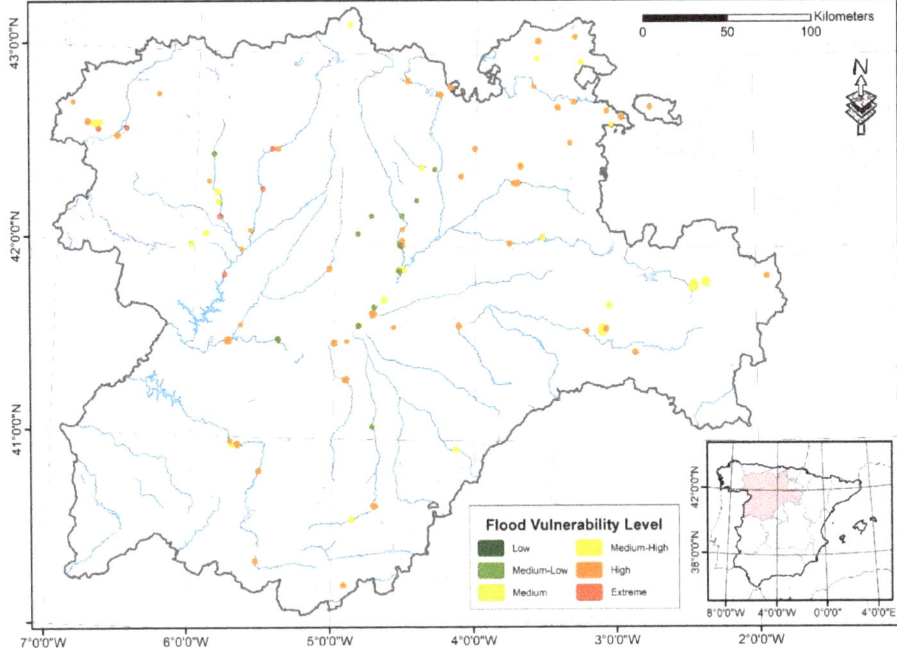

Figure 6. Spatial distribution of BICs in Castile and León region and their associated flood vulnerability level.

3.3. Results of Flood Potential Risk Assessment

Crossing the information related to flood hazards that may affect BICs with the BIC flood vulnerability has allowed us to define the level of flood potential risk with a return period of 10, 100, and 500 years for the Castile and León region.

The distribution of potential flood risk values based on the BIC typology is shown in Figure 7. This figure shows the existence of two types of BICs (historic cities and villages and monasteries) in which a level of Extreme potential risk is reached; rising to five types (the previous two plus palaces, churches, and towers) in those BICs that reach High or Extreme potential risk levels. In the opposite direction, eight types of BICs have levels of potential risk from Medium to Low (Medium, Medium-Low, and Low), which would be factories, towers, city walls, ancient sites, proto-historic sites, pre-historic sites, mills, bridges, and canals. Three of these eight typologies with medium or low levels are BICs that were originally associated with infrastructures related to water or rivers. This must condition their relatively low level of potential flood risk (associated with its low vulnerability).

From all the BICs considered, six of them present an Extreme level of potential flood risk; and another 44 show a High level of potential risk. This is around the 30% of the BICs considered as "affected by floods" in the present study. At the other extreme of the potential flood risk, 6 BICs present Low or Medium-Low risk levels (less than 5%). The spatial distribution of these potential BIC risk values can be seen in Figure 8, where the BICs with high-risk values are distributed around the peripheral municipalities and in the central part of the Castile and León region.

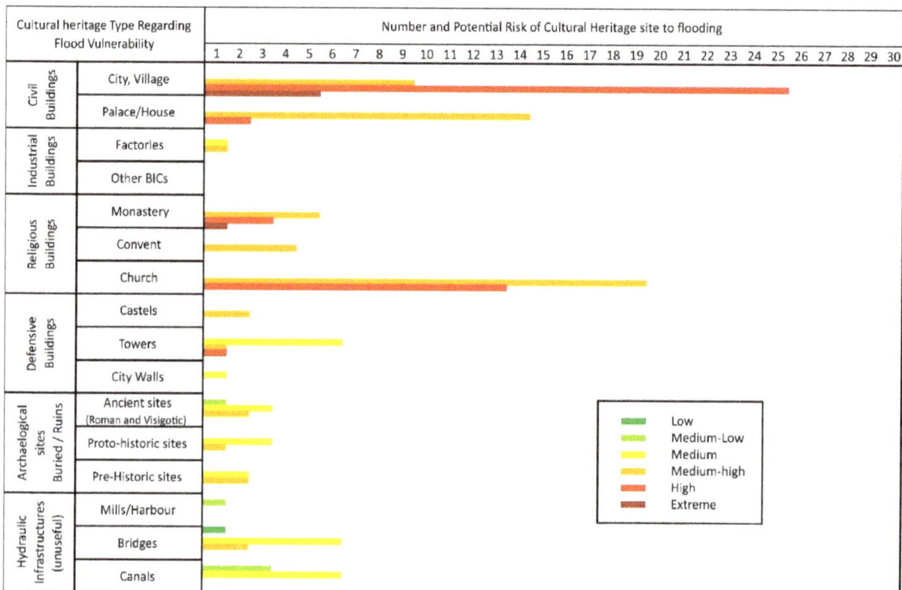

Figure 7. Distribution of flood potential risk levels for the different cultural heritage types defined for Castile and León region.

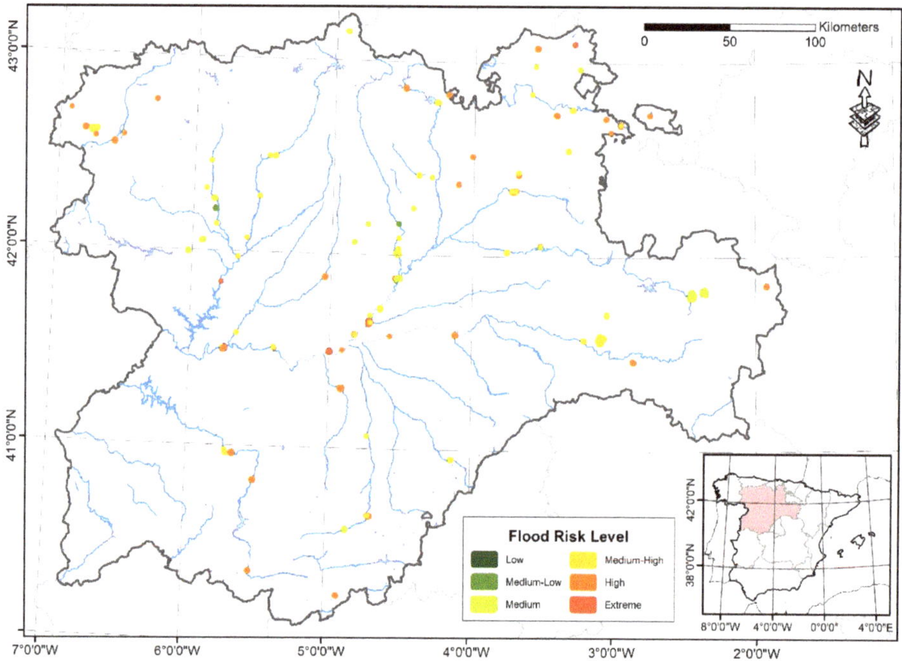

Figure 8. Spatial distribution of BICs in Castile and León region and their associated flood potential risk level.

4. Discussion

To estimate the potential flood risk in the elements classified as Cultural Heritage sites in the Castile and León region, three matrices have been developed to encompass the analysis of flood hazard, flood vulnerability, and potential flood risk. When considering the proposed potential flood risk model, it must be considered that the approach of the present methodological framework has a regional and qualitative focus, albeit with quantitative classes for the matrix. This regional approach will therefore determine the variables to be used in each of the risk components, prioritizing the use of easy estimation variables, regional scope, and the possibility of being easily integrated into a GIS environment, which has been used as a key tool for the analysis.

Thus, from the point of view of the flood hazard analysis, the variables used were the flood return period, the flow depth, and the time lag of the drainage basin in which the flood occurs: all of them easily derivable variables from DTMs and hydrodynamic models. In this selection of variables, those that are more difficult to obtain on a regional scale have been left out, even if their use at a micro-scale or in analyses of individual buildings could be feasible (analyses of dynamic loads in structures, distribution of pressures, and structural deformations due to the fluid-structure interaction), as shown by [29].

Another aspect to be taken into account is the value associated with the flow depth variable and its relationship with the flood hazard analysis (10, 100, and 500 years return period) at the locations of cultural heritage sites. From the different approaches for the representation of buildings in hydrodynamic models (e.g., [30,31]), the one used in this work uses the elevation of the cells where the buildings are located. Thus, in order to obtain the flow depth value, one polygon feature that delimits the extension of the cultural heritage site (with a neighborhood buffer of 1 m around it) has been used.

In the present analysis, the flow depth used is defined as the mean value for each Cultural heritage site. However, this average value does not represent the maximum possible flow depth, and the difference between those values will depend on the topographic variability inherent in the terrain. Therefore, it would be appropriate to consider the optimum value of flow depth to be used for the flood hazard assessment. The minimum and maximum values would take us to the extremes of security, and therefore, when dealing with patrimonial elements of difficult recovery or reproduction, the option of the minimum value does not seem optimal or desirable. At the same time, the use of the maximum value would lead us to work within the worst possible scenario, a situation that could be considered viable or optimal depending on the importance of the elements at risk. However, the use of the maximum value can also lead us to consider extreme flood hazards supported by errors in the hydraulic model or, what may be more common, the terrain topography represented through an DTM. In this sense, as indicated by Vaze et al. [32], the use of LiDAR data has significantly improved the terrain representation in the DTMs. However, the urban areas (the majority in the location of cultural heritage sites) remain the most problematic and have the highest uncertainty, both in the correct representation of urban areas (e.g., References [33–35]) and in the results of the hydrodynamic models used to estimate flood hazard (e.g., References [31,36]).

Due to the above, the estimation of the flood hazard from flow depth values should use average or modal values, which indicate the most frequent or probable value within the cultural heritage site under consideration. Under this approach, it is possible that the most optimal option was the modal value. However, this option clashes with the type of flow depth data (real numbers with two or three decimal digits), which does not allow a correct estimation of the modal value because it would require categorization and provide less accurate results. Furthermore, when the water depth data available for a BIC are scarce (due to the fact that the BIC is not affected by flood in all its extension), the use of average water depth value may reduce the uncertainty about a possible error in water depth estimation. All in all, the option of the average value has been viewed as the optimum indicator for flow depths value related to each cultural heritage site. This average value should be considered representative of the flow depth in most cases, and only in situations of clearly bi-modal distributions with the maximums located at the extremes of the distribution, could we regard the mean value as a non-representative measure of this variable.

Even when regarding the average flow depth value as optimal, we proceeded to analyze the change in the results that would occur if the flood hazard estimate was based on the maximum flow depth values for each cultural heritage site. Figure 9 shows the variations in the distribution of flood hazard and potential flood risk values for cultural heritage sites based on the use of the mean or maximum flow depth values. In general, and as it would be logical to use the maximum flow depth value instead of the average value, the flood hazard in cultural heritage sites increases. For flood hazard data, a clear change in BIC hazard level distribution is observed where the number of BICs in levels 5–6 (High and Extreme) increase against the number of BICs in hazard levels 1–4. In the case of potential flood risk, the shape of the distribution of BICs into risk levels remains similar, with the curve distribution for maximum flow depth always under the curve distribution for average flow depth. This distribution shape drastically changes for the highest flood risk level, where the curve associated to maximum flow depth sharply increased. Therefore, using average or maximum flow depth does not draw a completely different results scenario, but a clear difference can be pointed out between them. Maximum flow depth data leads us to an increase of high flood risk, although the possibility that these results are due to errors in the data sources used is also higher than if we use average flow depths.

These tendencies are in line with what was pointed out by [31], although here it is observed that the influence on the flood hazard estimate does not undergo modifications as significant as those shown in that study, which focused on the assignment of the flow depth to each building. In addition, the results obtained in the present work would support: (1) the consistency of the use of average flow depth as a key factor for flood hazard assessment; and (2) the negligible influence of possible anomalous flow depth point values in the study area.

Regarding the analysis of the flood vulnerability for cultural heritage sites, a qualitative approach has been proposed that takes into consideration, on the one hand, the building materials and structural relationship with the hydraulic parameters, and on the other hand, the different continent elements of each BIC. This is an approach that is undoubtedly less detailed than others (e.g., References [37,38]) previously used for the study of the vulnerability of buildings to flooding, which focuses on the analysis of vulnerability at a building scale (through the use of magnitude-damage functions in which the magnitude is associated with the variable flow depth and flow velocity or through physical models that analyze the impacts and fluid-structure interactions). The previously cited vulnerability models focus mainly on residential buildings, without the particular characteristics of cultural heritage sites.

A variable that has sometimes been introduced in the analysis of flood hazard and flood vulnerability is the flow velocity. In this study, it has not been considered relevant for two reasons: (1) in a regional study and using such generic categories of cultural heritage types, it is very difficult to estimate the incidence of the average speed on their structural stability; (2) almost all the analyzed cultural heritage sites are in the flood-prone area, although they can be affected by high flow depths, but not by flow velocity; since they are usually located outside the path of the floodway, especially the oldest ones.

In those previous assessments in which the flood risk analysis took into account elements of cultural heritage (e.g., [13,15]), the variables used are similar to those used in the present study (flood return period, potential vulnerability depending on the type of cultural heritage). However, and because the present study focuses specifically on the risk of heritage elements, the flood hazard and flood vulnerability matrices generated are more complete and complex, since they include other variables such as flow depth, flood time lag (time lag), or a more detailed classification of the cultural heritage elements and its relationship with the hydraulic environment. This means that the results obtained can be considered to be more detailed, even if the work scale is similar to the previous proposals (city or regional scale).

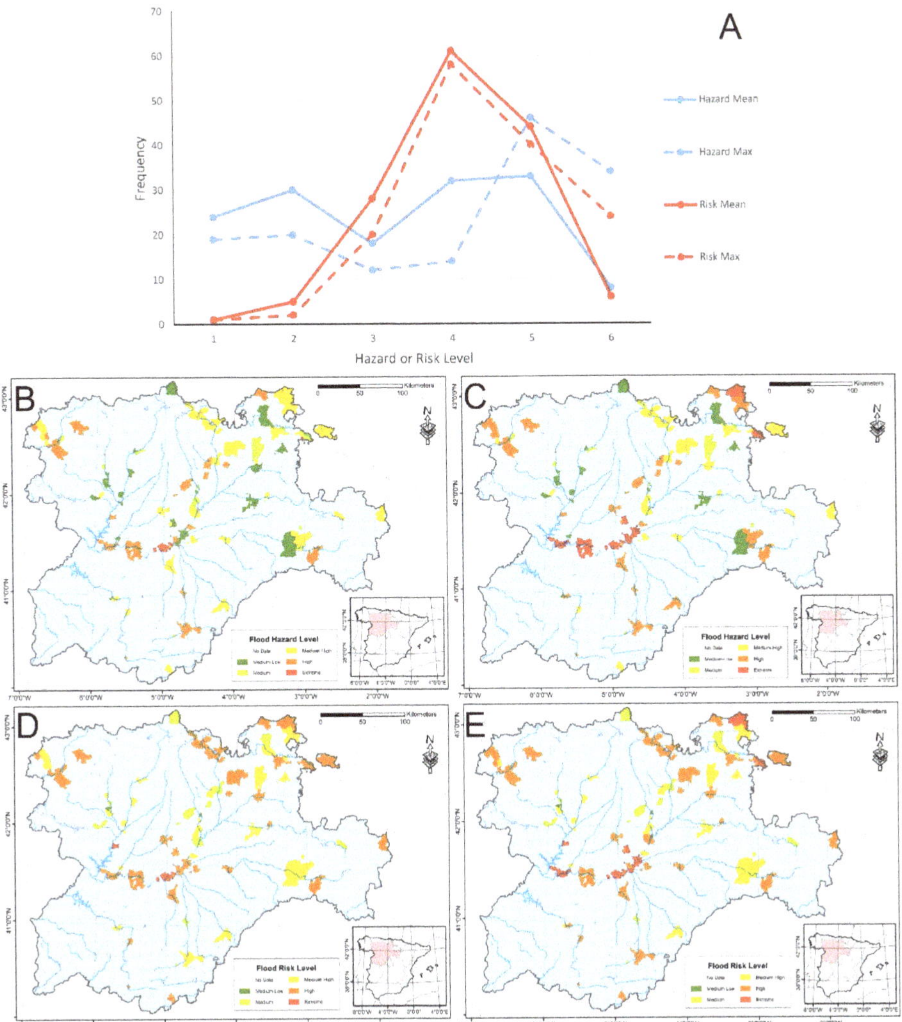

Figure 9. BICs frequency variations (**A**) for flood hazard and flood potential risk upon the use of mean or maximum flood depth values. Flood hazard spatial distribution at municipality level using average (**B**) and maximum (**C**) flow depth data. Potential flood risk spatial distribution at municipality level using average (**D**) and maximum (**E**) flow depth data.

An up-scaling process has been made, using the boundaries of the municipalities as the unit for display (Figure 9B–E). For both, flood hazard and potential risk, and increase in High and Extreme values is observed. Those changes are mainly concentrated along the Duero basin main river, and they are much less evident in the rest of the basin. Average and maximum flood hazard for each municipality (Figure 9B,C) do not show great differences, but the same patterns linked to the use of maximum values previously noted. Same situation is observed for flood risk (Figure 9D,E). Anyway, due to the variability of BICs types and vulnerabilities in a municipality, up-scaling process does not allow great advantages in analysis results beyond a simplification in its mapping.

For the working scale used here, it is possible, however, to include other variables, mainly related to the vulnerability component that would improve the estimation of the potential flood risk. These would be variables such as: (1) the structural conservation state of the cultural heritage, (2) the behavior of the construction material in contact with flood water, (3) the presence of mitigation measures against floods, etc. In relation to the experience in the region of Castile and León, however, this type of data is not included systematically in the databases of the Catalogs of Cultural Heritage Sites. To do so would require costly field and laboratory work that is not viable for analyses at the regional scale. Those data, however, are considered essential for local or micro-scale analysis.

At the assessment working scale, the limitations of the analysis come from data availability. Both hydrodynamic data (water depth, flow velocity, spatial resolution data, etc.) and vulnerability data (stored data about cultural heritage sites such as materials, structure, content, conservation, etc.) All those limitations are very similar to those pointed out by previous approach [39]. About hydrodynamic data, the scarce availability of data limits the possibility of regional analysis, and the data from European or Global projects have a spatial resolution not useful for cultural heritage sites flood risk analysis. From the point of view of vulnerability data, the limited number of tabulated variables (with much more information as plain text) makes difficult [39] a more complex and complete assessment of cultural heritage sites vulnerability to flooding.

We must keep in mind that the proposed methodological framework is a subjective proposal, based on the knowledge and authors' experience in both flood hazard and cultural heritage vulnerability. In this sense, the subjectivity of the proposal follows the line of previous proposals [40], in which the vulnerability of the cultural heritage and the flood hazard are also defined based on the authors' criteria.

The combination of the flood hazard and vulnerability matrices has made it possible to generate a matrix of potential flood risk that is easily implemented in a GIS environment and allows a quick visual analysis of the spatial distribution of the different levels of flood risk for cultural heritage elements and the identification of possible high flood risk "clusters" within them. However, a real clustering analysis may not be available due to the irregular spatial distribution of BICs and its dependence of the existence of associated flooding data. This visual analysis is also a perfect starting point for those flood risk and cultural heritage managers responsible for managing and safeguarding this irreplaceable heritage. These results can help to prioritize micro-scale flood risk analyses or directly implement flood mitigation measures in those places of greatest interest. For preventive conservation assessments at a regional scale, rather than use a high or extreme flood risk for an individual heritage site, it is convenient to look at high-risk clustering, which justifies a more detailed analysis at the local scale.

Finally, the methodological framework present here has the characteristic of being easy to replicate in other territories and settings. Flood hazard analysis may be replicated anywhere, and flood vulnerability too. About vulnerability, the matrix presented here may be used directly or may be adapted to the local characteristics of the cultural heritage sites where it will be applied. The vulnerability of each different building material, structure, or building content should be kept unchanged, but the vulnerability linked to each cultural heritage class could be different if the building material and structure or building contents have changed.

Calibration and Validation of the Methodological Proposal and the Obtained Results

The calibration and validation of the methodological proposal of this work, through its results, is very complex because it involves validating potential risk situations and not tangible and measurable scenarios, whose maps could be calibrated with the usual techniques for quality control in thematic cartographies. Therefore, in the international scientific literature, the calibration and validation of potential risk is usually addressed indirectly, through secondary risk variables and indicators [41,42]. However, in this case, the systematic quantitative validation of the results correlating them with indirect sources of past damage is also not feasible. This is due to the lack of correlation between the assigned risk and the volume and quality of documentary sources; having this lack of correlation a double cause: the scarcity and spatio-temporal bias of documentary information and the different spatial resolutions of the results and contrast information.

Regarding the scarcity of information, not even the *Spanish National Catalog of Historical Floods* collects a significant number of past events (only 308 historical records, between 181 BC and 2010, throughout the whole Duero basin), which allow representative statistical analysis at BIC scale. The spatial bias is manifested in the fact that BICs that are located in municipalities with smaller populations and in rural areas (villages), have less historical documentary information, transmitting a false sense of lower potential risk; while the opposite occurs in the large populations of historical importance (towns and cities), which were important political, religious, or military populations in the past and, therefore, with a great capacity for documentary production. The temporal bias is manifested in the greater information on recent floods (19th and 20th centuries) compared to the old ones (prior to 18th century), due to the increased recent documentary availability (historical press since the mid-19th century) and regarding the perception of risk by populations over time.

The hazard and risk results of this study (analyzed at the BIC scale) have different spatial resolution regarding the available documentary sources, ranging from the BIC (damage technical reports and repair works), to the location (village, town, or city), to the municipality or region affected by past events. Therefore, there is no certainty that past documented events will affect the BIC or only its environment ("location" and "municipality" spatial resolution documentary sources). This uncertainty affects calibration and validation analysis.

As an alternative to systematic validation using all documentary production in bulk, the only feasible strategy is the in-depth study of specific examples representing high and low risk situations among the results of this work.

To begin with the BIC with high values of flood hazard and risk, the Royal Old Mint of Segovia is an architectural complex that has an extensive documentary record on historical floods and their consequences, from descriptions of damages to machinery and facilities for more than 30 historical floods, collected in the manuscripts of archives and libraries (General Archive of Simancas), to technical reports of detailed analysis of flooding (historical floods, [43,44]) and possible mitigation solutions. There is also a good qualitative correlation between another of the higher risk BICs, as "certain areas of the city of Medina de Rioseco" and the existence of technical reports on damage conditions by floods. Another positive example is the San Francisco convent in the city of Medina de Rioseco (medium risk) and how the mitigation actions carried out have reduced the flooding of this BIC [45].

In the opposite situation, that is, BIC with lower flood hazard and risk and few conditions and past damages, it is complex to find reliable examples, due to the mentioned space-time bias of the documentation. In that way, there are some BICs, such as the Roman villages of Camarzana de Tera (Orpheus) or Quintana del Marco (Los Villares), which are located in the flood zone for T500 years, and apparently have not suffered significant flood damage in the last 1600–1700 years since their construction (as evidenced by the conservation of very vulnerable elements, such as mosaics), so their flood risk must have been low or very low, as they have been assigned in the results of this study.

In summary, it can be considered that the methodology used and the results obtained have an acceptable agreement with the historical and actual flood risk of Castile and León BICs, not so much for a complete systematic validation but for the timely analysis of the results for representative singular BICs, which makes them useful as a qualitative approach to risk.

5. Conclusions

An easy-to-implement and low-cost methodology is presented here to determine the potential flood risk of cultural heritage sites at a regional scale. The methodological framework improves previously proposed approaches to cultural heritage flood risk assessment.

The methodological framework (with a focus on analysis at a regional scale) has been implemented in the region of Castile and León (Spain), which has a *Catalog of Cultural Heritage Sites* composed of more than two thousand elements that have different degrees of administrative protection. Although the model calibration and validation are not fully available due to the scarce data on historic flood events and damages, cultural heritage sites with different potential flood risk have been used to check the reliability of the model results, having obtained a good agreement between them.

The implementation of the three proposed matrices (flood hazard, flood vulnerability, and potential flood risk) in a GIS environment allows the estimation of the flood risk level presented by the cultural heritage elements and its classification.

The matrices implemented in the present assessment are based on a series of variables related to flood risk that are generally available or can be easily derived from DTMs. These characteristics make the study and proposed methodological framework easy to replicate.

The qualitative characterization of the potential flood risk for the heritage site and its spatial distribution, both direct results of the methodological framework proposed here, can and should serve as a key tool for cultural heritage managers at a city, region, or state administrative level. This should be presented in a way that facilitates the tasks of prioritization and allocation of economic items for the protection of cultural heritage, through preventive conservation.

Finally, the results obtained through the methodological framework proposed can and must also make it easier to prioritize flood risk analysis on heritage elements at a local or micro-scale level. The micro-scale analysis, due to its inherent level of detail, may use specific or ultra-detailed magnitude damage functions (in which the magnitude is defined from the combination of flow depths–flow velocities–flooding times–flood sediment) or physical models that analyze the interactions between fluid and structure (distribution of pressures, deformations, thresholds of collapse of structures, etc.).

Author Contributions: All authors made a significant contribution to the final version of the paper. Conceptualization, J.G. and A.D.-H.; flood risk analysis, J.G. and A.D.-H.; BICs data, C.E. and I.G.; writing—original draft, J.G. and A.D.-H.; writing—review and editing, all authors. All authors have read and agreed to the published version of the manuscript.

Funding: This research was funded by the project DRAINAGE, CGL2017-83546-C3-R (MINEICO/AEI/FEDER, UE), and it is specifically part of the assessments carried out under the coverage of the sub-project DRAINAGE-3-R (CGL2017-83546-C3-3-R).

Acknowledgments: The authors would like to thank to the Planning, Research and Dissemination Service of the General Directorate of Cultural Heritage of the Castile and León regional government for granting access to BICs data; and finally, Juan Francisco Arazola (Directorate General of Water of the Spanish Ministry of the Environment and Rural and Marine Affairs) for granting access to flood depth data of SNCZI.

Conflicts of Interest: The authors declare no conflict of interest.

References

1. Carta del Rischio. Available online: www.cartadelrischio.it/index.asp (accessed on 12 November 2019).
2. NOAH's Ark Project. Available online: https://cordis.europa.eu/publication/rcn/11779_en.html (accessed on 12 November 2019).
3. Ghose, S. Protection against natural and man-made disasters. In *Risk Preparedness for Cultural Properties: Development of Guidelines for Emergency Response*; Saito, H., Ed.; Chuo-Koron Bijustu Shuppan Co., Ltd.: Tokyo, Japan, 1999; p. 159.
4. Jung, F.C. From "Blue Shield" to disaster management: The awareness and actions of risk preparedness from world's cultural property. *J. Cult. Property Conserv.* **2010**, *12*, 43–56.
5. Štulc, J. The 2002 Floods in the Czech Republic and their impact on built heritage. In *Heritage at Risk. Cultural Heritage and Natural Disasters*; Meier, H.-R., Petzet, M., Will, T., Eds.; ICOMOS TUDPress: Dresden, Germany, 2007; pp. 133–137.
6. Will, T. Integrating technical flood protection and heritage conservation planning for Grimma, Saxony. In *Heritage at Risk. Cultural Heritage and Natural Disasters*; Meier, H.-R., Petzet, M., Will, T., Eds.; ICOMOS TUDPress: Dresden, Germany, 2007; pp. 139–152.
7. Lanza, S.G. Flood hazard threat on cultural heritage in the town of Genoa (Italy). *J. Cult. Herit.* **2003**, *4*, 159–167. [CrossRef]
8. Holicky, M.; Sykora, M. Assessment of flooding risk to cultural heritage in historic sites. *J. Perform. Constr. Facil.* **2010**, *24*, 432–438. [CrossRef]

9. Daungthima, W.; Kazunori, H. Assessing the flood impacts and the cultural properties vulnerabilities in Ayutthaya, Thailand. *Procedia Environ. Sci.* **2013**, *17*, 739–748. [CrossRef]

10. Vojinovic, Z.; Hammond, M.; Golub, D.; Hirunsalee, S.; Weesakul, S.; Meesuk, V.; Medina, N.; Sanchez, A.; Kumara, S.; Abbott, M. Holistic approach to flood risk assessment in areas with cultural heritage: A practical application in Ayutthaya, Thailand. *Nat. Hazards* **2015**, *81*, 1–28. [CrossRef]

11. Naderi, M.; Raeisi, E.; Talebian, M.H. Effect of Extreme Floods on the Archaeological Sites of Persepolis and Naghsh-e-Rostam, Iran. *J. Perform. Constr. Fac.* **2014**, *28*, 502–510. [CrossRef]

12. Wang, J.J. Flood risk maps to cultural heritage: Measures and process. *J. Cult. Herit.* **2015**, *16*, 210–220. [CrossRef]

13. Ortiz, R.; Ortiz, P.; Martín, J.M.; Vázquez, M.A. A new approach to the assessment of flooding and dampness hazards in cultural heritage, applied to the historic centre of Seville (Spain). *Sci. Total Environ.* **2016**, *551–552*, 546–555. [CrossRef]

14. Li, H.; Zhang, J.; Sun, J.; Wang, J. A visual analytics approach for flood risk analysis and decision-making in cultural heritage. *J. Vis. Lang. Comput.* **2017**, *41*, 89–99. [CrossRef]

15. Arrighi, C.; Brugioni, M.; Castelli, F.; Franceschini, S.; Mazzanti, B. Flood risk assessment in art cities: The exemplary case of Florence (Italy). *J. Flood Risk Manag.* **2018**, *11*, S616–S631. [CrossRef]

16. Drdácký, M.F. Flood Damage to Historic Buildings and Structures. *J. Perform. Constr. Fac.* **2010**, *24*, 439–445. [CrossRef]

17. Díez-Herrero, A. Patrimonio e inundaciones: ¿una relación de riesgo? In *Actas del Congreso Internacional "Patrimonio Cultural y Catástrofes: Lorca como referencia", Lorca, Spain, October 3 2018*; Asociación Hispania Nostra: Lorca, Spain, 2018.

18. Herle, I.; Herbstová, V.; Kupka, M.; Kolymbas, D. Geotechnical Problems of Cultural Heritage due to Floods. *J. Perform. Constr. Fac.* **2010**, *24*, 446–451. [CrossRef]

19. Aroca-Jiménez, E.; Bodoque, J.M.; Antonio García, J.; Díez-Herrero, A. Construction of an integrated social vulnerability index in urban areas prone to flash flooding. *Nat. Hazards Earth Syst. Sci.* **2017**, *17*, 1541–1557. [CrossRef]

20. Aroca-Jiménez, E.; Bodoque, J.M.; García, J.A.; Díez-Herrero, A. A quantitative methodology for the assessment of the regional economic vulnerability to flash floods and implications for risk management. *J. Hydrol.* **2018**, *565*, 386–399. [CrossRef]

21. Lastra, J.; Fernández, E.; Díez-Herrero, A.; Marquínez, J. Flood hazard delineation combining geomorphological and hydrological methods: An example in the Northern Iberian Peninsula. *Nat. Hazards* **2008**, *45*, 277–293. [CrossRef]

22. MARM. *Guía Metodológica para el Desarrollo del Sistema Nacional de Cartografía de Zonas Inundables*; Ministerio de Medio Ambiente, y Medio Rural y Marino: Madrid, Spain, 2011; 349p.

23. Ortiz, P.; Antunez, V.; Martín, J.M.; Ortiz, R.; Vázquez, M.A.; Galán, E. Approach to environmental risk analysis for the main monuments in a historical city. *J. Cult. Herit.* **2014**, *15*, 432–440. [CrossRef]

24. USACE. *Flood Proofing Performance: Successes and Failures*; USACE (United States Army Corps of Engineers) National Flood Proofing Committee: Washington, DC, USA, 1998; 120p.

25. Creutin, J.D.; Borga, M.; Gruntfest, E.; Lutoff, C.; Zoccatelli, D.; Ruin, I. A space and time framework for analyzing human anticipation of flash floods. *J. Hydrol.* **2013**, *482*, 14–24. [CrossRef]

26. Borga, M.; Stoffel, M.; Marchi, L.; Marra, F.; Jakob, M. Hydrogeomorphic response to extreme rainfall in headwater systems: Flash floods and debris flows. *J. Hydrol.* **2014**, *518*, 194–205. [CrossRef]

27. Kelman, I.; Spence, R. An overview of flood actions on buildings. *Eng. Geol.* **2004**, *73*, 297–309. [CrossRef]

28. Kreibich, H.; Piroth, K.; Seifert, I.; Maiwald, H.; Kunert, U.; Schwarz, J.; Merz, B.; Thieken, A.H. Is flow velocity a significant parameter in flood damage modelling? *Nat. Hazards Earth Syst. Sci.* **2009**, *9*, 1679–1692. [CrossRef]

29. Lonetti, P.; Maletta, R. Dynamic impact analysis of masonry buildings subjected to flood actions. *Eng. Struct.* **2018**, *167*, 445–458. [CrossRef]

30. Bellos, B.; Tsakiris, G. Comparing Various Methods of Building Representation for 2D Flood Modelling In Built-Up Areas. *Water Resour. Management* **2015**, *29*, 379–397. [CrossRef]

31. Bermúdez, M.; Zischg, A.P. Sensitivity of flood loss estimates to building representation and flow depth attribution methods in micro-scale flood modelling. *Nat. Hazards* **2018**, *92*, 1633–1648. [CrossRef]

32. Vaze, J.; Teng, J.; Spencer, G. Impact of DEM accuracy and resolution on topographic indices. *Environ. Modell. Softw.* **2010**, *25*, 1086–1098. [CrossRef]

33. Bodoque, J.M.; Guardiola-Albert, C.; Aroca-Jimenez, E.; Eguibar, M.A.; Martinez-Chenoll, M.L. Flood damage analysis: First floor elevation uncertainty resulting from LiDAR derived digital surface models. *Remote Sens.* **2016**, *8*, 604. [CrossRef]

34. Molinari, D.; Scorzini, A.R. On the Influence of Input Data Quality to Flood Damage Estimation: The Performance of the INSYDE Model. *Water* **2017**, *9*, 688. [CrossRef]

35. Shen, D.; Qian, T.; Chen, W.; Chi, Y.; Wang, J. A Quantitative Flood-Related Building Damage Evaluation Method Using Airborne LiDAR Data and 2-D Hydraulic Model. *Water* **2019**, *11*, 987. [CrossRef]

36. Arrighi, C.; Campo, L. Effects of digital terrain model uncertainties on high-resolution urban flood damage assessment. *J. Flood Risk Manag.* **2019**, e12530. [CrossRef]

37. Mazzorana, B.; Simoni, S.; Scherer, C.; Gems, B.; Fuchs, S.; Keiler, M. A physical approach on flood risk vulnerability of buildings. *Hydrol. Earth Syst. Sci.* **2014**, *18*, 3817–3836. [CrossRef]

38. Custer, R.; Nishijima, K. Flood vulnerability assessment of residential buildings by explicit damage process modelling. *Nat. Hazards* **2015**, *78*, 461–496. [CrossRef]

39. Figueiredo, R.; Romão, X.; Paupério, E. Flood risk assessment of cultural heritage at large spatial scales: Framework and application to mainland Portugal. *J. Cult. Herit.* **2019**, in press. [CrossRef]

40. Romão, X.; Paupério, E.; Pereira, N. A framework for the simplified risk analysis of cultural heritage assets. *J. Cult. Herit.* **2016**, *20*, 696–708. [CrossRef]

41. De Moel, H.; Jongman, B.; Kreibich, H.; Merz, B.; Penning-Rowsell, E.; Ward, P.J. Flood risk assessments at different spatial scales. *Mitig. Adapt. Strateg. Glob. Chang.* **2015**, *20*, 865–890. [CrossRef] [PubMed]

42. Garrote, J.; Gutiérrez-Pérez, I.; Díez-Herrero, A. Can the quality of the potential flood risk maps be evaluated? A case study of the social risks of floods in Central Spain. *Water* **2019**, *11*, 1284. [CrossRef]

43. Génova, M.; Ballesteros-Cánovas, J.A.; Díez-Herrero, A.; Martínez-Callejo, B. Historical Floods and Dendrochronological Dating of a Wooden Deck in the Old Mint of Segovia, Spain. *Geoarchaeology* **2011**, *26*, 786–808. [CrossRef]

44. Génova, M.; Díez-Herrero, A.; Rodríguez-Pascua, M.A.; Moreno-Asenjo, M.A. Natural disasters written in historical woods: Floods, a thunderbolt fire and an earthquake. *J. Cult. Herit.* **2018**, *32*, 98–107. [CrossRef]

45. Boletín Oficial del Estado (BOE). Available online: http://www.boe.es/datos/pdfs/BOE/1955/188/A04088-04089.pdf (accessed on 13 May 2019).

Article

Flood Monitoring Based on the Study of Sentinel-1 SAR Images: The Ebro River Case Study

Francisco Carreño Conde [1,2,*] and María De Mata Muñoz [1]

[1]   Geology area, ESCET, Rey Juan Carlos University, c/ Tulipán s/n, 28933 Móstoles, Madrid, Spain; dematamunozmaria@gmail.com

[2]   IMDEA Water Institute, Avenida Punto Com 2, Parque Científico Tecnológico de la Universidad de Alcalá, 28805 Alcalá de Henares, Madrid, Spain

*   Correspondence: francisco.carreno@urjc.es

Received: 2 October 2019; Accepted: 16 November 2019; Published: 22 November 2019

Abstract: Flooding is the most widespread hydrological hazard worldwide that affects water management, nature protection, economic activities, hydromorphological alterations on ecosystem services, and human health. The mitigation of the risks associated with flooding requires a certain management of flood zones, sustained by data and information about the events with the help of flood maps with sufficient temporal and spatial resolution. This paper presents the potential use of the Sentinel-1 SAR (Synthetic Aperture Radar) images as a powerful tool for flood mapping applied in the event of extraordinary floods that occurred during the month of April 2018 in the Ebro (Spain). More specifically, in this study, we describe accurate and robust processing that allows real-time flood extension maps to be obtained, which is essential for risk mitigation. Evaluating the different Sentinel-1 parameters, our analysis shows that the best results on the final flood mapping for this study area were obtained using VH (Vertical-Horizontal) polarization configuration and Lee filtering 7×7 window sizes. Two methods were applied to flood maps from Sentinel-1 SAR images: (1) RGB (Red, Green, Blue color model) composition based on the differences between the pre- and post-event images; and (2) the calibration threshold technique or binarization based on histogram backscatter values. When comparing our flood maps with the flood areas digitalized from vertical aerial photographs, done by the Hydrological Planning Office of the Ebro Hydrographic Confederation, the results were coincident. The result of the flooding map obtained with the RADAR (Radio Detection and Ranging) image were compared with the layers with different return periods (10, 50, 100, and 500 years) for a selected zone of the study area of SNCZI (National Flood Zone Mapping System in Spain). It was found that the images are consistent and correspond to a flood between 10 and 50 years of return. In view of the results obtained, the usefulness of Sentinel-1 images as baseline data for the improvement of the methodological guide is appreciated, and should be used as a new source of input, calibration, and validation for hydrological models to improve the accuracy of flood risk maps.

Keywords: Ebro River; flood mapping; flood risk areas; RADAR SAR; Sentinel-1; RGB composition; calibrated thresholding

---

1. Introduction

A river flood occurs when the river flow can no longer be contained within its bed, and over spills its banks. Flooding is a natural and regular reality for many rivers, caused by any pulse of overflowing water (heavy rainfall, peak seasonal rains, or seasonal snow melt) that overwhelms a river channel, which supports the most naturally dynamic ecosystems, modeling soil and re-distributing nutrients in associated alluvial lands [1]. However, humans often perceive floods negatively due to damage and loss of life.

According to the United Nations [2], flooding is the most widespread hydrological hazard worldwide. According to this report, between 1995 and 2015, there were 3,062 flood disasters, representing 56% of all weather-related disasters, which affected 2 to 3 billion people, with a total of 157,000 fatal casualties. A report from the European Environment Agency (EEA) [3] found that from 1980 to 2010, European countries registered 3563 floods in total. The highest number of floods was reported for 2010 (321 floods), with 27 countries affected. The data analyses on these floods indicate significant increases in floods and envisage that flood losses will be multiplied by five as a consequence of climate change, the increase in urban development, and the price of land around floodplains. In Spain, there is an average of 10 serious flood episodes per year. Floods have caused the death of 312 people in the last 20 years and material damages worth about 800 million euros per year [4].

Extreme flooding is a hydrological event that affects water management, conservation efforts, economic activities, hydromorphological alterations on the ecosystem services, and human health. It presents a large number of technical challenges in terms of efficient management and public policy worldwide [5]. The mitigation of the risks associated with flooding requires certain management of flood zones sustained by data and information about the events. The estimation of damages caused by floods and their associated impacts must be supported by data on water levels, discharges, and flooded areas [6], and the integration of this data into a mathematical model for hydrological analysis of the related factors. When major progressive flooding occurs, it is essential to delimit the affected areas, and to be able to assess the damage caused, in order to provide a "ground truth" of great interest for better empirical knowledge of the water environment. These accurate mappings of the flooded areas are needed when monitoring and evaluating the status, impact, and effectiveness of efforts [7]. The delimitation of flooded areas through the collection of data in the field used to be complicated because access and mobility through the affected areas is difficult. Aerial flights to take vertical photographs used to generate mosaics of georeferenced images and High Definition (HD) videos are a common alternative. The flights are scheduled to fly over each of the sections at the peak of flooding; however, they have a high economic cost and are limited by weather conditions.

Remote sensing techniques are an effective source of information to discriminate bodies of water over large areas, and therefore, they can be used to map flooded surfaces with sufficient temporal and spatial resolution [8]. In the case of multispectral images, water bodies can be easily discriminated because of their spectral behavior in the visible and infrared spectrum because it is different from other land covers [9,10], but they only operate in natural daylight conditions and cloudless skies. Since severe flooding usually occurs as a result of heavy rainfall, they are not always useful for this purpose. In Spain, due to the physiographic characteristics of the basins and the Mediterranean climatic conditions, most floods are associated with periods of heavy rainfall. The validity of multispectral observations for this purpose is limited by clouds because unclear weather conditions are common during floods events [11]. However, the Synthetic Aperture Radar (SAR) is a form of radar that is used to create more useful images over a target region to provide fine spatial resolution [12,13], which can operate day and night and is unaffected by cloud cover, smoke, atmospheric water, or hydrometeors. Moreover, the signal can penetrate through foliage (wavelength dependent), and therefore can provide more complete information about the inundation state [14,15]. SAR based flood mapping methods are usually more complex than those based on optical sensors because of the process added to mitigate the error propagated from one or more of the error sources [11,16].

Accordingly, in the present study, we investigated the potential use of the Sentinel-1 SAR images as a powerful and complete tool for flood mapping, and as a crucial element of flood management [17]. This methodology was applied in the case of extraordinary floods that occurred during the month of April 2018 in the Ebro River due to the snow that was melting in the headwater basins. The specific goals were: (1) Analysis of the affected area and hydraulic parameters characteristic of the Ebro flood case; (2) selection of the optimal processing and correction options (orbital parameters, radiometric, noise reduction—speckle, and geometric geometric) to perform an accurate and thorough treatment of

Sentinel-1 SAR images, in order to optimize the detection and analysis of flooded areas; (3) application of two detection algorithm approaches (RGB—Red, Green, Blue color model - composition and calibration threshold technique) of free water surfaces to derive flood cartography from SAR images, which permits the delimitation of non-permanent water surfaces; and (4) final comparison and analysis between the flooded area maps derived from SAR images, the official orthophotography provided by the Ebro Hydrographic Confederation (CHE, acronym in Spanish), and flood risk areas by the National Flood Zone Mapping System (SNCZI acronym in Spanish) provided by the Spanish Ministry for the Ecological Transition (MITECO, acronym in Spanish).

2. Study Area and Data Source

The Ebro river runs through seven autonomous communities and its hydrographic basin is one of the largest in the Iberian Peninsula. Its great extent favors geological, topographic, climatic, and water balance variability. The study area was focused in the middle basin of the Ebro, from the villages of Novillas to El Burgo de Ebro, and this entire section of the Zaragoza province (Figure 1). It is characterized by an average and very smooth slope; it has a high sinuous index with a meandering morphology and wide floodplains [18]. Irrigated fields, such as wheat, alfalfa, corn, barley, or rice, and permanent crops, such as vineyards, olive groves, or apple trees, are prevalent on the banks of the river [19].

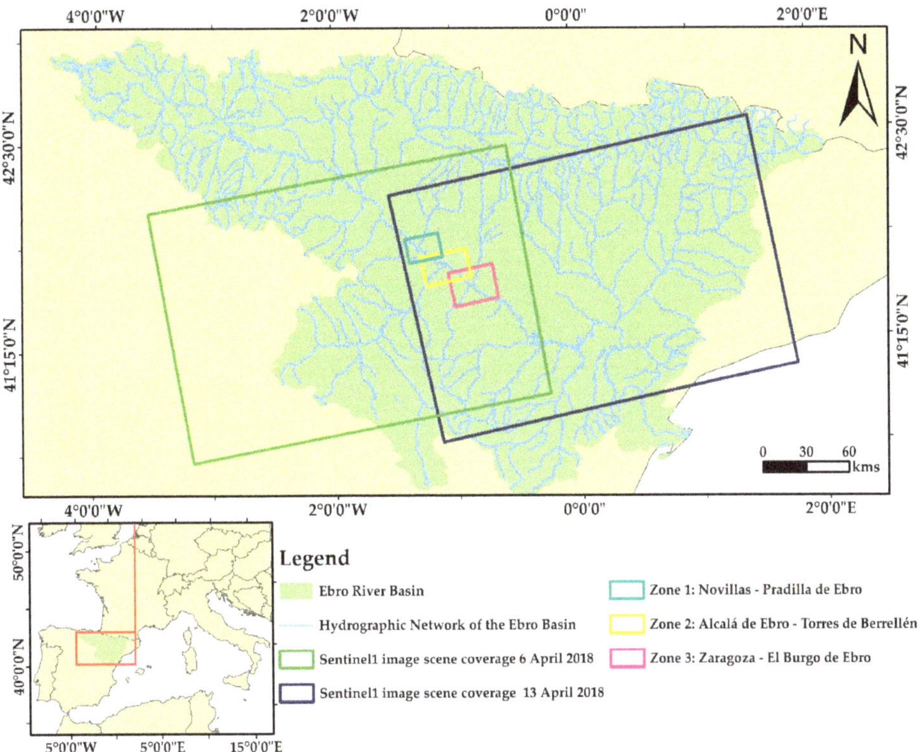

Figure 1. Location map and study areas.

The study of the great historical floods of the Ebro river, such as those that occurred in 1643, 1775, 1871, or 1961, and more recent ones, such as those of 2003 or 2015, provides information on the changing dynamics of the river's course and how the Ebro has tried to consolidate in the flood zone, encountering man-made barriers to prevent the flooding from occurring [20]. The Ebro is one of the most regulated rivers in the Iberian Peninsula and suffers from frequent flooding [21]. The data supplied by the Hydrological Information Systems (SAIH, acronym in Spanish), in operation since the end of 1997 at the CHE, has facilitated the correct management of the reservoirs of the basin and has helped to control the floods since then. They provide, among other data, information on the levels and water flows in the main rivers and affluents, as well as the level of dammed water in the reservoirs, through sensors or stations located therein. Table 1 shows the maximum values in the last three large floods of the Ebro River in 2003, 2015, and 2018 with data provided by the gauging stations of Castejón and Zaragoza [22]. The floods of 2003 and 2015 left maximum flow values higher than those of 2018. This is due, in part, to the fact that after the flood of 2015, measures were taken to improve the safety of the populations closest to the channel, restore the riverbed, create new deviation channels, and construct new containment infrastructure [23].

Table 1. Representative values on the annual levels and flows of the Ebro river in two gauging stations, Castejón and Zaragoza, in its course for the years 2003, 2015, and 2018. Historical data taken from the SAIH of the CHE, 2019.

Station Annual Gauging Average	Annual Average Level of Ebro (m)	Annual Average Flow of Ebro (m³/s)	Day	Hour	Annual Maximum Level of Ebro (m)	Annual Maximum Flow of Ebro (m³/s)
Castejón	2.31	226	6 February 2003	02:45	7.54	2847
	2.63	261	27 February 2015	00:00	7.78	2691
	2.82	265	13 April 2018	9:30	7.77	2682
Zaragoza	1.41	258.78	9 February 2003	03:00	5.76	2237
	1.38	268.95	2 March 2015	02:00	6.10	2448
	1.5	288.52	15 April 2018	19:45	5.36	2037

The winter of 2018 was one of the rainiest since records were first kept. In late February and in March, there were considerable episodes of rain and snowfall, accompanied by wind, throughout the Iberian Peninsula. This produced a high accumulation of water in the Ebro River Basin, especially in the upper and middle sections, and snow at the headwaters of the rivers. This was aggravated by the fact that extreme values of soil moisture had been reached, saturating it and increasing the runoff. Days before the event, some reservoirs in the basin had been preventively drained. Between 7 and 12 April, an extraordinary flood, which is the object of study of this article, took place. This was due to persistent, widespread, and very significant rainfall, and also snowfall at heights between 800 and 1600 m, with lower temperatures that were unusual for this time of the year. To this was added, in the previous days, snow thawing in the Pyrenees in the early weeks of spring. In the first half of April, it rained in the basin more than three times the normal amount [24].

2.1. Hydraulic Parameters Characteristic of the Flood

To know the maximum flow dates during the flood and to confirm the adequacy of the dates of the Sentinel-1 SAR images, data were taken from some of the gauging stations that exist along the course of the Ebro River (Figure 2). The values at the stations of Castejón and Tudela, in the province of Navarra and upstream of the study area, were taken into account, despite not being within the delimited study area. These give us an idea of the development of the flood and how the maximum flood points have been moving along down along the Ebro axis with time.

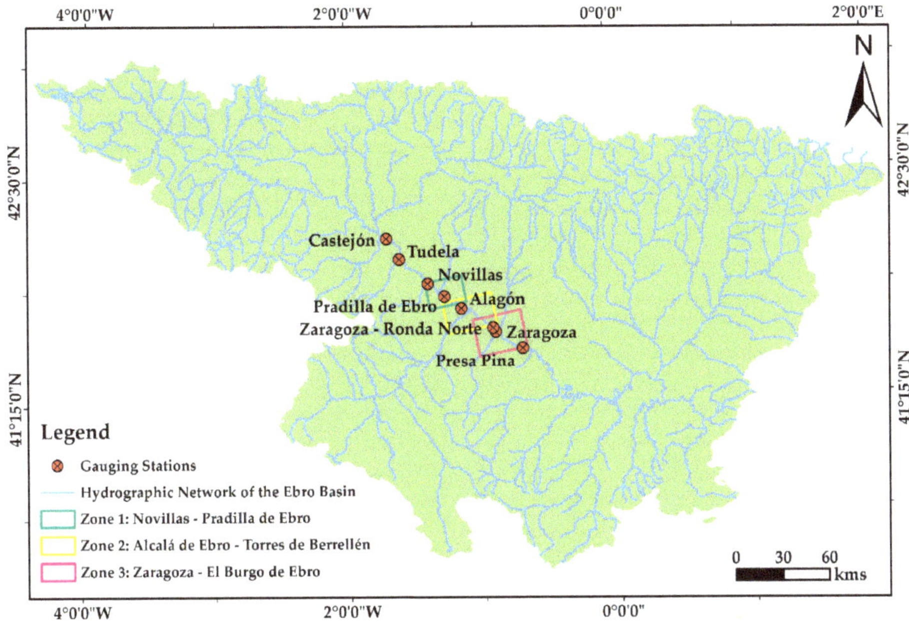

Figure 2. Location map of the gauging stations considered in the study.

Table 2, representative of the year 2018, shows the values collected at the stations. This reflects the extraordinary flood of the Ebro in the month of April when it reached the highest values of the year. These data show the evolution of the flood from 13 to 15 April, obtaining the maximum annual level of accumulation in the Pina Dam, below the village of El Burgo de Ebro, on 15 April at 9:30 pm at 6.03 m, having had an average annual level of 4.48 m [22].

Table 2. Representative values on the annual levels and flows of the Ebro river in different stations located in its course for the year 2018. Historical data taken from the SAIH of the CHE, 2019.

Gauging Station	2018		Day	Hour	Annual Maximum Level of Ebro (m)	Annual Maximum Flow of Ebro (m^3/s)
	Annual Average Level of Ebro (m)	Annual Average Flow of Ebro (m^3/s)				
Castejón	2.82	265	13 April 2018	9:30	7.77	2682
Tudela	1.22	270	14 April 2018	01:00	5.34	2413
Novillas	2.63	-	14 April 2018	11:45	8.24	-
Pradilla de Ebro	3.11	-	14 April 2018	20:30	8.51	-
Alagón	0.88	-	14 April 2018	07:30	2.26	-
Zaragoza—Ronda Norte	2.61	-	15 April 2018	17:15	8.02	2041
Zaragoza	1.5	288.52	15 April 2018	19:45	5.36	2037

2.2. Sentinel-1 SAR Data

Synthetic Aperture RADAR (Radio Detection and Ranging) is a powerful active remote sensing technology used for several applications, including flood monitoring. RADAR are active systems that illuminate the Earth's surface and therefore, images can be acquired by day during any lighting conditions, or at night, in the total absence of sunlight. In addition, these images are not affected by cloud cover, fog, or smoke because the RADAR signals have the ability to penetrate these covers [25,26].

Under the weather conditions in which floods occur, these types of images are the most suitable for study. Active radar systems incorporate an antenna that generates and transmits electromagnetic pulses (between 0.3 and 300 cm) to the Earth's surface, and after interacting with the objects, the echoes of the backscattered signal are collected and amplified by a receiver at certain frequencies and polarization modes. Currently, there is a wide variety of SAR missions that have improved spatial resolution, bands, revisiting times, and polarization modes that have increased the monitoring potential [27].

The backscattering coefficient, or sigma nought (σ_0), is the measure of the reflective strength of a radar target, defined as the per unit area on the ground, and is defined as the normalized measure of the radar return from a distributed target [28]. The amplitude and phase of the backscattered signal depends, among others, on the land surface properties, including the geometry, vegetation, roughness, and dielectric constant [29,30].

Continental waters, due to their low or null roughness in the absence of waves, behave like a specular surface, reflecting the radar return signal in a direction opposed to the sensor position, and therefore show very low backscatter values (negative) with respect to other land covers. This strong contrast in backscatter values allows the precise delimitation of flooded areas with SAR images. When a rough terrain has slopes with an angle that is more steep than the sensor depression angle, it generates shadow regions that appear without signal, very similar to the flooded areas, which can easily be confused. As flood areas tend to be flat, this effect is minor and is usually correctly delimited with detailed topographic information [31].

In this study, it is been used Sentinel-1 SAR images because of their features, configuration, and the free data set available online from Earth Observation (EO) data resources. Sentinel-1 SAR datasets, coming before and after the event, were downloaded via the Copernicus Open Access Hub platform [32]. The Sentinel-1 sensor is part of an ESA (European Space Agency) satellite constellation. It is composed of Sentinel-1A, launched on 3 April 2014, and Sentinel-1B, on 22 April 2016. Both satellites have a useful life of 7 years and a 12-day orbital repetition cycle (6 days between them). These temporal limitations of image capture mean that we did not have images of the day of the highest level of the Ebro at the study points. The available image that was closest to the moment of maximum flood was from 13 April 2018 and it was the one used for the development of the study.

Table 3 shows the characteristics of the two images used in the study. Usually, depending on the methodology used, the pre-event image serves as a reference and record of permanently flooded areas, while the post-event image serves to establish the differences with the initial image [33].

Table 3. Characteristics of the Sentinel-1 SAR images used in the study.

Image	Mission Identifier	Date of Capture	Hour of Capture
A	Sentinel-1A	6 April 2018	18:03
B	Sentinel-1B	13 April 2018	17:55

Each image has a coverage range of 250 km with a 5 × 20 m resolution. The images are Level-1 Ground Range Detected (GRD) products; this means that they were detected, multi-looked, and projected to the ground range using an Earth ellipsoid model [34]. The acquisition mode is Interferometric Wide Swath, and the images have a high resolution; the sensor pass is ascending, with incidence angles between 29° and 46°, and dual polarization, VV (Vertical-Vertical) and VH (Vertical-Horizontal). The study area is covered by VV and VH only according to the observation scenarios shown in Earth Observation Data Center for Water Resources Monitoring (EODC). In the free modes of IW (Interferometric Wide) Sentinel-1, only VV + VH polarization over land is available [35].

2.3. Additional Ebro Hydrographic Confederation Data

The Hydrological Planning Office of the CHE has been tracking air flights during flood episodes since the extraordinary flood that took place in February 2003. Flights are scheduled according to the time of the maximum flood peak as long as weather conditions allow it. Through these flights, oblique and vertical aerial photographs, as well as videos, are obtained, all in high resolution. This information provides the basis for the cartographic restitution of orthophotographs of the National Aerial Ortophotography Program (PNOA, acronym in Spanish) included in the General Spanish National Plan for Territory Observation (PNOT, acronym in Spanish) and to obtain, by digitization, the flood areas corresponding to each flood. This favors a greater understanding of the water behavior of the Ebro river over time.

For the study area, a vertical aerial photograph corresponding to a part of the section that the CHE called "Ribera Alta" was used (Figure 3). It covers a region that goes from the village of Novillas to Alagón and was captured on 14 April 2018.

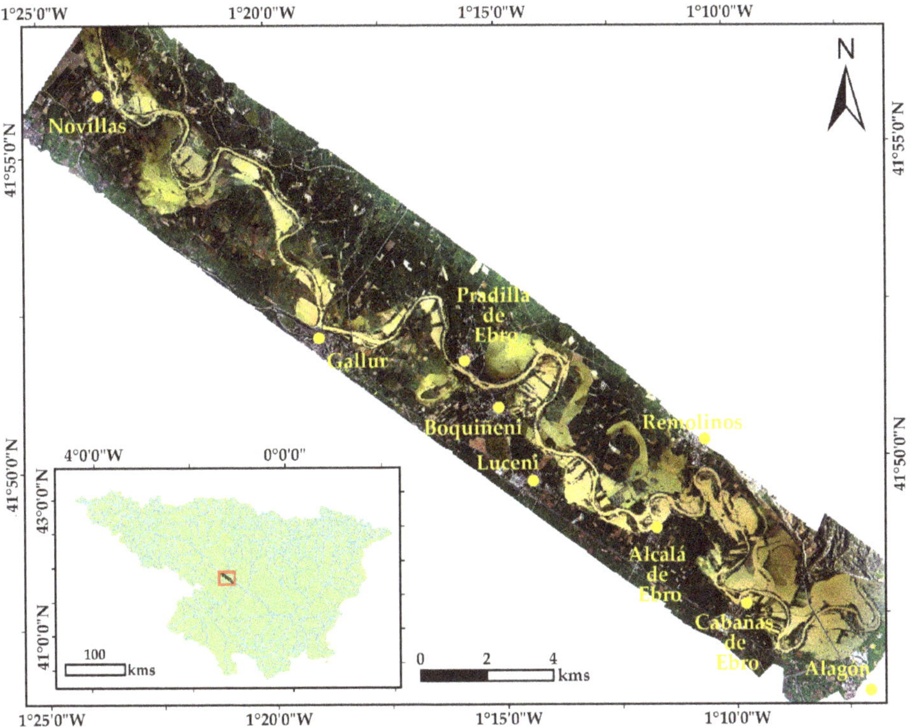

Figure 3. Ribera Alta orthophotography dated 14 April 2018. (Orthophotographs source: CHE, 2018 [36]).

3. Procedures and Methodologies

3.1. SAR Sentinel-1 Data Pre-Processing

The images were processed with the free software SNAP (Sentinel Application Platform) Tool version 6.0.0. [37], created specifically by ESA for the analysis of the data captured by Sentinel satellites. It is a very powerful program that has the right tools, Sentinel Toolboxes, to perform an accurate and thorough treatment of the images and subsequently run the analysis. In Figure 4, the sequence of the steps followed for the pre-treatment of the images is shown.

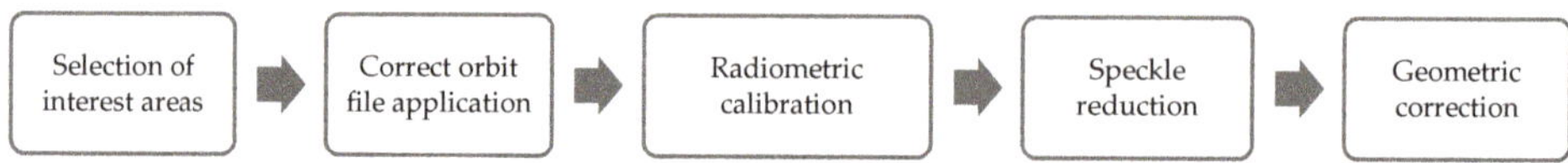

Figure 4. Flowchart showing the preprocessing chain used in RADAR SAR Sentinel-1 images.

In order to give continuity to the study, contribute to correct manipulation of images, and reduce the processing time, three subdivisions were made in each image, resulting in three areas of interest named after the most important towns and villages. These were zone 1—Novillas–Pradilla de Ebro, zone 2—Alcalá de Ebro–Torres de Berrellén, and zone 3—Zaragoza–El Burgo de Ebro (Figure 1).

The application of the right orbit helps to make the orbital data more accurate, improving geocoding and the subsequent processing results. With radiometric calibration, the value representing the radar backscatter intensity, in units of decibels, is taken for each pixel, and σ_0-bands are generated for the two polarizations that we provide, VH and VV [38,39]. There is the possibility of obtaining other bands, but the σ_0-band is the one that provides the best separation between terrestrial soil areas and those with water [40].

Permanent water bodies have temporal heterogeneity regarding backscattering [41]. The backscatter intensity values are usually below −20 dB and manifest themselves in dark, practically black tones. Applying the theoretical knowledge characteristic of the flat areas, the wave beams emitted by the sensor are reflected on the surface, which acts as a mirror, causing almost no return of energy towards the sensor; hence, its intensity value and representation in the radar image. There are surfaces that, at the scale of the measuring wavelength, behave as smooth and share almost identical scattering properties with water surfaces that can create "false positives" [42].

Speckle is represented in the images as mottled or granular noise. This causes degradation of the image quality and hinders its interpretation [43]. Mainly, noise is due to random oscillations of the signal returned to the sensor due to the interaction of the emitted wave with the rough terrain surfaces. There is no fixed procedure to suppress the speckle of RADAR images in flood extent mapping. The characteristics and configuration of each RADAR system, together with the conditions of the acquisition and characteristics of the land, establish the type and sequence of filters to be applied. The SNAP program has numerous filters, all of a statistical nature, to reduce this problem. A visual comparison was made with the filters Gamma Map, Lee, Refined Lee, and Lee Sigma, with different window sizes (Figure 5) in a small area, within the study area, which brings together flood areas, urban areas, and crop areas where the texture variability produced by the application of the different filters can be appreciated. The speckle filtering application does not modify the spatial detail and it is mostly used to achieve smoothing of the limits of the different forms represented, avoiding the loss of image details. The Gamma Map filter is based on the Bayes statistical analysis of the image and it works best for extensive areas, such as oceans, forests, or extensive fields [44]. The Lee filter uses statistical parameters, such as the mean and standard deviation, with a predetermined window size assessing different factors for smoothing. In homogeneous regions, such as flooded areas, the final pixel value is the linear average of neighboring pixels [45]. The Refined Lee filter provides a smoother texture in the image, minimizing radiometric and textural loss in the images [46,47]. The Lee Sigma filter uses a statistical distribution of the digital levels of a given area window chosen by the operator and estimates what pixel should be considered [48]. Taking into account mainly the topography, the geometry of the urban areas, the polarization, and the need to delimit those areas with smaller intensity values, the Lee filter with a window size 7 × 7 (m) was chosen as being most appropriate for the study.

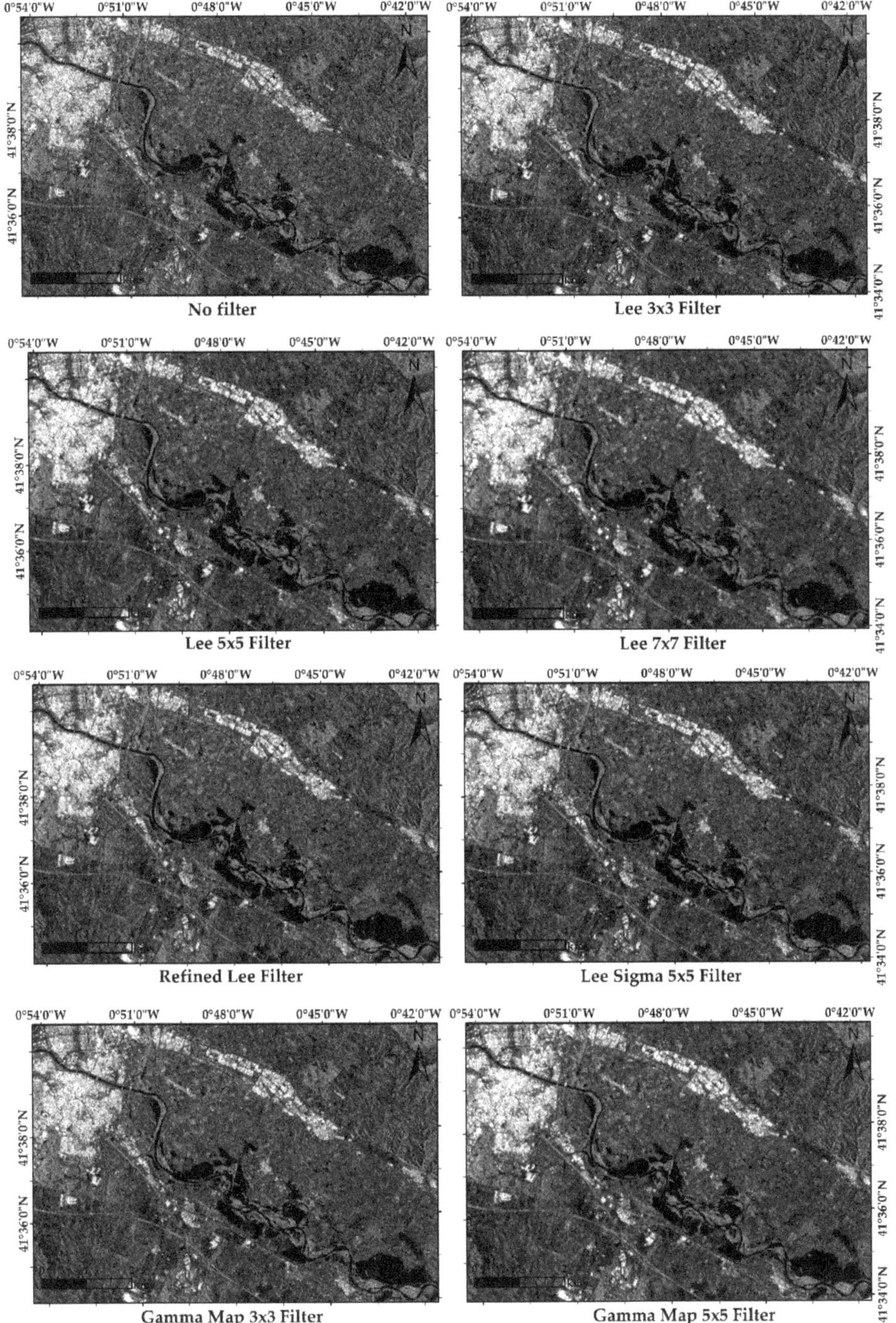

Figure 5. Detailed images of a section of zone 3 (Zaragoza–El Burgo de Ebro), 13 April 2018, after the flood event, with VH polarization in which the comparison is shown between the unfiltered image and the result of the application of different filters of different window sizes.

The last common step in the pre-treatment of images is to perform a correction of the terrain and orthorectification. This mainly eliminates distortions due to changes in the topography and the angle of incidence with the ground with respect to the nadir. The geometric calibration used in this study was range doppler terrain correction and the digital elevation model (DEM)–SRTM-3Sec.

3.2. SAR Sentinel-1 Data Processing

Many authors have used various algorithms for the detection of non-permanent water surfaces. Among these approaches, we can highlight the supervised classification [49], automatic non-supervised classification [46,50], RGB combination [38,51], or the detection or calibration threshold technique [41, 52–57]. In this study, the methodologies used were RGB composition and the calibrated thresholds technique. Before the application of any techniques and to perform a multitemporal analysis, it is necessary to unify in a single file the different bands of σ_0 of the images before and after the event using a stack.

The RGB composition is an appropriate method to visualize multi-temporal modifications and facilitates the detection of changes on the terrain surface through a temporal color image [58]. It is a method based on the differences between the pre- and post-event images, and in which a combination of bands is set for these differences to become visually remarkable. These differences are defined by the change in the intensity value for the same pixel on different dates. The result is a multitemporal image to which a band is assigned to each of the primary colors to form an image of RGB composition [38].

The other methodology used was the calibration threshold technique, also called binarization. It is a simple and quick process that differentiates flooded areas from those that are not by generating histograms. The use of this methodology is not recommended for fast flash flooding, where mapping is urgently needed, and the presence of noise causes some uncertainty [59]. Knowing that the backscatter values in permanent waters and in flooded areas are generally the most negative and differ markedly from the following radiometric values caused by other physical changes, such as, for example, the moisture content of the soil, it is possible to establish any limit [40]. It is necessary to emphasize that an equitable distribution of water and land is necessary so that the histogram has two clear distributions to more precisely limit the threshold. Through the characteristic polarization histogram, we obtained the number of pixels in the image for certain values of intensity or backscatter. In this way, the threshold could be set manually between flooded areas and those that were not. Later, using a mathematical equation of bands, a new bi-colour image separating both zones was achieved [54].

4. Results and Discussion

4.1. RGB Composition Results

The combination of bands used for the application of the method was the image before the event for the R-band and the image after the event for the G- and B-band. Different results were obtained for the two polarizations, VV and VH, as can be seen in Figure 6. There are many possible combinations, but the chosen combination favors the visualization of the flooded areas. Water surfaces that are stable on both dates, such as the Ebro river or reservoirs, are represented in black. The populations, urban and industrial areas with very high intensity values, are observed in white. Those areas, with intermediate intensity values and in which there have been hardly any variations, are consolidated in blue tones. The light pink tones are characteristic of areas with high humidity while the red ones represent the flood surfaces in which the water has completely flooded.

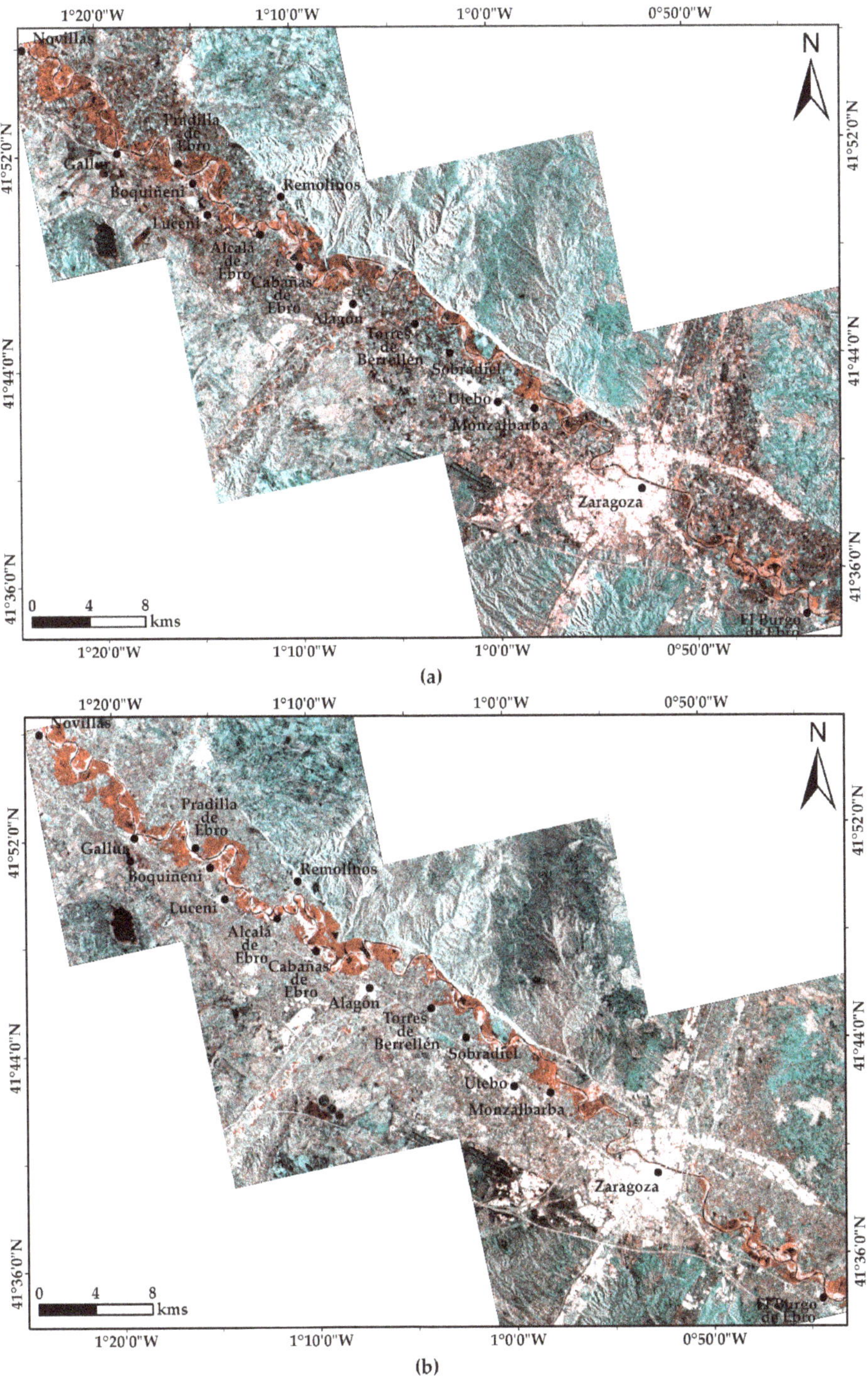

Figure 6. General flood map of the entire study area using RGB composition created from the RADAR images of 6 April 2018 and 13 April 2018: (**a**) in VV and (**b**) in VH polarization with a Lee 7×7 window size filter.

It should be noted that right on the riverbank, the presence of massive riparian vegetation and vertical growth generates a double rebound dispersion or a volume dispersion (Figure 7). The vegetation structure and submerged status are conditioning factors of different scattering mechanisms that can happen, and therefore a range of backscatter values [60]. This, combined with the fact that the sensor used works in the C-band, with a penetration range in bare soil or vegetation that has a maximum of 5 cm, results in the presence of water not being shown even though the area was flooded. These areas are seen in white or slightly pink tones. For these areas, the use of an L-band sensor (23.5 cm) would provide greater penetration because the wavelength is longer than the size of the leaves within the tree mass [61]. In fully submerged locations, scattering dampening occurs as in an open flood, while scattering enhancement is observed in partially or unsubmerged locations because of the dihedral scattering dominates.

Figure 7. Aerial views of the study area on 14 April 2018, showing the flood of the Ebro and the characteristic riverbank vegetation as highly flooded (Source: CHE, 2018 [36]).

A correct choice of polarization optimizes the discrimination of flooded areas [33]. The results obtained in our polarization configurations and the contributions of [41,62] in the comparative studies of polarizations to monitor flood areas reaffirm that the VH polarization is more suitable for delimiting flooded areas. It generates well-defined and correctly defined surfaces, results that VV polarization cannot offer. VV polarization is clearly influenced by the roughness and heterogeneity of the terrain [33,63]. We can see this fact in the floodplains, where the roughness of the soil of the crops fields becomes more evident in the VV polarization.

4.2. Calibration Threshold Technique Results

Before the application of this methodology for the extraction of several points from the radar images, specifically in zone 1 with VV and VH polarization, a comparison of how backscatter intensity values are modified in areas that have been flooded was considered (Figure 8). While the magnitude of intensity for the VV polarization, for both moments, is very variable and has a wide spectrum of values, the VH polarization is more defined within a certain range (Figure 9). The areas adjacent to the course of the river, mainly dedicated to cultivation, are characterized by partially rough surfaces and moderate vegetation. These move in values between −17 and −19 dB for the VH polarization, and between −11 and −15 dB for the VV polarization, with specific values that are out of this range. These same areas, subsequently flooded, reflect values between −21 and −24 dB approximately for VH polarization, and for VV, a wider range between −15 and −21 dB. The smooth surface generated by the sheet of water causes the backscatter intensity to decrease towards even more negative values.

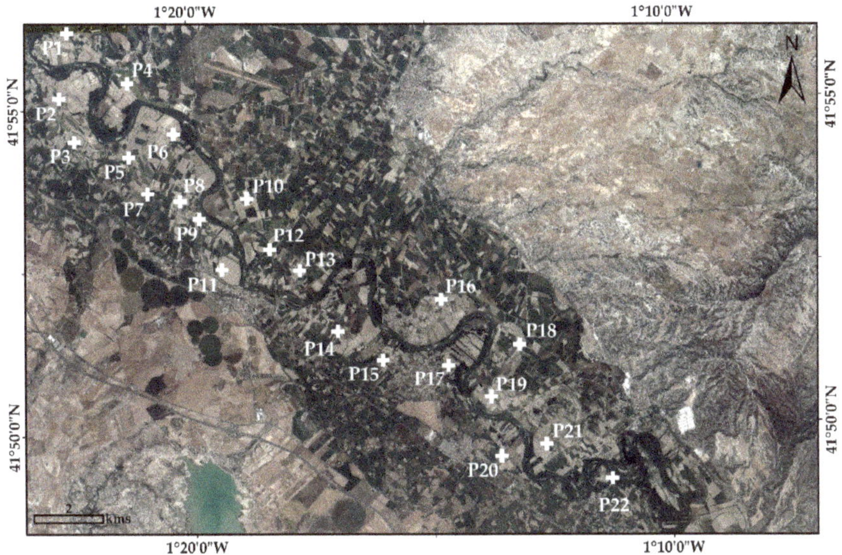

Figure 8. Location map of the points considered in the comparison of backscatter intensity before and after the flooding event.

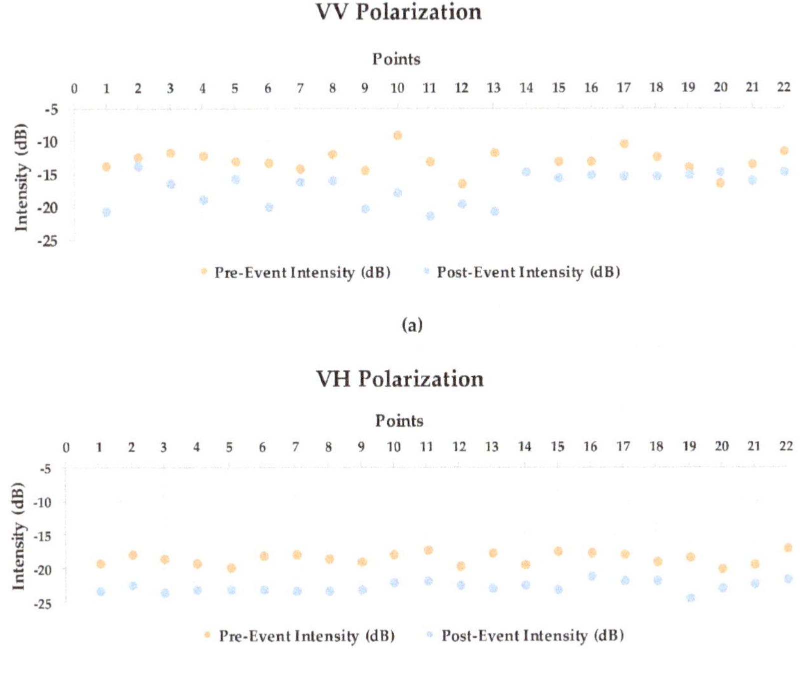

Figure 9. Graphical representation of the values of backscatter intensity before and after the event, for the VV (**a**) and VH (**b**) polarization, for the indicated points.

Through the graphics obtained by representation of characteristic points, we obtained an idea of the behavior of the radar on different surfaces. The calibration threshold methodology separated the dry zones from areas with the presence of a water sheet through the interpretation of histograms. In Figure 10, the relationship between polarizations and characteristic histograms for each of them is observed for the zone 1: Novillas–Pradilla de Ebro. The histogram of the VV polarization shows a very large curve representative of the intensity values ranging between −1.0 and −22 dB, a very wide spectrum characteristic of areas that range between very rough and moderately rough. While there is no clear projection of low backscatter values, the histogram of the VH polarization itself shows two clear populations where the specific intensity values of the image oscillate. This was chosen for further analysis. On the one hand, there is the curve that provides a greater number of pixels and that represents the areas without water, and on the other hand, a curve of smaller size and with smaller intensity values that represents the areas with water is shown. It is between both areas that the calibration limit can be set. Analyzing the histograms in greater depth, we observe that the cross polarization provides a wider range of backscatter values on surfaces with vegetation than the simple polarization, VV, which results in the latter omitting flooded areas [64].

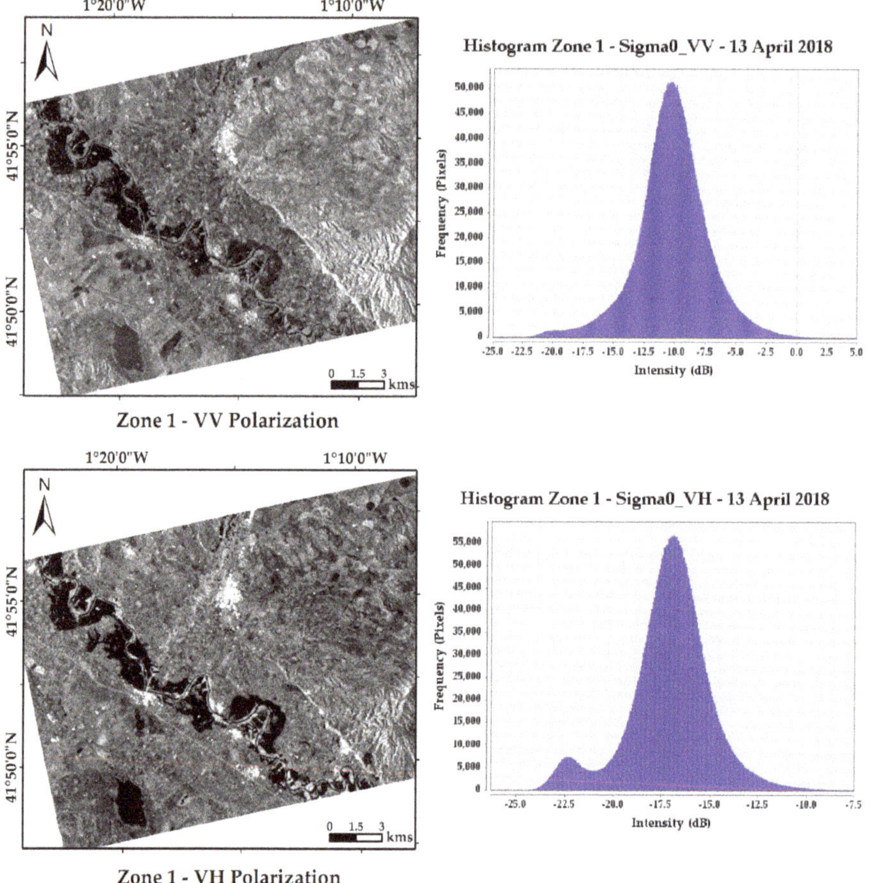

Figure 10. Representation of the characteristic histograms of zone 1: Novillas–Pradilla de Ebro for VV polarization (up) and for the VH (down) in the radar image of 13 April 2018 filtered with the Lee 7 × 7 window size filter.

To further specify the boundary point between the areas with water and without water, a water mask was generated on the VH polarization image. This layer was created by randomly capturing some flooded areas within zone 1 and subsequently generating a histogram. In this way, intensity values were obtained in which only areas with water or flooded areas are shown, independent of the frequency or number of pixels taken. It should be noted that this layer serves only to confirm and verify the suitability of the threshold taken with the total histogram of the zone. The comparison of the histogram of the water mask obtained and the polarization VH histogram is shown in Figure 11. A significant spatial linear correlation between both histograms is shown, something that does not happen with VV polarization [56]. After studying the results, we conclude that the most representative values of the presence of the water range mostly between −21 and −24 dB, so −21 dB was considered. The threshold taken was therefore set to −21 dB.

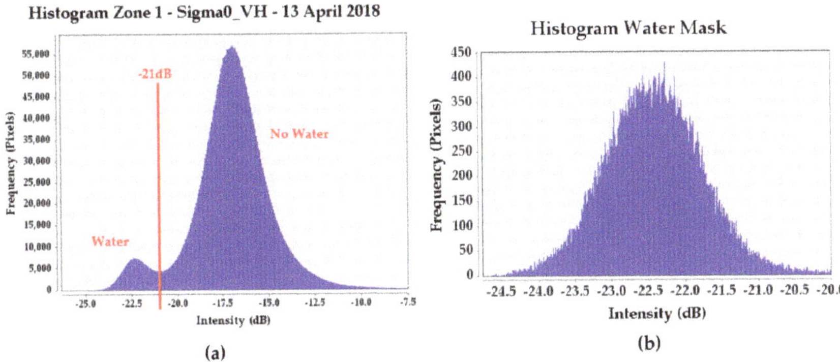

Figure 11. (**a**) Characteristic histogram of zone 1: Novillas–Pradilla de Ebro with VH polarization. (**b**) Representative histogram only for the water zones.

With the Raster-Band Maths tool of the SNAP program, a binary layer was generated in which, on the one hand, below −21 dB values are represented, representative of the water and flooded areas, and on the other, those above it that would show the dry zones. Figure 12 shows the areas with the presence of water that are observed in part of the study area for a radar image in which no filter was used and another for which the process of Lee 7 × 7 filtering was used. The unfiltered image has a large number of random pixels distributed by the image that give "false positives". All of them are points outside the flood area and the course of the river that have backscatter intensity values below −21 dB. On the other hand, the use of a filter minimizes this effect, favoring less overrepresentation errors.

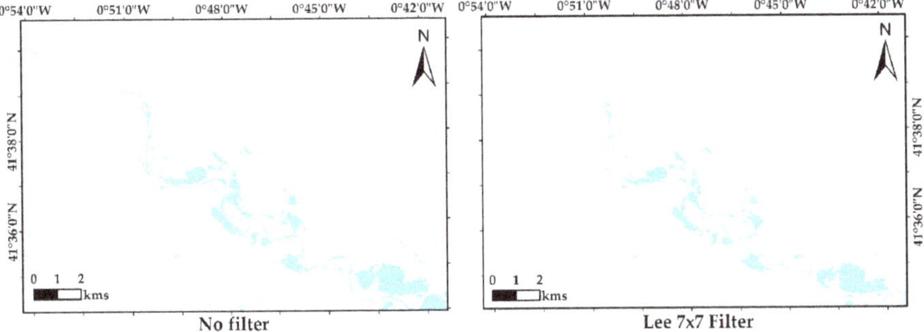

Figure 12. Representation of a specific part of zone 3: Zaragoza–El Burgo de Ebro (the same represented in Figure 5) of the areas that were captured as water in the unfiltered radar image and another with the Lee 7 × 7 filter.

The indicative layers of water presence, for the same specific area, were also generated for the other filters and different window sizes that were considered in the study (Refined Lee, Lee Sigma, and Gamma Map) and through which numerical data of the surfaces that show the presence of a water sheet were extracted (Table 4). It is quite significant that the use of one filter or another provides notable variations in the areas represented as water. Obviously, the unfiltered image is the one that generates more surface compared to any of the others that were filtered.

Table 4. Water covered area for a specific part of zone 3: Zaragoza–El Burgo de Ebro.

Filter	Total Area (ha)
No filter	1822.01
Lee 3 × 3	1122.15
Lee 5 × 5	880.53
Lee 7 × 7	762.90
Refined Lee	1115.54
Lee Sigma 5 × 5	910.17
Gamma Map 3 × 3	1093.30
Gamma Map 5 × 5	890.62

It is quite significant that the use of one filter or another provides notable variations in the areas represented as water. Obviously, the unfiltered image is the one that generates more surface compared to any of the others that were filtered. In comparison, the unfiltered image could provide even more than twice the area flooded than, for example, the image filtered with Lee 7 × 7. Therefore, the application of filtering is necessary not only for speckle reduction but to avoid errors when making quantitative calculations with images. The choice of the Lee 7 × 7 filter, which generates the least area, minimizes the presence of individual pixels outside the flood area and thus prevents overrepresentation in these areas.

The choice of the Lee 7 × 7 filter was verified as being the most appropriate for the results of the present study. The process represented on the PNOA results in the map of zone 1 is shown in Figure 13. It should be taken into account that factors such as the characteristics of the terrain or the angle of incidence, can produce areas of shadow (see this effect in Figure 13a in the upper right where there is a mountainous area with higher topography). These have backscatter values similar to those with surfaces with the presence of water, causing overrepresentation of flooded areas [64]. Therefore, in order to avoid delimitation errors, the threshold technique could be combined with photointerpretation to add areas not identified by default or eliminate areas identified in excess, thus improving the quality of the results obtained [47]. It should also be considered that this technique does not discriminate between bodies of surface water (lakes, lagoons, reservoirs, wetlands) flooded areas. In order to eliminate this overrepresentation, a layer of only permanent areas was generated. This was created from the RADAR image of 6 April 2018, in which the avenue period had not yet begun. By subtracting the flood layer on 13 April 2018 and the one generated on day 6, with the Band Maths tool, we obtained a display of the temporary flood areas that are represented in Figure 13b.

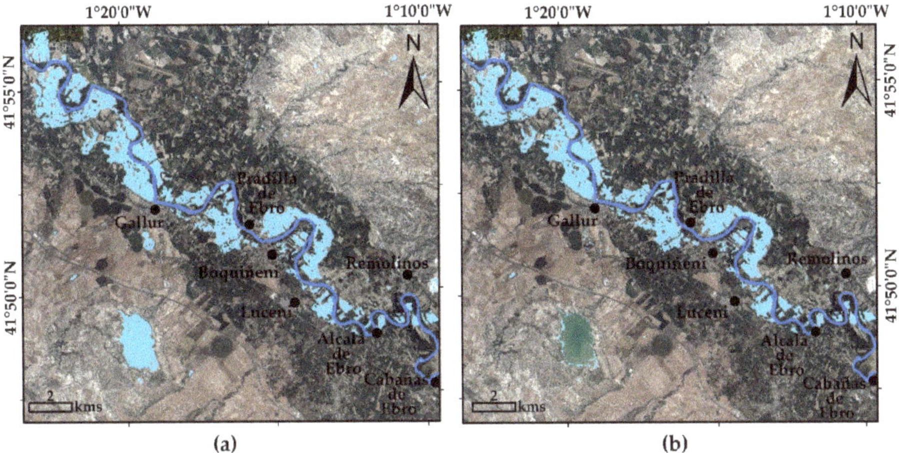

Figure 13. (**a**) Map of zone 1: Novillas–Pradilla de Ebro calculated using the calibration threshold methodology, with an intensity limit of −21 dB. (**b**) Representative map only of the flood zones of zone 1: Novillas–Pradilla de Ebro (Source: PNOA base map).

Following the same steps for the other two zones gave the total flood map for the calibration threshold technique, as represented in Figure 14.

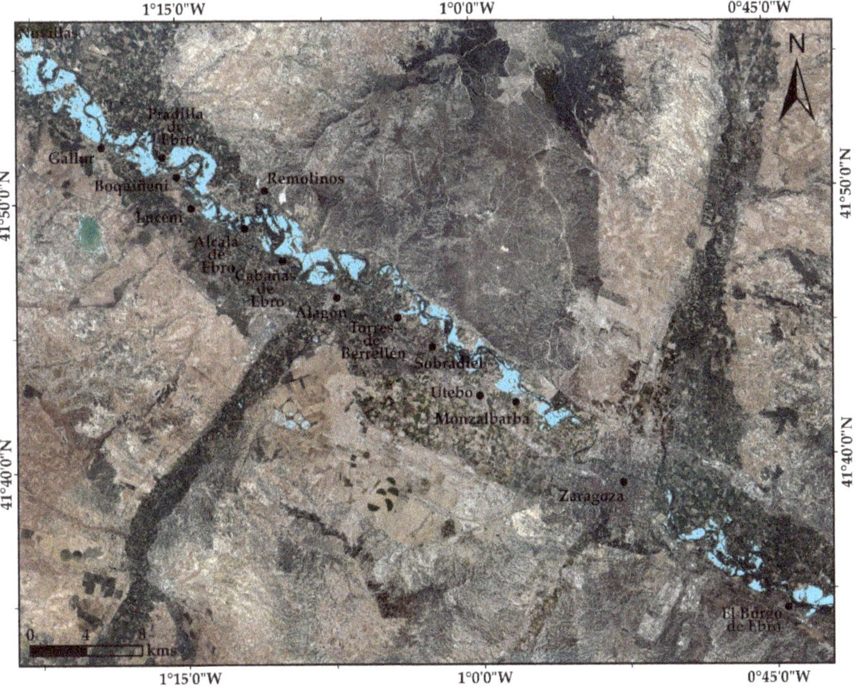

Figure 14. General flood map of the entire study area by calibration thresholds created from the RADAR image of 13 April 2018 and a scrubber with the permanently flooded water layer (Source: PNOA base map).

This technique applied to dual polarization images, VH in our case, gives good results in the extraction of water bodies, although these are small, but it has a number of limitations. These include, among other things, background noise. The calibration threshold is dependent on the characteristics of the surface to be captured and also on the specific parameters of the capture [65]. This implies that the threshold is not always fixed and cannot be generalized; it will have to be calculated in each study area.

4.3. Photointerpretation with Orthophoto Results

By analyzing the historical data of flows and levels in 2018, we can identify when the maximum annual flows of the Ebro were reached in the area of study. Specifically, they were reached in Novillas on 14 and 15 April at the Pina dam located downstream of the town of El Burgo de Ebro. In the villages between these two points, the maximum flow of the flood occurs between these two dates, migrating according to the direction of the river flow. In view of the data, it can be interpreted that the maximum flood day for the Ribera Alta was 14 April and that a comparison is possible between the RADAR image of 13 April and the orthophotography of the Ribera Alta. A comparison to determine both the flood areas was calculated through the result of the calibration threshold methodology and through photo interpretation of the flooded areas, and also calculation of the flooded surface for a specific day. Therefore, the areas in which the presence of water was perceived as flooding the fields were digitized, generating a vector layer on the orthophotography of the Ribera Alta. At the same time, the Ebro river was also digitized using the PNOA as a base map. For that purpose, the ArcGIS version 10.5 program was used to create the shape layers of the flood area and the Ebro river. In this way, the permanent river channel areas were discriminated from the temporarily inundated areas.

The flood areas of the RADAR image and the orthophotography are combined in Figure 15. In this way, common flood areas were obtained.

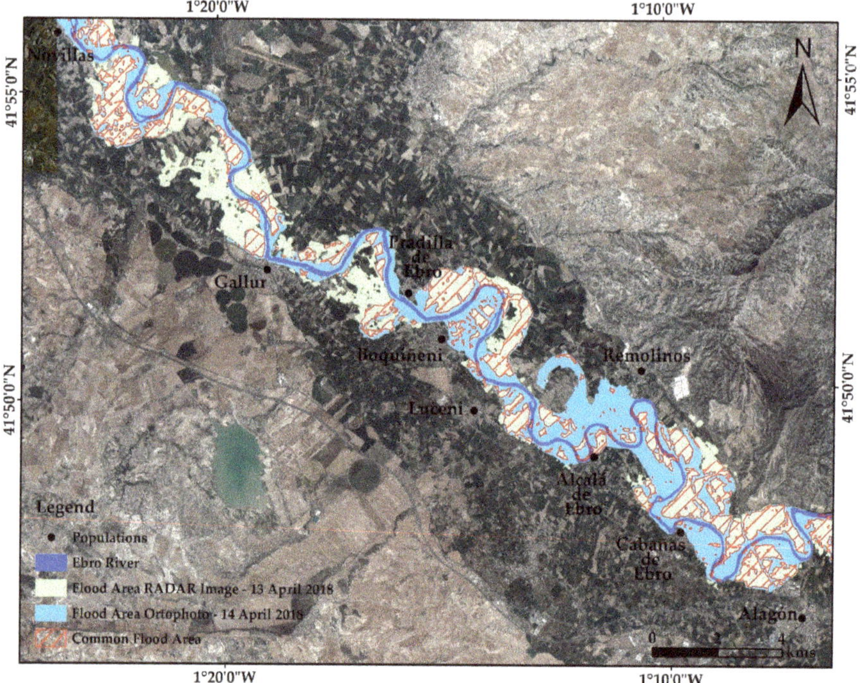

Figure 15. General map of comparative flooding between the RADAR image of 13 April 2018 and the orthophotography provided by the CHE on 14 April 2018 in Ribera Alta.

To establish the subsequent comparison, both layers were evaluated to calculate the flooded area overlapping the extent of flood areas mapped from both data sources (Table 5).

Table 5. Flooded surface values in the Ribera Alta.

Flood area RADAR 13 April 2018	2640.61 ha
Flood area Ortophotography 14 April 2018	4653.13 ha
Common flood area	1767.88 ha

It should be noted that in the RADAR image, the maximum flood areas are located in the upper part of that section of the river, and that in the orthophotography, the maximum flood is in the lower part of the same study area. Considering flood mapping obtained from Sentinel-1 on 13 April 2018, and that flood mapping obtained from ortophotography one day after on 14 April 2018, the fast evolution of the flood areas is clear.

The variability in the value of the flood area is mainly due to the fact that the quality of the RADAR image, which needed numerous treatments, was reduced. This generates a loss of resolution quality that, together with the fact that there were intensity values that were not perceived as flooded areas (tree areas close to the riverbed, mainly), causes a reduction in the calculated area.

In October 2007, the European Directive 2007/60/CE was approved. This directive highlights the need to create flood mapping to delineate areas with greater exposure to these events and to generate hazard and flood risk maps. In accordance with the European Directive 2007/60/CE, Spain's water governance system has made it possible to guarantee security and flood management.

The Flood Risk Management Plans are based on the flood hazard maps (calculation of the flood zone) of the SNCZI for the 10, 50, 100, and 500 year return periods. SNCZI is the official public source for flood hazard information produced by MITECO [66].

The tools used by Spain's government for the creation of the layers showing the flood return periods take into account the following: Hydrological studies that define the flows associated with the return periods considered in each moment; geomorphological studies, including a field cabinet analysis, LIDAR, or orthophotography images; historical studies; and a combination of the previous two and hydraulic studies, such as the generation of two-dimensional models [67]. However, they do not include SAR images as available base information to develop or validate the flood hazard maps.

In Figure 16, the result of the flooding map obtained with the RADAR image is compared with the layers with different return periods (10, 50, 100, 500 years) for a selected zone of the study area of SNCZI. On one hand, all flooded areas derived from Sentinel-1 SAR images are included inside the boundary for flood risk areas to 50, 100, and 500 years return period of SNCZI (Figure 16c–e. While, on the other hand, there are flooded areas that are outside the limits of the 10-year flood risk areas (Figure 16b). Therefore, we conclude that the results of flood area mapping derived from Sentinel-1 images are consistent with those of SNCZI, and correspond to a flood between 10 and 50 years of return. In view of the results obtained, the usefulness of Sentinel-1 images as baseline data for the improvement of the methodological guide is appreciated, and should be used as a new source of input, calibration, and validation for hydrological models to improve the accuracy of flood risk maps.

Flood maps obtained from Sentinel-1 SAR images complement and validate the information obtained by other means and help to understand the spatial and temporal evolution of the flood. Our results advise using flood maps derived from Sentinel-1 SAR images for the calibration and evaluation of hydrological models and as a data source that can complement and be compared with analysis obtained from other methodologies.

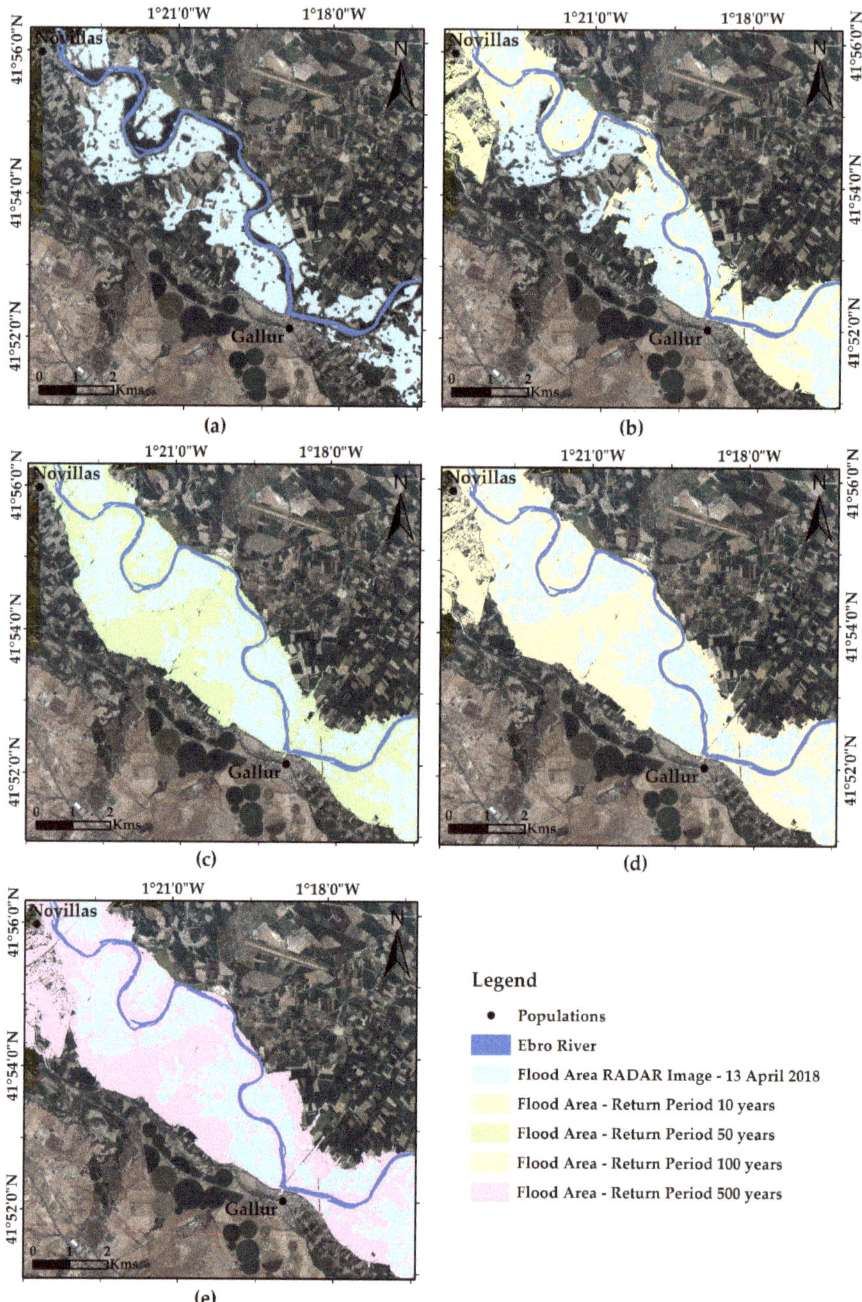

Figure 16. Comparative maps between (**a**) flood area RADAR image (13 April 2018) and (**b**) flood area high probability (10-year return period), (**c**) frequent flood zone (50-year return period), (**d**) flood zone of medium or occasional probability (100-year return period), and (**e**) flood zone of low or exceptional probability (500-year return period) (Source: MITECO, 2017, [66]).

5. Conclusions

In this study, a functional methodology was proposed for flood monitoring at a necessary spatial resolution derived from SAR images. Specifically, it aimed to highlight the potential use of the Sentinel-1 SAR data sets of the European Earth Observation mission for flood monitoring occurring in the middle course of the Ebro River (Spain) in April 2018.

In SAR images, continental waters have strong contrast in the backscatter values due to their low or null roughness in the absence of waves, behaving like a specular surface that reflects the radar return signal in a direction opposed to the sensor position. However, they need a correction and filtering process (orbital parameters, radiometric, noise reduction speckle, and geometric geometric). Evaluating the different Sentinel-1 parameters, our analysis showed that the best results were obtained using VH polarization configuration. Concerning the need to apply numerous filters to reduce speckle, which also reduces the resolution quality of the images and has an effect on the final flood mapping, Lee filtering with a 7×7 window was the most appropriate option and it was proven that it favors the elimination of "false positives".

Two methods were applied to detect and delimitate non-permanent water surfaces from Sentinel-1 SAR images. Method 1 RGB composition is based on differences between the pre- and post-event images, in which a combination of bands is set for identifying these differences as visually remarkable. The calibration threshold technique or binarization is based on a simple and quick process that differentiates flooded areas from those that are not from the characteristic polarization histogram using backscatter values in permanent waters and in flooded areas. The advantages of using the RGB composition methodology is that it guarantees clear differentiation between the areas of permanent flooding and the temporarily flooded ones, while achieving distinction through the technique of calibration thresholds requires other processes. Another advantage of the RGB methodology is that it detects the presence of massive riparian vegetation and vertical growth that can hide the presence of water, and it is essential to get a complete map of the flooded areas. In the calibration threshold technique, the most representative values of the presence of water were mainly between -21 and -24 dB; therefore, a central value of -21 dB was considered.

When comparing our flood maps with the flood areas digitalized from vertical aerial photographs planned by the Hydrological Planning Office of the Ebro Hydrographic Confederation, the results were consistent, even considering that the dates of both sources differed by a day. This confirmed and validated the use of Sentinel-1 SAR images to obtain flood maps with the required accuracy.

However, our flood maps derived from Sentinel-1 SAR images were consistent with flood risk areas provided by the SNCZI, and correspond to a flood between 10 and 50 years of return. In view of the results obtained, the usefulness of Sentinel-1 images as baseline data for the improvement of the methodological guide can be appreciated, and should be used as a new source of input, calibration, and validation for hydrological models to improve the accuracy of flood risk maps.

Finally, the development of methodologies that allow the synergy of different SAR missions (Sentinel-1, PAZ, Radarsat, TerraSAR-X, TanDEM-X, etc.) and multispectral optical systems will increase both the precision and the temporal resolution of flood monitoring, and should also be considered as powerful tools in the management and mitigation of the risks and defense against the adverse effects associated with floods.

Author Contributions: All authors have a significant contribution to final version of the paper. Conceptualization, F.C.C., and M.D.M.M.; Methodology, F.C.C., and M.D.M.M.; Supervision, F.C.C.; Processing SAR Sentinel-1 images, M.D.M.M.; Flood maps analysis, F.C.C., and M.D.M.M.; Investigation, F.C.C., and M.D.M.M.; Validation, F.C.C.; Writing—original draft, F.C.C., and M.D.M.M.; Writing—review and editing, F.C.C., and M.D.M.M.

Funding: This research received no external funding.

Acknowledgments: The authors would like to thank Hydrological Planning Office of the Ebro Hydrographic Confederation (Spanish Ministry for the Ecological Transition—MITECO) for data and photographs of the Ebro river flood reconnaissance aerial flights in April 2018, thank you to Cadonan for his diligent proofreading of this paper. Part of the present manuscript constituted the Master Thesis of the second of the authors, presented in the

Master in Hydrology and Water Resources Management interuniversity of Alcalá University and Rey Juan Carlos University (Spain).

Conflicts of Interest: The authors declare no conflict of interest.

References

1. Poff, N.L. Ecological response to and management of increased flooding caused by climate change. *Philos. Trans. A Math. Phys. Eng. Sci.* **2002**, *360*, 1497–1510. [CrossRef]
2. The United Nations Office for Disaster Risk Reduction (UNISDR). The Human Cost of Weather Related Studies. 2015. Available online: https://www.unisdr.org/2015/docs/climatechange/COP21_WeatherDisastersReport_2015_FINAL.pdf (accessed on 5 April 2019).
3. European Environment Agency. *Flood Risks and Environmental Vulnerability*; Exploring the Synergies between Floodplain Restoration, Water Policies and Thematic Policies; European Environment Agency: Copenhagen, Danmark, 2016; pp. 9–15.
4. Consorcio de compensación de seguros. Guía Para la Reducción de la Vulnerabilidad de los Edificios Frente a las Inundaciones. 2017. Available online: https://www.consorseguros.es/web/documents/10184/48069/guia_inundaciones_completa_22jun.pdf (accessed on 17 September 2018).
5. National Research Council. *Hydrologic Hazards Science at the U.S. Geological Survey*; The National Academies Press: Washington, DC, USA, 1999.
6. Merz, B.; Aerts, J.; Arnbjerg-Nielsen, K.; Baldi, M.; Becker, A.; Bichet, A.; Blöschl, G.; Bouwer, L.M.; Brauer, A.; Cioffi, F.; et al. Floods and climate: Emerging perspectives for flood risk assessment and management. *Nat. Hazards Earth Syst. Sci.* **2014**, *14*, 1921–1942. [CrossRef]
7. European Commission. Communication from the Commission to the European Parliament, the Council, the European Economic and Social Committee and the Committee of the Regions—An EU Strategy on Adaptation to Climate Change. 2013. Available online: https://ec.europa.eu/transparency/regdoc/rep/1/2013/EN/1-2013-216-EN-F1-1.pdf (accessed on 17 May 2018).
8. Bioresita, F.; Puissant, A.; Stumpf, A.; Malet, J.F. Fusion of Sentinel-1 and Sentinel-2 image time series for permanent and temporary surface water mapping. *Int. J. Remote Sens.* **2019**, *40*, 9026–9049. [CrossRef]
9. McFeeters, S.K. The Use of the Normalized Difference Water Index (NDWI) in the Delineation of Open Water Features. *Int. J. Remote Sens.* **1996**, *17*, 1425–1432. [CrossRef]
10. Ji, L.; Zhang, L.; Wylie, B. Analysis of Dynamic Thresholds for the Normalized Difference Water Index. *Photogramm. Eng. Remote Sens.* **2009**, *75*, 1307–1317. [CrossRef]
11. Shen, X.; Wang, D.; Mao, K.; Anagnostou, E.; Hong, Y. Inundation Extent Mapping by Synthetic Aperture Radar: A Review. *Remote Sens.* **2019**, *11*, 879. [CrossRef]
12. Tomer, S.T.; Al Bitar, A.; Sekhar, M.; Zribi, M.; Bandyopadhyay, S.; Sreelash, K.; Sharma, A.K.; Corgne, S.; Kerr, Y. Retrieval and Multi-scale Validation of Soil Moisture from Multi-temporal SAR Data in a Semi-Arid Tropical Region. *Remote Sens.* **2015**, *7*, 8128–8153. [CrossRef]
13. Filion, R.; Bernier, M.; Paniconi, C.; Chokmani, K.; Melis, M.; Soddu, A.; Talazac, M.; Lafortune, F.-X. Remote sensing for mapping soil moisture and drainage potential in semi-arid regions: Applications to the Campidano plain of Sardinia, Italy. *Sci. Total Environ.* **2016**, *573*, 862–876. [CrossRef]
14. Ulaby, F.T.; Moore, R.K.; Fung, A.K. *Microwave Remote Sensing: Active and Passive, Volume 2, Radar Remote Sensing and Surface Scattering and Emission Theory*; Addison-Wesley: Reading, MA, USA, 1983; pp. 1840–1852.
15. Kelly, R.; Davie, T.; Atkinson, P. Explaining temporal and spatial variation in soil moisture in a bare field using SAR imagery. *Int. J. Remote. Sens.* **2003**, *24*, 3059–3074. [CrossRef]
16. Smith, L. Satellite remote sensing of river inundation area, stage, and discharge: A review. *Hydrol. Process.* **1997**, *11*, 1427–1439. [CrossRef]
17. European Exchange Circle on Flood Mapping (EXCIMAP). Handbook on Good Practices for Flood Mapping in Europe. Available online: https://ec.europa.eu/environment/water/flood_risk/flood_atlas/pdf/handbook_goodpractice.pdf (accessed on 10 September 2019).
18. Sánchez Fabre, M.; Ballarán Ferrer, D.; Mora, D.; Ollero, A.; Serrano-Notivoli, R.; Saz, M. *Las Crecidas del Ebro Medio en el Comienzo del Siglo XXI*; XXIV Congreso de la Asociación de Geógrafos Españoles. Análisis espacial y representación geográfica: Innovación y aplicación; Universidad de Zaragoza y Asociación de Geógrafos Españoles: Zaragoza, Spain, 2015; pp. 1853–1862.

19. Confederación Hidrográfica del Ebro, CHE. Available online: http://www.chebro.es (accessed on 15 February 2019).

20. Galván Plaza, R. Cuatro grandes inundaciones históricas del Ebro en la ciudad de Zaragoza: 1643, 1775, 1871 y 1961. *Pap. De Geogr.* **2018**, *64*, 7–25. [CrossRef]

21. Pueyo Anchuela, Ó.; Revuelto, C.; Casas Sainz, A.; Rajamo Cordero, J.; Pocovi, A. Las crecidas del Ebro de febrero/marzo de 2015. ¿Qué hemos aprendido y qué falta por aprender? *Geogaceta* **2016**, *60*, 119–122.

22. Sistema Automático de Información Hidrológica de la Cuenca Hidrográfica del Ebro (SAIH of the CHE). Available online: http://www.saihebro.com/saihebro/index.php (accessed on 12 March 2019).

23. Polanco Fernández, L. Obras de restauración fluvial en el ámbito del Plan PIMA Adapta. In Proceedings of the Conference: La Gestión del Riesgo de Inundación Fluvial en el Contexto del Cambio Climático, Madrid, Spain, 10 December 2018.

24. Rodríguez Marcos, F.J. Principales episodios de inundaciones de 2018. In Proceedings of the Conference: La Gestión del Riesgo de Inundación Fluvial en el Contexto del Cambio Climático, Madrid, Spain, 10 December 2018.

25. Martínez, J.M.; Toan, T. Mapping of flood dynamics and vegetation spatial distribution in the Amazon floodplain using multitemporal SAR data. *Remote Sens. Environ.* **2007**, *108*, 209–223. [CrossRef]

26. Cunjian, Y.; Yiming, W.; Siyuan, W.; Zengxiang, Z.; Shifeng, H. Extracting the flood extent from satellite SAR image with thesupport of topographic data. In Proceedings of the International Conferences on Info-Tech and Info-Net. Networks (ICII 2001), Beijing, China, 29 October–1 November 2001; IEEE: Piscataway, NJ, USA, 2001; Volume 1, pp. 87–92.

27. Argenti, F.; Lapini, A.; Alparone, L. A tutorial on speckle reduction in synthetic aperture radar images, IEEE Geosci. *Remote Sens. Mag.* **2013**, *1*, 6–35.

28. European Space Agency (ESA). The ASAR User Guide. Available online: https://earth.esa.int/handbooks/asar/toc.html (accessed on 10 July 2019).

29. Gorrab, A.; Zribi, M.; Baghdadi, N.; Mougenot, B.; Fanise, P.; Chabaane, Z.L. Retrieval of both soil moisture and texture using TerraSAR-X images. *Remote Sens.* **2015**, *7*, 10098–10116. [CrossRef]

30. Zribi, M.; Dechambre, M. A new empirical model to retrieve soil moisture and roughness from C-band radar data. *Remote Sens. Environ.* **2002**, *84*, 42–52. [CrossRef]

31. Twele, A.; Cao, W.X.; Plank, S.; Martinis, S. Sentinel-1-based flood mapping: A fully automated processing chain. *Int. J. Remote Sens.* **2016**, *37*, 2990–3004. [CrossRef]

32. European Space Agency (ESA). Copernicus Open Access Hub. Available online: https://scihub.copernicus.eu/dhus/#/home (accessed on 2 May 2019).

33. Klemas, V. Remote sensing of floods and flood-prone areas: An overview. *J. Coast. Res.* **2015**, *31*, 1005–1013. [CrossRef]

34. ESA Sentinel Online. User Guides and Technical Guides of Sentinel-1 SAR. Available online: https://sentinel.esa.int/web/sentinel (accessed on 8 August 2019).

35. ESA Sentinel Online. Product Types and Processing Levels. Available online: https://sentinel.esa.int/web/sentinel/user-guides/sentinel-1-sar/product-types-processing-levels (accessed on 8 August 2019).

36. Confederación Hidrográfica del Ebro, CHE. *Ministerio Para la Transición Ecológica*; Gobierno de España. Vuelos aéreos de reconocimiento para la inundación del río Ebro en abril de 2018; CHE: Zaragoza, Spain, 2018.

37. SNAP Software Version 6.0.0. Available online: https://step.esa.int/main/download/snap-download (accessed on 22 December 2018).

38. Tavus, B.; Kocaman, S.; Gokceoglu, C.; Nefeslioglu, H.A. Considerations on the use of Sentinel-1 data in flood mapping in urban areas: Ankara (Turkey) 2018 floods. *Int. Arch. Photogramm. Remote Sens. Spat. Inf. Sci.* **2018**, *XLII-5*, 575–581. [CrossRef]

39. Ban, H.-J.; Kwon, Y.-J.; Shin, H.; Ryu, H.-S.; Hong, S. Flood monitoring using satellite-based RGB composite imagery and refractive index retrieval in visible and near-infrared bands. *Remote Sens.* **2017**, *9*, 313. [CrossRef]

40. Bioresita, F.; Puissant, A.; Stumpf, A.; Malet, J.-P. A method for automatic and rapid mapping of water surfaces from Sentinel-1 imagery. *Remote Sens.* **2018**, *10*, 217. [CrossRef]

41. Martinis, S.; Rieke, C. Backscatter analysis using multi-temporal and multi-frequency SAR data in the context of flood mapping at river Saale, Germany. *Remote Sens.* **2015**, *7*, 7732–7752. [CrossRef]

42. Henry, J.-B.; Chastanet, P.; Fellah, K.; Desnos, Y.-L. Envisat multipolarized ASAR data for flood mapping. *Int. J. Remote Sens.* **2006**, *27*, 1921–1929. [CrossRef]

43. Kudahetty, C. *Flood Mapping Using Synthetic Aperture Radar in the Kelani Ganga and the Bolgoda Basins, Sri Lanka*; Master of Science in Geo-information Science and Earth Observation; University of Twente: Enschede, The Netherlands, 2012.

44. Senthilnath, J.; Handiru, V.; Rajendra, R.; Omkar, S.N.; Mani, V.; Diwakar, P. Integration of speckle de-noising and image segmentation using Synthetic Aperture Radar image for flood extent extraction. *J. Earth Syst. Sci.* **2013**, *122*, 559–572. [CrossRef]

45. Park, J.-M.; Song, W.J.; Pearlman, W.A. Speckle filtering of SAR images based on adaptive windowing. *IEE Proc. Vis. Image Signal Process.* **1999**, *146*, 191–197. [CrossRef]

46. Borah, S.B.; Sivasankar, T.; Ramya, M.N.S.; Raju, P.L.N. Flood inundation mapping and monitoring in Kaziranga National Park, Assam using Sentinel-1 SAR data. *Environ. Monit. Assess.* **2018**, *190*, 520. [CrossRef]

47. Ezzine, A.; Darragi, F.; Rajhi, H.; Ghatassi, A. Evaluation of Sentinel-1 data for flood mapping in the upstream of Sidi Salem dam (Northern Tunisia). *Arab. J. Geosci.* **2018**, *11*, 170. [CrossRef]

48. García, R.; González, C.; De la Vega, R.; Valverde, A.; Seben, E. *Análisis del Comportamiento de Filtros de Reducción de Speckle en Imágenes ERS2-SAR*; Teledetección y Desarrollo Regional; X Congreso de Teledetección: Cáceres, Spain, 2003; pp. 325–328.

49. Chapman, B.; McDonald, K.; Shimada, M.; Rosenqvist, A.; Schroeder, R.; Hess, L. Mapping regional inundation with spaceborne L-Band SAR. *Remote Sens.* **2015**, *7*, 5440–5470. [CrossRef]

50. Martinis, S.; Twele, A.; Voigt, S. Unsupervised extraction of flood-induced backscatter changes in SAR data using Markov image modeling on irregular graphs. *IEEE Trans. Geosci. Remote Sens.* **2011**, *49*, 251–263. [CrossRef]

51. Perrou, T.; Garioud, A.; Parcharidis, I. Use of Sentinel-1 imagery for flood management reservoir-regulated river basin. *Front. Earth Sci.* **2018**, *12*, 506–520. [CrossRef]

52. Cao, H.; Zhang, H.; Wang, C.; Zhang, B. Operational Flood Detection Using Sentinel-1 SAR Data over Large Areas. *Water* **2019**, *11*, 786. [CrossRef]

53. Martinis, S.; Twele, A.; Voigt, S. Towards operational near real-time flood detection using a split-based automatic thresholding procedure on high resolution TerraSAR-X data. *Nat. Hazards Earth Syst. Sci.* **2009**, *9*, 303–314. [CrossRef]

54. Psomiadis, E. Flash flood area mapping utilising Sentinel-1 radar data. In Proceedings of the SPIE 10005, Earth Resources and Environmental Remote Sensing/GIS Applications, Dresden, Germany, 11 November 2013; VII 100051G. SPIE: Edimburgh, UK, 2016.

55. Dumitrascu, N.; Oniga, E.; Florian, S.; Marcu, C. Floods damage estimation using sentinel-1 satellite images. Case study: Galati County, Romania. RevCAD. *J. Geod. Cadas.* **2017**, *22*, 115–122.

56. Pham-Duc, B.; Prigent, C.; Aires, F. Surface Water Monitoring within Cambodia and the Vietnamese Mekong Delta over a Year, with Sentinel-1 SAR Observations. *Water* **2017**, *9*, 366. [CrossRef]

57. Zhang, B.; Wdowinski, S.; Oliver-Cabrera, T.; Koirala, R.; Jo, M.J.; Osmanoglu, B. Mapping the extent and magnitude of severe flooding induced by hurricane Irma with multi-temporal Sentinel-1 SAR and InSAR observations. *Int. Arch. Photogramm. Remote Sens. Spat. Inf. Sci.* **2018**, *XLII-3*, 2237–2244. [CrossRef]

58. Amitrano, D.; Guida, R.; Ruello, G. Multitemporal SAR RGB Processing for Sentinel-1 GRD Products: Methodology and Applications. *IEEE J. Select. Top. Appl. Earth Obs. Remote Sens.* **2019**, *12*, 1497–1507. [CrossRef]

59. Westerhoff, R.S.; Kleuskens, M.P.H.; Winsemius, H.C.; Huizinga, H.J.; Brakenridge, G.R.; Bishop, C. Automated global water mapping based on wide-swath orbital synthetic-aperture radar. *Hydrol. Earth Syst. Sci.* **2013**, *17*, 651–663. [CrossRef]

60. Shen, X.; Hong, Y.; Qin, Q.; Chen, S.; Grout, T. A backscattering enhanced canopy scattering model based on mimics. In Proceedings of the American Geophysical Union (AGU) 2010 Fall Meeting, San Francisco, CA, USA, 13–17 December 2010.

61. Townsend, P.A. Relationships between forest structure and the detection of flood inundation in forested wetlands using C-band SAR. *Int. J. Remote Sens.* **2002**, *23*, 443–460. [CrossRef]

62. Matgen, P.; Schumann, G.; Henry, J.-B.; Hoffmann, L.; Pfister, L. Integration of SAR-derived river inundation areas, high-precision topographic data and a river flow model toward near real-time flood management. *Int. J. Appl. Earth Obs. Geoinf.* **2007**, *9*, 247–263. [CrossRef]

63. Manjusree, P.; Prasanna Kumar, L.; Bhatt, C.M.; Srinivasa Rao, G.; Bhanumurthy, V. Optimization of threshold ranges for rapid flood inundation mapping by evaluating backscatter profiles of high incidence angle SAR images. *Int. J. Disaster Risk Sci.* **2012**, *3*, 113–122. [CrossRef]

64. Clement, M.; Kilsby, C.; Moore, P. Multi-temporal SAR flood mapping using change detection. *J. Flood Risk Manag.* **2017**, *11*, 152–168. [CrossRef]

65. Nguyen, D. Automatic detection of surface water bodies from Sentinel-1 SAR images using Valley-Emphasis method. *Vietnam Earth Sci.* **2016**, *37*, 328–343.

66. Ministerio para la Transición Ecológica, MITECO. 2017. Available online: https://www.miteco.gob.es/es/cartografia-y-sig/ide/descargas/agua/zi-lamina.aspx (accessed on 6 May 2019).

67. Ministerio para la Transición Ecológica, MITECO. *Guía Metodológica Para el Desarrollo del Sistema Nacional de Cartografía de Zonas Inundables*, 1st ed.; Ministerio de Medio Ambiente y Medio Rural y Marino: Madrid, Spain, 2011; pp. 19–52.

Article

Real-Time Data and Flood Forecasting in Tagus Basin. A Case Study: Rosarito and El Burguillo Reservoirs from 8th to 12th March, 2018

Ignacio Menéndez Pidal [1,*], José Antonio Hinojal Martín [2], Justo Mora Alonso-Muñoyerro [3] and Eugenio Sanz Pérez [1]

1 Department of Engineering and Morphology of the Terrain, Civil Engineering Insitute, Polytechnic University of Madrid, 28040 Madrid, Spain; eugenio.sanz@upm.es
2 Tajo Hydrographic Confederation, 28011 Madrid, Spain; joseantonio.hinojal@chtajo.es
3 Applied Geology Research Group, Polytechnic University of Madrid, 28040 Madrid, Spain; jmora@ciccp.es
* Correspondence: ignacio.menendezpidal@upm.es

Received: 27 December 2019; Accepted: 27 March 2020; Published: 1 April 2020

Abstract: The hydrological regime of the Iberian Peninsula is characterized by its extreme irregularity, including its propensity for periodic floods, which cause severe floods. The development of suitable cartographies and hydraulic models (HEC-RAS, IBER, etc.) allows for defining, with sufficient precision, the areas flooded by a determined return period, and for elaborating maps of danger and areas at risk of flooding, making it possible to adopt the corresponding preventive measures of spatial planning. These preventive measures do not avoid the need for contingent plans, such as the Civil Flooding Protection Plans. Many of the Peninsula's watercourses and rivers are regulated by reservoirs built to ensure water supply and to smooth floods by releasing water in extreme hydrological climates. Hydrological modeling tools (rain/run-off) and Decision Support Systems DSS have been developed for the optimal operation of these dams in flood situations. The objective of the article is to study and prove the effectiveness of the integrated data provision in real time, while the event occurs—a circumstance that was not possible from the limited available meteorological stations available from Official Weather Services. The development of the Automatic Hydrological Information System (SAIH) in the Spanish River Basin Authorities (Confederaciones Hidrográficas), which includes a dense network of thermo-pluviometric stations and rain-river flow gauges, has allowed for new perspectives in order to realize an effective forecast method of flows during episodes of extreme precipitation. In this article, we will describe the integration of a hydrological modeling system, developed by the Confederación Hidrográfica del Tajo (River Authority), to meet the described objectives; the results of this methodology are novel. This allows for the processing of the 15-minute data provided, including the simulation of the snow accumulation/melting processes and the forecast of inflows to the reservoir to help in the establishment of safeguards and preventive waterflow releases. Finally, the methodology described is shown in a real case of study at Rosarito and El Burguillo Reservoirs.

Keywords: flood forecasting; snow cover control; flood management; Tagus Basin

---

1. Introduction

Floods are natural phenomena that cause significant damage. In Spain, 61.7% of the indemnities paid by the Insurance Compensation Consortium for claims in the years 1971–2017 were due to floods [1]. Their impact on goods, people, and, in general, on the economy and the organization of society, are related to both the intensity and frequency of the events generated by these phenomena, and the degree of exposure of human activity to flood risk. While floods are natural phenomena, the progressive occupation and alteration of the territory, as well as the increasing magnitude and recurrence of floods, induced by climate change [2], have increased our vulnerability to them, raising international concern about their danger.

In Europe, a new regulatory framework was developed to deal with the situation, formalized by Directive 2007/60/EC of the European Parliament and of the Council of 23 October 2007, on the assessment and management of flood risks (transposed to the Spanish legal order by RD 903/2010), in order for member countries of the EU to prepare Flood Risk Assessment and Management Plans (FRMPs). The elaborations of the cartography corresponding to the hazard and risk maps are fundamental elements in establishing the objectives and measures to be adopted by the plan. These flood risk management plans should encompass all aspects of flood risk management, focusing on prevention, protection and preparedness, including flood forecasting procedures that reduce uncertainty when it occurs [3].

The preparation of the FRMPs does not exclude the need for other contingency plans, designed for those cases in which the previous prevention and protection measures cannot guarantee that certain extraordinary episodes affect people and property. They include the actions and measures to be adopted to reduce damages in these situations in which, despite the precautions anticipated, the episode is imminent and inevitable. The existence of Early Warning Systems must allow for the adequate forecasting of phenomenon of this magnitude occurring in a predictable time, making it possible to warn the population and to mobilize the means and resources established in the plan.

In Spain, these contingent plans are formalized through the Civil Protection Plans against the Risk of Floods (Agreement of the Council of Ministers of December 9, 1994 and the Basic Guideline for Planning of Civil Protection against the Risk of Floods) to which two early warning systems are incorporated: the meteorological alert and hydrological alert. The Prediction and Meteorological Surveillance System, in charge of AEMET (Spanish Meteorological Agency), communicates, to the corresponding Civil Protection Unit, the warnings concerning the forecasts of extraordinary events, in this case, rains that exceed a certain threshold.

With regard to the Hydrological Prediction and Surveillance System, this function is facilitated by the Hydrographic Confederations by providing the relevant information, for the purposes of flooding, provided by the SAIH, whose characteristics are described below. Among the data provided to Civil Protection, when the corresponding authority orders the implementation of the Plan, it includes data on the flows and water levels in certain sections of rivers, as well as the evolution of the volumes stored and the inputs/outputs of the reservoirs.

These Plans dedicate an important part to the management of critical situations of dams at risk of falling-out, establishing communication protocols with Civil Protection and warning the population, contained in their respective Emergency Plans.

In Spain, due to the number and volume of the same rain, the reservoirs play a very important role in the control and reduction in flood damage due to their storage capacity and smooth rolling effect during floods. However, the statistics indicate that a significant percentage of serious accidents [4] occur precisely during the events, when there are stress demands on its structure, and its stability conditions sub pressures and the operability of its mechanical elements (valves, spill gates, etc.) are extreme. Among these situations, the worst is when overtopping occurs, and is especially dangerous in the case of earth dams, meaning the structure will certainly be ruined and result in the generation of a flood wave that propagates downstream and generally has catastrophic effects.

Likewise, the existence of information systems capable of providing hydrometeorological data in real time and the availability of hydrological simulation models for forecasting inlet flows in reservoirs allows for the optimization of the operation of the reservoirs in order to minimize damage downstream of the dams, thus maintaining their structural safety. That is essential in a country, like Spain, where the irregularity of its hydrological regime has determined the existence of more than 1300 large dams.

2. Objectives

Hydrological forecasting is especially interested in the headwaters of the rivers where precipitations are usually more intense, generating floods where propagation downstream can cause flooding on the banks. In general, this circumstance is singularly compromised at the end of spring, when it is necessary to ensure the storage of a sufficient volume of water for the next summer's irrigation campaign, in the middle of the river low season, which obliges the minimization in the guards of the dams.

In the Iberian Peninsula, many of them are located in mountainous areas (rivers in the Ebro, Duero, Tajo basins, etc.) where the presence of snow is hydrologically significant, so its modeling must reproduce the processes of accumulation/fusion of snow—a circumstance that does not meet the conventional applications commonly used in the SAIH environment. However, these areas have been studied through the study of Water Resources Assessment from the Snowing (ERHIN) [5], including the application of specific models. Its calibration was carried out using official meteorological data (precipitation and temperature, fundamentally) from the current State Meteorological Agency (AEMET) and flow rates from the Official Gauging Stations Network (ROEA)—achieving sufficiently adjusted results in the cases considered.

Obviously, the procedure followed to obtain these results, although valuable for learning and improving the rules of exploiting reservoirs, based on the analysis of their behavior in the face of historical flood events, is not valid as a predictive methodology. For these applications to be effective, it is necessary to supply data in real time, while the event occurs—a circumstance that was not possible due to the few available meteorological stations.

The objective pursued by the work described in this article is to develop a process that allows for the construction of hydrological forecasts of inlet flows in reservoirs located in mountainous areas of the Tagus Basin (Spain) in order to help establish shelters and preventive imbalances in dams during flood periods at risk of flooding.

The research to achieve the proposed objective consists of the coordinated application of the instruments available in the Tajo Hydrographic Confederation, adapting the hydrological modeling system of the ERHIN program for its continued operation, coupled with SAIH, as a source of data supply in real time. This allows for fifteen-minute data processing, including the simulation of snow accumulation/melting processes. Its systematic application on a daily basis, starting from the month of November of each hydrological year, reproduces the virtual construction of the snow cover, which allows the ERHIN-Tajo modeling system to ascertain the water reserves in the form of snow, which are potentially available for fusion and incorporation into ordinary water runoff.

This methodology has been successfully applied to forecast entrances into the Rosarito (Tiétar River) and El Burguillo (Alberche River) reservoirs, and help establish safeguards and preventive imbalances therein. For these purposes, the results of applying this methodology during the episode of heavy rains in March 2018 are shown below.

Additionally, the application of this methodology has contributed to a more rational and efficient use of the tools that are already available in the Hydrographic Confederation, with practically no additional investment. This process, extended to the rest of the Confederations where the necessary elements of the SAIH and ERHIN are operational, has been completely innovative in Spain, providing useful information for the improvement of the flood forecasting system.

3. Available Means and Tools

3.1. The Automatic Hydrological Information System (SAIH) and the Program for the Evaluation of Resources from Innivation (ERHIN)

In Spain, the then General Directorate of Hydraulic Works (now the General Directorate of Water) promoted, from 1982, the implementation of the Automatic Hydrological Information System (SAIH) [6] in the main peninsular hydrographic basins. It is a system for the provision of real time data, based on the capture, transmission and processing of the adopted values by the most significant hydrometeorological and hydraulic variables, at certain geographical points of the monitored river basins. In particular, the data are from their own network of:

- Automatic pluviometers and nivometers.
- Gauging stations in rivers, canals and impulsion systems.
- Reservoirs level, position of floodgates, evacuated flows (latter, spillways), which allow estimating other calculated data (reservoir volume, inputs, etc.).
- Complete and thermopluviometric weather stations.
- Telenivometers (Ebro, Tagus and Duero Basins).

These data are transmitted by HISPASAT satellite or by the radio network, which owns the system, to a central position in the corresponding Hydrographic Confederation, namely the River Authority, and other secondary management. Central headquarters manages the entire procedure of data processing, visualization and distribution, periodic reports, as well as the development of programs and applications for the management of hydraulic infrastructures, or, where appropriate, for the transmission of data to Civil Protection services and for advice to its Central Command Post of the plan when it is activated.

The data update is normally done every fifteen minutes, and it can be considered as a real time or quasi-real system due to temporal variation of the magnitudes of the physical variables involved (rain, temperature, etc.), hydrological (rainfall, runoff, and flow) and hydraulic processes (river propagation, etc.)

3.2. Evaluation of Resources from Snow (ERHIN)

However, the same General Directorate mentioned, pushed through, in 1986, the production of studies in the sub-basins of the Pyrenees watershed to the Ebro River to determine the Evaluation of Water Resources from Snow (ERHIN). The snow phenomenon produces a natural regulation of water resources, generating an accumulation of these at high levels to subsequently, in certain circumstances, proceed to their melting point through a complex process governed by various meteorological variables, among which are temperature, rain, radiation, humidity and wind speed. This circumstance means that a specific treatment is required for the quantification of nival resources, as well as the forecasting of their evolution over time and the corresponding flow contributions that come from their melting process. This specific treatment is not reproduced by the ordinary models of hydrological simulation. For this, a method was initially developed to correlate snow thickness and density measurements, taken from a steel pole network that was outspread in the Spanish Pyrenees, with flow data observed in the Official Gauging Stations Network (ROEA) of the Hydrographic Confederation of the Ebro and of the entrances to the downstream reservoirs [7] This resulted in the first simplified tool being used for the estimation of the volume of water stored in the form of snow.

Going deeper into its analysis, it was decided to adapt the methodology applied in other pre-existing models to the Spanish casuistry. For the treatment of the snow phenomenon, the SNOW-17 model of the North American National Weather Service was selected [8]—being implemented on a distributed model, in which discretization of the study area is allowed for its adaptation to the orography existing in the Peninsula and its climatic characteristics. Despite the numerous modern approaches dealing with hydrological models that are useful for flood prediction and flood frequency estimation, the complexity of the implementation for management purposes suggested work with a specific algorithm—which was constructed for this purpose—from the start, which was many years ago [9–12]. For modelling the flow of liquid water on the ground the calculation algorithm, which is the core of the hydrologic model called ASTER, was constructed [13]. The model uses a usual hydrological balance in each of the cells of the distributed model, which is similar to the one used by the CEQUEAU model developed by the National Institut de la Recherche Scientifique (INRS) of Canada [14].

This ASTER model has been used the most by water management agencies in Spain (Hydrographic Confederations), in the field of the ERHIN program, and is conceptually composed of the integration of a distributed rain-runoff model and a specific module to simulate the evolution of water storage simulating snow.

Progressively, these works were extended to other basins with similar characteristics. Finally, the areas of the Iberian Peninsula where the presence of snow turned out to be hydrologically significant were analyzed, identifying and delimiting the sub-basins of mountain headlands to be integrated into the ERHIN program. [15,16].

The ASTER hydrological model, initially developed in the Aragón basin and Yesa reservoir, was refined for the simulation of the main snow-capped basins leading to reservoirs, using, for the calibration of the same, the readings of the poles installed for the measurement of thicknesses and densities of snow and hydrological data, provided by the former National Institute of Meteorology (INM)—today the State Meteorological Agency (AEMET)—and the Official Network of Gauging Stations (ROEA) of the respective Hydrographic Confederations [17–19]

4. Methodology

As previously stated, the objective of the proposed methodology is to design and apply a process that allows for the construction of hydrological forecasts of inlet flows in reservoirs located in mountainous areas of the Tajo basin (Spain) in order to help establish preventive safeguards and imbalances in dams during flood periods in situations of flood risk. This would facilitate the optimization of the operation of reservoirs when snow processes upstream of their sites affect them.

The application of hydrological simulation models to predict flood hydrographs at the entrance of reservoirs is common. However in cases, such as those discussed, in which, due to its orographic and climatic characteristics, the watershed can store significant amounts of water as snow cover, in addition to the conventional runoff fraction generated by rainfall, it is necessary to consider the effect of melting or snow accumulation. This requires the use of distributed models, such as ASTER, that contain an accumulation/melting subroutine that determines, in each cell, whether solid rainfall will be received or if the conditions for the liquefaction of part or all of the snow cover are logged.

The application of these types of models as tools to help decision-making in the management of a reservoir in a normal or exceptional hydrological situation (floods) requires at least:

- Having information in real time regarding the main hydrometeorological variables.
- Knowing the characteristics of the snow cover: distribution, thickness, and equivalent in water at all times.

To solve this issue, the following methodology was adopted, as shown in Figure 1, and was applied to the Tagus basin:

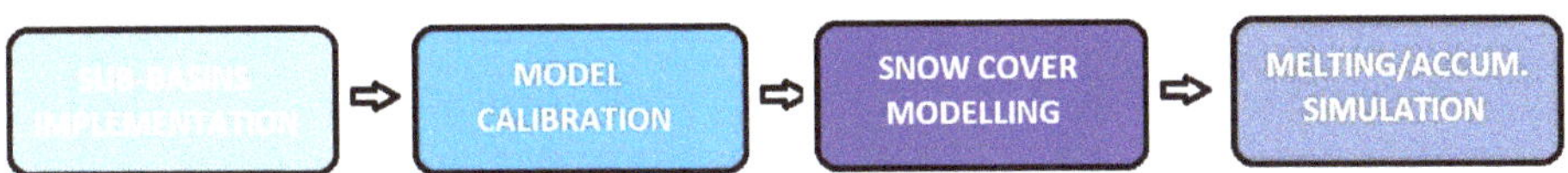

Figure 1. Methodological scheme applied to the Tagus Basin.

4.1. ASTER Tagus Sub-basins Implementation

The ASTER model was implemented in all the sub-basins identified in the ERHIN program, as headers, where the presence of snow is hydrologically significant. Among them, it was implemented in the ERHIN zones of the Tagus basin—its geographical location can be seen in Figure 2 and Table 1.

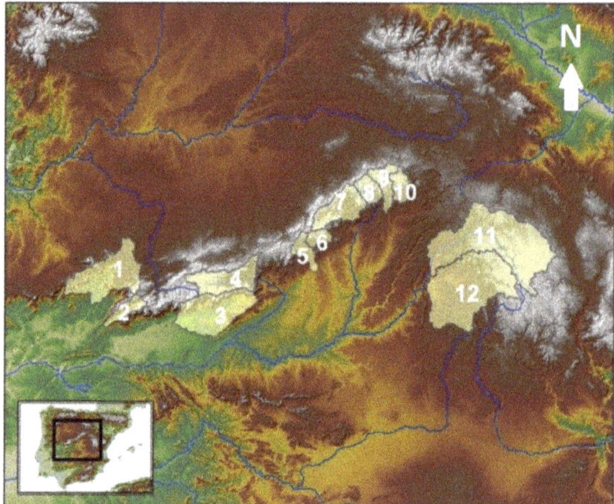

Figure 2. Headland sub-basins included in the ERHIN program in the Tagus basin.

Table 1. ERHIN zones in the Tagus Basin, where ASTER is operational.

Basin	Control Point	Surface (km^2)
1. Alagón	Gabriel y Galán Reservoir	1.841,60
2. Jerte	Plasencia	367,60
3. Tiétar	Rosarito Reservoir	1.730,30
4. Alberche	Burguillo Resevoir	1.052,70
5. Guadarrama	Picotejo	357,20
6: Manzanares	Santillana Reservoir	247,10
7. Lozoya	El Atazar Reservoir	925,00
8. Jarama	El Vado Reservoir	378,00
9. Sorbe	Beleña Reservoir	475,60
10. Bornova	Alcorlo Reservoir	362,30
11. Alto Tajo	Entrepeñas Reservoir	3.825,40
12. Guadiela	Buendía Reservoir	3.355,70

4.2. Model Calibration

Calibration of the model with the rainfall and temperature data supplied by the SAIH and was adjusted to the system flow control points, since the data from AEMET and the official ROEA capacity network are not available in real time. ASTER supports different extrapolation curves of the temperature data recorded by the SAIH stations in order to estimate their value in each cell, taking into account the meteorological situation and the corresponding thermal gradients. Figure 3 shows the geographical location of the SAIH data supply points. Among them, the river gaps and reservoirs serve to compare the flow values measured and supplied in quasi-real time (every 15 minutes) with those calculated by the model for adjustment and calibration.

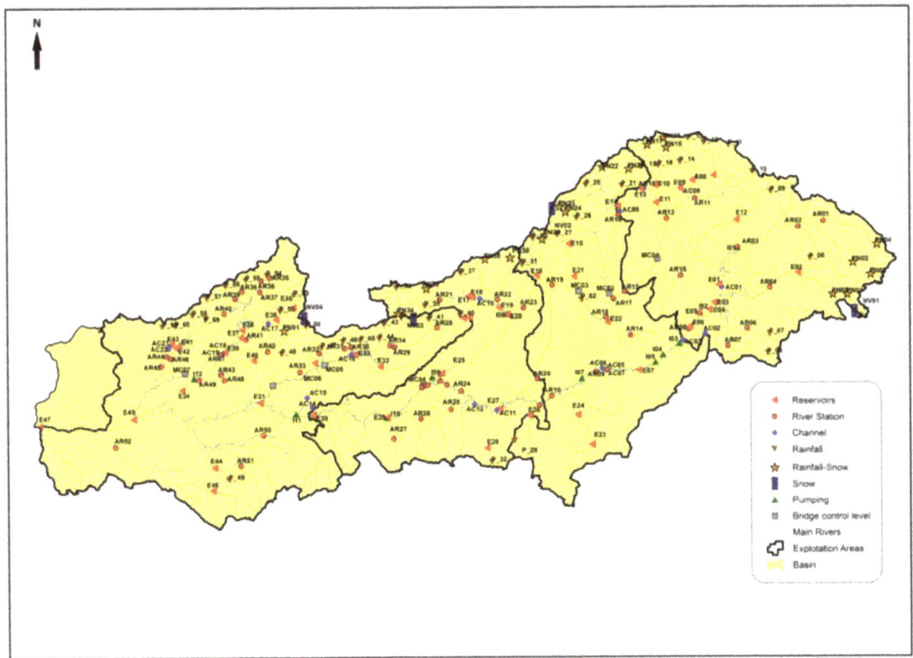

Figure 3. SAIH network control points—Tagus Basin.

The model, fed daily (during the months November to June) with the data provided by the SAIH meteorological stations, allows for the virtual construction of the snow cover. Its thickness and density are contrasted with field measurements (usually twice a year). The extent of snow cover is estimated by the model using cells, in which the existence of stored snow is determined, and is compared with satellite observations (Figure 4)

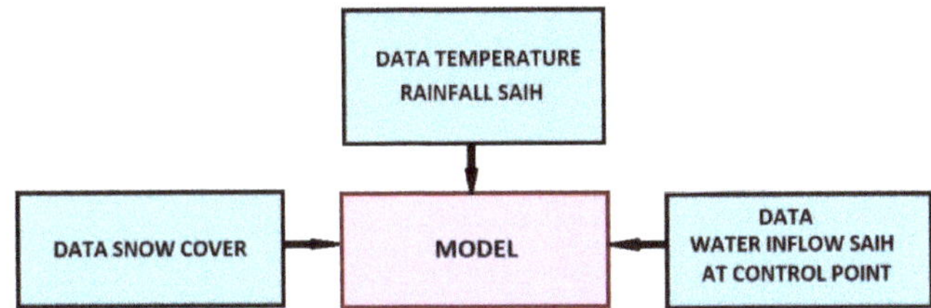

Figure 4. Calibration of the ASTER model with data from the SAIH network control points.

More recently, the ECMW (European Center for Medium-Range Weather Forecasts) numerical prediction model facilitates seven-day snow accumulation forecasts, allowing an additional precision in the results of the evolution of the snow cover obtained by ASTER

4.3. Snow Cover Modelling

The periodic (daily or weekly) and systematic application of the model from the beginning of November is displayed in order to reproduce the generation of the snow cover, represented by the snow storage conditions in each cell. Periodically, the gap of the snow cover, obtained with ASTER, is controlled against its real limits observed by remote sensing (TERRA/AQUA MODIS sensor images), and when possible, by collecting in situ snow thicknesses and densities.

The model is able to determine the cells in which the liquefaction conditions are reached and activate the corresponding simulation module of the melting process, but it needs to ascertain whether there was actually snow previously stored in it, and if applicable, the thickness and equivalence in water. These data and the month in which the snow storage process can start from certain levels are known thanks to the historical monitoring of the evolution of the mantle that has been carried out every year since November. The numerical results are reflected in the corresponding monitoring graphs for each zone or sub-basin, obtaining, by aggregation, the evolution of the total snow accumulated in the whole of the basin. See Figures 5 and 6.

River Basin				(*) 02/23/2014 State			
Order	Sistem	Code	Control Point	Modelled surface (km^2)	VAFN (hm^3)	Contribution (hm^3)	Snow Reserves
1		D.II.6.1/I	Alagón at Gabriel and Galán	1.848,2	7,2	723,5	Very high
2		D.II.6.2/I	Jerte at Plasencia	370,0	3,2	175,7	Low
3		D.II.5/I	Tietar at E. de Rosarito	1.743,2	12,8	501,8	5 year Maximum
4		D.II.4/I	Alberche at the Burguillo	1.052,7	19,0	130,6	High
5		D.II.3/I	Guadarrama at Picotejo	355,9	2,3	33,8	5 year Maximum
6	Central Sistem	D.II.2.3/I	Manzanares at Santillana	247,1	7,8	46,1	5 year Maximum
7		D.II.2,2.B/I	Lozoya at the Atazar	925,0	30,4	180,4	5 year Maximum
8		D.II.2.2.A/I	Jarama at the Vado	378,0	7,3	88,0	Very high
9		D.II.2.1.B/I	Sorbe at Beleña	475,6	0,0	81,4	Normal
10		D.II.2.1.A/I	Bornova at Alcorló	362,3	0,0	50,9	Normal
11		D.II.1.1/I	Alto Tajo at Entrepeñas	3.825,6	0,0	162,5	Low
12		D.II.1.2/I	Guadiela at Buendia	3.355,8	0,0	161,8	Normal
			Total:	14.939,4	90,0	2336,6	

(*) Obtained by the ASTER model.
VAFN: Water volume in snow form

Figure 5. Example of monthly snow report (ERHIN) in the Tagus Basin (Tagus Hydrographic Confederation).

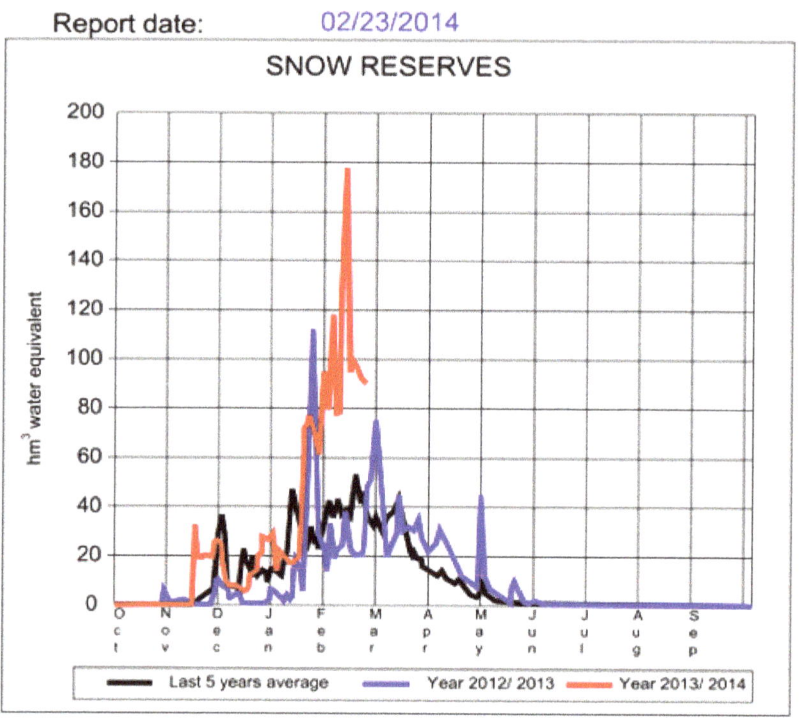

Figure 6. Graphic type of monitoring of snow reserves in the ERHIN area of the Tagus basin (Tagus Hydrographic Confederation).

4.4. Melting and Accumulation Snow Cover Simulation

Finally, there is a hydrological model that works with a subroutine that allows for the simulation of snow accumulation/melting processes in the mountainous headwaters of the rivers that flow to the reservoir in operation, and with a system for providing basic data (temperature and rainfall), in real time, on the information at any time about the stored snow and its geographical distribution. Therefore, it is a system capable of automatically recording rainfall over the basin, upstream of the reservoir, and anticipating the inflow generated by the runoff, and where appropriate, by the melting of the snow cover.

In addition, the system is fed with the data corresponding to the weather forecasts for several days (3 to 7 days). This significantly extends the time available to carry out the necessary preventive actions, establishing a safety margin in line with the forecasts of incoming flows that allow for the management of the reservoir in the case of floods, see Figure 7. Therefore:

- Minimizing the peak value of the outflow, and therefore, downstream damage.
- Maximizing the volume of water stored and the guarantee of supply, avoiding excesses in preventative release cases.
- Ensuring the safety of hydraulic infrastructure by reducing the chances of overflow by crowning (overtopping).

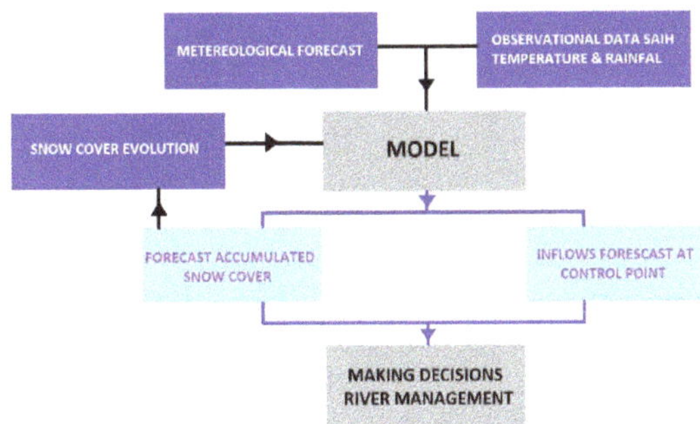

Figure 7. Scheme of operation of the model in real time.

The ASTER-SAIH system aims to obtain the maximum performance from the ASTER application, taking as input the data provided in real time by the SAIH. The system execution requires the observed thermopluviometric values available at that time at SAIH stations as input data. If required, it can propose forecast scenarios with different meteorological hypotheses, adjusting the model with real data of flows from SAIH and real data snow cover.

The data obtained, as a result of the forecast, are the inflow flows to the reservoir and the equivalent volume of water, stored as snow in the watershed. These are very helpful values in decision making (reservoir management).

5. Results: Rosarito and El Burguillo Reservoirs Practical Cases

The actions described in the previous section need to be completed in order to develop the proposed methodology and to make it fully operational—this has been applied by the technical services of the Tagus Hydrographic Confederation under different flood episodes. Some recent results are described below; specifically, those corresponding to the period between 8th and 12th March, 2018, during the storm caused by the "Felix" cyclone. At that time, a progressive and intense cyclogenesis swept the Iberian Peninsula from West to East, and involved the Canary Islands. This episode caused the activation of Civil Protection Plans in all the Spanish Autonomous Communities. In the first few days, there was intense rainfall throughout the Spanish territory.

In the Tagus basin, the most important rains were recorded on 9th and 10th March, while the floods in the main rivers, as well as the flood control in the reservoirs, were extended until 12th March.

The monitoring of the meteorological forecasts that are continuously received in the SAIH room board at the Hydrographic Confederation, allowed for the launching of the Reservoir Release Standing Committee. For this reason, the first AEMET rainfall forecasts were developed at the scale of the Tagus basin, see Figure 8.

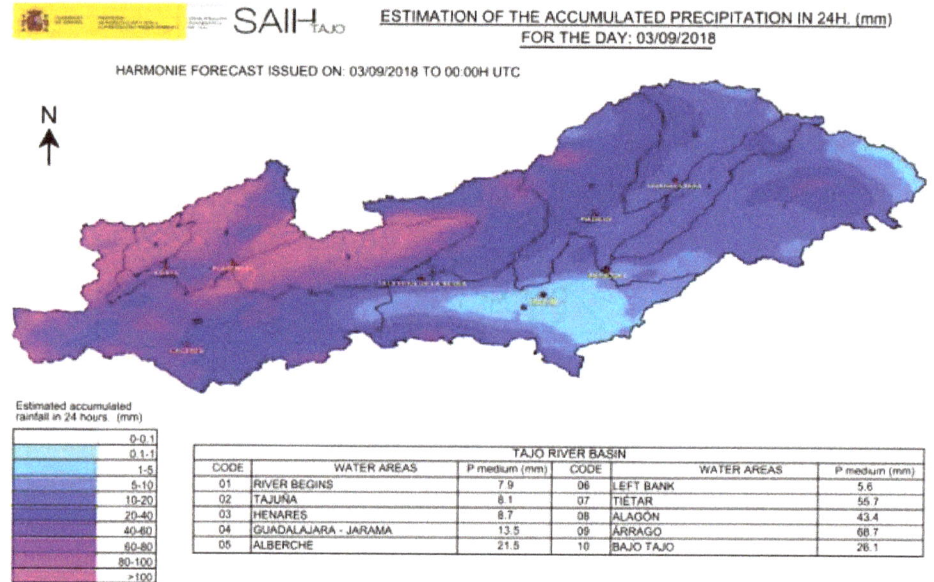

TAJO RIVER BASIN					
CODE	WATER AREAS	P medium (mm)	CODE	WATER AREAS	P medium (mm)
01	RIVER BEGINS	7.9	06	LEFT BANK	5.6
02	TAJUÑA	8.1	07	TIÉTAR	55.7
03	HENARES	8.7	08	ALAGÓN	43.4
04	GUADALAJARA - JARAMA	13.5	09	ÁRRAGO	68.7
05	ALBERCHE	21.5	10	BAJO TAJO	26.1

Figure 8. Rainfall forecasts on the day of the starting episode.

These predictions show the special intensity of the episode in the western part of the basin, particularly in the river basins of the Árrago River, Alagón River and Tiétar River, and less intensively, in the Alberche River. On 10th March, different yellow warnings were received (referred to as a three steps scale: red, yellow, and green) from the AEMET, State Meteorological Agency, which affected the Autonomous Communities of Madrid and Castilla la Mancha, and alerted to them heavy rainfalls, snowfalls and winds, see Figure 8.

The Tiétar basin deserves special attention since it is located the Rosarito reservoir (Figure 9), and its operation is uniquely delicate. With only 8190 hm^3 of capacity compared to an average contribution of 800 hm^3, it must meet a demand of 240 hm^3, the majority for irrigation, concentrated in the months between April and September. The altitude of the Rosarito dam is 283.7 m (minimum level) and it is located at the bottom of the southern part of Sierra de Gredos (2500 m maximum altitude), where the annual average rainfall regime, at its summits, often exceeds 2000 mm/year. The flow contributions of its watershed are characterized by a steep average slope and a short concentration period. Therefore, there is a high risk of sudden floods, with very high peak flow values, in which flows from the snow melting can contribute significantly. Therefore, anticipating the forecast of the flood hydrograph, including the additional fraction of snow melting, is extraordinarily important.

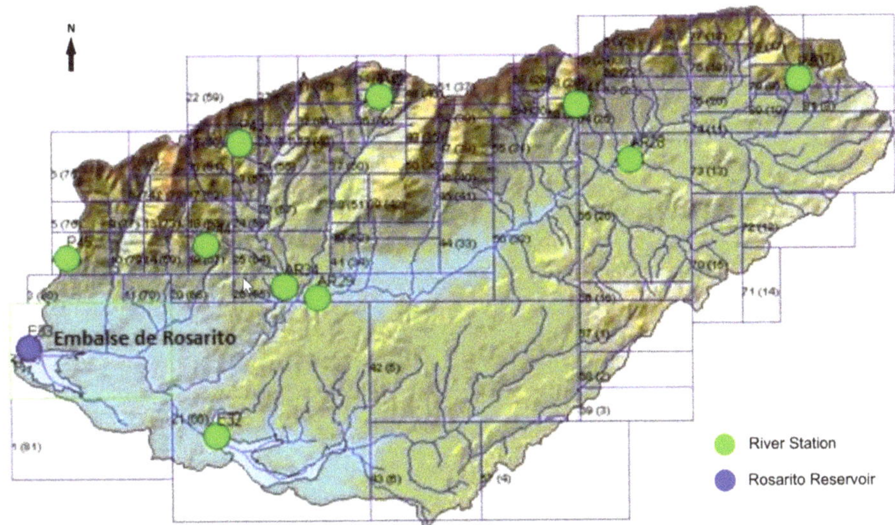

Figure 9. Location of the Rosarito Reservoir in the lower part of the Tiétar Basin.

Figure 10 shows the results that correspond to the forecast of flows and the volume of water stored in the form of snow in the period between 9th and 18th March, 2018. In the first and second part of the graph, the real values (blue) of temperature and precipitation appear on the previous dates (provided by SAIH), and these parameters forecasted by AEMET(green). Between them, the more unfavourable meteorological values were chosen and correspond to the most probable and expected precipitation and temperature values.

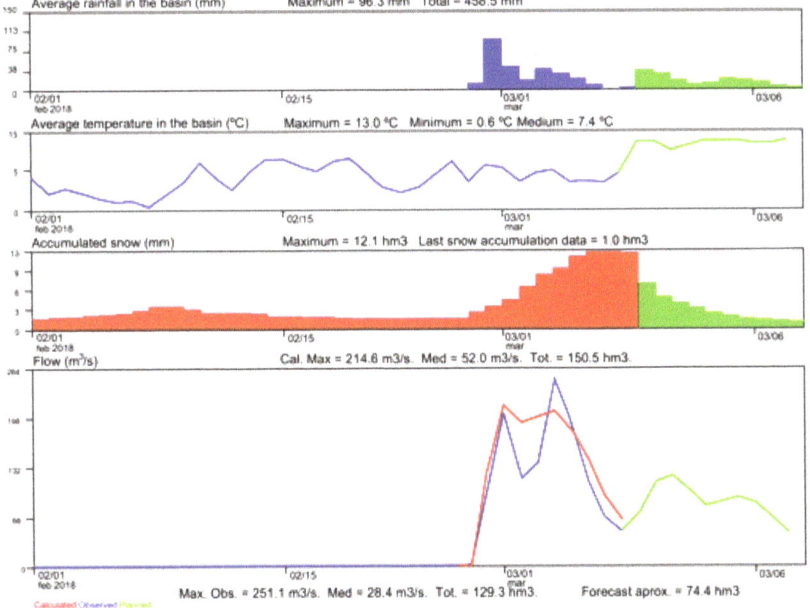

Figure 10. Flow forecast graph (green) in the Rosarito reservoir, with the AEMET (most unfavorable) temperature and rainfall predictions.

In the third part, the volumes of water that accumulated in the form of snow—estimated by ASTER-ERHIN (red) on the date preceding the beginning of the episode (9th March, 2018)—are shown. It is observed how the storm, one day before, contributed a notable increase in snow storage, with potential fusion, and added to the runoff due to the foreseeable rains, which added an additional risk of flows. Finally, in green, the total snow values accumulated during that period of time which reflect a clear melting process of almost 11 hm^3, due to the expected increase in temperature

Finally, in the fourth part, the red line corresponds to the inlet flows to the reservoir, calculated by the ASTER-ERHIN model, with the precipitation and temperature data provided by the SAIH; the blue line shows the measured input values (input/output balance/volume increase) by SAIH and the green line represents the flow forecast provided by the model with the prediction of temperatures and rainfall delivered by AEMET.

The calculated values of the Volume (hm^3) of Water in Snow (VAFN), the predicted values (ASTER-ERHIN) of average daily and cumulative flows (m^3/s) into the Rosarito reservoir, with AEMET (most unfavourable) predictions of temperature and rain and its accumulated volume inflow (hm^3), are shown in Table 2.

Table 2. Expected values of average daily flows entering the Rosarito reservoir from AEMET (most unfavourable) temperature and rainfall predictions. Between 8 March, 2018 and 18 March, 2018, near 11 m^3, are coming from snowmelt.

Date (at 00:00 h)	Snow Water Volume (hm^3)	Flow (m^3/s)	Accumulated Volume Inflow (hm^3)
8 March 2018	11.93		
9 March 2018	7.13	71.92	6.21
10 March 2018	4.83	112.14	15.90
11 March 2018	3.88	120.95	26.35
12 March 2018	3.12	101.60	35.13
13 March 2018	2.40	81.60	42.18
14 March 2018	1.85	86.77	49.68
15 March 2018	1.48	92.01	57.63
16 March 2018	1.19	84.45	64.93
17 March 2018	1.06	65.94	70.62
18 March 2018	0.97	44.29	74.45

A similar case, corresponding to the contributions to the El Burguillo reservoir (Figure 11), in the Alberche basin during the same episode in 2018, is shown below.

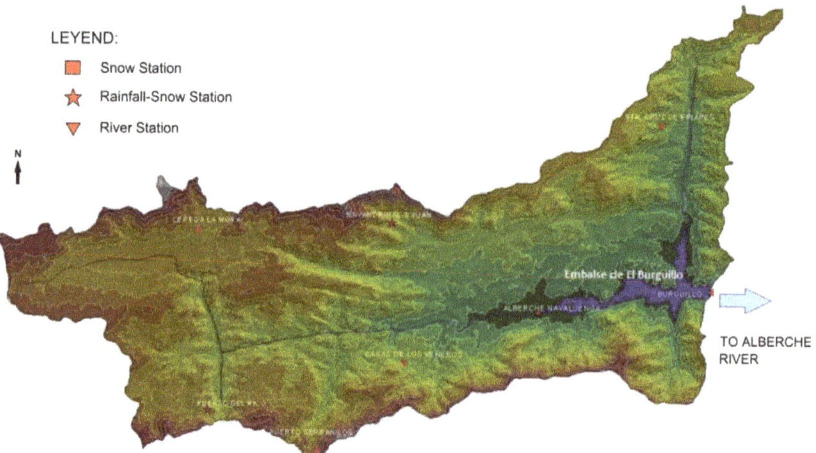

Figure 11. Location of the El Burguillo Reservoir in the upper part of the Alberche Basin.

In Table 3, it can be seen that the storm produced a merger process of more than 13 hm^3 of Snow Water Volume (VAFN), previously accumulated between 8 March, 2018 and 18 March, 2018, which demonstrates the need to use hydrological models with a snow subroutine.

Table 3. Expected values of average daily flows, entering the El Burguillo reservoir from AEMET (most probable) temperature and rainfall predictions.

Date (at 00:00 h.)	Snow Water Volume (hm^3)	Flow (m^3/s)	Accumulated Volume Inflow (hm^3)
8 March 2018	16.24		
9 March 2018	7.60	30.65	3.17
10 March 2018	4.31	79.37	10.02
11 March 2018	3.91	107.41	19.30
12 March 2018	3.84	65.11	24.93
13 March 2018	3.67	53.35	29.54
14 March 2018	3.20	53.40	34.15
15 March 2018	2.91	57.13	39.09
16 March 2018 00:00	2.87	56.32	43.95
17 March 2018 00:00	2.92	51.64	48.42
18 March 2018 00:00	2.90	45.09	52.31

As in the previous case of Tiétar (closure in the Rosarito reservoir), the storm before the episode that began on 9 March 2019, brought with it a large accumulation of snow in the basin (Figure 12). Its potential melt was considered in the calculation of future contributions to the El Burguillo reservoir. This gives consistency to the optimal management of reservoirs and their release that minimizes downstream damage without jeopardizing the safety of the dam or the guarantees of water supply.

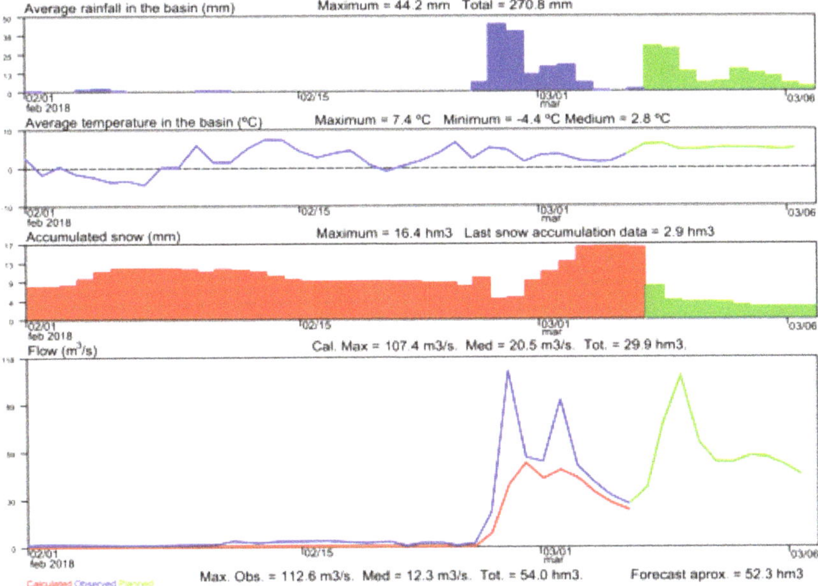

Figure 12. Flow forecast graph (green) in the El Burguillo reservoir, with the AEMET (most probable) temperature and rainfall predictions.

In both cases—Rosarito and El Burguillo—a very positive consequence can be obtained from the results achieved, since they have meant a notable advance with respect to the previous predictive methods, which could not take into account the additional flows generated by the snow layer fusion. The integrated SAIH-ASTER application of providing real time data to the ASTER model allowed for information on the accumulated snow reserves at the beginning of the episode to be obtained. This also helped to make a forecast of incoming flows in the El Burguillo reservoir. This forecast was calculated over time, using weather forecasts from AEMET. All this facilitated the management of the reservoirs and water release from the reservoir with the lowest possible flow peaks downstream.

6. Conclusions

Floods are among the most damaging natural disasters economically, but also causing serious disorder in human activity and endangering people's safety. Its effects are aggravated by anthropogenic intervention, through exposure to the risk of human transformation in the territory, and alteration of the natural regime of the rivers.

In the case of Spain, its special hydrological characteristics and the structure of water demand—with an important agricultural use that concentrates its consumption in the dry months—have promoted the construction of a significant number of dams to regulate its rivers. These infrastructures are an essential element for flood control and lamination. However, at the same time, they constitute a potential risk in the event of breakage due to overtopping, making it a priority to guarantee their safety.

The complex management of such a high number of hydraulic infrastructures has required the development of a technologically advanced Automatic Hydrological Information System (SAIH), which provides quasi-real-time data on the main hydrometeorological variables of the basin and dam hydraulics—information that facilitates the safe operation of reservoirs and establishing preventive safeguards during floods. However, the operation of reservoirs in mountainous areas, where the presence of snow is hydrologically significant, was not adequately resolved due to the lack of methods and tools to forecast inlet flows that took those resulting from melting snow layers into account.

The methodology developed by the Tajo Hydrographic Confederation integrates tools already available in the Hydrographic Confederations into the SAIH environment. This allows for the coupling of that system to supply input data to the ASTER hydrological modelling of the Water Snow Resources Assessment program (ERHIN). Thus, an appreciable improvement has been obtained in the forecasts of incoming flows in reservoirs of mountainous headwaters. The joint and coordinated application of SAIH and ERHIN has facilitated a more efficient use of resources and the means of both.

The data obtained as a result of the forecast are the inflows to the reservoir and the equivalent volume of water stored as snow in the watershed. These values help decision-making in real time (reservoir management), as its application by the Tajo Hydrographic Confederation has repeatedly demonstrated.

In the cases shown regarding the application of this process (Rosarito and El Burguillo reservoirs), it has been possible to verify the possibility of incorporating weather forecasts provided by the State Meteorological Agency (AEMET).

With this, it is possible to schedule, on time, preventive dam releasing and anticipate the implementation of more reliable safeguards.

The improvement of the described methodology and the generalization of its application to other hydrographic basins represent a great advance in the reduction in flood damage and improving the safety of dams located in mountainous areas.

Author Contributions: Conceptualization, I.M.P., J.A.H.M. and E.S.P.; Data curation, J.M.A.-M.; Formal analysis, J.M.A.-M.; Investigation, I.M.P. and J.M.A.-M.; Methodology, J.A.H.M. and J.M.A.-M.; Project administration, I.M.P.; Resources, J.A.H.M.; Validation, E.S.P.; Writing—original draft, I.M.P.; Writing—review & editing, I.M.P. All authors have read and agreed to the published version of the manuscript.

Funding: This research received no external funding

Acknowledgments: Authors acknowledge support given by Tagus Hydrographic Confederation by providing detailed data.

Conflicts of Interest: The authors declare no conflict of interest.

References

1. Consorcio de Compensación de Seguros. Estadística de Riesgos Extraordinarios Serie 1.971-2.017. Available online: www.consorseguros.es (accessed on 15 October 2019).
2. IPCC, W.G. AR5 Climate Change 2014: Impacts, Adaptation, and Vulnerability. 2014. Available online: https://www.ipcc.ch/report/ar5/wg2/ (accessed on 5 October 2019).
3. EC. circabc.europe.eu. Available online: https://circabc.europa.eu/webdav/CircaBC/env/wfd/Library/water_directors/documents_presidency/meetingsdocuments/9%20-%20Best%20Practices%20Flood%20final%20version.pdf (accessed on 5 September 2019).
4. Comité Español de Grandes Presas SPANCOLD. *Rotura de Presas. Análisis Estadístico, 1996*; Comité Español de Grandes Presas SPANCOLD: Madrid, Spain, 1996.
5. Arenillas Parra, M.; Cobos Campos, G.; Navarro Caraballo, J. *Datos sobre la Nieve y los Glaciares en las Cordilleras Españolas: El programa ERHIN 1984–2008*; Monreal Teodoro, E., Francés Mahamud, M., Eds.; Ministerio de Medio Ambiente y Medio Rural y Marino: Madrid, Spain, 2008.
6. Available online: www.miteco.gob.es/es/cartografia-y-sig/ide/descargas/agua/saih.aspx (accessed on 10 September 2019).
7. Arenillas Parra, M.; Martínez Costa, R. *El método Hidrológico-Estadístico. La Nieve en el Pirineo Español*; MOPU: Madrid, Spain, 1988; pp. 99–126.
8. Anderson, E.A. National Weather Service river forecast system: Snow accumulation and ablation model (Vol. 17). In *Technical Memorandum NWS Hydro-17*; US Department of Commerce, National Oceanic and Atmospheric Administration, National Weather Service: Silver Spring, MD, USA, 1973.
9. Gioia, A.; Manfreda, S.; Iacobellis, V.; Fiorentino, M. Performance of a Theoretical Model for the Description of Water Balance and Runoff Dynamics in Southern Italy. *J. Hydrol. Eng.* **2014**. [CrossRef]
10. Haggett, C. An integrated approach to flood forecasting and warning in England and Wales. *J. Water Environ. Manag.* **1998**, *12*, 425–432. [CrossRef]
11. Gioia, A.; Iacobellis, V.; Manfreda, S.; Fiorentino, M. Comparison of different methods describing the peak runoff contributing areas during floods. *Hydrol. Process.* **2017**, *31*, 2041–2049. [CrossRef]
12. Liu, J.; Han, D. On selection of the optimal data time interval for real-time hydrological forecasting. *Hydrol. Earth Syst. Sci.* **2013**, *17*, 3639–3659. [CrossRef]
13. Cantarino, I. Anuario Aster 98 UPV. 1998. Available online: http://www.upv.es/dit/ASTERanuario.html (accessed on 15 June 2019).
14. Morin, G.; Fortin, J.P.; Lardeau, J.P.; Sochanska, W. *Modèle CEQUEAU: Manuel d'utilisation*; INRS-Eau, Rapport Scientifique: Quebec, QC, Canada, 1981; p. 93.
15. Gascoin, S.; Hagolle, O.; Huc, M.J.; Dejoux, J.-F.; Szczypta, C.; Marti, R.; Sanchez, R. A snow cover climatology for the Pyrenees from MODIS snow products. *Hydrol. Earth Syst. Sci.* **2015**, *19*, 2337–2351. [CrossRef]
16. Arenillas, M.; Cobos, G.; Mora, J. *El Programa ERHIN y la Gestión de Embalses en Áreas con Alta Precipitación Nival*; Ingeniería Civil (144): Madrid, Sapin, 2006; pp. 89–96.
17. Cobos, G. Cuantificación de las Reservas Hídricas en Forma de Nieve y Previsión en Tiempo Real de los Caudales Fluyentes Derivados de la Fusión. Aplicación al Pirineo Español: Cuenca alta del río Aragón. Ph.D. Thesis, Universidad Politécnica de Valencia, Valencia, Spain, 2004.
18. Mora, J.; Ferrer, C.; Arenillas, M.; Cobos, G. Hydrological Peculiarities of Mountain Basins. The Case of the Spanish Pyrenees. In Proceedings of the International Conference on Water Observation and Information System for Decision Support, Ohrid, Macedonia, 25–29 May 2004.
19. Ferrer, C.; Romeo, R.; Arenillas, M.; Cobos, G. El Sistema ASTER-SAIH Aplicado a la Explotación de Embalses en Cuencas con Marcado Comportamiento Nival. In Proceedings of the VII Jornadas Españolas de Presas, Zaragoza, Spain, 29–31 May 2002.

Article

Modelling the 2012 Lahar in a Sector of Jamapa Gorge (Pico de Orizaba Volcano, Mexico) Using RAMMS and Tree-Ring Evidence

Osvaldo Franco-Ramos [1,*], Juan Antonio Ballesteros-Cánovas [2,3], José Ernesto Figueroa-García [4], Lorenzo Vázquez-Selem [1], Markus Stoffel [2,3,5] and Lizeth Caballero [6]

[1] Instituto de Geografía, Universidad Nacional Autónoma de México, Ciudad Universitaria Coyoacán, México 04510, Mexico; lselem@igg.unam.mx

[2] Dendrolab.ch, Department of Earth Sciences, University of Geneva, 13 rue des Maraîchers, CH-1205 Geneva, Switzerland; juan.ballesteros@unige.ch (J.A.B.-C.); markus.stoffel@dendrolab.ch (M.S.)

[3] Climate Change Impacts and Risks in the Anthropocene (C-CIA), Institute for Environmental Sciences, University of Geneva, 66 Boulevard Carl-Vogt, CH-1205 Geneva, Switzerland

[4] Posgrado en Geografía Universidad Nacional Autónoma de México, Ciudad Universitaria Coyoacán, México 04510, Mexico; ernestfigue.7@gmail.com

[5] Department F.-A. Forel for Environmental and Aquatic Sciences, University of Geneva, 66 Boulevard Carl-Vogt, CH-1205 Geneva, Switzerland

[6] Facultad de Ciencias, Universidad Nacional Autónoma de México, Ciudad Universitaria Coyoacán, México 04510, Mexico; lcaballero@ciencias.unam.mx

* Correspondence: ofranco@igg.unam.mx

Received: 18 December 2019; Accepted: 21 January 2020; Published: 23 January 2020

Abstract: A good understanding of the frequency and magnitude of lahars is essential for the assessment of torrential hazards in volcanic terrains. In many instances, however, data on past events is scarce or incomplete, such that the evaluation of possible future risks and/or the planning of adequate countermeasures can only be done with rather limited certainty. In this paper, we present a multiidisciplinary approach based on botanical field evidence and the numerical modelling of a post-eruptive lahar that occurred in 2012 on the northern slope of the Pico de Orizaba volcano, Mexico, with the aim of reconstructing the magnitude of the event. To this end, we used the debris-flow module of the rapid mass movement simulation tool RAMMS on a highly resolved digital terrain model obtained with an unmanned aerial vehicle. The modelling was calibrated with scars found in 19 *Pinus hartwegii* trees that served as paleo stage indicators (PSI) of lahar magnitude in a sector of Jamapa Gorge. Using this combined assessment and calibration of RAMMS, we obtain a peak discharge of 78 m^3 s^{-1} for the 2012 lahar event which was likely triggered by torrential rainfall during hurricane "Ernesto". Results also show that the deviation between the modelled lahar stage (depth) and the height of PSI in trees was up to ±0.43 m. We conclude that the combination of PSI and models can be successfully used on (subtropical) volcanoes to assess the frequency, and even more so to calibrate the magnitude of lahars. The added value of the approach is particularly obvious in catchments with very scarce or no hydrological data at all and could thus also be employed for the dating and modelling of older lahars. As such, the approach and the results obtained can be used directly to support disaster risk reduction strategies at Pico de Orizaba volcano, but also in other volcanic regions.

Keywords: lahar; Pico de Orizaba; paleostage indicators; tree-ring analysis; RAMMS program

---

1. Introduction

Lahars are torrential mass movements on volcanic slopes and have been documented in various volcanic environments around the world. Being mixtures of water and sediment, they can mobilize large amounts of weakly consolidated debris, rocks, and wood. These processes are know for their high flow velocities, and for their potential to cause major destruction and loss of lives [1–3]. Lahars are commonly triggered by heavy rainfall, condensate gas coming from volcanoes, excessive pore water, or glacial and snow meltdown, but the process has also been reported to occur as a result of river erosion or dam overflow [4,5].

On active volcanoes, so-called syn-eruptive lahars may be formed after heavy downpours onto unconsolidated volcanic sediment or by the sudden melting of snow and ice during volcanic eruptions. Similar processes can occur in the absence of volcanic activity; in that case, they are referred to as intra-eruptive or post-eruptive lahars [6,7]. This last type of lahars can occur on active, dormant or extinct volcanic complexes and can persist long after the last volcanic activity [8–10]. Intra-eruptive and post-eruptive lahars are frequently triggered by enhanced hurricane activity in subtropical environments or by cyclonic events (i.e., torrential rainfall) [8,11], but seismic activity has been shown to trigger lahars as well [4,12]. Different approaches have been used in the past to understand the occurrence and triggering of lahars. Past work included rainfall–runout analyses [8,11,13–15], real-time monitoring of lahars with rainfall, seismic (geophones) and observational data [16,17], but also analyses of disturbances recorded in tree-ring series [18–21].

Lahar magnitude can be approximated by lahar discharge, i.e., the volume of water and sediment moving down a gorge per unit of time. It depends on the initial volume of water and sediment, the erosion and incorporation of secondary, exotic debris by lahars as they move downstream (bulking) [5], the incorporation of water from tributaries, and sediment deposition (debulking). Lahar discharges have been reported in the literature, and one of the largest contemporary lahars occurred in 2007 at Nevado del Huila (Colombia) [22]. Since they are widely applied to calibrate numerical models used in lahar hazard evaluation, accurate volume estimations are necessary to strengthen their results.

The magnitude of lahar events has been assessed frequently with models using different bidimensional hydraulic codes and input data; among the more commonly used modeling tools are LAHARZ [23], TITAN2D [24], FLO-2D [25] or RAMMS [26]. Thanks to these models, the magnitude of several lahars has been calculated, including, among others, the very large 2007 lahar at Nevado del Huila with an estimated maximum peak discharge of ~13,000 m^3 s^{-1} [22]. Other studies investigated the process dynamics of the lahar caused by the 2010 Merapi eruption [27], but also looked at hypothetical lahars on the base scenario of historical lahar deposits from an event that occurred in 1877 on Cotopaxi volcano [28]. Additionally, Nocentini et al. [29] have used FLO-2D to perform debris-flow modelling in volcanic terrains of Italy.

On the most active volcanoes of central Mexico, several hydraulic models have been used in the past to study lahars, mostly with the aim of obtaining better insights into sedimentology and/or rheological parameters (discharge, velocity, lahar depth). The LAHARZ code has been used on Popocatépetl [30], Pico de Orizaba [31], Nevado de Toluca [32], San Martin at Tuxtlas [33], Tacaná [34] and Ceboruco [35] volcanoes. Subsequently, the FLO-2D code has been applied to analyze Popocatépetl's 2001 lahar [36] and in the Nexpayantla Gorge located on the northwestern slope of Popocatépetl volcano [37] to model the 2010 lahar. At Colima volcano, lahar peak discharges have been estimated for recent events as well using the FLO-2D code [16,38].

In general, all of the above-mentioned studies have contributed considerably to delimit the reach and lateral extension of lahar events and to understand the hazards and risks induced by lahars. The value and accuracy of modelling approaches will depend substantially on the availability of (post-event) field data or observations for the calibration and validation of models; however, such data are normally absent [39–41]. In particular, paleo-deposits (lahar terraces or deposits) are only rarely present on Mexican volcanoes due to omnipresent erosion in the high-gradient torrents. Furthermore, historical records of volcanic or geomorphic events are not generally available, especially in areas with limited population densities. In this context, the reconstruction of past events with tree-ring series can contribute significantly to an improved knowledge of the frequency and/or magnitude of past volcanic or geomorphic processes [19–21,42]. Hence, scarred trees have proven very useful for the dating and frequency analysis of past geomorphic events [43] and for the determination of paleostage indicators (PSI; [39]) as the level of scars on trees is considered to indicate the minimum flood level during a given event [44]. Several studies have used scar heights to calibrate two-dimensional (2D) hydraulic models and to estimate the flow discharge (magnitude) of paleofloods [45–47]. This methodology, originally used for torrential and river floods, has been adapted on volcanic terrains to estimate the discharge of a lahar on Canary Island, Spain, in 1997 [48].

In this paper, we aim at estimating the magnitude of a lahar that occurred in 2012 on the slopes of Pico de Orizaba volcano (Mexico) in a sector of Jamapa Gorge by combining botanical evidence (i.e., scars on tree trunks) and hydraulic modelling performed on a highly-resolved topography acquired with an unmanned aerial vehicle (UAV). To this end, we used RAMMS, a dynamic numerical model that allows simulation of debris flows (including water depth, discharge and flow velocity) in mountain basins [49–51].

2. Study Site

2.1. Volcanic Setting and Recorded Lahar Activity

The Jamapa Gorge is located on the northern slopes of Pico de Orizaba volcano (5640 m above sea level (a.s.l.)), also known as Citlaltépetl ("Mountain of the Star") in the Náhuatl language. Pico de Orizaba is an active, yet quiescent Quaternary stratovolcano of andesic–dacitic composition located in the eastern sector of the Trans-Mexican Volcanic Belt (Figure 1). The volcanic complex is characterized by a long history of eruptive episodes with repeat growth and collapse of domes and other volcanic structures. According to Carrasco-Núñez [52,53], the first volcanic paleo-structure of Pico de Orizaba can be dated to ~0.65 Ma BP (= Before Present) and is referred to as "Torrecillas". This volcanic structure was destroyed between ~0.29 and ~0.21 Ma by a large collapse associated to debris avalanche deposits on the northeastern slope of the volcano. The second paleo-structure was dated to ~0.21 Ma BP and is known as "El Espolón de Oro". The cone belonging to this episode was destroyed by a large collapse and debris avalanche associated with hydrothermal alteration of the volcanic edifice dated to ~20,000 years (or 20 kyr; the abbreviation will be used hereafter). The current cone (Pico de Orizaba), is ~20 kyr in age. Two Plinian eruptions have been dated to the late Pleistocene and the Holocene, between ~13 kyr and ~8.5 kyr ago. These events covered the flanks of the volcano with thick pyroclastic flow and tephra fall deposits known as "Citlaltépetl" ignimbrite [54–56]. Other volcanic events during the Holocene were recorded between 7.0–6.2 and 3.4 kyr BP [57]. Block and ash flows and lahar deposits were produced at the summit crater around 4.1–4.0 kyr BP [54] in relation to a large explosive event in the western sector of the volcano. On the southern slope of Pico de Orizaba, two lava flows were dated to between 3 and 1 kyr BP [58]. Furthermore, historical effusive and explosive volcanic events were reported in 1537, 1545, 1566 and 1613 AD [59].

Holocene and historical lahar deposits can be found in different gorges and rivers around the Pico de Orizaba volcano. For example, Carrasco-Núñez et al. [52] dated a large lahar (~16.5 kyr BP) associated with the collapse of the Espolón de Oro cone; the resulting lahar deposits were emplaced in the Tetelzingo River where they still have thicknesses of 12–20 m on average and a maximum of 100 m. Other lahar deposits were identified inside Huitzilapan Gorge and dated to ~5260 yr BP [12]. In the Huitzilapan river (north-northeastern slope of Pico de Orizaba), one can observe an important lahar deposit from an event that occurred in 1920 ("Huitzilapan lahar"). The 1920 lahar was a typical compounded event, as an earthquake (magnitude 6.5) triggered several landslides from slopes saturated with water after 10 days of torrential rainfall [12]. Iverson et al. [59] reported high water marks 40 to 60 m above the channel for the 1920 event, which allowed for estimation of its volume at 4.4×10^7 m^3 [31].

More recently, on 5 June 2003, a catastrophic debris flow occurred on the southern slope of Pico de Orizaba after a torrential rainfall, resulting in a peak discharge of 350 m^3 s^{-1} [60]. On the north-northeastern slopes of Pico de Orizaba volcano (i.e., Jamapa and Seca Gorges), intra-eruptive lahars occurred in 2012, again triggered by heavy rainfalls that can be linked to the passage of Hurricane Ernesto over the Atlantic Ocean [61].

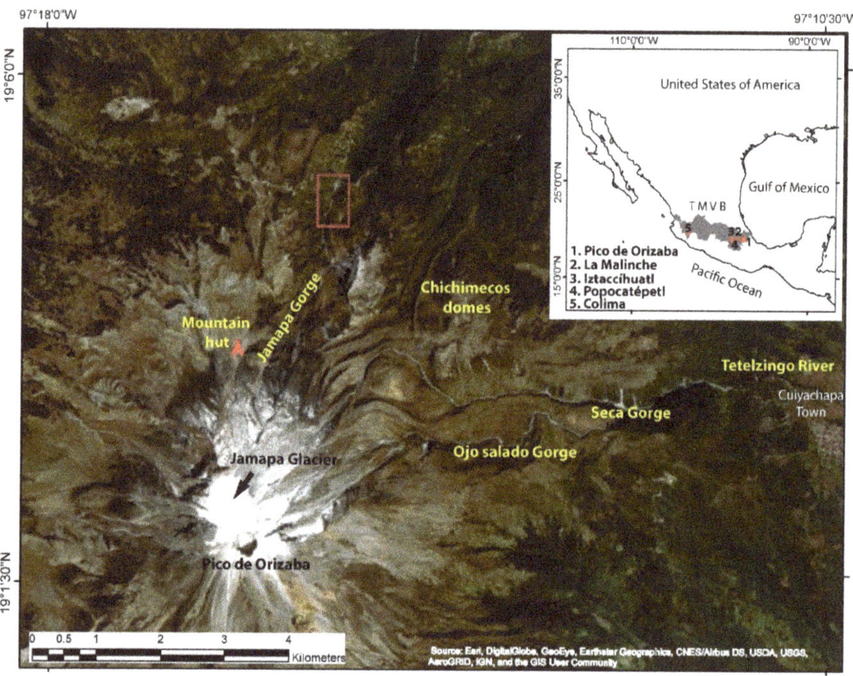

Figure 1. The Pico de Orizaba volcano is located in the eastern sector of the Trans-Mexican Volcanic Belt (TMVB). Satellite imagery illustrating the volcano as well as the gorges mentioned. The Jamapa Gorge, where this study was realized, is located on the northern slope of the volcano. The study site is indicated with a red box.

2.2. Geomorphic Features on Jamapa Gorge

The Jamapa Gorge drains the northern flank of Pico de Orizaba volcano; its headwaters are at ~5000 m a.s.l. near the terminus of Jamapa glacier (Figure 1). In that area, glacial landforms from the late Pleistocene and Holocene have been covered by Holocene pyroclastic deposits and lava flows. The glaciers of the Little Ice Age (LIA) covered the top of the Jamapa Gorge drainage basin and built terminal moraines ridges around 4700 m a.s.l. [62]. In addition, several paraglacial and postglacial geomorphic features can be found in the area, such as debris flows or rockfall deposits and sectors characterized by fluvial erosion [63].

The study site investigated in this paper is located on the central sector of Jamapa Gorge, between 3535 and 3620 m a.s.l. (Figures 1 and 2A). Several old laharic terraces can be recognized with thicknesses of 1–4 m (Figure 2B). Buried, tilted and scared trees are common in the laharic deposits of the low terraces (Figure 2C). Superficial debris-flow deposits with block sizes of ~3 m in diameter are embedded in a sandy matrix. Recent laharic deposits (from debris flows and/or hyperconcentrated flows) have filled the main channel with ~2 m of sediments in this sector of Jamapa. Some of these lahars have reached the road connecting the federal states of Puebla and Veracruz (Figure 2D).

Figure 2. Geomorphic characteristics of Jamapa Gorge. (**A**) Active channel with recent lahar deposits. (**B**) Lahar terraces formed by historical events. (**C**) Trees of *Pinus hartwegii* buried and impacted by blocks and sediment. (**D**) Recent lahar deposits burying the road at 3546 m a.s.l.

2.3. The 2012 Hydrometeorological Event and Rainfall Triggering

On 6 August 2012, tropical storm "Ernesto" attained a hurricane strength at 1200 UTC (= Coordinated Universal Time) in the Caribbean, east of Chetumal, Mexico. The hurricane then turned westward and continued to strengthen as it approached the coast of the southern Yucatan Peninsula. The hurricane made landfall at Cayo Norte in the Banco Chinchorro Islands of Mexico around 0100 UTC on 8 August [64]. On 9 August, it caused torrential rainfalls in the range of 112 to 300 mm 24 h^{-1} in the states of Oaxaca, Veracruz, Hidalgo and Puebla. On 10 August, rainfall decreased from 126 to 180 mm 24 h^{-1} in Puebla, Guerrero, Chiapas, and Veracruz [65]. Based on records from the Huatusco meteorological station (located on the NE slopes of Pico de Orizaba, 1186 m a.s.l.) the maximum rainfall was recorded on 9 August 2012 with 133 mm, and reached 153 mm in two days (9–10 August 2012), 159 mm in three days (8–10 August 2012), and 212 mm in five days (9–13 August 2012) [66]. This hydrometeorological event triggered regional landslide and debris-flow events on the northern slopes of the volcano, i.e., in the Jamapa and Tliapa Gorges where they caused massive structural damage [61]. Field recognition showed several deposits left by the geomorphic events, namely in the form of terraces, debris and block deposits, scarps, and scars on several trees (Figure 2A–C).

3. Methods

3.1. Digital Terrain Model

A UAV was used to construct a digital superficial model (DSM) and a digital terrain model (DTM) based on the structure-from-motion (SfM) photogrammetry of the landforms with high-resolution topography [67]. For the modelling of the 2012 lahar discharge in the study reach within Jamapa Gorge, we constructed a DTM based on 306 digital images taken with a DJI Phantom 4 and a Mavic Pro Platinum (Figure 3A). Flight height was 70 m on average above the terrain surface with an image resolution of 4000 × 3000 dpi. Side overlap between images was 75% and front overlap was 40%. The processing and DTM extraction were performed using Pix4DMapper software (Pix4D SA, Switzerland) with a cell size of 3.6 cm (Figure 3B).

Figure 3. (**A**) The digital terrain model shown was obtained from imagery taken with an unmanned aerial vehicle (UAV). (**B**) Digital terrain model (DTM) and location of trees sampled for dendrogeomorphic analysis.

3.2. Dendrogeomorphic Analysis and PSI

Within the study sector in Jamapa Gorge, we identified and sampled all trees with visible scars inflicted by the 2012 lahar between ~3535 to 3620 m a.s.l., at a point located ~7 km away from the Pico de Orizaba crater (Figure 1). To this end, we used increment borers and an electric chainsaw to obtain cores or wedges to properly date the injuries [43]. A total of 47 *Pinus hartwegii* trees were sampled, of which 38 were disturbed trees (with scars) and 9 were undisturbed trees for the construction of a reference chronology (Figure 3B). In addition to their dating, scars were used as paleostage indicators (PSI) by measuring the height of impacts with respect to the channel bed [40]. In the case that a tree had scars located at different heights, different samples were taken to distinguish between different lahar events. We considered the uppermost point of the highest scar as the peak of lahar discharge (Figure 4) [45,46]. In addition, for each tree, we also recorded the geomorphic and geographic position with a Global Positioning System (GPS) device. Additional measures were taken in the field to delineate the distance, direction and inclination of trees with respect to topographic features (such as big blocks, old terraces, or volcanic slopes) using a compass, tape measure and inclinometer.

After fieldwork, samples were prepared in the laboratory according to the standard procedures described in Bräker [68] and Stoffel and Bolslchweiler [69]. After the dating of samples, ring widths were measured by means of a microscope and a sliding Velmex [70] measurement plate connected to a computer using the time series analysis software TSAPWin^{TM} (Time Series and Analysis Program for Windows) [71].

In addition to the dating of scars to the year [45,46], we also aimed to identify the seasonal timing of lahars based on the position of growth anomalies (i.e., injuries and callus tissue) within the tree ring according to Stoffel et al. [72]. The growing season was adapted to central Mexico conifers [20].

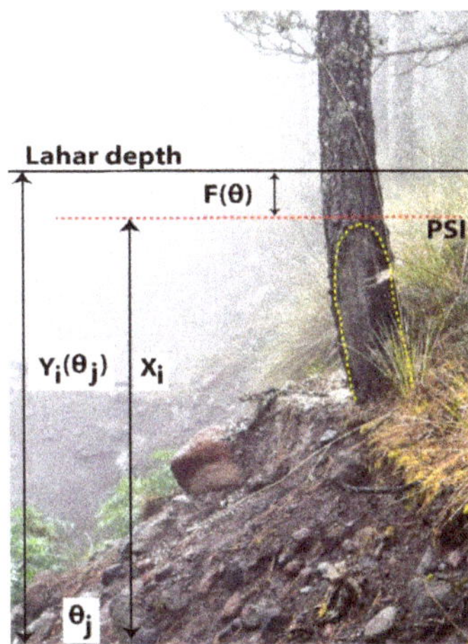

Figure 4. Assessment of deviations between observed paleostage indicators (PSI) and lahar depths modelled with rapid mass movement simulation (RAMMS) for the 2012 lahar: X_i represents observed values (i.e., impact scars on trees) at position i; and $Y_i(\theta_j)$ gives the computed water depth for a given roughness θ_j. Deviation between both values is represented by $F(\theta)$.

3.3. Numerical Simulations

RAMMS has been developed in the Swiss Federal Institute for Forest, Snow and Landscape research (WSL) [73] to simulate the runout of muddy and debris-laden flows in complex terrain. The model uses the Voellmy–Salm fluid flow continuum model [74] based on the Voellmy fluid flow law [75]. In the model, the frictional resistance is divided into two parts: a dry-Coulomb type friction, proportional to the normal stress at the flow bottom (coefficient μ) and a viscous resistance turbulent friction depending on the square of the velocity (coefficient ξ (m/s^2)). The model yields maximum water depth, sediment deposition, flow velocity, and pressure. For further details about RAMMS and the equations used, see [76,77].

In our study, the input parameters of the model were set up as follows. The lahar density was initially fixed to 1400 g/m^3. The μ coefficient was defined on the basis of field observations at a value of 0.15, representing the average slope of the deposition zones [75]. The ξ coefficient was set to 400 m/s^2 according to the nature of the flux observed in similar contexts in Mexico [36]. The lahar peak discharge was obtained by applying an iterative step-backwater procedure [44]. For each of the simulation runs, we released a triangular hydrograph at the beginning of the river reach in which we defined the maximum peak discharge to occur after 60 s and the event to stop after 850 s [36]. Then, in order to estimate the peak discharge of the 2012 lahar, the maximum water depth modelled by RAMMS was compared with the maximum scar height observed in the 19 trees analyzed (Figures 3B and 4). The magnitude of the lahar event was then defined as the peak discharge that minimized the deviation between the observed scar height and the modelled lahar stage. Once the peak discharge of the 2012 lahar was fixed, we performed a sensitivity model analysis by varying the density, μ and ξ coefficients by ±25% with respect to the initial values and evaluating the changes in the deviation between observed and modelled water heights.

4. Results

4.1. Tree Ring Evidence of the 2012 Lahar Event

In this study, a total of 38 trees with ages of up to 150 years have been sampled, all showing disturbances related to the 2012 lahar event. Most disturbances were in the form of scars, stem tilting and decapitation. Among these disturbances, 19 *Pinus hartwegii* trees had scars that were suitable to be used as PSI. The passage of the lahar left by far the largest amount of severe damage in trees that were growing in the main channel as well as in those individuals standing on the lower terraces (13 scars at heights 22–120 cm), whereas in six cases, scarred trees were located in a secondary channel (with scar heights of 3–70 cm). Scars were recorded on the upslope side of trunks, facing the direction of the lahar flow (Figure 4). The analysis of scar seasonality showed that the injuries were located in the early latewood, corresponding to the late summer of 2012 (Figure 5) and likely of the passage of hurricane "Ernesto".

4.2. Magnitude Reconstruction of the 2012 Lahar

We simulated the 2012 lahar in a trial-and-error approach with maximum peak discharges ranging from 25 to 500 m^3 s^{-1} to approximate and ultimately reduce the deviation between scar heights found in the analyzed trees and maximum lahar depths obtained in the model. Our results suggest a mean deviation ranging from −0.43 m for 25 m^3 s^{-1} to 3.08 m for 500 m^3 s^{-1}. Refining the approach further, the minimum absolute mean deviation was reached for a peak discharge of 78 m^3 s^{-1}, translating into a lahar volume of 33,000 m^3 based on the assumed shape of the hydrograph (see Figure 6D). Varying lahar density and the ξ coefficients by ±25% indicate that the mean deviation can change by up to ~±0.43 m and ~±0.47 m, respectively (Table 1).

Figure 6A shows the relationship between the maximum water depth and scar heights, as well as flow velocity and stream power for the estimated peak discharge. Deviations between the lahar model and scar heights were between −0.68 m and 1.18 m. Relatively small deviations (±0.25 m; 50th percentile) were observed in 52% of the cases. Moderate deviations (ranging between ±0.25 and ±0.35 m; 50th–75th percentile) could be found in 26% of the cases, whereas 22% of the scars found in trees showed deviations exceeding ±0.35 m (>75th percentile). Only one tree showed deviations exceeding 1 standard deviation (σ = 0.77 m; Table 2). Noteworthy, most of the trees for which deviations were more relevant were growing at sites where they tend to be quite exposed to the flow, i.e., especially in those parts of the reach with the steepest slopes and unruffled bedrock and where specific kinetic energy tends to be higher as well. The maximum lahar velocity simulated by RAMMS (Figure 6B) occurred in the central part of the main channel with ~18 m/s.

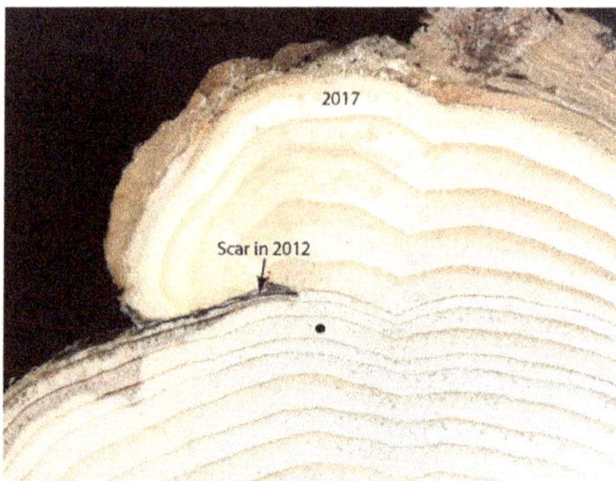

Figure 5. Cross section of a *Pinus hartwegii* stem wedge injured by the 2012 lahar event. The injury is located in late latewood, which allows dating of the event to August 2012.

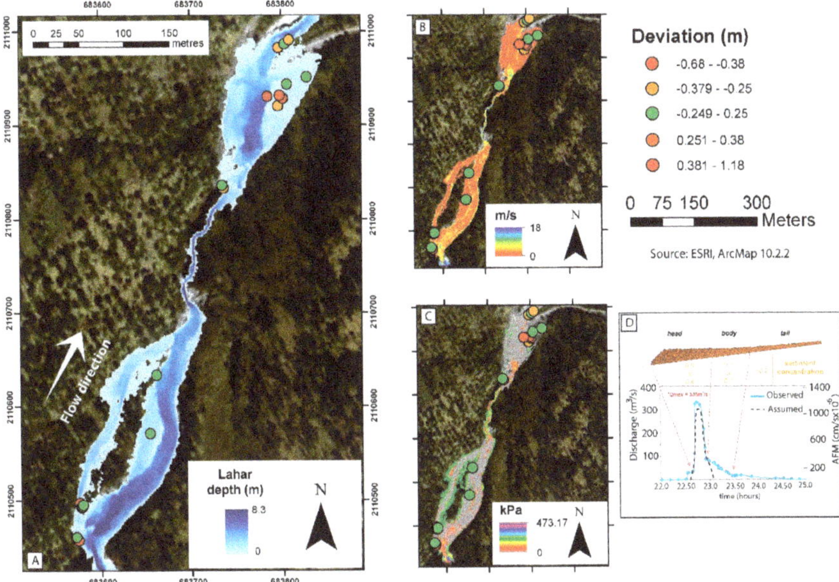

Figure 6. Absolute deviation (in m) of PSI as observed on the trunks of *Pinus hartwegii* trees in a sector of Jamapa Gorge and their relationship with (**A**) lahar depth, (**B**) flow velocity, and (**C**) pressures modeled for the 2012 lahar. (**D**) The hydrograph data correspond to the 2001 lahar in Popocatépetl volcano by Caballero and Capra [36]. The dashed black line represents the assumed shape of the hydrograph used in this study.

Table 1. Sensitivity analysis of lahar modelling.

Simulation	Mean (m)	Standard Deviation
density: 1400 kg/m³; chi: 400	0	0.43
density: +25%	0.43	0.72
density: −25%	0.43	0.72
Chi: +25%	0.39	0.72
Chi: −25%	0.47	0.73

Table 2. Tree locations and absolute deviation (in m) between PSI as observed in trees and modelled flow heights. X-UTM and Y-UTM refer to the x and y values within the Universal Transverse Mercator coordinate system.

Tree Code	X-UTM	Y-UTM	Maximum Scar Height (m)	Deviation from Modelled Lahar Sediment (m)
0	683,574.4	2,110,457	0.1	0.39
1	683,572.1	2,110,461	0.1	−0.04
2	683,575.9	2,110,497	0.032	0.648
3	683,577.4	2,110,493	0.05	0.32
4	683,736.7	2,110,833	0.4	−0.31
5	683,578.5	2,110,494	0.7	−0.6
6	683,579.3	2,110,495	0.3	−0.25
7	683,797.1	2,110,982	0.41	0.02
8	683,803.7	2,110,986	0.22	−0.21
9	683,808.9	2,110,991	0.52	0.13
10	683,796.3	2,110,920	0.36	0.2
11	683,801.8	2,110,928	0.69	−0.37
12	683,798.2	2,110,932	0.66	−0.34
13	683,784.7	2,110,930	0.7	1.18
14	683,806.2	2,110,943	0.25	0.19
15	683,828	2,110,951	0.3	−0.08
16	683,735	2,110,836	0.15	−0.14
17	683,655.1	2,110,571	1.2	0.0
18	683,662.4	2,110,633	0.6	−0.6

5. Discussion and Conclusions

In this paper, we provide insights into a tree-ring based lahar discharge reconstruction for an event that occurred in the Jamapa Gorge (northern slopes of Pico de Orizaba volcano, Mexico) during August 2012, likely as a result of the passage of hurricane "Ernesto". To this end, we used scar heights found in 19 *Pinus hartwegii* trees to calibrate simulation runs realized with the two-dimensional RAMMS model. Even if the combined use of tree-ring evidence and model outputs has been used previously for peak discharge estimation of fluvial processes [40], this study likely is the first of its kind to reconstruct peak discharge of a lahar event with botanical evidence and models.

The intra-annual dating of 19 scars found in trees (Figure 3B) indicate that the lahar event must have taken place in early latewood which locally corresponds to late summer (Figure 5); this timing is consistent with the passage of hurricane "Ernesto" on 9 August 2012 at the study site. Evidence of lahars triggered by the 2012 hurricane can be found elsewhere in central Mexico, namely on Popocatépetl [20]; Iztaccíhuatl [78] and La Malinche [41] volcanoes. The timing of the 2012 lahar at our study site is further supported by evidence from local newspapers [79]. However, the 2012 lahars in Jamapa Gorge could also have been favored by water added to the system as a result of progressive ice-melting at Jamapa glacier in recent years [61].

In Mexico, dendrogeomorphic approaches have been used to date and reconstruct lahars [18–20] and debris flows [21,41] in the past. By contrast, studies focusing on the magnitude of hydrogeomorphic events are still critically lacking. The approach that we present here has allowed for retrospective estimation of the peak discharge of the 2012 lahar at 78 m^3 s^{-1} at the level of the study reach with a ~3 km^2 catchment area. For this modelled peak discharge, the difference between the maximum height of the scars and lahar stage modelled were minimized. The reconstructed peak discharge is somehow lower than the lahars of Nevado del Huila [22], which was estimated at 13,500 m^3 s^{-1} (maximum peak discharge) with a ~26 km^2 catchment area (Páez River). By contrast, the calculated peak discharge is higher than the 48 m^3 s^{-1} estimated for "Patrio" lahar at Colima volcano [16] (Montegrande catchment area: ~6.2 km^2). In addition, the shape of the channel [30], lahar rheologic characteristics [36] and resolution of the digital terrain model [48] can affect the lahar scenarios.

Documented differences between observed scar heights and modeled flow heights are comparable with results obtained in past work focusing on the fluvial domain. The largest deviations (exceeding ±0.35 m or the >75th percentile of deviations) were found in trees located on cut banks and/or in straight and exposed positions along the investigated river reach at the lower end of the channel. By contrast, moderate and smaller deviations were typically found in trees located in intermediate and low-energy landforms, such as alluvial terraces located in the central and lower end of the channel. These results are in agreement with previous studies carried out in the fluvial domain [40,80–83], suggesting that intermediate-energy landforms such as alluvial terraces could be considered appropriate geomorphic locations for tree-ring-based reconstructions of lahar magnitude. Here, instead of using a non-Newtonian flow-like model for high concentrated flows [48], we employed a single-phase model to simulate inertial flows of solid–fluid mixtures such as debris flows [84]. In this study, the calibration of RAMMS has been based exclusively on scar heights and the extension of fresh deposits observed during field recognition. Despite the fact that we performed a sensitivity analysis on the main parameters, i.e., ±25% on lahar density and ξ coefficients (μ was fixed according to field recognition), uncertainties remain regarding the total volume of the event.

The combination of tree-ring based lahar reconstruction and process modelling with RAMMS is the first of its kind and has helped considerably to estimate the magnitude of the 2012 lahar in a sector of Jamapa Gorge. As such, we conclude that the combination of 2D dimensional dynamic modeling and tree-ring analysis can be very useful to assess and calibrate the magnitude of lahars in other temperate forests growing on Mexican volcanoes, especially in areas with scarce or inexistent hydrologic data and/or for the dating and assessment of older lahars.

Author Contributions: All authors have contributed to the final version of the article. Conceptualization, O.F.-R. and J.A.B.-C.; Methodology, J.A.B.-C. and O.F.-R.; Software, J.A.B.-C.; Writing—Original Draft, O.F.-R. and J.A.B-C; Writing—Review and Editing, O.F.-R., J.A.B.-C., J.E.F.-G., L.V.-S., M.S., and L.C. All authors have read and agreed to the published version of the manuscript.

Funding: This research was funded by DGAPA-PAPIIT, UNAM project number IA100619.

Acknowledgments: We acknowledge Andres Prado for his support during fieldwork and for the digital elevation model with an orthomosaic map based on drone images. We acknowledge Salvador Ponce and Mireya Vazquez for their help during fieldwork.

Conflicts of Interest: The authors declare no conflict of interest.

References

1. Smith, G.A.; Fritz, W.J. Volcanic influences on terrestrial sedimentation. *Geology* **1989**, *17*, 375–376. [CrossRef]
2. Vallance, J.W. Volcanic debris flow. In *Debris-Flows Hazards and Related Phenomena*; Jakob, M., Hungr, O., Eds.; Springer: Berlin/Heidelberg, Germany, 2005; pp. 247–274.
3. Lavigne, F.; Thouret, J.C. Sediment transportation and deposition by rain-triggered lahars at Merapi Volcano, Central Java, Indonesia. *Geomorphology* **2002**, *49*, 45–69. [CrossRef]
4. Scott, K.; Macías, J.L.; Naranjo, J.A.; Rodríguez, S.; McGeehin, J.P. Catastrophic debris flows transformed from landslide in volcanic terrains: Mobility, hazard assessment and mitigation strategies. In *U. S. Geological Survey Professional Paper*; U.S. Geological Survey: Reston, VA, USA, 2001; Volume 1630, p. 59.
5. Vallance, J.W.; Iverson, R.M. Lahars and their deposits. In *Encyclopedia of Volcanoes*, 2nd ed.; Academic Press: London, UK, 2015; pp. 649–664. [CrossRef]
6. Capra, L.; Macías, J.L. The cohesive Naranjo debris flow deposit (10 km^3): A dam breakout flow derived from the pleistocene debris-avalanche deposit of Nevado de Colima volcano (Mexico). *J. Volcanol. Geoth. Res.* **2002**, *117*, 213–235. [CrossRef]
7. Cortes, A.; Macías, J.L.; Capra, L.; Garduño-Monroy, V.H. Sector collapse of the SW flank of Volcán de Colima, México. The 3600 yr BP La Lumbre-Los Ganchos debris avalanche and associated debris flows. *J. Volcanol. Geoth. Res.* **2010**, *197*, 52–66. [CrossRef]
8. Saucedo, R.; Macías, J.L.; Sarocchi, D.; Bursik, M.I.; Rupp, B. The rain-triggered Atenquique volcaniclastic debris flow of October 16, 1955 at Nevado de Colima Volcano, Mexico. *J. Volcanol. Geoth. Res.* **2008**, *173*, 69–83. [CrossRef]
9. Bisson, M.; Pareschi, M.T.; Zanchetta, G.; Sulpizio, R.; Santacroce, R. Volcaniclastic debris flow occurrences in the Campania region (southern Italy) and their relation to Holocene—Late Pleistocene pyroclastic fall deposits: Implications for large-scale hazard mapping. *Bull Volcanol.* **2007**, *70*, 157–167. [CrossRef]
10. Bisson, M.; Zanchettab, G.; Sulpizio, R.; Demib, F. A map for volcaniclastic debris flow hazards in Apennine areas surrounding the Vesuvius volcano (Italy). *J. Maps* **2013**, *9*, 230–238. [CrossRef]
11. Vallance, J.W.; Schilling, S.P.; Devoli, G.; Reid, M.E.; Howell, M.M.; Brien, D.L. *Lahar Hazards at Casita and San Cristóbal Volcanoes, Nicaragua*; Open-File Report 01-468; U.S. Geological Survey: Reston, VA, USA, 2004; pp. 1–18.
12. Carrasco-Núñez, G.; Díaz-Castellón, R.; Siebert, L.; Hubbard, B.; Sheridan, M.F.; Rodríguez, S.R. Multiple edifice-collapse events in the Eastern Mexican Volcanic Belt: The role of sloping substrate and implications for hazard assessment. *J. Volcanol. Geoth. Res.* **2006**, *158*, 151–176. [CrossRef]
13. Newhall, C.; Stauffer, P.H.; Hendley, J.W. *Lahars of Mount Pinatubo, Philippines*; Fact Sheet 114; U.S. Geological Survey: Reston, VA, USA, 1997.
14. Lavigne, F.; Thouret, J.C.; Voight, B.; Suwa, H.; Sumaryono, A. Lahars at Merapi volcano, Central Java: An overview. *J. Volcanol. Geoth. Res.* **2000**, *100*, 423–456. [CrossRef]
15. Capra, L.; Borselli, L.; Varley, N.; Gavilanes-Ruiz, J.C.; Norini, G.; Sarocchi, D.; Caballero, L.; Cortes, A. Rainfall-triggered lahars at Volcán de Colima, Mexico: Surface hydro-repellency as initiation process. *J. Volcanol. Geoth. Res.* **2010**, *189*, 105–117. [CrossRef]
16. Vázquez, R.; Capra, L.; Caballero, L.; Arámbula-Mendoza, R.; Reyes-Dávila, G. The anatomy of a lahar: Deciphering the 15th September 2012 lahar at Volcán de Colima, Mexico. *J. Volcanol. Geoth. Res.* **2014**, *272*, 126–136. [CrossRef]

17. Procter, J.N.; Cronin, S.J.; Zernack, A.V. Landscape and sedimentary response to catastrophic debris avalanches, western Taranaki, New Zealand. *Sediment. Geol.* **2009**, *220*, 271–287. [CrossRef]

18. Bollschweiler, M.; Stoffel, M.; Vázquez-Selem, L.; Palacios, D. Tree-ring reconstruction of past lahar activity at Popocatépetl volcano, México. *Holocene* **2010**, *20*, 265–274. [CrossRef]

19. Franco-Ramos, O.; Stoffel, M.; Vázquez-Selem, L.; Capra, L. Spatio-temporal reconstruction of lahars on the southern slopes of Colima volcano, Mexico—A dendrogeomorphic approach. *J. Volcanol. Geoth. Res.* **2013**, *267*, 30–38. [CrossRef]

20. Franco-Ramos, O.; Castillo, M.; Muñoz–Salinas, E. Using tree-ring analysis to evaluate the intra-eruptive lahar activity in the Nexpayantla Gorge, Popocatépetl Volcano (Central Mexico). *Catena* **2016**, *147*, 205–215. [CrossRef]

21. Franco-Ramos, O.; Stoffel, M.; Vázquez Selem, L. Tree-ring based record of intra-eruptive lahar activity: Axaltzintle valley, Malinche volcano, Mexico. *Geochronometria* **2016**, *43*, 74–83. [CrossRef]

22. Worni, R.; Huggel, C.; Stoffel, M.; Pulgarin, B. Challenges of modelling recent, very large lahars at Nevado del Huila Volcano, Colombia. *Bull. Volcanol.* **2012**, *74*, 309–324. [CrossRef]

23. Schilling, S. *LAHARZ: GIS Programs for Automated Mapping of Lahar-Inundation Hazard Zones*; OFR 98-638; USGS: Vancouver, WA, USA, 1998.

24. Williams, R.; Stinton, A.J.; Sheridan, M.F. Evaluation of the Titan2D two-phase flow model using an actual event: Case study of the 2005 Vazcún Valley Lahar. *J. Volcanol. Geoth. Res.* **2008**, *177*, 760–766. [CrossRef]

25. O'Brien, J.; Julien, P.; Fullerton, W. Two-dimensional water flood and mudflow simulation. *J. Hydraul. Eng.* **1993**, *119*, 244–261. [CrossRef]

26. Cesca, M.; D'Agostino, V. Comparison between FLO-2D and RAMMS in debris-flow modelling: A case study in the Dolomites. *WIT Trans. Eng. Sci.* **2008**, *60*, 197–206. [CrossRef]

27. Lee, S.K.; Lee, C.W.; Lee, S. A comparison of the Landsat image and LAHARZ-simulated lahar inundation hazard zone by the 2010 Merapi eruption. *Bull. Volcanol.* **2015**, *77*, 46. [CrossRef]

28. Pistolesi, M.; Cioni, R.; Rosi, M.; Aguilera, E. Lahar hazard assessment in the southern drainage system of Cotopaxi volcano, Ecuador: Results from multiscale lahar simulations. *Geomorphology* **2014**, *207*, 51–63. [CrossRef]

29. Nocentini, M.; Tofani, V.; Gigli, G.; Fidolini, F.; Casagli, N. Modeling debris flows in volcanic terrains for hazard mapping: The case study of Ischia Island (Italy). *Landslides* **2014**, *12*, 831–846. [CrossRef]

30. Muñoz-Salinas, E.; Castillo-Rodríguez, M.; Manea, V.; Manea, M.; Palacios, D. Lahar flow simulation using LAHARZ program: Application for Popocatépetl volcano, Mexico. *J. Volcanol. Geoth. Res.* **2009**, *182*, 13–22. [CrossRef]

31. Hubbard, B.E. Volcanic Hazard Mapping Using Aircraft, Satellite, and Digital Topographic Data: Pico de Orizaba (Citlaltépetl), Mexico. Ph.D. Thesis, University at Buffalo, Buffalo, NY, USA, 2001; p. 354.

32. Capra, L.; Norini, G.; Groppelli, G.; Macías, J.L.; Arce, J.L. Volcanic hazard zonation of the Nevado de Toluca volcano, México. *J. Volcanol. Geoth. Res.* **2008**, *176*, 469–484. [CrossRef]

33. Sieron, K.; Capra, L.; Rodríguez-Elizararrás, S. Hazard assessment at San Martín volcano based on geological record, numerical modeling, and spatial analysis. *Nat. Hazards* **2014**, *70*, 275–297. [CrossRef]

34. Vázquez, R.; Macías, J.L.; Cisneros, G. A first approach to the numerical modeling of lahars at Tacaná Volcanic Complex, Chiapas, México. In *EGU General Assembly Conference Abstracts*; EGU2018-905; Geophysical Research Abstracts: Vienna, Austria, 2018; Volume 20.

35. Sieron, K.; Ferrés, D.; Siebe, C.; Constantinescu, R.; Capra, L.; Connor, C.; Connor, L.; Groppelli, G.; González Zuccolotto, K. Ceboruco hazard map: Part II—Modeling volcanic phenomena and construction of the general hazard map. *Nat. Hazards.* **2019**. [CrossRef]

36. Caballero, L.; Capra, L. The use of FLO2D numerical code in lahar hazard evaluation at Popocatépetl volcano: A 2001 lahar scenario. *Nat. Hazards Earth Syst. Sci.* **2014**, *14*, 3345–3355. [CrossRef]

37. Zaragoza Campillo, A.G. Caracterización y Dinámica del Lahar Secundario Ocurrido en 2010, Volcán Popocatépetl: Mecanismos Disparadores, Características Texturales y Simulación Numérica. Bachelor's Thesis, Facultad de Ciencias, Universidad Nacional Autónoma de México, Coyoacán, Mexico, 2017; p. 92. (In Spanish).

38. Capra, L.; Coviello, V.; Borselli, L.; Márquez-Ramírez, V.H.; Arámbula-Mendoza, R. Hydrological control of large hurricane-induced lahars: Evidence from rainfall-runoff modeling, seismic and video monitoring. *Nat. Hazards Earth Syst. Sci.* **2018**, *18*, 781–794. [CrossRef]

39. Baker, V.R.; Webb, R.H.; House, P.K. The scientific and societal value of paleoflood hydrology. In *Ancient Floods, Modern Hazards: Principles and Applications of Paleoflood Hydrology*; Water Science and Application; House, P.K., Webb, R.H., Baker, V.R., Levish, D.R., Eds.; American Geophysical Union: Washington, DC, USA, 2002; Volume 5, pp. 1–19.

40. Ballesteros-Cánovas, J.A.; Stoffel, M.; St George, S.; Hirschboeck, K. A review of flood records from tree rings. *Prog. Phys. Geogr.* **2015**, *39*, 794–816. [CrossRef]

41. Franco-Ramos, O.; Stoffel, M.; Ballesteros-Cánovas, J.A. Reconstruction of debris-flow activity in a temperate mountain forest catchment of central Mexico. *J. Mt. Sci.* **2019**, *16*, 2096–2109. [CrossRef]

42. Stoffel, M.; Corona, C. Dendroecological dating of geomorphic disturbance in trees. *Tree-Ring Res.* **2014**, *70*, 3–20. [CrossRef]

43. Webb, R.H.; Jarrett, R.D. One-dimensional estimation techniques for discharges of paleofloods. In *Ancient Floods, Modern Hazards: Principles and Applications of Paleoflood Hydrology: Water Science and Application*; House, P.K., Webb, R.H., Baker, V.R., Levish, D.R., Eds.; American Geophysical Union: Washington, DC, USA, 2002; pp. 111–126.

44. Ballesteros, J.A.; Bodoque, J.M.; Díez, A.; Sanchez-Silva, M.; Stoffel, M. Calibration of floodplain roughness and estimation of palaeoflood discharge based on tree-ring evidence and hydraulic modelling. *J. Hydrol.* **2011**, *403*, 103–115. [CrossRef]

45. Ballesteros, J.A.; Eguibar, M.; Bodoque, J.M.; Díez, A.; Stoffel, M.; Gutiérrez, I. Estimating flash flood discharge in an ungauged mountain catchment with 2D hydraulic models and dendrogeomorphic paleostage indicators. *Hydrol. Process.* **2011**, *25*, 970–979. [CrossRef]

46. Ballesteros-Cánovas, J.A.; Stoffel, M.; Spyt, B.; Janecka, K.; Kaczka, R.J.; Lempa, M. Paleoflood discharge reconstruction in Tatra Mountain streams. *Geomorphology* **2016**, *272*, 92–101. [CrossRef]

47. Bodoque, J.M.; Díez-Herrero, A.; Eguibar, M.A.; Benito, G.; Ruiz-Villanueva, V.; Ballesteros-Cánovas, J.A. Challenges in paleoflood hydrology applied to risk analysis in mountainous watersheds—A review. *J. Hydrol.* **2015**, *529*, 449–467. [CrossRef]

48. Garrote, J.; Díez-Herrero, A.; Génova, M.; Bodoque, J.M.; Perucha, M.A.; Mayer, P.L. Improving Flood Maps in Ungauged Fluvial Basins with Dendrogeomorphological Data. An Example from the Caldera de Taburiente National Park (Canary Islands, Spain). *Geosciences* **2018**, *8*, 300. [CrossRef]

49. Hussin, H.Y.; Quan Luna, B.; Westen, C.J.; Christen, M.; Malet, J.-P.; Van Asch, T.W. Parameterization of a numerical 2-D debris flow model with entrainment: A case study of the Faucon catchment, Southern French Alps. *Nat. Hazards Earth Syst. Sci.* **2012**, *12*, 3075–3090. [CrossRef]

50. Fan, L.; Lehmanna, P.; McArdell, B.; Or, D. Linking rainfall-induced landslides with debris flows runout patterns towards catchment scale hazard assessment. *Geomorphology* **2017**, *280*, 1–15. [CrossRef]

51. Frank, F.; McArdell, B.W.; Oggier, N.; Baer, P.; Christen, M.; Vieli, A. Debris-flow modeling at Meretschibach and Bondasca catchments, Switzerland: Sensitivity testing of field-data-based entrainment model. *Nat. Hazards Earth Syst. Sci.* **2017**, *17*, 801–815. [CrossRef]

52. Carrasco-Núñez, G.; Vallance, J.W.; Rose, W.I. A voluminous avalanche-induced lahar from Citlaltépetl volcano, Mexico: Implications for hazards assessment. *J. Volcanol. Geoth. Res.* **1993**, *59*, 35–46. [CrossRef]

53. Carrasco-Núñez, G. Structure and proximal stratigraphy of Citlaltépetl volcano (Pico de Orizaba), Mexico. *Geol. Soc. Am.* **2000**, *334*, 247–262. [CrossRef]

54. Carrasco-Núñez, G.; Rose, W.I. Eruption of a major Holocene pyroclastic flow at Citlaltépetl volcano (Pico de Orizaba), México, 8.5-9.0 ka. *J. Volcanol. Geoth. Res.* **1995**, *69*, 197–215. [CrossRef]

55. Rossotti, A.; Carrasco-Núñez, G. Stratigraphy of the 8.5–9.0 ka B.P. Citlaltépetl pumice fallout sequence. *Rev. Mex. Ciencias Geol.* **2004**, *21*, 353–370.

56. Hoskuldsson, A.; Robin, C. Late Pleistocene to Holocene eruptive activity of Pico de Orizaba, Eastern Mexico. *Bull. Volcanol.* **1993**, *55*, 571–587. [CrossRef]

57. Siebe, C.; Abrams, M.; Sheridan, M.F. Major Holocene block-and-ash fan at the western slope of ice-capped Pico de Orizaba volcano, México: Implications for future hazards. *J. Volcanol. Geoth. Res.* **1993**, *59*, 1–33. [CrossRef]

58. Alcalá-Reygosa, J.; Palacios, D.; Schimmelpfennig, I.; Vázquez-Selem, L.; García-Sancho, L.; Franco-Ramos, O.; Villanueva, J.; Zamorano, J.J.; ASTER Team. Dating late Holocene lava flows in Pico de Orizaba (Mexico) by means of in situ-produced cosmogenic ^{36}Cl, lichenometry and dendrochronology. *Quat. Geochron.* **2018**, *47*, 93–106. [CrossRef]

59. Iverson, R.M.; Schilling, S.P.; Vallance, J.W. Objective delineation of lahar hazard zones. *Geol. Soc. Am. Bull.* **1998**, *110*, 972–984. [CrossRef]

60. Rodríguez, S.; Mora-González, I.; Murrieta-Hernández, J.-L. Flujos de baja concentración asociados con lluvias de intensidad extraordinaria en el flanco sur del volcán Pico de Orizaba (Citlaltépetl), México. *Boletín Soc. Geológica Mex.* **2006**, *58*, 223–236. [CrossRef]

61. Martínez, M.A.M.; Rodríguez, C.M.W.; Monjardín, L.C.R.; Weissling, B.; Sieron, K.; Martínez, C.A.O. Afectaciones por posible asociación de eventos hidrometeorológicos y geológicos en los municipios de Calcahualco y Coscomatepec, Veracruz. *Teoría y Praxis* **2016**, *12*, 31–49. [CrossRef]

62. Palacios, D.; Vazquez-Selem, L. Geomorphic effects of the retreat of Jamapa glacier, Pico de Orizaba volcano (Mexico). *Geogr. Ann.* **1996**, *78*, 19–34. [CrossRef]

63. Palacios, D.; Parrilla, G.; Zamorano, J.J. Paraglacial and postglacial debris flows on a Little Ice Age terminal moraine: Jamapa Glacier, Pico de Orizaba (Mexico). *Geomorphology* **1999**, *28*, 95–118. [CrossRef]

64. Brown, D.P. Tropical Cyclone Report. Hurricane Ernesto. National Hurricane Center. NOAA, 2013. Available online: https://www.nhc.noaa.gov/data/tcr/AL052012_Ernesto.pdf (accessed on 22 November 2019).

65. SMN-CNA. Reseña Huracán Ernesto del Océano Atlántico. Temporada 2012 de Ciclones y Huracanes. Servicio Metrológico Nacional-Comisión Nacional de Agua, 2012. (In Spanish). Available online: http://smn1.conagua.gob.mx/ciclones/tempo2012/atlantico/Ernesto-a2012.pdf (accessed on 22 November 2019).

66. *Archivo de Datos Climatológicos del Servicio Meteorológico Nacional*; CNA-SMN: Mexico, 2019. Available online: https://smn.conagua.gob.mx/tools/RESOURCES/Diarios/30342.txt (accessed on 22 November 2019).

67. Westoby, M.J.; Brasington, J.; Glasser, N.F.; Hambrey, M.J.; Reynolds, J.M. 'Structure-from-Motion'photogrammetry: A low-cost, effective tool for geoscience applications. *Geomorphology* **2012**, *179*, 300–314. [CrossRef]

68. Bräker, O. Measuring and data processing in tree-ring research—A methodological introduction. *Dendrochronologia* **2002**, *20*, 203–216. [CrossRef]

69. Stoffel, M.; Bollschweiler, M. Tree-ring analysis in natural hazards research—An overview. *Nat. Hazards Earth Syst. Sci.* **2008**, *8*, 187–202. [CrossRef]

70. Robinson, W.J.; Evans, R. A microcomputer-based tree-ring measuring system. *Tree-Ring Bull.* **1980**, *40*, 59–64.

71. Rinn, F. TSAP-Win. Time Series Analysis and Presentation for Dendrochronology and Related Applications. Version 4.64 for Microsoft Windows. User Reference. Rinntech: Heidelberg, Germany, 2003; p. 22. Available online: http://www.rinntech.de (accessed on 22 January 2020).

72. Stoffel, M.; Lièvre, I.; Monbaron, M.; Perre, S. Seasonal timing of rockfall activity on forested slope at Täschgufer (Swiss Alps)—A dendrogeomorphological approach. *Z. Geomorphol.* **2005**, *49*, 89–106.

73. Bartelt, P.; Buehler, Y.; Christen, M.; Deubelbeiss, Y.; Graf, C.; McArdell, B.W. *RAMMS—A Modelling System for Debris Flows in Research and Practice—User Manuel v.1.01/Hillslope Debris Flow*; WSL Institute for Snow and Avalanche Research SLF: Davos, Switzerland, 2011; p. 91.

74. Salm, B. Flow, flow transition and runout distances of flowing avalanches. *Ann. Glaciol.* **1993**, *18*, 221–226. [CrossRef]

75. Voellmy, A. Über die Zerstörungskraft von Lawinen. *Schweiz. Bauztg.* **1955**, *73*, 159–162.

76. Christen, M.; Bartelt, P.; Kowalski, J. Back calculation of the in den Arelen avalanche with RAMMS: Interpretation. *Ann. Glaciol.* **2010**, *51*, 161–168. [CrossRef]

77. Christen, M.; Kowalski, J.; Bartelt, P. RAMMS: Numerical simulation of dense snow avalanches in three-dimensional terrain. *Cold Reg. Sci. Technol.* **2010**, *63*, 1–14. [CrossRef]

78. Prado-Lallande, A. Geomorfología, Dendrocronología y Análisis Sedimentológico en la Cuenca Alta del Valle Alcalican, SW del Volcán Iztaccíhuatl. Bachelor's Thesis, Colegio de Geografía, Facultad de Filosofía y Letras, Universidad Nacional Autónoma de México, Mexico City, Mexico, 2017; p. 106. (In Spanish).

79. La Jornada. The Newspaper of Mexico. 2012. (In Spanish). Available online: https://www.jornada.com.mx/2012/08/09/estados/033n1est (accessed on 18 September 2019).

80. Gottesfeld, A.S. British Columbia flood scars: Maximum flood-stage indicator. *Geomorphology* **1996**, *14*, 319–325. [CrossRef]

81. Yanosky, T.M.; Jarrett, R.D. Dendrochronologic evidence for the frequency and magnitude of paleofloods. In *Ancient Floods, Modern Hazards: Principles and Applications of Paleoflood Hydrology*; Water Science and Application; House, P.K., Webb, R.H., Baker, V.R., Levish, D.R., Eds.; American Geophysical Union: Washington, DC, USA, 2002; Volume 5, pp. 77–89.

82. Victoriano, A.; Díez-Herrero, A.; Génova, M.; Guinau, M.; Furdada, G.; Khazaradze, G.; Calvet, J. Four-topic correlation between flood dendrogeomorphological evidence and hydraulic parameters (the Portainé stream, Iberian Peninsula). *Catena* **2018**, *162*, 216–229. [CrossRef]

83. Quesada-Román, A.; Ballesteros-Cánovas, J.A.; Granados, S.; Birkel, C.; Stoffel, M. Dendrogeomorphic reconstruction of floods in a dynamic tropical river. *Geormophology*. in press.

84. Naef, D.; Rickenmann, D.; Rutschmann, P.; McArdell, B.W. Comparison of flow resistance relations for debris flows using a one-dimensional finite element simulation model. *Nat. Hazards Earth Syst. Sci.* **2006**, *6*, 155–165. [CrossRef]

Article

Analysis of Flood Fatalities–Slovenian Illustration

Maruša Špitalar [1,2,*], Mitja Brilly [1], Drago Kos [2] and Aleš Žiberna [2]

[1] Faculty of Civil and Geodetic Engineering, University of Ljubljana, 1000 Ljubljana, Slovenia; mitja.brilly@fgg.uni-lj.si

[2] Faculty of Social Science, University of Ljubljana, 1000 Ljubljana, Slovenia; drago.kos@fdv.uni-lj.si (D.K.); ales.ziberna@fdv.uni-lj.si (A.Ž.)

* Correspondence: marusaspitalar@yahoo.co.uk; Tel.: +386-4164-1735

Received: 31 October 2019; Accepted: 15 December 2019; Published: 23 December 2019

Abstract: Floods not only induce vast economic damages but also pose a great danger to human life. In Slovenia, floods rank number one on the scale of damage magnitude. Different factors external to the hazard of flooding influence the gravity and extent of the impacts. A comprehensive collection and analysis of the information related to the understanding of causative factors of human impacts can substantially contribute to the mitigation and the minimisation of fatalities and injuries. In this work, historical analysis was performed for flood fatalities in the years between 1926 and 2014, with 10 flood events that induced 74 casualties considered. A detailed collection and review on human impacts was made. Victims in cars have seemed to increase in recent years and rural areas tend to be more vulnerable and susceptible to having victims of flooding. With regards to gender, the majority of victims are male. The focus was on demographic aspects (age and gender) of fatalities and analysis of the circumstances of loss of life. Based on a description of the activities of victims during flood events and repetitive patterns, groups were made based on the type of flood fatality. Eight interviews were performed with rescuers and people who were affected by floods in order to obtain more extensive information on preventative measures, received help, and flood risk perception.

Keywords: floods; fatalities; circumstances of loss of life; historical analysis

---

1. Introduction

Floods, especially flash floods, cause extensive damage to property, disrupt human lives, and may result in irreparable losses. The HANZE (Historical Analysis of Natural Hazards in Europe) database notes that the previously mentioned flood type is the most frequent in the Europe, as out of 1564 events, 56% were flash floods [1]. The majority of work is built around flash floods, as these are the most prevailing. Analysis that focuses on social aspects of floods mainly deals with circumstances of death with the intersection of parameters like population density, time of the event, and hydrological parameters (catchment size). In order to work towards loss prevention, there has been valuable work performed on flood hazard modelling, which includes flood prone areas and flood damage maps [2]. The focus in research has also been on the importance of population growth, infrastructure decay, increasing urbanisation, and effective risk communication [3]. It is imperative to have an interdisciplinary approach as multiple factors may play a vital role.

Based on public perception, loss of life is ranked on the top of the spectrum of damage severity. Thus, it is considered the most important type of loss. The scale of impacts varies between different locations and social environments as it does among different forms and sizes of floods. Hence, each flood event can be considered a unique hazard with specific characteristics [4]. Diversity and specificity of the flood events and uniqueness of the environments bring different, individual vulnerability factors but patterns can be observed with a larger sample. The available literature on loss of life focuses on different aspects of the topic [5]. Different angles of loss of life analysis contribute to the understanding of important factors, which influence the vulnerability and exposure of individuals and groups on a global level with cultural diversity. Furthermore, these different angles represent the variety of contextual emphasis in the areas of flood fatalities that have been studied.

Existing analyses focus on a global scale [6–8], discuss fatalities in the context of health impacts [9], investigate circumstances of loss of life [10–15], study mortality in flash flood events in the US [16], and analyse vehicle flood fatalities [17].

The complexity of the factors that contribute to loss of life probability implies a need for a different, interdisciplinary approach that is emphasised by Gruntfest and Handmer [18] and has been applied in studies [10,19]. Furthermore, Creutin et al. [20] have pointed out the importance of collaboration between hydrologists and social scientists and have set the groundwork that is needed for solving complex problems and for effective decision-making. The results of analyses increasingly display the relationship of distribution and correlation of flood fatalities with physical vulnerability as well as with social vulnerability. Studies of flood fatalities in general relate to demographic data (gender, age, and ethnic groups), circumstances of loss of life, and evaluation of geographic, temporal, and hydrological parameters and their interplay with societal factors [7,10,19,21]. For example, Vinet et al. [22] studied flash floods over the period 1988–2015 and showed that 50% of fatalities happened in a small watershed (<150 km^2). Recent analyses have focused more on models for predicting the human impacts of floods [4,23,24].

Circumstances of loss of life represent an important part of works because of the detailed sources of information. There are studies on the topic globally. Jonkman and Kelman [7] have analysed 13 flood events and a total of 247 of fatalities that have happened in the US and Europe with a goal to bring about a better understanding of circumstances and improvement of prevention measures. The highest portion of victims (67, 6%) was due to drowning that occurred while the victims crossed flooded bridges and roads. Analysis of loss of life in the US between 1969 and 1981 has also shown a higher share of victims in cars [16].

Furthermore, a study carried out by Ashley and Ashley [10] also supports previous findings in which a 47-year-long database (1959–2005) has shown that 63% of victims in the US were car-related. Sharif et al. [17] focused on flood fatalities in Texas (from 1959 till 2008), and the majority were found to happen in cars. Texas is the only state that had, in the era from 1960 till 2009, flood victims every year [25]. In the latest analysis performed by Sharif et al. [14] it was pointed out again that Texas is the leading state with regard to the number of flood fatalities. Gissing et al. [26] analysed flood fatalities in Australia between 1960 and 2015. The results showed, similarly, that 49% of flood fatalities were vehicle-related. The findings of Diakakis et al. [27], which were based on an analysis of 60 flood fatalities in Greece (all vehicle-related), concluded that in most cases loss of life happened during the night and in rural areas.

This paper focuses on Slovenia (which has an area of around 20,000 km^2), which is located in Central Europe and borders Italy, Croatia, Hungary, and Austria (Figure 1). It has a coastline on the Adriatic Sea and a population of 2 million people. In terms of geographical regions, it is possible to divide countries' land into four major natural units, namely, the Mediterranean, the Dinaric mountains, the Pannonian basin and the Alps [28], which contribute to a diverse climate. Floods rank number three on a scale of damage severity after droughts and hail. Flash floods are the prevailing type of flood because of the structure of most river basins (relatively steep slopes that cause rapid runoff) [29]. Floods in Slovenia pose a danger to around 15% of the country's surface area. Nearly half the flood zone is in the Sava basin, 40% in the Drava basin, and 4% in the Soča basin.

The importance of information on the circumstances of death and the lack of an existing database in Slovenia which could store information on human impacts in floods was the reason for this research. Good practises existing in the US which have comprehensive collections on fatalities and injuries in flood events was also an initiative.

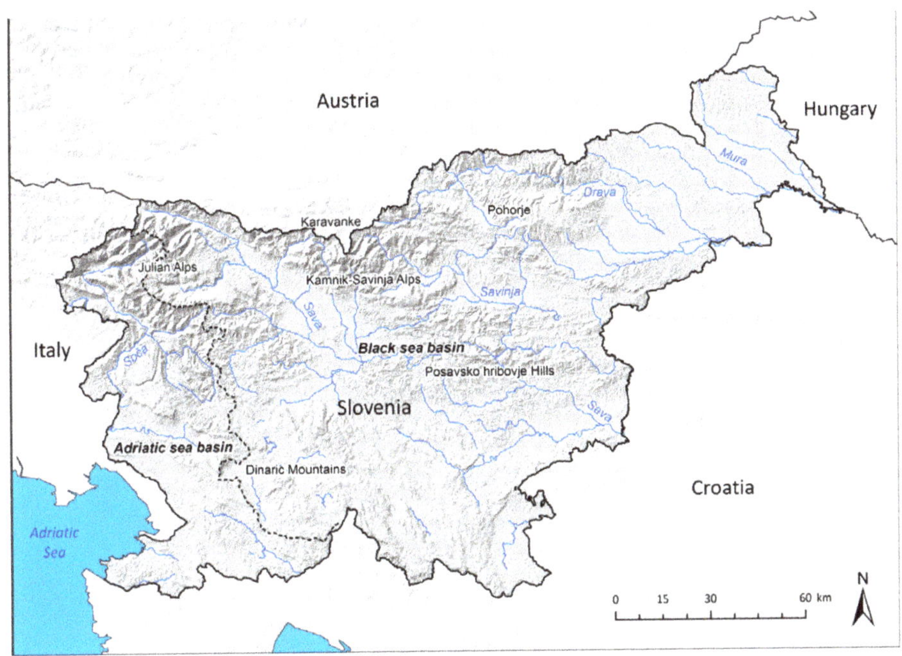

Figure 1. Map of Slovenia [28].

Furthermore, findings in the analysis of flash floods in the US have been used at the end solely for illustration [21]. Employing historical analysis, this study addresses circumstances of flood fatalities in Slovenia using primary and secondary sources (available papers, reports from institutions, and mass media). Available data in the sources about the victims enabled us to address three aspects, namely, circumstantial (vulnerable groups in Table 1), demographic (distribution of victims across gender and age) and temporal (did fatalities occur during the night or in the daytime) aspects. Our work represents a comprehensive collection and analysis of fatalities in flood events based on a long time scale, which was lacking in our system. This research does not incline towards making generalisations of any sort, but the main goal and purpose was to investigate and address human impacts, analyse demographic data of fatalities, and interplay it with the nature of hazards. Based on that information it is possible to determine vulnerable groups, which is an important tool for creating preventative measures. The essential goal of flood risk management is management "without fatalities". In order to develop flood risk management with a victim prevention approach, there needs to be an understanding of factors that contribute to loss of life during flood events. Measures of social impact management can be carried out efficiently based on the information of the circumstances of death and knowing which parts of the population are more vulnerable. An existing dataset of previous impacts can provide a base ground of needed information for preventative actions in the future.

Analysis brings a new perspective to the field of natural hazards in Slovenia. The paper is organised in the following way. The following part focuses on the data analysis framework, and then we present the results, which include variability of human impacts in flood events from 1926–2014, analysis of demographic information, analysis of loss of life circumstances, types of flood fatalities and findings from the interviews, and a discussion, followed by conclusions.

Table 1. Types of flood fatalities.

Group	General Description	Number of Flood Fatalities in Slovenia
Victims in cars	People losing their lives due to incorrect evaluations of situations on the road, unawareness of cars' instability in flood flows, and the substantial feeling of safety in large, heavy cars.	6
Victims in collapsed buildings	People trapped in buildings that collapse during flooding due to bad constructing standards. A lack of transparency in informing people of upcoming hazards. Problems with people refusing to leave their homes and individuals returning to dangerous areas/homes to rescue their property.	20
Professional rescuers	All people that are members of rescue teams while on duty. This group includes fire workers, policeman, and soldiers.	1
Voluntary rescuers	Individuals who are not part of any rescue team and get involved in helping people who are affected by hazardous events by rescuing them or their property, or helping in any other kind of way of their own will.	1
Visitors	People unfamiliar with the area, local processes, and conditions due to a temporary visit. Communication channels for warning messages are not as effective as with local people.	1
Observers	Individuals that consciously and subconsciously expose themselves to danger with risk–taking behaviour. Their drive is to see the event as closely as possible and observe what is happening. The problem of rising disaster tourism exists.	10
Victims when structural measures collapse	Structural measures are related to any kind of physical construction used to achieve resistance, and when construction standards are low, people's lives are jeopardised in the case of a hazardous event. For example, an abundance of rainfall may lead to levees overtopping and/or dam failures, and consequently fatalities.	21
Victims of non–structural failure	Non–structural measures include all the communication channels used to warn people, educate, raise awareness of risks, and basically work on tools for hazard prevention. It also covers other fields like law, policy, agreements, and trainings. Grounds for victims in this group are low risk perception, which leads to a willingness to risk more, and on the other hand bad communication and warning messages (false non-alarms). People's interpretation of messages and signs and the question of whether information has reached them and whether it has happened on time.	13
Others	This covers indirect victims that had identified medical causes of death. Floodwaters carry large debris, cars, and trees that can result in trauma, orthopaedic injuries, and lacerations. The occurrence of contaminated local water supply and sewage systems through floodwaters. There have been cases of hypothermia and heart attack.	5

2. Data Analysis Framework

The approach used in our research of the human impacts of floods in Slovenia was historical analysis of fatalities. Primary and secondary sources were analysed to collect data on the victims. Based on this data we addressed circumstantial, demographic, and temporal aspects. Descriptive statistics were used to analyse fatalities and categorise them and address especially demographic aspects. Structured interviews were executed to obtain more information on rescuing during floods and how people evaluated the help received. Questions were not related to a specific flood event but more so an overall view on what they had experienced so far. Three of the interviews were carried out with people, who have been rescuers during flooding and five were with people who had experienced flood events. Based on the two groups, two different sets of questions were constructed. This means that three interviewees were asked the same questions (questioner 1) and the rest (five interviewees) were interviewed based on questioner 2. Questions in questioner 1 were related to the following areas: description of the flood events that the interviewees had experienced, description of sources of information regarding the hazard, what they felt when experiencing the flood, what would contribute to their feeling of safety, what surprised them the most, description of the work of the rescuers, how long the redevelopment of the area took, what their preventative measures will be, and how much more attention they pay to the weather channel. Questioner 2 covered the following points: description of the flood events, the location of intervention, work flow of the protection and rescuing, all the sources of flood information and forecasts, problems that they faced during intervention, reactions of the people, how can each individual's events be useful for future (preventative) measures, and how common was evaluation at the end of the events.

Injuries were also a part of the framework, but through the data collections it was unfortunately evident that there was no detailed information available except some numbers that were not consistent for all flood events. Furthermore, the time frame of the flood events analysed was 88 years, starting from 1926 and continuing all the way to 2014. Flood events used in the analysis were chosen based on their production of at least 1 human impact. We found 10 flood events and a total of 74 fatalities in the data, which had been collected from several different available sources (news channels, Administration of the Republic of Slovenia for civil protection and disaster relief, and the Slovenian environment agency). Years 1926, 1933, and 1954 were particularly difficult since the only sources of information were the few existing general newspapers (Slovenec, Savinjski vestnik, and Jutro) that mainly covered economic and political news but sometimes also wrote about natural hazards and the damage inhabitants suffered in the affected areas. After the year 1954, human impact data was mentioned in the reports on flood events among main hydrological details performed by the Slovenian Environment Agency, which is a branch of the Ministry of Agriculture and Environment and which primarily focuses on forecasting, analysing and monitoring natural phenomena, and mitigating damage and loss of life. Other sources included were the Acta Geographica Slovenica, the Administration for Civil Protection and Disaster Relief, and the mass media. Our main focus and goal was to investigate all the available factors of flood fatalities which contribute to an understanding of the circumstances and the level of influence. Gender and age, if available, were also included in the analysis to see which group tended to be more exposed.

Additionally, based on the story narrative of individual casualties, the circumstances of death were determined, through which types of flood fatalities were derived. In order to obtain more information on how people perceive floods and what seem to be the main problems when flood events happen, several structured interviews were conducted. Three were with people, who have been rescuers during flood events and five were with people, who have experienced floods.

Considering the lack of available information on casualties in flood events in Slovenia and the low emphasis that is put on collecting and storing data of this kind, a comprehensive analysis represents a step towards implementing a systematic data storing. The assessment of these influential parameters is a way to improve the understanding of human impacts and a contribution to flood managers when planning preventative measures. Flooded roads seem to have consistently posed a great threat over the years to drivers and have caused fatalities. As a result of numerous roads being flooded during every flood event, questions about the relationship between watercourses and roads arose. This issue itself is too complex to be analysed in this article but will be the subject of future work.

3. Results

3.1. Variability of Human Impacts in Flood Events from 1926–2014

Information regarding the number of flood fatalities in each individual year in Slovenia was collected and has been presented in Figure 2. There are 10 flood events and 74 flood victims. It is important to point out that there were more flood events between the years 1926 and 2014 but that they did not cause loss of life. For a comparison, the total number of more extensive floods (return period >50 years in at least three river basins) between 1926 and 2010 was 17, as presented in Figure 3 [30]. Based on Figure 2, there is a trend consisting of a decrease in flood victims through the years despite hydrological reports showing an increase in the frequency of extensive floods. There is a large contrast due to a higher number of flood fatalities in the years 1926, 1933, and 1954. The use of a long timeframe enables us to mark a threshold at the year 1954 for the purpose of data analysis. There is a trend of consistent increase in loss of life from 1926 to 1954 and fluctuations in loss of life after the year 1990. It is important to point out the 36 year period (between 1954 and 1990) in which no flood fatalities were recorded. As already mentioned, there were 17 extensive floods which occurred between 1926 and 2010, with eight of them causing loss of life and nine flood events occurring without fatalities which happened in the years 1948, 1965, 1974, 1979, 1980, 1982, 1986, 2008, and 2009 [30].

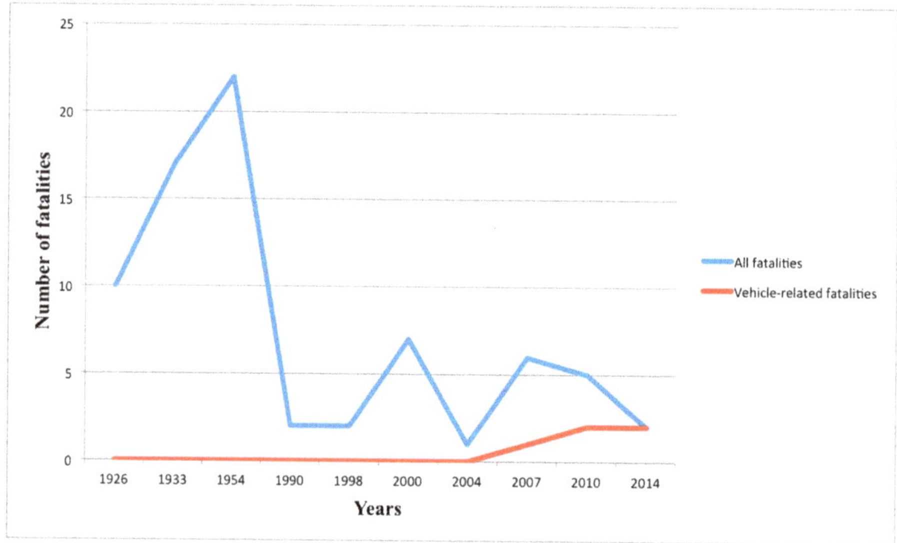

Figure 2. Number of fatalities in Slovenia from 1926 to 2014.

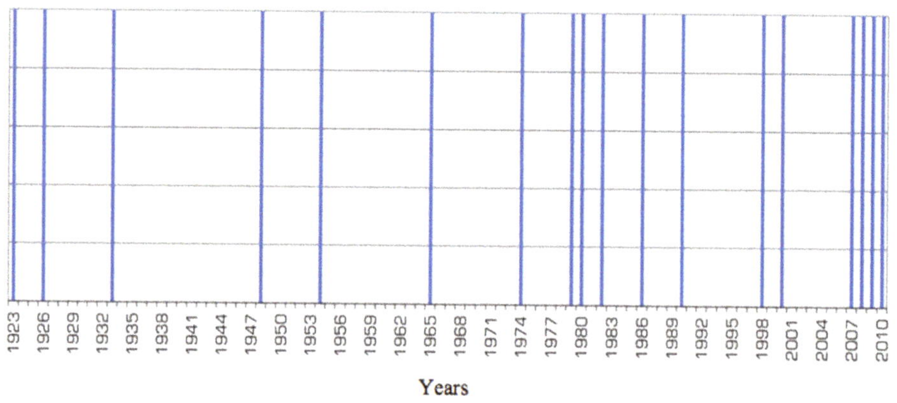

Figure 3. Occurrence of extensive floods in Slovenia [30].

Extensive data of maximum annual discharges were used for the period 1895 to 2013 for the river Sava. In the fourth quartile there were a total of 7 years with flood fatalities (1926, 1933, 1990, 1998, 2000, 2007, and 2010) and 4 years without victims (1965, 1979, 1982, and 2009). The flood event in the year 2010 ranks as one of the most extreme floods recorded, with discharges equal to those of a 100-year return period. The largest floods recorded are still from the year 1990 and they exceeded a 100-year return period. There are 5 years worth pointing out (1936, 1964, 1973, 1975, and 1985) in the fourth quartile which fall in the period 1954 to 1990 which had no documented flood events with fatalities. Furthermore, in the first quartile there were 8 years within the period 1954 to 1990 and 7 years falling within the period 1933 to 1954 in which there were no flood fatalities. Discharge data for the river Sava indicate a very dry season between 1940 and 1960 [30]. Furthermore, data show a 50-year return period for the year 1998 and a close to 100-year return period for the year 1933.

Discharge data for the river Savinja point out a more than 100-year return period of the floods in the years 1990 and 1998 In the years 1933 and 1980 there was close to a 50-year return period. The river Krka approached a 1000-year return period, with floods in the year 1933 having a 200-return period and a 100-return period occurring in the year 1948. Discharge data for the river Ljubljanica exceeded a 100-year return period. After this, the highest recorded discharge rates were from the years 1975 and 1998 (a 50-year return period).

Lastly, the river Vipava showed discharges with a 1000-return period in the year 2010, a 100-return period in 2009, and a 50-year return period in 1965 [30]. Information on the return period indicates some correlation with severity of flood event and human impacts, with the exception of the year 1954, which counts as the most fatal event in the entire period covered and does not coincide with a high return period.

3.2. Analysis of Demographic Information

Gender and age were used in the demographic segment in order to determine which groups are the most vulnerable in the case of flooding. From the total number of 74 fatalities, we were able to collect gender information for 43 people. The frequency of male fatalities (60%/26 people) was higher than that of female fatalities (40%/17 people).

Data regarding age were divided into three groups for the analysis. The Statistical Office of the Republic of Slovenia has two types of age classification. The first approach divides people by age into children (0–14), the working–age population (15–64), and the elderly (65 and over) while the second approach divides people by age into the young (0–19), adults (20–59), and the elderly (60 and over). For the purpose of this analysis we used the second approach of age classification. The age information sample included 21 people, making it a very small sample.

The distribution of fatalities shows that the most vulnerable groups are adults (10 people) and the elderly (8 people). Three people fell in the young people group.

3.3. Analysis of Loss of Life Circumstances

Many factors influence the degree of vulnerability of people in the affected areas. While the focus is too often solely targeted on compositional variables (hydrological parameters) it must be noted that contextual factors (situational and temporal parameters) also contribute significantly to the fluctuation of the distribution of human impacts and their scale. Apart from external influences, some authors [8] believe that risk–taking behaviours should not be underestimated because a substantial proportion of flood fatalities are believed to be due to those unnecessary types of risk–taking behaviours. Based on the collection of flood fatalities and their activities across a long time scale, we were able to see how and if people's activities changed throughout the years and if these activities are really attributable to unnecessary risk–taking behaviours. Contextual information helped us to determine how or if people's behaviours reduce personal safety.

The most severe floods in terms of human impacts happened in the year 1926 (10 fatalities), 1933 (17 fatalities), and 1954, when 22 people lost their lives. The data revealed that three families were trapped and drowned in their own houses in the year 1954. Further analysis also showed that professional and volunteer rescuers lost their lives while trying to help others. People's attachment to their personal belongings led them back to their houses and they lost their lives while attempting to salvage their belongings. Wrong judgement of their current situation and unnecessary exposure to danger resulted in several deaths.

For example, a male subject was trying to reach a piece of wood in the flood flow while a group of 13 people was standing on a wooden bridge, observing. The latter later collapsed and resulted in all the people drowning. Conformity behaviour is where a substantial number of observers cause new to join.

A few people also died in their vehicles, which at the time were most likely wagons pulled by horses. Moreover, in two cases people died at home during a flood event due to medical reasons. These medical circumstances could have been induced from the event but there was not enough information to make a final conclusion. There have been several reports of a high number of people with serious and minor injuries but unfortunately no definitive actual number was stated.

Furthermore, in the last two decades there has been a trend of an increase of landslides as a consequence of flash flooding, taking 7 lives in 2000 and 3 lives in 2007. The background of the event in 2000 shows a failure of non–structural measures. Inhabitants exposed to the hazard were evacuated but sent back in the evening, thinking the area was safe. During the night, an abrupt landslide destroyed several structures and killed 7 people. The flood event that happened in 2007 also triggered landslides that caused 3 fatalities. A couple was found dead in their house, which had been partly destroyed by a landslide. In this case, structural measures collapsed and caused fatalities since back in 1990 the area had not been properly restored and protected with barriers after the landslide had already happened. Another fatality happened when somebody tried to go back to their house to save personal belongings. The circumstances of three other fatalities that happened in the flood events in 2007 were related to drowning, electrocution (of a volunteer fire fighter), and an attempt to save a car. Five people lost their lives in the flood event in 2010.

Among these fatalities, two were vehicle-related. The first victim ignored a closed and marked flooded road and drove on. The second fatality happened while the victim lost control over their car on a flooded road and ended up in the stream. In addition, situations existed in which people exposed themselves unnecessarily, resulting in two more victims that were found on the side of the stream and railway underpass. Another victim died in a flooded house while sleeping.

The latest flood event that took place in September 2014 caused 2 vehicle-related fatalities which once again occurred due to a failure of structural measures and low visibility. There was a situation involving unmaintained stream crossing and poor road conditions in which high flow velocities led to destruction of the road surface, causing the car to slide into the stream.

3.4. Types of Flood Fatalities

The story narratives of each individual fatality gave us an overview of the circumstances. Through our analysis, we started to see repetitive patterns that presented a basis for classification. The time frame of the data of human impacts in flood events we used was from 1926 to 2014. Based on information regarding flood casualties that occurred in Slovenia, flood fatality types (illustrated in Figure 4) were constructed. Understanding the circumstances can be useful to several different areas, such as early warning systems, risk prevention, and preparedness, and to voluntary and professional rescuers as well. This kind of categorisation emphasises the most vulnerable groups of people in the case of flooding and should be considered by flood managers. Furthermore, some clarification of how these groups were formed is necessary. As shown in Table 1, medical causes of death were taken into consideration but not as a main focus.

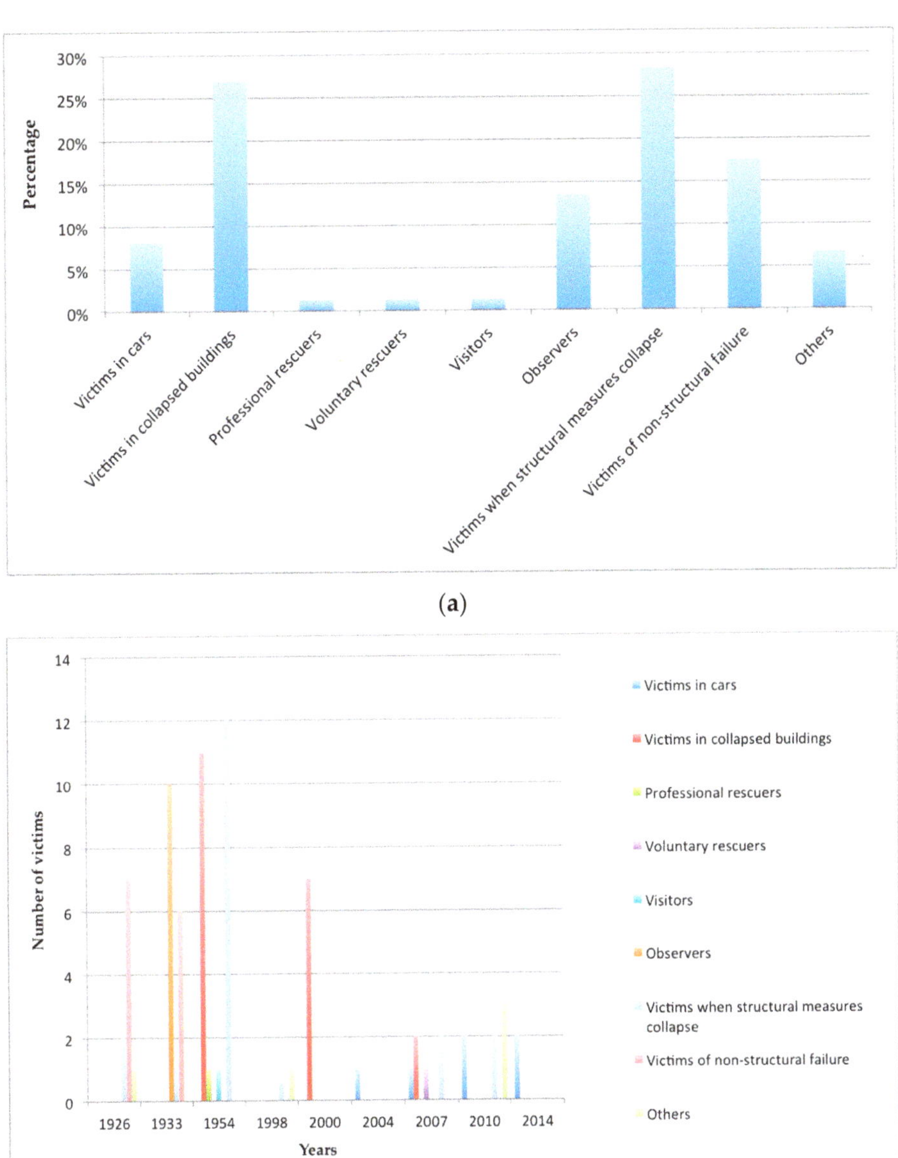

Figure 4. (**a**) Percentage of each type of flood fatality. (**b**) Number of victims per year by each type of flood fatality.

This is the reason for which these people were categorised as the group "Others". Latter was created based on analysis of various situations: people involved of their own will, people on duty, or people being in the wrong place at the wrong time. This categorisation can also be applied with small alterations if needed to other hazardous events and countries. The sum of victims in Slovenia for each fatality type does not represent the total sample because with certain victims we did not have enough circumstantial information, which would enable us to classify them, and certain fatalities were placed in more than one group. The majority of loss of life in floods in Slovenia happened in collapsed buildings and due to the structural and non–structural failures (Table 1, Figure 4a). Each type per year is on the one hand cumulated with victims of non–structural failure, observers, victims in collapsed buildings, and victims when structural measures collapsed in the period 1926 to 1954; on the other hand, we see overrepresentation of car victims in the period 2004–2014 (Figure 4b).

3.5. Findings from the Interviews

Two of the rescuers in the flood events were fire fighters and one was a member of the Administration of the Republic of Slovenia for civil protection and disaster relief. They were all male and between the ages of 27 and 51. People who had been affected by floods were of ages between 24 and 51; four were male and one was female. All the rescuers pointed out that no matter what the predictions were about the flooded area where they had to intervene, the situation with water was always worse when they arrived to the actual place. When the conditions were critical, the equipment with which they worked was useless. Fire fighters often face a lack of staff and equipment. Employment as a fire fighter in Slovenia is still, in the majority of cases, based on volunteering. When water quickly exceeds a certain level and starts flooding it causes a domino effect, meaning multiple disasters are experienced at the same time, including failure of the infrastructure (destroyed roads and collapsed bridges) and landslides, which can severely change the situation and make intervention more difficult or impossible. Decision-making and preventative measures (evacuations and sand bags) when the first warnings are received are crucial. Malfunctions of the Administration of the Republic of Slovenia for civil protection and disaster relief were also mentioned. People sometimes make situations much worse when they do not want to leave their homes and when they underestimate the power of the water. After the event, analysis of the intervention is performed. The latter is really important because this type of intervention is very specific and demands a lot of knowledge.

People that were affected by the floods all described how unimaginable the destruction of the water could be. Regardless, for them it was still a priority to take care of their personal belongings first over the safety of their own life. It was pointed out how little local authorities warned or taught them about the possible danger of floods in the area in which they live. Since they have experienced floods they pay much more attention to the weather news. All of them expressed disappointment over the poor preventative measures and insufficient restoration of the area after the flood event.

4. Discussion

There have been other studies that have shown males as the prevailing group among victims [10, 11,27]. This vulnerability could be explained in two ways: due to their nature, males are more willing to risk their lives, and also, there are more males than females in emergency and rescue teams. For the purposes of comparison, in the year 2019 49% of the population was male and 51% was female (Statistical Office of Slovenia), meaning there was not an overrepresentation of either group.

Trends in population development have been obtained by the Statistical Office of the Republic of Slovenia and these trends show a continuous growth of the elderly population and a decrease in the number of young people. This is an increasing problem in Slovenia [31]. Furthermore, a constrictive pyramid serves as an important tool for flood manager to prioritise elderly in safety and evacuation measures.

Concerns over structural measures failing should be addressed with preventative measures. The reality of the danger posed that comes from insufficiently repaired damage of past flood events is continuously repeating and is shown during new flood hazards. This issue has not only caused fatalities but also increases the level of damage made from one flood event to another. A new approach in management is needed to efficiently repair damage and to find ways to mitigate watercourses that are prone to flooding. Mapping of the exposed areas is needed, as well as an assessment of the influential parameters that contribute to the areas' flood exposure [2]. Maps, as an informative tool, can contribute in an important manner to people's awareness of danger and how powerful and destructive floods can be. We need to make sure of the availability of these maps to the public and that people understand and know how to read them. Furthermore, roads have been extensively affected in terms of damage by every flood event. In the last decade there were also 5 vehicle-related fatalities: 2 were due to risk-taking behaviour and 3 due to a collapse of structural measures. Each year there are more registered vehicles and in the year 2018 the number was as high as 1.143,150 registered vehicles [31]. It is necessary to address the relationship between roads and watercourses and create a map of road flood hazards and adequate road flood preventions. There needs to be an emphasis on structural prevention, for example having building standards that can sustain floods with higher return period. This could mean that roads would sustain more and would get less flooded, which would enable evacuation on the road.

This review of the circumstances of loss of life through a longer time scale indicates a reduction in the number of casualties, different means of transportation, and fewer channels of warning. At the same time, it points out the ongoing existence of problems with the collapse of structural and non–structural measures. Slovenia has also faced various situations where landslides occurred in conjunction with flood events, which should be considered the rule rather than mere coincidence.

In the future, problems which arise could be victims in cars, people refusing evacuation, and/or people returning back to their homes to save their property, which is an unintentional risk–taking behaviour. Also, unsafe conditions for professional and volunteer rescuers are to be worried about as much as the issue of people not obtaining information about an upcoming danger (dying while sleeping at home). There is a growing importance of good communications channels [3].

It is very difficult to compare Slovenia and the US due to geographical and cultural differences. For illustration purposes we want to point out some findings. There is a seasonality factor. Monthly variability of flood events in the US shows the domination of summer months [21], whereas Slovenia has an autumn maximum. Based on information regarding the circumstances of death, we can see cultural differences. The US as an automotive society has more than 50% of its fatalities in vehicles [21], even though data for Slovenia shows an increasing problem of victims in cars in recent years. For example, in the last ten years Slovenia had nearly 50% of vehicle-related fatalities among the total number of flood deaths.

The lifestyle of living in trailers is not so much present in Slovenia, and thus Slovenia does not have victims related to this circumstance. Rural areas have been shown to be more problematic, as structural measures collapse and more fatalities tend to happen in these areas in Slovenia. A similar trend has been pointed out for the US [21].

Inconsistencies between flood plain areas and spatial planning (granted building permits) and a lack of preventative measures also result in fatalities. Analysis has shown that people do not learn from mistakes. For example, a comparison of flood damage between the years 1954 and 1990 has pointed out that certain buildings have been damaged in the same way.

Furthermore, newly constructed buildings have suffered more damage than older objects, indicating a lack of building standards for damage prevention in the case of natural disasters [32]. Information on the return period of flood events gives a certain correlation between floods producing fatalities and events with no loss of life. Events with a high return period tend to result in casualties but there are certain exceptions with extreme events and no fatalities. It is interesting that the year 1990 accounts for the most extreme floods recorded and has one of the lowest numbers of fatalities. Comprehensive preventative measures can only be performed with sufficient and holistic data, and thus it is important to continue this systematic collection of information and link it with changes in education, warning signs, and systems and building standards that would sustain higher return periods.

5. Conclusions

In this work, historical analysis of casualties in flood events was performed for the 1926 to 2014, employing a dataset of 10 floods that ended in 74 fatalities. However, there were 21 severe floods in total that happened in this time frame and 48% of them caused loss of life. Despite a trend which shows a decrease in fatalities through the years, the percentages of floods with casualties and floods without casualties are similar, since 52% of the flood events occurred without any casualties. Data show a phenomenon of car victims in recent years.

The primary focus in this study was floods with fatalities, since there was a lack of a comprehensive view on human impacts in Slovenia. Loss of life was addressed through different angles. Through the collected data on flood victims, demographic, circumstantial, and temporal aspects were addressed. The main points for each aspect are summarised below.

The demographic distribution showed a higher vulnerability of males, which could be due to their being overrepresented as volunteer and professional rescuers and their nature of being willing to take more risks.

The age analysis of fatalities pointed out that adults and the elderly tend to be more vulnerable in the case of hazardous events. However, population development shows a growth of the elderly group and a decrease in the young population. This is important information for future mitigation strategies to make sure the elderly in hazardous areas have received warnings about upcoming dangers and if necessary have been evacuated from the affected areas.

Data collection and analysis of the circumstances of loss of life gave an important insight into the activities of individuals during flood events. Usage of a long time scale showed how circumstances have changed. We noticed a reduction in the number of people falling in the group of observers and an increase in victims in cars in the last decade.

Based on the types of loss of life (victims in cars, victims in collapsed buildings, professional rescuers, voluntary rescuers, visitors, observers, victims when structural measures collapsed, victims of non-structural measures, and others) which were derived from the circumstances, a platform for the vulnerable groups was established which could be a useful tool for flood managers, especially since the main indicated issues were structural and non–structural failures and people exhibiting risk behaviours.

Our analysis highlighted problems regarding small municipalities (rural areas) that lack mitigating structures, the severity of events with heavy precipitation, and high flows and high return periods. The severity of events was difficult to predict, and this opens up a new challenge on how to improve prediction systems, and, even further, make these specific to a location.

Furthermore, in two events (in the years 2010 and 2014) fatalities in cars happened early in the morning and in one event (2014) low visibility was crucial. There is also the importance of small catchment size (in flood events in the years 2007, 2010, and 2014), which is more unpredictable and usually causes more damage. Existing good practices such as "safety education", including "Turn around, don't drown", as performed by the National Weather Service, could be implemented in Slovenia. This could include placing warning signs on frequently flooded roads, educating drivers on how dangerous flooded roads can be and how fast water can float vehicles. Social media could be used to educate people about the dangers of floodwater and as a channel for people to obtain information from.

In many emergencies, evacuations are obstructed due to people refusing to leave their homes. Residents are often unfamiliar with the flood dangers in their area. Investments need to be carried out in the equipment which rescuers work with. Evaluations have shown an increasing need for improvements in communication strategies between fire fighters and the Administration of the Republic of Slovenia for civil protection and disaster relief.

This study advances understanding of flood fatalities in Slovenia. It gives an overview of demographic, temporal, and circumstantial aspects of victims. By addressing the problem of loss of life from different angles, it advances understanding and highlights the mitigation and preventative measures that can be taken in the future and the remaining challenges regarding how to reduce vulnerability (delving deeper into social dimensions like perception) and improve mitigation structures and post-flood damage restoration.

Author Contributions: This study was conceptualized by M.Š., M.B. and D.K., data analysed by M.Š. and A.Ž., original manuscript written by M.Š., manuscript was reviewed and edited by M.Š. All authors have read and agreed to the published version of the manuscript.

Funding: This research received no external funding.

Conflicts of Interest: The authors declare no conflict of interests.

References

1. Paprotny, D.; Sebastian, A.; Napoles, O.M.; Jonkman, S. Trends in flood losses in Europe over the past 150 years. *Nat. Commun.* **2018**. [CrossRef] [PubMed]
2. Dandapat, K.; Panda, G.K. A geographic information system-based approach of flood hazards modelling, Paschim Medinipur district, West Bengal, India. *Jàmbá J. Disaster Risk Stud.* **2018**, *10*. [CrossRef] [PubMed]
3. Salman, A.M.; Li, Y. Flood rick assessment, future trend modelling and rick communication: A review of on going research. *Nat. Hazards Rev.* **2018**, *19*, 04018011. [CrossRef]
4. Jonkman, S.N.; Vrijling, J.K. Methods for the estimation of loss of life due to floods: A literature review and a proposal for a new method. *Nat. Hazards* **2008**, *46*, 353–389. [CrossRef]
5. Jonkman, S.N.; Vrijling, J.K. Loss of life due to floods. *J. Flood Risk Manag.* **2008**, *1*, 43–56. [CrossRef]
6. Doocy, S.; Daniels, A.; Murray, S.; Kirsch, T.D. The Human Impact of Floods: A Historical Review of Events 1980–2009 and Systematic Literature Review. Currents Disasters. 2013. Available online: http://currents.plos.org/disasters/article/the-human-impact-of-floods-a-historical-review-of-events-1980-2009-and-systematic-literature-review/ (accessed on 23 December 2014).
7. Jonkman, S.N.; Kelman, I. An analysis of the causes and circumstances of flood disaster deaths. *Disasters* **2005**, *29*, 75–97. [CrossRef]
8. Jonkman, S.N. Global perspectives on loss of human life caused by floods. *Nat. Hazards* **2005**, *34*, 151–175. [CrossRef]
9. Lowe, D.; Ebi, K.; Forsberg, B. Factors increasing vulnerability to health effects before, during and after floods. *Int. J. Environ. Res. Public Health* **2013**, *10*, 7015–7067. [CrossRef]
10. Ashley, S.T.; Ashley, W.S. Flood fatalities in the United States. *J. Appl. Meteorol. Climatol.* **2008**, *47*, 806–818. [CrossRef]
11. Coates, L. Flood fatalities in Australia 1788–1996. *Aust. Geogr.* **1999**, *30*, 391–408. [CrossRef]
12. Haynes, K.; Coates, L.; Honert, R.; Gissing, A.; Bird, D.; Oliveira, F.D.; D'Archy, R.; Smith, C.; Radford, D. Exploring the circumstances surrounding flood fatalities in Australia 1900–2015 and the implications for policy and practice. *Environ. Sci. Policy* **2017**, *76*, 165–176. [CrossRef]
13. Salvati, P.; Petrucci, O.; Rossi, M.; Bianchi, C.; Pasqua, A.A.; Guzzetti, F. Gender, age and circumstances analysis of floods and landslide fatalities in Italy. *Sci. Total Environ.* **2017**, *610*, 867–879. [CrossRef] [PubMed]
14. Sharif, H.O.; Jackson, T.L.; Hossain, M.M.; Zane, D. Analysis of flood fatalities in Texas. *Nat. Hazards Rev.* **2015**, *19*, 311–323. [CrossRef]
15. Staes, C.; Orengo, J.C.; Malilay, J.; Rullan, J.; Noji, E. Death due to flash floods in Puerto Rico, January 1992: Implications for prevention. *Int. J. Epidemiol.* **1994**, *23*, 968–975. [CrossRef]
16. French, J.G.; Ing, R.; Von Allmen, S.; Wood, R. Mortality from flash floods: A review of the National Weather Service reports, 1969–1981. *Public Health Rep.* **1983**, *98*, 584–588.
17. Sharif, H.O.; Hossain, M.M.; Jackson, T.; Bin–Shafique, S. Person–place–time analysis of vehicle fatalities caused by flash floods in Texas. *Geomat. Nat. Hazards Risk* **2012**, *3*, 311–323. [CrossRef]
18. Gruntfest, E.; Handmer, J.W. *Coping with Flash Floods*; Kluwer Academic Publishers: Dordrecht, The Netherlands, 2001.
19. Ruin, I.; Creutin, J.D.; Anquetin, S.; Lutoff, C. Human exposure to flash floods—Relation between flood parameters and human vulnerability during a storm of September 2002 in Southern France. *J. Hydrol.* **2008**, *361*, 199–213. [CrossRef]
20. Creutin, J.D.; Borga, M.; Gruntfest, E.; Lutoff, C.; Zoccatelli, D. A space and time framework for analysing human anticipation of flash floods. *J. Hydrol.* **2013**, *482*, 14–24. [CrossRef]
21. Špitalar, M.; Gourley, J.J.; Lutoff, C.; Kirstetter, P.E.; Brilly, M.; Carr, N. Analysis of flash flood parameters and human impacts in the US from 2006–2012. *J. Hydrol.* **2014**, *519*, 863–870. [CrossRef]

22. Vinet, F.; Boissier, L.; Saint-Martin, C. Flash flood-related mortality in southern France: First results from new database. In Proceedings of the FLOODrisk 2016—3rd European Conference on Flood Risk Management, Lyon, France, 17–21 October 2016. [CrossRef]

23. Terti, G.; Ruin, I.; Gourley, J.J.; Kirstetter, P.; Flaming, Z.; Blanchet, J.; Arthur, A.; Anquetin, S. Toward probabilistic prediction of flash flood human impacts. *Risk Anal.* **2019**, *39*, 140–161. [CrossRef]

24. Zhai, G.; Fukuzono, T.; Ikeda, S. An empirical model of fatalities and injuries due to floods in Japan. *J. Am. Water Resour. Assoc.* **2006**, *42*, 863–875. [CrossRef]

25. Jackson, T.L. The Impacts of Increasing Rainfall: Flood Fatalities in Texas. Ph.D. Thesis, The University of Texas at San Antonio, San Antonio, TX, USA, 2009.

26. Gissing, A.; Opper, S.; Tofa, M.; Coates, L.; McAneney, J. Influence of road characteristics on flood fatalities in Australia. *Environ. Hazards* **2019**, *18*, 434–445. [CrossRef]

27. Diakakis, M.; Deligiannakis, G. Vehicle-related fatalities in Greece. *Environ. Hazards* **2013**, *12*, 278–290. [CrossRef]

28. Trobec, T. Frequency and seasonality of flash floods in Slovenia. *Geogr. Pannonica* **2017**, *21*, 198–211. [CrossRef]

29. Brilly, M.; Mikoš, M.; Šraj, M. *Vodne Ujme—Varstvo Pred Poplavami, Erozijo in Plazovi*; Fakulteta za Gradbeništvo in Geodezijo: Ljubljana, Slovenia, 1999.

30. Kobold, M. Comparison of floods in September 2010 with registered historic flood events. *Ujma* **2011**, *25*, 48–56.

31. SURS, Prebivalstvena Piramida: 2014. Available online: http://www.stat.si/PopPiramida/Piramida2.asp (accessed on 23 January 2015).

32. Komac, B.; Natek, K.; Zorn, M. *Geografski Vidiki Poplav v Sloveniji*; Založba ZRC: Ljubljana, Slovenia, 2008.

Article

Can the Quality of the Potential Flood Risk Maps be Evaluated? A Case Study of the Social Risks of Floods in Central Spain

Julio Garrote [1,*], Ignacio Gutiérrez-Pérez [2] and Andrés Díez-Herrero [3]

[1] Department of Geodynamics, Stratigraphy and Paleontology, Complutense University of Madrid, E-28040 Madrid, Spain
[2] Ferrovial Agroman, US Corp., North America Headquarters, Austin, TX 78759, USA; ignaciogutierrezperez@gmail.com
[3] Geological Survey of Spain (IGME), Ríos Rosas 23, E-28003 Madrid, Spain; andres.diez@igme.es
* Correspondence: juliog@ucm.es; Tel.: +34-913-944-850

Received: 9 May 2019; Accepted: 18 June 2019; Published: 20 June 2019

Abstract: Calibration and validation of flood risk maps at a national or a supra-national level remains a problematic aspect due to the limited information available to carry out these tasks. However, this validation is essential to define the representativeness of the results and for end users to gain confidence in them. In recent years, the use of information derived from social networks is becoming generalized in the field of natural risks as a means of validating results. However, the use of data from social networks also has its drawbacks, such as the biases associated with age and gender and their spatial distribution. The use of information associated with phone calls to Emergency Services (112) can resolve these deficiencies, although other problems are still latent. For example, a bias does exist in the relationship between the size of the population and the number of calls to the Emergency Services. This last aspect determines that global regression models have not been effective in simulating the behavior of related variables (calls to Emergency Services–Potential Flood Risk). Faced with this situation, the use of local regression models (such as locally estimated scatterplot smoothing (LOESS)) showed satisfactory results in the calibration of potential flood risk levels in the Autonomous Community of Castilla-La Mancha (Spain). This provides a new methodological path to the calibration studies of flood risk cartographies at national and supra-national levels. The results obtained through LOESS local regression models allowed us to establish the correct relationship between categorized potential risk levels and the inferred potential risk. They also permitted us to define the cases in which said levels differed ostensibly and where potential risk due to floods assigned to those municipalities led to a lower level of confidence. Therefore, based on the number of calls to the Emergency Service, we can categorize those municipalities that should be the subject of a more detailed study and those whose classification should be revised in future updates.

Keywords: flood risk; LOESS model; risk map calibration; 112 emergency service; central Spain; PRICAM project

---

1. Introduction and Objectives: The Assessment of Quality of Maps and the Peculiarities of Flood Risk Maps

Floods are probably the natural process with the greatest temporal and spatial recurrence affecting society throughout the world. Data compiled by "The international disasters database" for the period 1900–2018 [1], for example, show that maximum annual flooding was reached during the year 2008. That is why flood risk management has become an essential tool from both social and economic perspectives to reduce both economic losses and loss of human lives. However, the extension of the territory conditions the flood risk analysis approach [2]; each scale of work requires the use of different methods of analysis, and the results must satisfy different uses. Depending on the spatial extent of the analysis, de Moel et al. [2] propose four scales (supra-national, macro-scale or national, meso-scale or regional, and micro-scale or local). This does not mean they are isolated, because some of the analysis methodologies are valid for different scales (in many cases by using grouping techniques or by simplifying calculation processes). Approaches, analysis techniques, results, uncertainties, and processes used for validation of the results associated with each of these four working scales are included in de Moel et al. [2]. The processes used to validate results are key points that will require a greater effort in the future (regardless of the scale of work of the analysis), since they also determine the utility for end users [3]. Thus, de Moel et al. [2] raise the need to focus efforts on obtaining post-disaster information (with as much detail as possible) as a fundamental tool to improve the calibration, the validation, and the representativeness of flood risk models.

When the study area is large (macro-scale analysis and above), one of the main problems we face in validation and calibration of the results is the lack of post-event information uniformly distributed throughout the territory. This lack of validation, as has been shown in several previous works [4–6], can lead to an over-estimation of risk from a quantitative analysis point of view. However, other authors showed that this quantitative overestimation did not have a negative influence in terms of prioritization between localities for future, more detailed work [6,7]. The quantitative magnitude of the risk varies but not its scalar ordering and ranking. This problem of validation and calibration of the results was also considered by Dottori et al. [8]; they state that the validation of supra-national models of flood hazard is limited by the scarce availability of reference maps of flooded areas. This is largely the case in Africa, Asia, and South America, and in developing countries in general [9].

In cases of both macro-scale (national) and supra-national scale models, validation of the results is affected by the scarce availability of contrast information—an aspect favored by the spatial and the temporal discontinuity in the occurrence of floods. The particularities of flood risk maps contribute to this situation [10] in which the use of information referring to flooded areas in river floodplains (through, for example, the use of aerial or satellite images) does not necessarily produce good results in locations of great interest for flood risk analysis; for example, the purpose of validation of flood risk models based on data from insurance policies [11]. Although the extension of the affected areas remains homogeneous (with respect to the validation through data from flooded areas), a tendency to reduce the damage associated with the flood events considered is observed.

Another form of risk map calibration used mainly in the analysis of flood hazards (by hydro-dynamics modeling) is one that was proposed at the beginning of 2010s. It suggested using data contributed through citizen collaboration [12–16]. It is worth highlighting the state of the art of these citizen observation techniques [17] carried out by Assumpção et al. [18], which includes not only data provided by direct observation but also those provided by social networks. The use of data from social networks has gained importance in recent years, with a proliferation of works in which these data are used for flood modeling and monitoring [19–24], the spatial planning of settlements for displacement by floods [25], as well as other natural disasters [26]. However, the use of data from citizen collaboration (in its different modalities) is not trivial despite the great interest in using such sources in the assessment of natural hazards [27]. As pointed out by Rosser et al. [28], the data have generally not undergone any type of validation or quality assessment [29] and may contain deliberate or unintentional biases [30].

In addition, as noted by Xiao et al. [30], socio-economic factors play a major role (of greater importance than the size of populations or the magnitude of associated natural disasters) in the "natural disasters–social networks" relationship. This means that factors such as age, sex, educational and economic level, or location in urban or rural areas can condition this relationship, since they also condition the access and the use of social networks. This combination of factors and their relationship with the use of social networks can be a negative handicap for the calibration of risk analysis in rural areas (with an ageing population and, in general, lower educational and economic levels).

This study tries to overcome some of the limitations previously raised. First, a flood risk analysis model was developed in the region of Castilla-La Mancha (located in the center of the Iberian Peninsula), which mainly focused on the social analysis of flood risk. Although the analysis could be administratively placed on a regional scale, the autonomous community of Castilla-La Mancha covers a very large area (greater than over 50% of the countries in Europe), meaning that the study could be placed within the macro or the national scale. The flood risk analysis (RICAM project) was developed for the Civil Protection Service (regional government of Castilla-La Mancha), thus the validation of the results obtained was of vital importance. To this end, a validation/calibration model of potential flood risk analysis was proposed (associated with different flood return periods or recurrence probabilities), and this analysis was based on the information gathered in the emergency telephone (112) of the Castilla-La Mancha region. This approach was based on the hypothesis of a higher accessibility to this service (emergency telephone) than to other social networks and a less biased calibration due to the socio-economic factors of the population.

Therefore, two main objectives were achieved. The first objective was to propose a methodological approach for the analysis of potential flood risks (with a predominantly social focus) with the objective of obtaining a flood risk level ranking (using a multi-criterion analysis-geographic information system approach, MCA-GIS). The second objective was to propose a validation method for the results obtained in the previous phase (with the combination of Emergency Service 112–locally estimated scatterplot smoothing (LOESS)). The ultimate goal was to prioritize mitigation actions and maximize the efficiency of economic investments.

2. Study Site

2.1. Environmental Description

The region of Castilla-La Mancha (Figure 1) is located in the southern half of the Iberian Peninsula (between latitudes 38°01′ and 41°20′ N and longitudes 0°55′ to 5°24′ W), occupying the greater part of the South Sub-plateau with an area of 79,226 km^2.

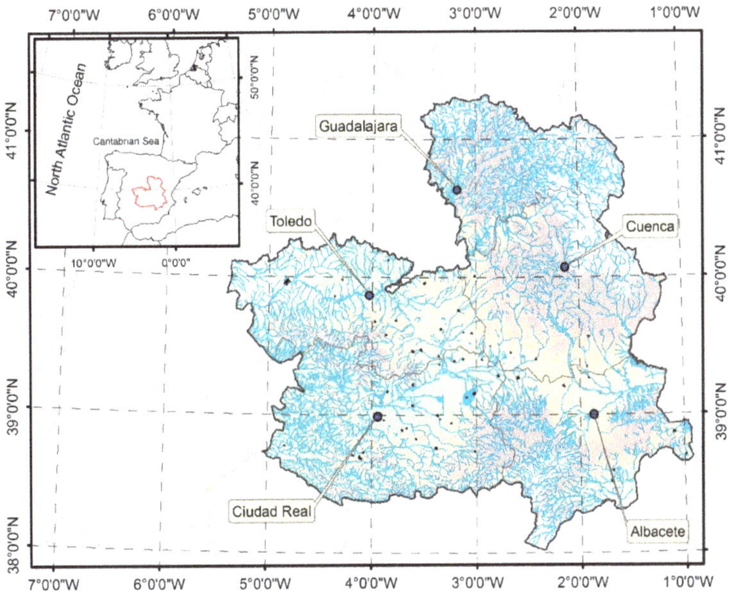

Figure 1. Location map of the autonomous community of Castilla-La Mancha. Dark blue circles show the location of each province capital city, and the black dots show the location of major towns.

The region has an average altitude between 600 and 700 m above sea level (a.s.l.). The maximum and the minimum elevations can be found in the Pico Lobo (2262 m) and in the floodplain of the Tagus River (289 m) at the western boundary of the province of Toledo. The autonomous community of Castilla-La Mancha has three major relief groups. The first of them corresponds to a plain that covers the entire central area of the autonomous community, and it is thus called "La Mancha plain". The relief of this area fluctuates between 500–1000 m high, and it makes up close to 70% of the total territory.

The second group is associated with the most important reliefs, which are mainly located in the northeastern boundary of the region (where there are mountains peaks above 2000 m a.s.l.). The foothills of other major reliefs of the Iberian Peninsula (Cordillera Bética, Cordillera Ibérica, Sierra Morena, and Central System) are in the region of Castilla-La Mancha where altitudes can range between 1000 and 1500 m a.s.l. Finally, the third orographic group includes the valleys of the main rivers that drain the territory of the autonomous community. The largest area of this group is associated with flood plains and adjacent areas of the Tagus and the Guadiana Rivers and their main tributaries.

Castilla-La Mancha has a Mediterranean climate, although the special latitudinal and geographical configuration of the region makes it very continentalized [31]. The weather is characterized by the seasonality of its temperatures, a clear period of summer drought (usually very severe in both duration and intensity), and the irregularity of annual rainfall. The distribution of precipitation is conditioned by the sequence of anticyclonic or cyclonic conditions throughout the year. The weather is anticyclonic in summer and winter (warm and cold anticyclonic) and cyclonic in autumn and spring equinoxes. In terms of annual rainfall [32], the total volume is 41,000 hm^3, which is equivalent to an average of 510 mm m^{-2} $year^{-1}$, with maximum values associated with the Tagus and the Guadalquivir river basins (590 mm m^{-2} $year^{-1}$) and a minimum value for the Segura river basin (410 mm m^{-2} $year^{-1}$). Rainy days can vary between 53 and 78 days, with very few days of snow except in the northeast quadrant, which clearly stands out from the rest (19 days). Storms, however, affect the entire territory in a similar way (between 15 and 25 days a year).

The waters of the autonomous community of Castilla-La Mancha are distributed among eight hydrographic basins of the Iberian Peninsula. The most important are the Tagus, the Guadiana, and the Júcar basins, while the area covered by the remaining five can be considered marginal. In general, the main rivers and their tributaries have the usual characteristics of Mediterranean rivers with a very low water level in summer, a maximum in spring, a secondary maximum in autumn, and a secondary minimum in winter. Only the rivers with their headwaters in the central system have a nivo-pluvial supply; most are supplied with rainwater, often through large avenues. This pattern translates into an annual distribution of floods that is concentrated from September to April.

2.2. Social Description

The autonomous community of Castilla-La Mancha is administratively divided into five provinces in which approximately 1.8 million inhabitants live (with an average population density of 22.25 inhabitants per km^2). These five provinces are subdivided into 919 municipalities with 1487 population centers. The largest population center of the autonomous community corresponds to the city of Albacete (capital of the homonymous province) with 149,000 inhabitants. Next are other provincial capitals (Ciudad Real, Cuenca, Guadalajara, and Toledo, also with names homonymous to those of their provinces) and the city of Talavera de la Reina (the second largest population in the autonomous community). All of them have populations that fluctuate between 50,000 and 75,000 inhabitants. In only 54 of the remaining 913 municipalities does the number of inhabitants exceed 5000. These data show that the population is spread out over a large number of small population centers. Thus, the average population values per municipality stands at around 2000 inhabitants, while the average inhabitants per population centers would be reduced to 1200.

Population distribution by age and gender is very similar in all the provinces, and the population pyramid reaches a maximum (independent of sex) between the ages of 40 and 60 with a second peak in the age range of 80 and 90. While the male population is slightly higher in the young and the middle-aged groups, the trend is reversed for the older population in which female inhabitants dominate.

This trend in population distribution has been preferentially observed in the most populated municipalities. However, smaller municipalities (those with a total population of less than 2000 inhabitants) do not fit very well into the age distribution pattern noted above. In these small urban centers, population distribution remains stable throughout all of the age ranges used (five year intervals), and the maxima, which is poorly defined, moves upward, which points us towards aging populations. This situation is important and should be taken into account in a flood risk analysis with a social approach, as older population groups may be more vulnerable. To conclude, the population scenario of the autonomous community of Castilla-La Mancha indicates that 635 of its 919 municipalities have a population of less than 1000 inhabitants (representing 69% of municipalities) in which aging is more evident.

Finally, from a socio-economic point of view, services are considered the main productive sector of the autonomous community, accounting for almost 60% of its gross domestic product (GDP). This sector is followed by industry (23% of GDP), whose major activities include energy production, agro-food industry, wood and furniture manufacturing, leather and footwear, production of non-metallic minerals, and oil refining. Finally, the influence that the agricultural sector has on the autonomous community (7.5% of GDP), where rain-fed cereals, vines, and olive trees are the main crops, cannot be ignored.

3. Data Sources and Methodologies

3.1. The Flood Risk Maps of the PRICAM Project (MCA and GIS)

The RICAM project [33] was developed between 2005 and 2007 to comply with the Spanish civil protection legislation in terms of risk prevention against floods in the national territory [34]. Specifically, the project formed the basis of knowledge for the development of the Special Plan of Civil Protection against the Flood Hazard of the autonomous community of Castilla-La Mancha (PRICAM, level II plan, regional scope) for which the autonomous community itself is responsible.

The objective of the PRICAM project was to determine the social risk of floods and the categorization of the different population centers of Castilla-La Mancha. These administrative entities were therefore established as units of analysis in the project. PRICAM was initially a flood risk study that shared characteristics associated with macro- and meso-scale studies [2] on the 79,000 km^2 extension of the autonomous community of Castilla-La Mancha. The PRICAM project considered floods that were produced from the typologies listed below:

- River flooding,
- In situ rainfall flooding,
- Dam failure flooding.

For flood risk analysis, MCA techniques supported in a GIS environment were used. As stated by Meyer et al. [35], the use of spatial MCA techniques began to spread in different fields of study from the last quarter of the 20th century and the first years of the 21st century, including the analysis of flood risk [35–38]. This analysis approach could be classified into the parametric approaches to the study of flood risks [39–41]. One of the main factors in the spatial MCA analysis is the allocation of weights for the different variables used. In this case, a simplified variant of the so-called Delphi method consensus was applied [42,43]. Simplification consisted of eliminating the second circulation of the results of the first round of surveys, considering it inoperative with adequate convergence of answers in the first round. The importance assigned by the experts to each of the variables was therefore considered in the MCA model (Table 1), reducing the subjectivity of the analysis and the results obtained [44].

Table 1. List of variables used in PRICAM project and associated values of DELPHI analysis.

Variable	Mean Value	Skewness	Kurtosis
Flooding probability: Qt to Qb ratio	3.5;4.2;4.9	−1.30	0.72
Time of concentration (Tc)	3.9;4.5;5.0	−0.48	−1.65
Distance to river reach	2.9;3.6;4.3	−0.25	−0.90
Urban center–river reach elevation difference	3.1;3.9;4.7	−1.29	2.10
Lithology and geomorphology sediments	3.4;4.0;4.6	−0.48	−1.65
Terrain slope	3.6;4.1;4.6	−0.26	−1.12
Terrain morphology (concave or convexity)	3.7;4.2;4.7	−0.44	−1.23
24 h rainfall	3.9;4.4;4.8	−0.69	−0.25
Dam or reservoir class (volume)	2.5;3.2;3.9	−0.73	1.03
Downstream distance from reservoir	3.2;4.0;4.8	−1.85	4.13
N° of historical flood events	4.0;4.5;4.9	−0.69	−0.25
Historical vs actual reservoirs scenario	3.2;3.8;4.4	−0.67	−0.10
Toxic of danger materials industries	3.4;4.1;4.8	−1.07	0.37
Industrial estate	2.6;3.5;4.3	−0.04	−1.65
Total population (number)	3.8;4.5;5.1	−2.05	4.83
Population clustering by age	2.8;3.4;3.9	−0.85	−0.76
Unemployment index	1.2;1.8;2.4	0.57	−0.86
Presence of educational institutions	2.5;3.1;3.7	0.71	0.53
Presence of hospital centers	3.4;3.9;4.5	−1.06	1.93
Housing type	3.0;3.9;4.8	−1.25	1.26
Under 6 year population ratio	3.6;4.2;4.8	−1.24	1.75
Up 65 year population ratio	3.8;4.4;4.9	−1.58	2.78
Presence of people with disabilities	3.7;4.3;4.8	−1.54	2.90
Knowledge of the local language ratio	2.0;2.8;3.6	0.20	−0.81
Presence of sick or convalescent people	3.5;4.1;4.7	−0.98	0.93
Degree of buildings accessibility	2.6;3.2;3.8	−0.35	−1.45
State of buildings preservation	2.9;3.6;4.3	−0.27	−0.65
N° of stores above ground surface	3.5;4.1;4.7	−0.76	0.16
N° of stores below ground surface	4.3;4.7;5.2	−1.64	1.13
Distance to roads or evacuation paths	2.8;3.5;4.3	−0.51	−0.92
Distance to urban centers	3.5;4.0;4.5	−1.62	5.50
Existence of defined evacuation paths	4.4;4.7;5.0	−0.81	−1.65

Within the spatial MCA, up to 44 variables were evaluated (17 associated with flood hazard, 11 with flood exposure, and 16 with flood vulnerability), as can be seen in Figure 2. The sources of information and data for the estimation of the different variables considered within the MCA were mostly official. This information was in both vector and raster format. The topography came from raster datasets with a spatial resolution of 100 × 100 m. Meteorological variables were contained from raster datasets with a resolution of 500 × 500 m [45]. Furthermore, geology, lithology, and geomorphology data came from two cartographies at a scale of 1:200 k and 1:400 k, both in vector format [46]. All socio-economic information had its origin in the Ministry of Public Administration of the Spanish Government. All datasets were supported by a GeoDatabase in ArcGIS (ESRI Geosystems).

Results for each criterion (variable), factor, and total integrated risk were obtained from the MCA and the whole of Castilla-La Mancha. The scales were quantitative and qualitative (normally with values between one and five) or three classes (low, medium, and high). In the case of the integrated total risk map, the scale of the Basic Directive for Civil Protection against Flood Hazards [34] was used, which established classes A1, A2, A3, B, and C (Table 2). The spatial units for estimating the flood risk were the 1487 population centers of Castilla-La Mancha, to which we added the protected natural areas and the campsites. Categorization of population centers into risk classes based on the statistical analysis of flood risk values (flood frequency histogram) and binary graphs (risk vs. hazard) resulted in the interval classes reflected in Figure 3.

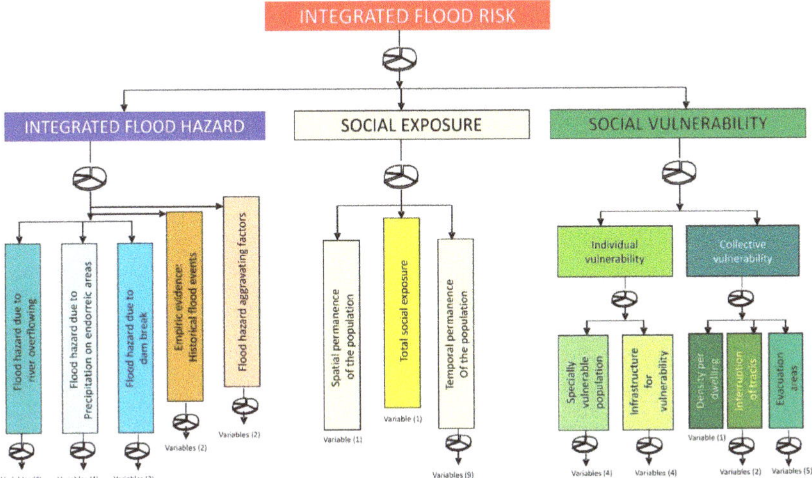

Figure 2. Diagram of the main variable groups used for the flood risk analysis of PRICAM project.

The results of the spatial MCA process allowed us to differentiate those population centers with a greater potential social flood risk. Thus, of the 1489 entities analyzed, 49 (3.4%) of them were assigned a maximum risk (A1), 371 (24.9%) were considered high risk (category A2), 605 (40.6%) were medium risk (A3), 417 (28%) were low risk (B), and 47 (3.1%) were assigned a very low risk (C).

Table 2. Flood risk categories based on the Basic Directive for Civil Protection against Flood Hazards.

Risk Level	Description
A1	High Frequency—High Flood Risk; 50-year flood produces significant damage to urban center
A2	Medium Frequency—High Flood Risk; 100-year flood produces significant damage to urban center
A3	Low Frequency—High Flood Risk; 500-year flood produces significant damage to urban center
B	Medium Flood Risk; 100-year flood produces significant damage outside the urban center
C	Low Flood Risk; 500-year flood produces significant damage outside the urban center

Based on the results from the flood risk analysis, a proposal was made for population centers to identify hot-spots with the objective of prioritizing the development of local flood risk mitigation plans. It should be kept in mind that there will always be a priority for action by those population centers classified within risk groups A1 and A2. However, based on proximity and spatial concentration, five areas were delimited (Figure 4):

- The so-called "La Mancha" area,
- Southeast area of the province of Albacete,
- East Zone of the province of Guadalajara,
- Axis of the Tagus River in the province of Toledo,
- Areas surrounding the city of Cuenca.

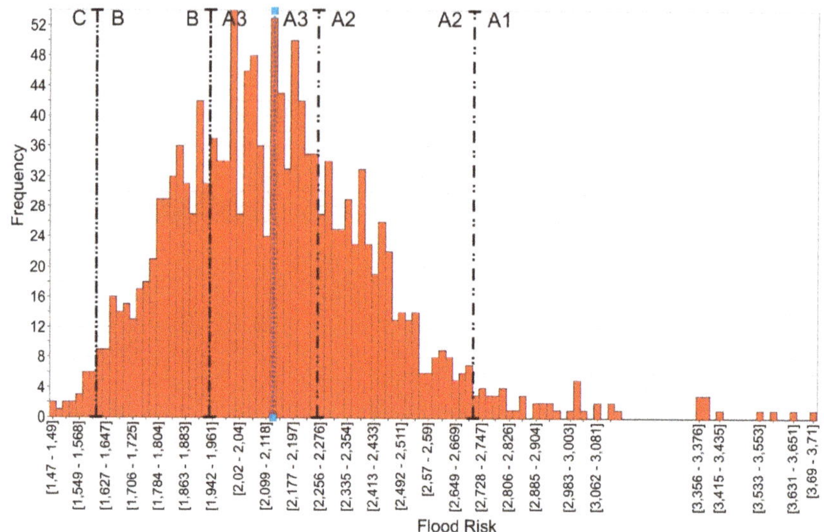

Figure 3. Flood risk values histogram used for risk class thresholds delimitation into the flood risk analysis of the PRICAM project.

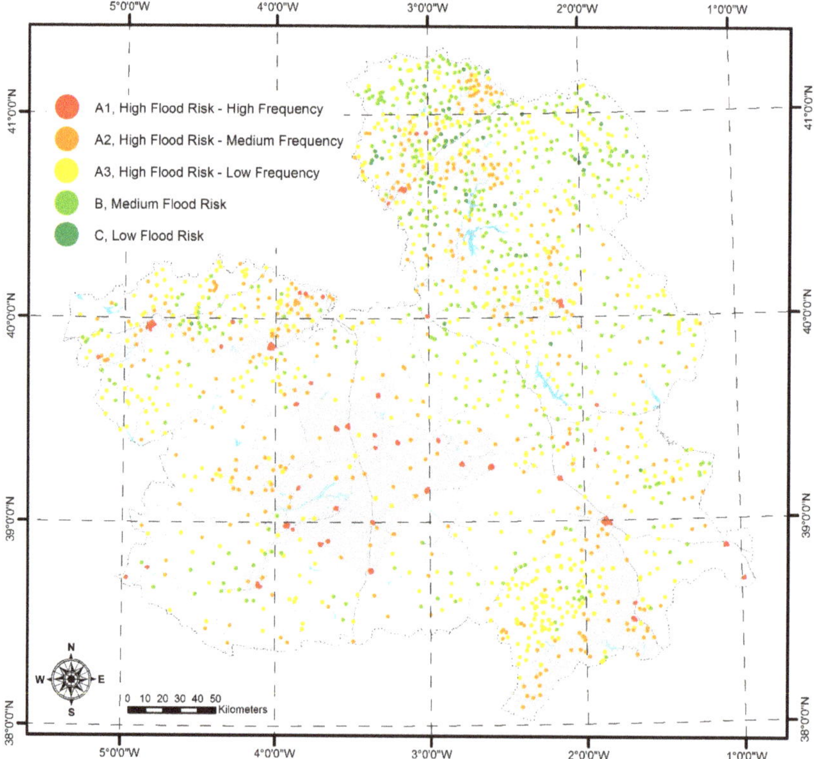

Figure 4. Urban centers flood risk classification for autonomous community of Castilla-La Mancha. Grey shadow areas show the flood risk high values concentration areas.

Another important result was the high flood risk level of the main population centers of the autonomous community of Castilla-La Mancha. Out of the 62 populations that exceed the 5000 inhabitants threshold, almost 70% were classified with a maximum risk level (A1).

3.2. The Quality Assessment of Flood Risk Analysis (Phone Calls and LOESS)

As stated in the Introduction, the quality assessment of risk analysis and cartography as a potential situation is not easy. In the present study, we opted for the use of telephone calls to the Emergency Services during the seven years after the risk analysis ended (2007–2013). The data were analyzed using advanced statistical methods.

The set of baseline data used in this study was that provided by the Public Emergency Service 112 of the Castilla-La Mancha region. The service's mission is to provide effective and coordinated help through the emergency number 112 in situations of emergency that may endanger people's lives, the environment, property, rights, and the heritage of this region. The information contained in the records of the 112 Service is very basic for each phone call: identifier, date and time, municipality and province, and cause of the emergency (fire, flood, etc.). The dataset used corresponded to the calls received (only related to flooding processes) in the seven years after the completion of the risk analysis (year 2007), which amounted to around 17,000 phone calls distributed as shown in Table 3.

Table 3. Summary of phone calls received by the Emergency Service 112 of Castilla-La Mancha regarding floods.

Year	Semester	N° Phone Calls
2007	1	1943
2007	2	1101
2008	1	1062
2008	2	1730
2009	1	817
2009	2	2190
2010	1	1448
2010	2	1500
2011	1	1239
2011	2	584
2012	1	586
2012	2	951
2013	1	740
2013	2	825
Total:		**16,828**

The statistical function LOESS [47,48] consists of a locally weighted regression to moderate a set of baseline data. The function attempts to smooth data to generate scatter diagrams.

The purpose of local regression is to create a trend line starting with a point cloud and taking into account the vicinity of the data to minimize the error. The algorithm is iterative and does not generate a function. It has the following parameters: bandwidth (h), which acts as a smoothing coefficient, and the polynomial degree used in interpolation. When performing a local regression, a parametric family is used as in a global regression adjustment, but the adjustment is only performed locally.

In practice, certain assumptions are made: (i) regression, continuity, and derivability are assumed, thus it can be well approximated by polynomials of a certain degree; (ii) for the variability of Y around the curve, for example, a constant variability is assumed.

The estimation methods that result from this type of models are relatively simple: (i) for each point x, an environment is defined; (ii) within that environment, we assume that a regressor function is approximated by some member of the parametric family that would be quadratic polynomials; (iii) the parameters are then estimated using contour data; (iv) the local adjustment is that of the adjusted function evaluated at x.

Generally, a weight function, W, is incorporated to give greater weight to the x_i values that are close to x. Estimation criteria depend on the assumptions made about the distribution of the Y's. If, for example, we assume that they have a Gaussian distribution with constant variance, it makes sense to use the least squares method.

The point cloud for which one wants to use LOESS is expressed as:

$$A = \begin{bmatrix} x_0 & y_0 \\ \dots & \dots \\ x_n & y_n \end{bmatrix}$$

For example, use a third-degree polynomial that is expressed as:

$$u(x, x_i) = a_0 + a_1(x_i - x) + \frac{1}{2}a_2(x_i - x)^2 + \frac{1}{6}a_3(x_i - x)^3$$

In general, a polynomial of degree p can be expressed as:

$$u(x, x_i) = \sum_{k=0}^{p} a_k \frac{(x_k - x)^k}{k!}$$

where a_k is the coefficient that must be solved to interpolate the value of x through local regression. To calculate the value of the coefficient a_k, the following formula can be used:

$$\begin{bmatrix} a_0 \\ \dots \\ a_p \end{bmatrix} = \left(X^T W X\right)^{-1} X^T W Y$$

in which X is the design matrix, where:

$$X = \begin{bmatrix} 1 & \dots & \frac{(x_0 - x)^n}{n!} \\ \vdots & \ddots & \vdots \\ 1 & \dots & \frac{(x_n - x)^n}{n!} \end{bmatrix}$$

In the case of the third-degree polynomial that is being used as an example:

$$X = \begin{bmatrix} 1 & x_0 - x & \frac{(x_0 - x)^2}{2} & \frac{(x_0 - x)^3}{6} \\ 1 & x_1 - x & \frac{(x_1 - x)^2}{2} & \frac{(x_1 - x)^3}{6} \\ \vdots & \vdots & \vdots & \vdots \\ 1 & x_n - x & \frac{(x_n - x)^2}{2} & \frac{(x_n - x)^3}{6} \end{bmatrix}$$

W is a diagonal matrix where each value assigns a weight:

$$w(s) = \begin{cases} \left(1 - /s/^3\right)^3, & -1 \leq x \leq 1 \\ 0, & -1 > x > 1 \end{cases}$$

$$W = \begin{bmatrix} w\left(\frac{x_0 - x}{h}\right) & \dots & 0 \\ \vdots & \ddots & \vdots \\ 0 & \dots & w\left(\frac{x_0 - x}{h}\right) \end{bmatrix}$$

The matrix Y simply represents the y coordinates of the input data:

$$Y = \begin{bmatrix} y_0 \\ \vdots \\ y_n \end{bmatrix}$$

The LOESS model in this study (source code for LOESS model is available as a Supplementary File) was applied using the R working environment. R is a free software environment for statistical computing that is widely used in academic and scientific fields (i.e., [49]) and which was created by Robert Gentleman and John Chambers in the Bell laboratories [50]. The degree of adjustment between the risk values and the number of phone calls defined both the statistical capacity of simulating the phenomena of the LOESS model and its usefulness for the validation of the analysis and the flood risk cartographies. Additionally, a positive correlation between the risk level of each municipality (according to MCA) and the number of calls to the Emergency Services (112) would show the usefulness of the analysis of 112 calls as a method for calibrating potential risk cartographies.

4. Results: Correlation Risk Categories vs. Number of Phone Calls

The combination of information and data sources used in this study allowed us firstly to describe the relationship between these variables as flood risk–population–emergency calls. As would be expected (and as shown in Figure 5), a positive relationship could be established between the variables considered both in the flood risk–emergency calls relationship and in the population–emergency calls relationship. However, even if there was a positive (and expected) relationship between these variables, the presence of values (corresponding to municipalities) that did not fit perfectly in this relation was observed. Those populations stood out that seemed to present a number of calls greater than what was expected for their assigned risk and population.

In addition to this descriptive analysis, the LOESS model also provided a detailed analysis of the relationship between the pairs of variables considered. This analysis allowed us to relate the flood risk value established from the MCA analysis (PRICAM project) with respect to the inferred risk value established from the LOESS statistical model. When applying the LOESS model using R, a table was obtained in which, for each municipality, a new risk value was inferred using the population of each locality as a major reason for the adjustment. In this table, one could see the municipalities for which there was a greater difference between the categorized risk and the inferred risk. A municipality appearing above or below the LOESS trend line in the LOESS graph indicated the need for a categorized risk review. The points above the trend line indicated that the categorized risk was greater than the inferred one, and the opposite situation was indicated for those that were below it.

Although the regression analysis gave good results, the presence of municipalities with a very high number of records (phone calls) controlled the shape of the local regression function in such a way that the municipalities with the highest number of calls presented extreme risk values that were much higher than the values from the PRICAM model. The highest value of the inferred risk obtained through the local regression function was 4.96, whereas it was only 3.64 from the PRICAM model.

A non-systematic search of incidents and information (from newspapers, reports, and field visits) on these municipalities showed that they were places in which episodes of floods were periodically recorded with important consequences for people and assets. Thus, high flood risk values were justified. It could be concluded that these inferred flood risk values should not be understood as a new flood risk assignment but as a starting point for the review of the PRICAM model flood risk database and its cartography on a risk map.

The inferred flood risk value for municipalities with a lower number of phone calls was very similar to those collected in the PRICAM. For municipalities with a record of incidences below 300 calls, the LOESS graph had a very high positive slope, thus it was understood that the inferred risk grew according to the calls, which indicated a good correlation between the calls and the value of the risk.

An inferred flood risk value lower than the categorized risk (PRICAM model) could also be used as a starting point in the PRICAM update process, since it indicated that the consequences of floods may have been neglected in these urban centers.

As the PRICAM model classified each urban center into one of the five risk categories used by Civil Protection regulations (A1, A2, A3, B, and C), a local regression trend could be established for each class using LOESS. In this way, the degree of correlation between the variables (emergency phone calls and flood risk) could be inferred for each level of risk considered (Figure 6). Figure 6 shows that the best correlation was established for the urban centers of flood risk group A1, since the higher the number of calls corresponded to higher risk value. In any of the other groups, the slope of the graph remained mainly constant, indicating a low correlation between the number of calls and the risk, which may have also indicated that the number of flood events collected was insufficient to establish the relationship between phone calls and risk.

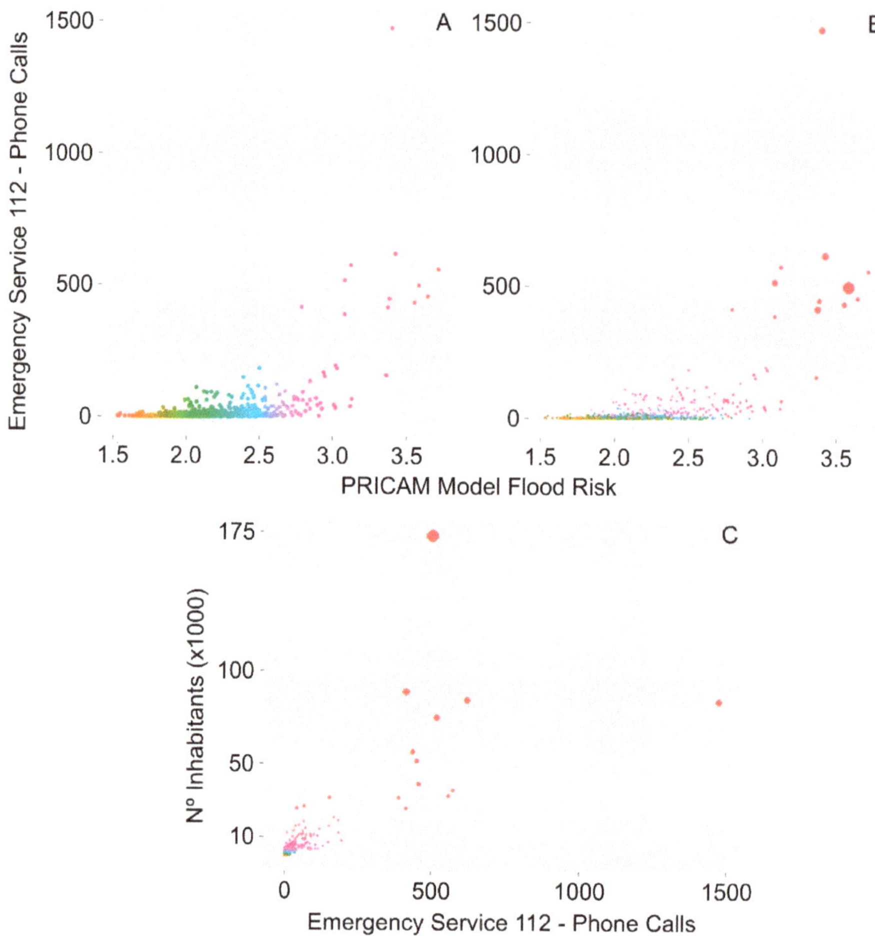

Figure 5. Emergency Service 112 (phone calls) vs. PRICAM model flood risk values (**A**), with color dots size (**B**) showing urban centers population size. N° of Inhabitants vs. Emergency Service 112 phone calls (**C**), with color dots size showing urban centers population size.

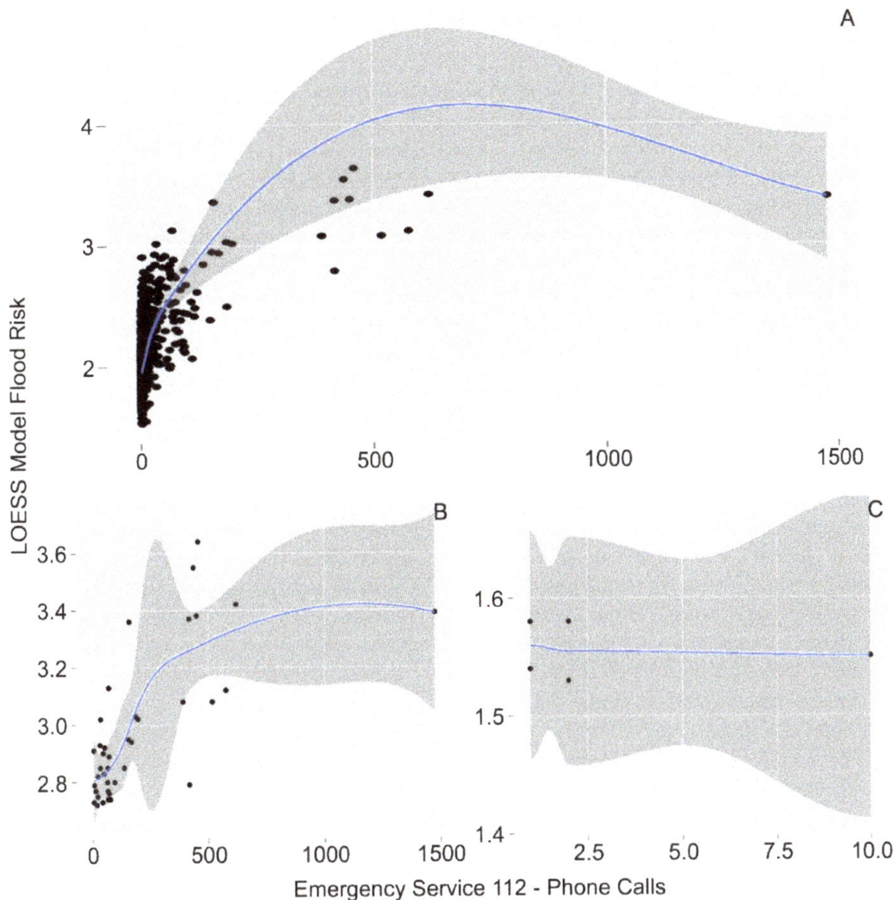

Figure 6. Locally estimated scatterplot smoothing (LOESS) model fits for flood risk values vs. phone calls number (Emergency Service 112) for all flood risk classes (**A**), flood risk class A (**B**), and flood risk class C (**C**). Flood risk classes are from PRICAM project results.

For the regression trend by flood risk level, the local regression function no longer generated inferred risk values greater than those contemplated in the PRICAM model and even showed values below these.

For the municipalities of flood risk classes B and C, there were two situations to consider— the number of urban centers in these classes was very small, and the number of phone calls was also scarce. In these cases, LOESS assigned all municipalities a very similar risk value (which occurred in class C), or the dispersion between the value of the risk categorized and the one inferred was very small, thus there was hardly any difference.

5. Discussion

5.1. Data Sources and Methodological Limitations

The methodological proposal developed for the analysis of flood risk in the civil protection plan of Castilla-La Mancha solved, in an imaginative or an innovative way, the initial limitations and the difficulties derived from the limited availability of both cartographic and alphanumeric data. This point limited the variables that could be considered and the conditions implemented using the Delphi Fuzzy method for selection and weighting. Therefore, for its application to other spatial areas, the variables that could be used should be reconsidered by adapting them to the particularities of the study area.

Flood risk analyses that present a macro-scale approach generally imply some simplification in the analysis model, either in the hazard estimation [2] or the exposed elements [51], in order to make their development viable. In the potential flood risk PRICAM analysis model, floods affect the population as a whole, not subsets of it. This is due to not knowing the extension of the water surface and because the spatial resolution of socio-economic information would not allow it.

Therefore, the results derived from the PRICAM project would define a maximum potential flood risk for urban centers, which could differ (being generally greater) from the real flood risk obtained from micro-scale approaches [4–6,52]. However, according to the main objective of the study (potential flood risk ranking of urban centers as a criterion for the development of their local plan against flood risk), the deviations in the estimation of the real flood risks due to analysis simplification should not condition the validity of the analysis [2].

Beyond these limitations and simplifications, any risk analysis implies a certain subjectivity that depends on the method chosen or the variables analyzed. In the PRICAM project, a combination of MCA and GIS was chosen for flood risk analysis [53]. The choice of this model was based on considerations such as the type and the quality of the data available, the extent of the study area, or the type of result expected by the model. This choice was made despite the fact that, in most case studies, the MCA-GIS combination has been used more to evaluate flood risk mitigation measures than for risk mapping itself [35]. Perhaps the main source of subjectivity in the results obtained from an MCA is associated with the choice and the weighting of the variables used [35,54]. To limit this subjectivity, a Delphi Fuzzy model was used in such a way that the criteria that governed the MCA were based on the consensus among a national and multidisciplinary team of flood risk experts and the fact that the convergence of the experts' criterion was reasonable.

The level of aggregation of socio-economic data can condition flood risk assessments (through its exposure and vulnerability components) based on MCA models and can thus bring about the assignment of high-risk levels that are associated with main urban areas. This effect can be found fundamentally when aggregation reaches municipality or urban center levels. This relationship would be based on considerations such as:

- High exposure linked to the total number of people in the urban center,
- High exposure linked to population density,
- High exposure linked to the presence of educational institutions, tourist camps, nursing homes, hospitals, or industrial estates,
- High vulnerability linked to the presence of children,
- High vulnerability linked to the presence of patients or convalescents (hospitals),
- Most of these major urban areas are historically developed close to rivers, thus the hazard factor (according to the characteristics of the model used) will also show a high value.

The situation described above could be considered as a bias in the results obtained. The clearest example of this effect is found if we analyze those urban centers classified as an A1 flood risk. Out of the total 1489 urban centers, only 49 reached that level of risk, and 43 of these 49 urban centers could be considered as "major towns". Therefore, the following conclusion could be reached—the results of the MCA model of flood risk analysis magnify the level of risk in the main urban centers, underestimating the role of the flooding hazard in the obtained results. However, after a more in-depth analysis, this statement loses its validity. Let us start with one data item; from the 1489 urban centers, only 31 were classified with the maximum level of flood hazard. Out of those 31 populations, only one obtained a flood risk assessment lower than A2. As regards the other 30 urban centers, 13 were included in flood risk level A1, and the remaining 17 were in flood risk level A2. From these results, another (perhaps more realistic) conclusion could be drawn regarding the results of the analysis—the flood hazard played an equal role in the whole flood risk analysis, as can be seen in the figure relating both variables (Figure 7). Those urban centers that presented high values of flood hazard were finally included in the high or the extreme flood risk classes.

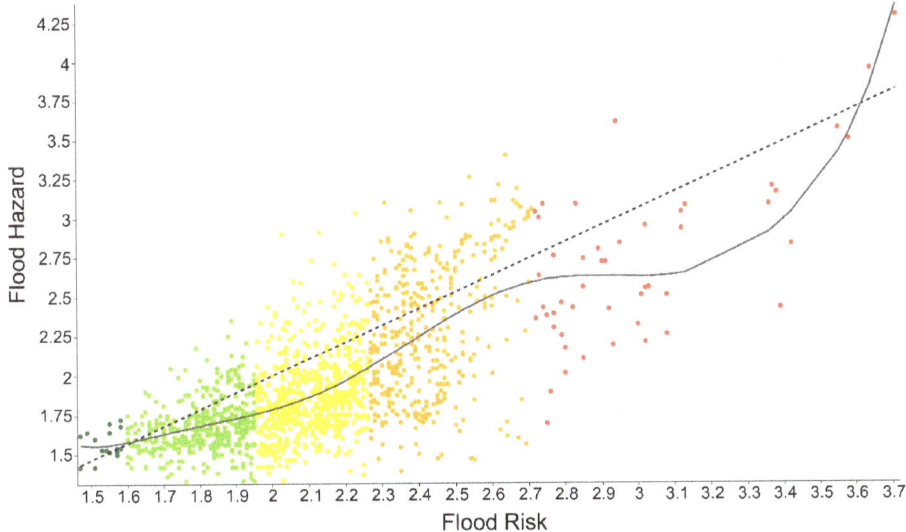

Figure 7. Flood risk vs. flood hazard values from results of the PRICAM project. Dot colors (from dark green to red) show the flood risk class (from C to A1). Dashed line shows the linear regression fit for values. Solid line shows the 10th order polynomial regression fit for flood risk and flood hazard values.

Another point of the PRICAM model with which a certain degree of subjectivity could be associated was the urban center flood risk classification into the Basic Civil Protection Directive flood risk levels of Spain. This classification was based on the opinion of the experts involved in the development of the PRICAM model from descriptive and graphical statistical analyses (Figures 3 and 7) of the results of the project. Similar approaches to the assessment and the management of flood risk have been adopted for the flood risk management policy in countries such as the United Kingdom, Austria, and Germany [55].

Regarding the LOESS function used to approach a calibration method for the analysis of potential flood risk from phone calls to the Emergency Service (112), it was observed that an individual flood risk level analysis was necessary. The conclusion came from visual analysis of the graphs showing the number of calls and the flood risk. When all flood risk levels were included without any discrimination, unexpected results were obtained in such a way that the greater value of the inferred risk (the one obtained through the local regression fit) was much higher than the maximum value of flood risk collected for any urban center in the PRICAM model. Therefore, the error estimated by the regression model was very high.

In a second approach, it was decided to use the flood risk classes of the Basic Civil Protection Directive of Spain—those that differentiated urban centers into A1, A2, A3, B, and C levels. As expected, the largest number of phone calls was concentrated on the first classes, which suffered more flooding events. By focusing on the analysis of these flood risk levels, more accurate analyses were achieved, the local regression trend worked much better, and estimated errors were smaller.

For any of the risk levels, analysis of flood risk validation required filtering the phone calls to the 112 service. This was due to the fact that, in urban centers for which only one or two flood events were registered in the time period analyzed (2007–2013), very few phone calls could imply "noise" that altered the regression model and did not contribute valuable data. In this sense, as proposed by

Calle et al. [49] for the generation of longitudinal profiles of rivers and the adjustment of historical precipitation trends, the application of the LOESS function could limit the "noise" generated by small anomalies in the variable analyzed.

Although the above could be considered as a bias in the spatio-temporal distribution of data associated with 112 phone calls, the magnitude and the importance of this bias were not comparable to those currently associated with the use of citizen or crowd-source data [30,56], which is higher by one or two orders of magnitude. On the other hand, the use of citizen sources is more common in studies focused on natural hazards assessment (disasters detection and real time monitoring), not for integrated risk assessment (including damages and losses). This circumstance may be due to the need for element vulnerability (elements fragility) when we consider the existence of risk. Additionally, in these situations, the population tends to contact those agencies in charge of emergency management more than social or citizen projects and networks.

Moreover, the study time period (seven years) seemed insufficient, since the frequency of flood events in many of the urban centers analyzed was less than the time range covered by the used phone call dataset. However, this methodological approach could be a good starting point to determine which urban centers should be considered for PRICAM model reviews. The criteria for doing the reviews can be considered as follows:

- Urban centers for which the value of the inferred flood risk (LOESS regression trend) is greater than that of the categorized risk (PRICAM model); in these urban centers, flooding events have registered consequences on people and/or assets that result in a high number of phone calls to the Emergency Service 112.
- Urban centers with a value of inferred flood risk lower than the categorized risk; in these situations, there have not been enough flooding events to justify flood risk classifications greater than those inferred by LOESS.

5.2. Future Applications

Due to the impossibility of calibrating and validating the results of an analysis of potential flood risk over large areas due to the innumerous exposed elements and their different vulnerabilities, it is necessary to search for methodologies that allow us to implement post-analysis calibration measures. In order to achieve these goals, the use of information derived from social (or citizen) networks is generalized in the process of calibrating the hazard level (less for risk characterization) posed by different natural phenomena [19–27]. However, it has been observed that the use of data from social networks has its disadvantages [28–30]. From the results of other recent studies [56], a bias in crowd-source data availability and spatial distribution could be denoted, even in projects that enjoyed an important degree of maturity, as in the case of the CrowdHydrology project in the USA. In this sense, the use of information from Emergency Services solves some of these weak points, such as the different degree of implementation and the use of social networks based on age, gender, and location (rural or urban environment). In spite of the above, it is evident that the joint use of official and crowd-source data must be developed and improved in the coming years, and it can and should be the future trend for the validation of natural hazard and risk assessments.

However, data collected from phone calls to Emergency Services are also limited (and require a multi-year period of data collection to be considered significant), as they do not collect information regarding the severity of the flood event, the exposed elements affected, or the degree of damage suffered by them. In this sense and with the aim of achieving the best possible calibration of potential flood risk cartographies, we may have to focus our efforts on the implementation and/or the detailed study of the reports generated by Territorial Intervention Units in the territory with the flood risk, mainly formed in Spain by the Civil Guard (SEPRONA), local police, fire departments, and Civil Protection Units. Thus, reports generated in the field by the intervention units can respond to the uncertainties associated with the severity of the phenomenon or the elements affected by it. The use of this type of information together with local regression models (such as the LOESS model used in the present study) should allow better calibration of the studies of potential flood risk as well as prioritization based on scientific criteria of the more detailed studies (meso- and micro-scale).

5.3. Map Update

Regarding the flood risk analysis, the most important updates and those that should be made more frequently are those related to the determination of exposure and vulnerability of the population. The variables that take into account characteristics of the buildings, as well as some related to the population (e.g., civil status, economic activity) are only updated every ten years in the case of the Spanish official Census of the Population. The variables linked to the basic characteristics of the population (age, sex, nationality, level of studies) may have a much higher update frequency, since they are included in the Municipal Register, which is constantly updated. However, because the record-keeping systems are non-standard, using the 919 available municipal registers seems to be unfeasible from the point of view of methodology and the time needed.

Regarding the variables associated with flood hazard, the optimal update frequency tends to prefer a 10-year review rather than a continuous update. The updating of peak flows (or rainfall) associated with different return periods based on a series of instrumental measures regarding annual maximum peak flows should not show significant variations (for the proposed frequency update), even in the global climate change scenario.

The development or the general acceptance of new techniques or methodologies for the validation and the calibration of flood risk should at least be taken into account for their use in the next scheduled update. In any case, in the field of flood risk analysis at a regional or a national scale, the implementation of a program of regular model updates should be considered as the fourth axis on which the risk analysis is based.

6. Conclusions

A new proposal was presented for the calibration and the validation of flood risk maps at a regional, a national, or even a supra-national scale based on information derived from phone calls to the Emergency Service (112) of the autonomous community of Castilla-La Mancha (Spain) and the use of local regression models such as LOESS. Within this methodological framework, the LOESS model establishes the relationship between the number of calls to the Emergency Service and the categorized flood risk level derived from the PRICAM project.

The results obtained in the study show the usefulness of the data provided by citizen collaboration (calls to the emergency telephone number) in the process of calibration of flood risk analysis and maps. Since this calibration process is not biased by the socio-economic factors of the population, the limiting calibration factor depends on the volume of data available (which are limited by the period of time elapsed after completion of the flood risk analysis).

When the local regression model (LOESS) used the total number of calls to the Emergency Service, it was observed that the model overestimated the potential risk value defined within the PRICAM project, mainly in those populations with the highest number of phone calls. However, these differences between categorized and inferred risk values did not modify the hierarchical order of municipalities according to the level of risk, thus they remained valid in their prioritization. When regression trends were calculated on subsets of calls (defined by the flood risk level of urban centers in the PRICAM model), the differences disappeared between categorized and inferred flood risk; moreover, no change in the layout of urban centers with respect to their flood risk level was observed.

These results highlight two main ideas; on the one hand, the estimation of the flood risk level carried out in the PRICAM project (combining Fuzzy DELPHI, MCA, and GIS techniques) can be considered optimal for a national scale analysis. On the other hand, the application of statistical models of local regression on data collected in the regional or the national Emergency Services will serve to improve the processes of calibration and validation of potential flood risk cartographies. It also allows us to prioritize urban centers for the development of flood risk analysis at regional or local scales.

Supplementary Materials: The following are available online at http://www.mdpi.com/2073-4441/11/6/1284/s1.

Author Contributions: All authors have a significant contribution to final version of the paper. Conceptualization, J.G. and A.D.-H.; Flood risk analysis, J.G. and A.D.-H.; LOESS analysis, I.G.-P. and A.D.-H.; Writing—original draft, J.G. and A.D.-H.; Writing—review & editing, all authors.

Funding: This research was funded by the project DRAINAGE, CGL2017-83546-C3-R (MINEICO/AEI/FEDER, UE), and it is specifically part of the assessments carried out under the coverage of the sub-project DRAINAGE-3-R (CGL2017-83546-C3-3-R).

Acknowledgments: The authors would like to thank the following administrations for their collaboration: the Civil Protection Service, 112, and the General Directorate for Citizen Protection of the Regional Government of Castilla-La Mancha; the Statistical Institute of Castilla-La Mancha; to the Geological Survey of Spain (IGME); and to all the staff of experts who collaborated in the Fuzzy DELPHI analysis. Part of the present manuscript constituted the Master Thesis of the second of the authors, presented in the Master in Evaluation and Management of the Geographic Information Quality of the University of Jaén (Spain).

Conflicts of Interest: The authors declare no conflict of interest.

References

1. Centre for Research on the Epidemiology of Disasters. The International Disaster Database. Available online: http://emdat.be/emdat_db/ (accessed on 10 October 2018).

2. De Moel, H.; Jongman, B.; Kreibich, H.; Merz, B.; Penning-Rowsell, E.; Ward, P.J. Flood risk assessments at different spatial scales. *Mitig. Adapt. Strateg. Glob. Chang.* **2015**, *20*, 865–890. [CrossRef] [PubMed]

3. Sayers, P.; Lamb, R.; Panzeri, M.; Bowman, H.; Hall, J.; Horritt, M.; Penning-Rowsell, E. Believe it or not? The challenge of validating large scale probabilistic risk models. In Proceedings of the FLOODrisk 2016, Lyon, France, 17–21 October 2016.

4. Evans, E.P.; Johnson, P.J.; Green, C.H.; Varsa, E. Risk assessment and programme prioritisation: The Hungary flood study. In Proceedings of the 35th Annual MAFF Conference of River and Coastal Engineers, London, UK, July 2000.

5. Cook, A.; Merwade, V. Effect of topographic data, geometric configuration and modeling approach on flood inundation mapping. *J. Hydrol.* **2009**, *377*, 131–142. [CrossRef]

6. Penning-Rowsell, E.C. A realistic assessment of fluvial and coastal flood risk in England and Wales. *Trans. Inst. Br. Geogr.* **2015**, *40*, 44–61. [CrossRef]

7. Penning-Rowsell, E.C. A 'realist' approach to the extent of flood risk in England and Wales. In *Comprehensive Flood Risk Management: Research for Policy and Practice*; Klijn, F., Schweckendiek, T., Eds.; Taylor and Francis: London, UK, 2013.

8. Dottori, F.; Salamon, P.; Bianchi, A.; Alfieri, L.; Hirpa, F.A.; Feyen, L. Development and evaluation of a framework for global flood hazard mapping. *Adv. Water Resour.* **2016**, *94*, 87–102. [CrossRef]

9. McCallum, I.; Liu, W.; See, L.; Mechler, R.; Keating, A.; Hochrainer-Stigler, S.; Mochizuki, J.; Fritz, S.; Dugar, S.; Arestegui, M.; et al. Technologies to support community flood disaster risk reduction. *Int. J. Disaster Risk Sci.* **2016**, *7*, 198–204. [CrossRef]

10. Pappenberger, F.; Beven, K.; Frodsham, K.; Romanowicz, R.; Matgen, P. Grasping the unavoidable subjectivity in calibration of flood inundation models. A vulnerability weighted approach. *J. Hydrol.* **2007**, *333*, 275–287. [CrossRef]

11. Zischg, A.P.; Mosimann, M.; Bernet, D.B.; Röthlisberger, V. Validation of 2D flood models with insurance claims. *J. Hydrol.* **2018**, *557*, 350–361. [CrossRef]

12. Alfonso, L.; Lobbrecht, A.; Price, R. Using mobile phones to validate models of extreme events. In Proceedings of the 9th International Conference on Hydroinformatics, Tianjin, China, 7–11 September 2010; pp. 1447–1454.

13. Poser, K.; Dransch, D. Volunteered geographic information for disaster management with application to rapid flood damage estimation. *Geomatica* **2010**, *64*, 89–98.

14. Hung, K.C.; Kalantari, M.; Rajabifard, A. Methods for assessing the credibility of volunteered geographic information in flood response: A case study in Brisbane, Australia. *Appl. Geogr.* **2016**, *68*, 37–47. [CrossRef]

15. Le Coz, J.; Patalano, A.; Collins, D.; Guillén, N.F.; García, C.M.; Smart, G.M.; Bind, J.; Chiaverini, A.; Le Boursicaud, R.; Dramais, G.; et al. Crowd-sourced data for flood hydrology: Feedback from recent citizen science projects in Argentina, France and New Zealand. *J. Hydrol.* **2016**, *541*, 766–777. [CrossRef]

16. Rollason, E.; Bracken, L.J.; Hardy, R.J.; Large, A.R.G. The importance of volunteered geographic information for the validation of flood inundation models. *J. Hydrol.* **2018**, *562*, 267–280. [CrossRef]

17. Montargil, F.; Santos, V. Citizen observatories: Concept, opportunities and communication with citizens in the first EU experiences. In *Beyond Bureaucracy: Towards Sustainable Governance Informatisation*; Paulin, A.A., Anthopoulos, L.G., Reddick, C.G., Eds.; Springer International Publishing: Cham, Switzerland, 2017; pp. 167–184.

18. Assumpção, T.H.; Popescu, I.; Jonoski, A.; Solomatine, D.P. Citizen observations contributing to flood modelling: Opportunities and challenges. *Hydrol. Earth Syst. Sci.* **2018**, *22*, 1473–1489. [CrossRef]

19. Schnebele, E.; Cervone, G. Improving remote sensing flood assessment using volunteered geographical data. *Nat. Hazards Earth Syst. Sci.* **2013**, *13*, 669–677. [CrossRef]

20. Triglav-Čekada, M.; Radovan, D. Using volunteered geographical information to map the November 2012 floods in Slovenia. *Nat. Hazards Earth Syst. Sci.* **2013**, *13*, 2753–2762. [CrossRef]

21. Smith, L.; Liang, Q.; James, P.; Lin, W. Assessing the utility of social media as a data source for flood risk management using a real-time modelling framework. *J. Flood Risk Manag.* **2015**, *10*, 370–380. [CrossRef]

22. Cervone, G.; Sava, E.; Huang, Q.; Schnebele, E.; Harrison, J.; Waters, N. Using Twitter for tasking remote-sensing data collection and damage assessment: 2013 Boulder flood case study. *Int. J. Remote Sens.* **2016**, *37*, 100–124. [CrossRef]

23. Li, Z.; Wang, C.; Emrich, C.T.; Guo, D. A novel approach to leveraging social media for rapid flood mapping: A case study of the 2015 South Carolina floods. *Cartogr. Geogr. Inf. Sci.* **2017**, *45*, 97–110. [CrossRef]

24. Starkey, E.; Parkin, G.; Birkinshaw, S.; Large, A.; Quinn, P.; Gibson, C. Demonstrating the value of community based ("citizen science") observations for catchment modelling and characterization. *J. Hydrol.* **2017**, *548*, 801–817. [CrossRef]

25. Kusumo, A.N.L.; Reckien, D.; Verplanke, J. Utilising volunteered geographic information to assess resident's flood evacuation shelters. Case study: Jakarta. *Appl. Geogr.* **2017**, *88*, 174–185. [CrossRef]

26. Kryvasheyeu, Y.; Chen, H.; Obradovich, N.; Moro, E.; Van Hentenryck, P.; Fowler, J.; Cebrian, M. Rapid assessment of disaster damage using social media activity. *Sci. Adv.* **2016**, *2*, e1500779. [CrossRef] [PubMed]

27. Goodchild, M.F.; Glennon, J.A. Crowdsourcing geographic information for disaster response: A research frontier. *Int. J. Digit. Earth* **2010**, *3*, 231–241. [CrossRef]

28. Rosser, J.F.; Leibovici, D.G.; Jackson, M.J. Rapid flood inundation mapping using social media, remote sensing and topographic data. *Nat. Hazards* **2017**, *87*, 103–120. [CrossRef]

29. Goodchild, M.F.; Li, L. Assuring the quality of volunteered geographic information. *Spat. Stat.* **2012**, *1*, 110–120. [CrossRef]

30. Xiao, Y.; Huang, Q.; Wu, K. Understanding social media data for disaster management. *Nat. Hazards* **2015**, *79*, 1663–1679. [CrossRef]

31. Fernández-García, F. *El Clima de la Meseta Meridional. Los Tipos de Tiempo*; Universidad Autónoma de Madrid: Madrid, Spain, 1985; 215p.

32. Instituto Nacional de Meteorología. *Guía Resumida del Clima en España 1971—2000*; INM: Madrid, Spain, 2001.

33. Díez-Herrero, A.; Garrote, J.; Baillo, R.; Laín, L.; Mancebo, M.J.; Pérez-Cerdán, F. Análisis del riesgo de inundación para planes autonómicos de protección civil: RICAM. In *El Estudio y la Gestión de Los Riesgos Geológicos*; Jiménez, I.G., Huerta, L.L., Isidro, M.L., Eds.; Instituto Geológico y Minero de España: Madrid, Spain, 2008; pp. 53–70.

34. Ministerio de Justicia e Interior del Gobierno de España. Directriz básica de planificación de protección civil ante el riesgo de inundaciones. *Boletín Oficial del Estado* **1995**, *38*, 4846–4858.

35. Meyer, V.; Scheuer, S.; Haase, D. A multicriteria approach for flood risk mapping exemplified at the Mulde river, Germany. *Nat. Hazards* **2009**, *48*, 17–39. [CrossRef]

36. Brouwer, R.; van Ek, R. Integrated ecological, economic and social impact assessment of alternative flood control policies in the Netherlands. *Ecol. Econ.* **2004**, *50*, 1–21. [CrossRef]

37. Akter, T.; Simonovic, S.P. Aggregation of fuzzy views of a large number of stakeholders for multiobjective flood management decision-making. *J. Environ. Manag.* **2005**, *77*, 133–143. [CrossRef]

38. Raaijmakers, R.; Krywkow, J.; van der Veen, A. Flood risk perceptions and spatial multi-criteria analysis: An exploratory research for hazard mitigation. *Nat. Hazards* **2008**, *46*, 307–322. [CrossRef]

39. Huang, D.; Zhang, R.; Huo, Z.; Mao, F.; Youhao, E.; Zheng, W. An assessment of multidimensional flood vulnerability at the provincial scale in China based on the DEA method. *Nat. Hazards* **2012**, *64*, 1575–1586. [CrossRef]

40. Li, C.H.; Li, N.; Wu, L.D.; Hu, A.J. A relative vulnerability estimation of flood disaster using data envelopment analysis in the Dongting Lake region of Hunan. *Nat. Hazards Earth Syst. Sci.* **2013**, *13*, 1723–1734. [CrossRef]

41. Balica, S.F.; Popescu, I.; Beevers, L.; Wright, N.G. Parametric and physically based modelling techniques for flood risk and vulnerability assessment: A comparison. *Environ. Model. Softw.* **2013**, *41*, 84–92. [CrossRef]

42. Linstone, H.A.; Turrof, M. *The Delphi Method. Techniques and Applications*; Addison-Wesley Educational Publishers Inc.: Boston, MA, USA, 1975; 621p.

43. Landeta, J. *El Método Delphi. Una Técnica de Previsión Para la Incertidumbre*; Ariel: Barcelona, Spain, 1999; 223p.

44. Nasiri, H.; Yusof, M.J.M.; Ali, T.A.M.; Hussein, M.K.B. District flood vulnerability index: Urban decision-making tool. *Int. J. Environ. Sci. Technol.* **2018**, *16*, 2249–2258. [CrossRef]

45. Ministerio de Fomento. *Máximas Lluvias Diarias en la España Peninsular*; Ministerio de Fomento de España: Madrid, Spain, 1999; 54p, (include software program MaxPluWin).

46. Instituto Geológico y Minero de España. Geological and Geomorphological Map of Spain. Available online: http://info.igme.es/cartografiadigital/portada/default.aspx?mensaje=true (accessed on 20 September 2018).

47. Cleveland, W.S. Robust locally weighted regression and smoothing scatterplots. *J. Am. Stat. Assoc.* **1979**, *74*, 829–836. [CrossRef]

48. Cleveland, W.S.; Devlin, S.J. Locally weighted regression: An approach to regression analysis by local fitting. *J. Am. Stat. Assoc.* **1988**, *83*, 596–610. [CrossRef]

49. Calle, M.; Alho, P.; Benito, G. Channel dynamics and geomorphic resilience in an ephemeral Mediterranean river affected by gravel mining. *Geomorphology* **2017**, *285*, 333–346. [CrossRef]

50. R Development Core Team. *R: A Language and Environment for Statistical Computing*; R Foundation for Statistical Computing: Vienna, Austria, 2008.

51. Merz, B.; Kreibich, H.; Schwarze, R.; Thieken, A. Review article "Assessment of economic flood damage". *Nat. Hazards Earth Syst. Sci.* **2010**, *10*, 1697–1724. [CrossRef]

52. Garrote, J.; Alvarenga, F.M.; Díez-Herrero, A. Quantification of flash flood economic risk using ultra-detailed stage—Damage functions and 2-D hydraulic models. *J. Hydrol.* **2016**, *541*, 611–625. [CrossRef]

53. Malczewski, J. GIS-based multicriteria decision analysis: A survey of the literature. *Int. J. Geogr. Inf. Sci.* **2006**, *20*, 703–726. [CrossRef]

54. Lee, G.; Jun, K.S.; Chung, E.S. Integrated multi-criteria flood vulnerability approach using fuzzy TOPSIS and Delphi technique. *Nat. Hazards Earth Syst. Sci.* **2013**, *13*, 1293–1312. [CrossRef]

55. Thaler, T.; Hartmann, T. Justice and flood risk management: Reflecting on different approaches to distribute and allocate flood risk management in Europe. *Nat. Hazards* **2016**, *83*, 129–147. [CrossRef]

56. Lowry, C.S.; Fienen, M.N.; Hall, D.M.; Stepenuck, K.F. Growing pains of crowdsourced stream stage monitoring using mobile phones: The development of crowdhydrology. *Front. Earth Sci.* **2019**, *7*, 128. [CrossRef]

Article

Analysis of Climate Change's Effect on Flood Risk. Case Study of Reinosa in the Ebro River Basin

Eduardo Lastrada *, Guillermo Cobos and Francisco Javier Torrijo

Department of Geotechnical Engineering, Universitat Politècnica de València, 46022 Valencia, Spain;
gcobosc@trr.upv.es (G.C.); fratorec@trr.upv.es (F.J.T.)
* Correspondence: edlasmar@aaa.upv.es

Received: 27 February 2020; Accepted: 7 April 2020; Published: 14 April 2020

Abstract: Floods are one of the natural hazards that could be most affected by climate change, causing great economic damage and casualties in the world. On December 2019 in Reinosa (Cantabria, Spain), took place one of the worst floods in memory. Implementation of DIRECTIVE 2007/60/EC for the assessment and management of flood risks in Spain enabled the detection of this river basin with a potential significant flood risk via a preliminary flood risk assessment, and flood hazard and flood risk maps were developed. The main objective of this paper is to present a methodology to estimate climate change's effects on flood hazard and flood risk, with Reinosa as the case study. This river basin is affected by the snow phenomenon, even more sensitive to climate change. Using different climate models, regarding a scenario of comparatively high greenhouse gas emissions (RCP8.5), with daily temperature and precipitation data from years 2007–2070, and comparing results in relative terms, flow rate and flood risk variation due to climate change are estimated. In the specific case of Reinosa, the MRI-CGCM3 model shows that climate change will cause a significant increase of potential affected inhabitants and economic damage due to flood risk. This evaluation enables us to define mitigation actions in terms of cost–benefit analysis and prioritize the ones that should be included in flood risk management plans.

Keywords: climate model projections; flood risk; flood hazard; Reinosa; climate change prioritization

---

1. Introduction

Floods are natural hazards that produce great material damage and human losses worldwide [1]. The increasing population density and infrastructure on river banks contribute to increased floodplain vulnerability, which can result in severe social, economic and environmental damage [2].

The fourth of the nine essential rules of flood risk management indicates that it should be taken into account that "The future will be different from the past. Climate and societal change as well as changes in the condition of structures can all profoundly influence flood risk" [3].

For all these reasons, it is essential to design methodologies that enable one to estimate flow regime variations and evaluate flood risk modification due to climate change. [4–6].

The technical document of the IPCC IV forecasts a probable increase in the frequency and intensity of precipitation episodes, as well as a decrease in average values in summer for mid-latitudes countries as Spain. The Fifth Assessment Report (AR5) of the IPCC (2013–14), points out that it is probable that the frequency or intensity of intense rainfall has increased in Europe, and in relation to future changes, that extreme precipitation events over most of the mid-latitude lands will most likely be more intense and more frequent [7]. Additionally, climate change can specially affect those regions in which the snow phenomenon is relevant in hydrological behavior.

As a matter of these facts, the 2019 flash flood European notifications beat the last 7 years of notifications clearly [8], and Reinosa (Cantabria, Spain) suffered two important floods on January and

December 2019, the second one being one of the worst floods in memory. This flood was caused by the Elsa storm, on December 19 and 20, which came from the southwest and was accompanied by very strong winds and very high temperatures for that time of the year, causing the melting of the snow-cover in the highest part of the catchment. A203 gauging station, recorded on 20 December 2019 at 1:00 a.m., 246 m^3/s instant (15 minutes period) flow, while the maximum ever recorded was 100 m^3/s (27 February 2010). The daily flow rate recorded by A203 for this event was 83 m^3/s and the return period has been estimated approximately as Tr = 300 years. This flood caused great damage in Reinosa, affecting residential areas (Figure 1).

Figure 1. Reinosa (Cantabria, Spain). 20th December 2019.

Reinosa (Cantabria, Spain) is located in the north of the Iberian Peninsula, in the junction of Hijar, Ebro and Izarilla rivers, immediately upstream the Ebro reservoir. Hijar catchment, with 27.6 km length and 147.6 km^2 surface, drains the highest area of Cantabrian mountain range that belongs to the Ebro river basin. The catchment height ranges between 865 and 2023 m above sea level, with a mean height of 1351 m. A203 gauging station is located in Hijar River, immediately upstream of Reinosa, with a time of concentration of approximately 15 hours.

The average rainfall of the head of the Ebro for the period 1920–2002 was 1190 mm/year, concentrated in autumn and winter, characteristic of an Atlantic climate, with some continental patterns producing 33 L/s/km^2. The average annual temperature is 8.9 °C in Reinosa. The main land uses in the catchment are forest and scrub. Figure 2 shows Hijar river basin overall and site map.

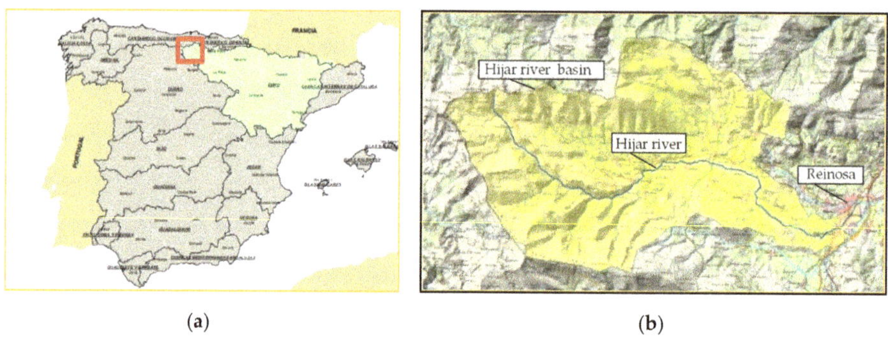

(a) (b)

Figure 2. Hijar river basin. (**a**) Overall Site map. (**b**) Detailed Site map.

Reinosa belongs to the Hijar river basin, and had potential significant flood risk (ES091_HIJ_01_02_04_05_06) during the implementation of DIRECTIVE 2007/60/EC, as shown in Figure 3.

Figure 3. Hijar potential significant flood risk (ES091_HIJ_01_02_04_05_06). http://iber.chebro.es/SitEbro/sitebro.aspx.

DIRECTIVE 2007/60/EC implies an objective quantification of flood risk, for low (500 years return period), medium (100 years return period) and high probabilities (10 years return period), that will allow the efficient development of flood risk management plans by implementing a set of combined actions to reduce floods' consequences [9].

To evaluate different climate change scenarios, the Spanish Meteorological Agency (AEMET) regionalized a set of global climate models, by using two statistical downscaling methods. The usefulness of this regionalization was assessed by their fitting to the observed data in the control period (1961–2000). A comparison based on a set of statistics show that although the fit is good for annual mean values, annual maximum values for both regionalization methods are not adequately simulated, since they provide lower extremes with a smaller variability. However, different fitting was observed depending on the Spanish region [10].

Although results might show a decrease in the magnitude of extreme floods for climate model projections downscaled by AEMET [11], a different pattern might be concluded, when taking into account snowmelt, if the melting flow rate's increase due to higher temperatures is bigger than the decrease caused by less snow accumulation.

The main aim of the work is to estimate flow and flood risk variation due to climate change by using different climate models regarding a scenario of comparatively high greenhouse gas emissions (RCP8.5), with daily temperature and precipitation data.

2. Materials and Methods

2.1. Overall Methodology

According to DIRECTIVE 2007/60/EC, Spanish hazard and flood risk maps can be downloaded at https://sig.mapama.gob.es/snczi/, regarding the following scenarios: (a) floods with a low probability, or extreme event scenarios (return period ≥ 500 years); (b) floods with a medium probability (likely return period ≥ 100 years); (c) floods with a high probability (return period ≥ 10 years).

By relating flow rates in A203 gauging station with hazard and risk maps, correlations with flooding surface, economic damage and casualties can be achieved.

In order to evaluate climate change, projections of a series of climatic variables are entered as input data in a calibrated hydrological model in the study basin. The ASTER®model is a distributed hydrological model that calculates snowmelt and accumulation regarding energy balance [12,13]

Two global climate models (GCM), downscaled by the Spanish Meteorological Agency (AEMET), have been selected; we validated them with real flow rates from the control period (1961 to 2000) and used under the highest greenhouse gas emissions pathway (RCPs 8.5) from the Fifth Assessment Report of the Intergovernmental Panel on Climate Change.

The relationship between calculated daily flow rates and instant (15 minutes period) flow rates is not necessary for these downscaled climate models, as climate change's effect will be calculated in relative terms comparing climate models in the calculation period in two equal length stages (2007–2038) and (2039–2070).

Once the flow rate variation for the climate projection is estimated and applied on real flow, new impacts on hazard and risk maps can be established. Figure 4 shows the overall methodology scheme.

Figure 4. Overall methodology scheme.

2.2. Hazard and Risk Maps

Hazard maps are the result of two-dimensional hydraulic models, indicating flooded area, flood depths and flood velocities. Return periods were given by Spanish Authority in the framework of DIRECTIVE 2007/60/EC implementation by "Maximum Flow Maps" software, based on a regionalized gauging station study [14].

Flood risk maps shall show the potential adverse consequences associated with flood scenarios, and express, among other things, terms of the indicative numbers of inhabitants potentially affected and the type of economic activity of each area potentially affected. Additionally, economic damage can be estimated.

According to DIRECTIVE 2007/60/EC, these maps serve as a tool for establishing priorities and for making additional technical, economic and political decisions related to flood risk management. Measures to be implemented in the at-risk areas can be prioritized, depending on the results of the cost-benefit analysis. They also allow us to evaluate climate change's effects. The risk associated with flood events is established based on the vulnerability of the threatened element and the hazard to which it is exposed. The estimation of indicative number of inhabitants potentially affected depends on updated census data. The Spanish Water Authority has also established a methodology to estimate the economic value of flood damage partially based on the PREEMPT project [15,16].

Finally, a relationship, as shown in Table 1, between flow rates for different return periods in Hijar, A203, and hazard and risk maps can be estimated, and subsequently a trend line, as shown in Figure 5, can be fitted:

Table 1. Instant (15 minute) flow rate, flood surface, inhabitants potentially affected and economic damage [1].

Instant Flow Rate (m³/s)	Return Period (years)	Surface (km²)	Inhabitants	Economic Value (€)
138	10	2.783	346	29,194,901
214	100	3.843	982	56,806,383
273	500	4.620	1484	90,791,074

[1] Obtained for overall ES091_HIJ_01_02_04_05_06 potential significant flood risk area, including the Spanish Public Water Domain (RDPH).

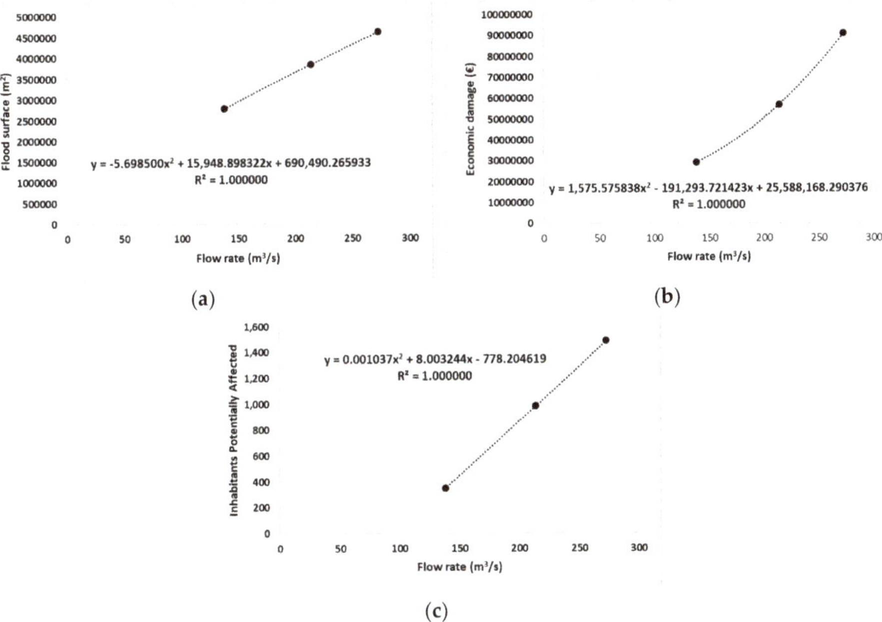

Figure 5. (**a**) Flood surface, (**b**) inhabitants potentially affected and (**c**) economic damage value in terms of flow rate in A203 gauging station.

2.3. Hydrological Model

2.3.1. Hijar Basin Model

A 500 by 500 m grid and daily time resolution distributed hydrological model has been built for the Hijar river basin (Figure 6), with A203 gauging station in Reinosa as the outlet and calibration point, 2014–2016 as the calibration period and 2017–2019 as the validation period.

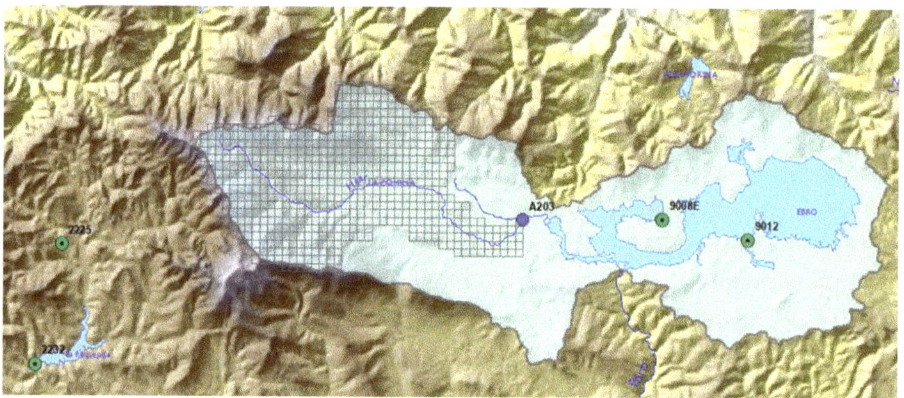

Figure 6. ASTER hydrological model. Hijar river basin mesh (grey grid), Gauging Station (A203), rainfall stations (9008E, 9012) and temperature stations (2225, 2232).

ASTER®is a distributed hydrological model that calculates streamflow in a river using rainfall, temperature, windspeed and radiation as climate inputs. It has been applied in several countries and implemented by Spanish water authorities as an accurate model in river basins with strong snow phenomena, as in our case study (Hijar catchment) [17,18]. Figure 7 shows the snow–water equivalent for two consecutive days in Hijar model.

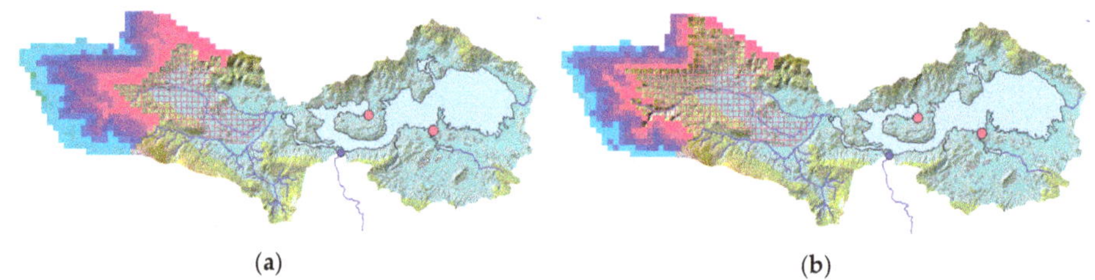

(a)   (b)

Figure 7. ASTER hydrological model snow accumulation (snow–water equivalent) (**a**) Day 1, (**b**) Day 2. Hijar model (ranging from 1 mm–pink to 215 mm–cyan).

2.3.2. Snow Accumulation Routine

Precipitation will be snow or water form at different temperatures depending on several atmospheric conditions. This range of temperatures can vary between −2 and 2 °C. The model sets a temperature, named rain/snow temperature, at which 50% of precipitation is snow, with a bilineal increase and decrease of snow form up to −2 and 2 °C.

2.3.3. Snowmelt Routine

As aforementioned, snowmelt is calculated using the energy balance equation. Melting takes place when air temperature is below a specific temperature (T_{mi}—snowmelt temperature). Snowmelt can be divided into two types: (a) rainfall snowmelt; (b) no rainfall snowmelt.

Rainfall Snowmelt

It is the snowmelt produced by rainfall energy. Assuming that the snow surface temperature is equal to 0 °C (273 K), snowmelt caused by energy from rainfall is [19]:

$$M_r = 0.0125 {\cdot} P {\cdot} f_r {\cdot} T_r, \tag{1}$$

where:

P = Precipitation (mm)
f_r = fraction of precipitation in the form of rain
T_r = temperature of rain (°C)

In addition to energy from rainfall, other secondary terms as radiation and condensation energy must be added to calculate snowmelt when rain takes place.

No Rainfall Snowmelt

It is the snowmelt on a dry day. Due to difficulties estimating wind speed, hydrological models usually use non-forced convection or other empirical relations depending on air temperature. The following equation is used to estimate no rainfall melt M_c (mm) [20]:

$$M_c = M_f {\cdot} (T_a - T_{mi}) \tag{2}$$

where:

M_f = melt factor (mm/°C)
T_a = air temperature (°C)
T_{mi} = snowmelt temperature (°C)

This melt factor (M_f) indicates the seasonal variation, and can be represented with a sine function [21,22].

$$M_f = \frac{\text{MFMAX} + \text{MFMIN}}{2} + \sin\left(\frac{n{*}2\pi}{366}\right){*}\frac{\text{MFMAX} + \text{MFMIN}}{2} \tag{3}$$

where:

MFMAX = maximum melt factor on June 21st (mm/°C/24hrs)
MFMIN = minimum melt factor on December 21st (mm/°C/24hrs)
n = number of days from March 21st.

2.3.4. Model Calibration and Validation

Calibration and validation have been carried out, comparing real daily flow rates, in A203 Hijar in Reinosa Gauging Station with the ASTER model daily flow rates being calculated with real meteorological data. The available continuous period for the observed daily flow rate ranges from 2014 to 2019, defining the calibration period as October 1st 2014 to September 30th 2016 and the validation period from October 1st 2017 to December 22nd 2019, including the aforementioned extraordinary event.

Figure 8, shows results during the calibration period, with a correlation coefficient of 0.90 and a Nash index (NSE) of 0.78. The maximum observed daily flow rate was 23.6 m^3/s, and the maximum calculated daily flow rate was 23.5 m^3/s. Total observed runoff was 189 hm^3 and calculated runoff was 200.7 hm^3, with 27 hm^3 of overall snow–water equivalent accumulation.

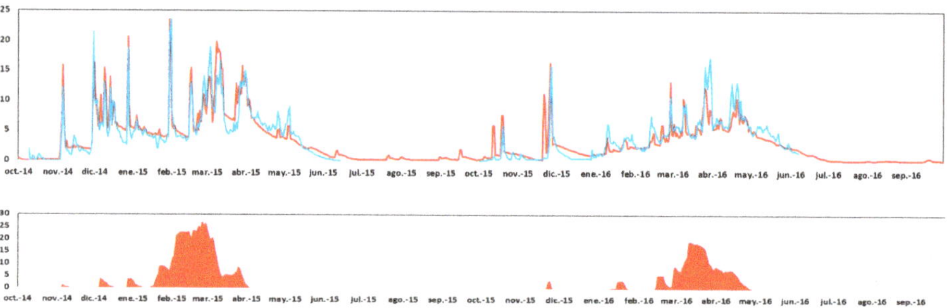

Figure 8. Upper level. Real (blue) and calculated (red) daily flow rates in A203. Lower level. Snow water equivalent in the catchment for the calibration period.

The main calibration parameters of the model are:

- MFMAX = 5.5 (mm/°C/24 h);
- MFMIN = 1.46 (mm/°C/24 h);
- T_{mi} = −0.35 (°C);
- Rain/Snow temperature = −0.53 (°C);
- Altimetry temperature gradient = −3.1 (°C/1000 m).

Figure 9 shows results during the validation period, with a correlation coefficient of 0.90 and a Nash index (NSE) of 0.80. The maximum observed daily flow rate was 83 m^3/s, and the maximum calculated daily flow rate was 69 m^3/s. Total observed runoff was 231 hm^3 and calculated runoff was 289 hm^3, with 20 hm^3 of overall snow–water equivalent accumulation.

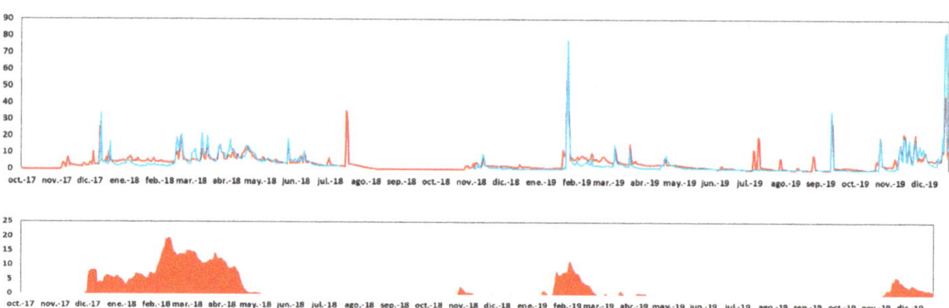

Figure 9. Upper level. Real (blue) and calculated (red) daily flow rates in A203. Lower level. Snow water equivalent in the catchment for the validation period.

2.4. Climate Projection

Daily rainfall and temperature data from two regionalized climate models (analog method—ANA and statistical regionalization method—SDSM) can be obtained from AEMET website (http://www.aemet.es/es/serviciosclimaticos/cambio_climat) for RCP 8.5.

Two climate models have been selected from 24, to analyze the combination of rainfall and temperature:

- ANA—BCC-CSM1-1-m (BCC);
- ANA—MRI-CGCM3 (MRI).

These two models have been pre-selected as generating the highest rainfalls. To evaluate their suitability, daily flow rates for the control period (1961–2000) have been calculated with ASTER for real meteorological data and both climate models. Figure 10a shows that annual maximum real daily flow rates are above climate models, and the trend line shows a higher increase as well. Figure 10b shows that annual mean real daily flow rates fit the MRI better at the beginning of the control period, while a fit better with BCC is shown at the end of the control period. Supplementary Materials for rainfall stations (9008E, 9012) and temperature stations (2225, 2232) is given.

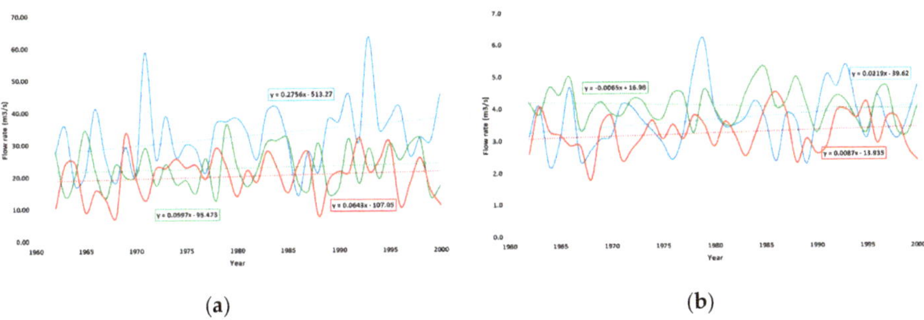

(a)   (b)

Figure 10. Daily flow rate evolution during control period (1961–2000) for real data (blue), BCC (green) and MRI (red). (**a**) Annual maximum, (**b**) annual mean.

2.5. Calculation of Flow Rates for Different Return Periods

Daily flow rates for the calculation period (2007 to 2070) have been estimated with ASTER model for both climate projections (BCC and MRI). Calculation period has been divided in two equal length stages (2007 to 2038) and (2039–2070) and a probability distribution of extreme values has been estimated using the Gumbel distribution, presenting the best fit goodness:

$$X = u - \beta \cdot Ln\left(-Ln\left(\frac{Tr-1}{Tr}\right)\right) \tag{4}$$

where:

u = location parameter;
β = scale parameter;
Tr = Return period.

Finally, daily flow rates and instant (15 minute period) flow rates observed in A203 can be fitted in a linear relationship. As climate projections flow rates are being studied in relative terms with observed flow rates, transformation from daily flow rates into instant flow rates is not necessary.

3. Results

3.1. Flow Rate Evolution (RCP 8.5)

Table 2 and Figure 11 show results and trend lines from percentile analysis for both calculated climate models. For BCC-CSM1-1-m a small decrease in flow rate projection due to climate change is estimated, with small variability for different probabilities, while for MRI-CGCM3 bigger increase is predicted, showing a smooth variability for different probabilities, describing a more extreme climate.

Table 2. Return period, daily flow rate projection, 1st Stage (2007–2038); daily flow rate projection ratio, 2nd Stage (2039–2070); and flow rate ratio (2nd Stage/1st Stage).

ANA—BCC-CSM1-1-m			
Tr (years)	**Flow Rate Projection 2007–2038 (m³/s)**	**Flow Rate Projection 2039–2070 (m³/s)**	**2039–2070/ 2007–2038 [1] (%)**
10	28.13	27.00	96.0%
25	31.92	30.62	95.9%
50	34.73	33.30	95.9%
100	37.52	35.96	95.9%
500	43.96	42.11	95.8%

ANA—MRI-CGCM3			
Tr (years)	**Flow Rate Projection 2007–2038 (m³/s)**	**Flow Rate Projection 2039–2070 (m³/s)**	**2039–2070/ 2007–2038 [1] (%)**
10	27.36	31.05	113.5%
25	31.99	36.51	114.1%
50	35.43	40.57	114.5%
100	38.85	44.59	114.8%

[1] Column 3 divided by column 2.

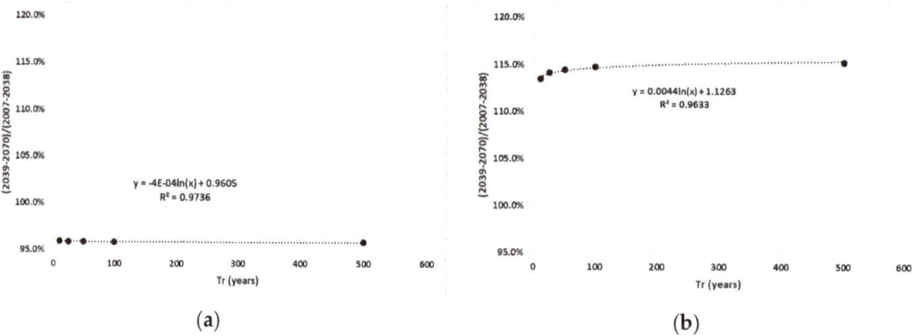

Figure 11. Flow rate projection ratio (2nd Stage (2039–2070)/1st Stage (2007–2038)) trend line in terms of return periods. (a) Analog method (ANA)—BCC-CSM1-1-m; (b) ANA—MRI-CGCM3.

3.2. Hazard and Risk Maps

After calculating the projected flow rate by entering for each return period (2nd Stage (2039–2070)/1st Stage (2007–2038)) the flow rate ratio in the trend line fitted for flood surface, economic damage and inhabitants potentially affected (Figure 5), new climate change estimations show similar results as for flow rate (Table 3).

As for flow rate, the MRI-CGCM3 climate model shows an increase in flood risk effects due to climate change, while BCC-CSM1-1-m shows a decrease. Results are enlarged for economic value and inhabitants potentially affected, while flooded surface decreases.

A logarithmic trend line is the one that best fits, and represents how the atmosphere behaves that shows in Figures 12–14.

Table 3. Projection ratio 2nd Stage (2039–2070)/1st Stage (2007–2038) for the flow rate, flood surface, inhabitants potentially affected and flood damage's economic value.

ANA—BCC-CSM1-1-m

Flow Rate (%)	Instant Flow Rate (m³/s)	Return Period (years)	Flood Surface (%)	Inhabitants (%)	Economic Value (%)
−4.0	132.4	10	−2.9	−13.3	−4.5
−4.1	205.1	100	−3.1	−7.6	−7.3
−4.2	261.5	500	−3.2	−6.6	−8.2

ANA—MRI-CGCM3

Flow Rate (%)	Instant Flow Rate (m³/s)	Return Period (years)	Flood Surface (%)	Inhabitants (%)	Economic Value (%)
13.5	156.7	10	9.6	44.8	17.4
14.8	245.6	100	11.0	27.3	29.7
15.3	314.7	500	11.4	24.2	33.7

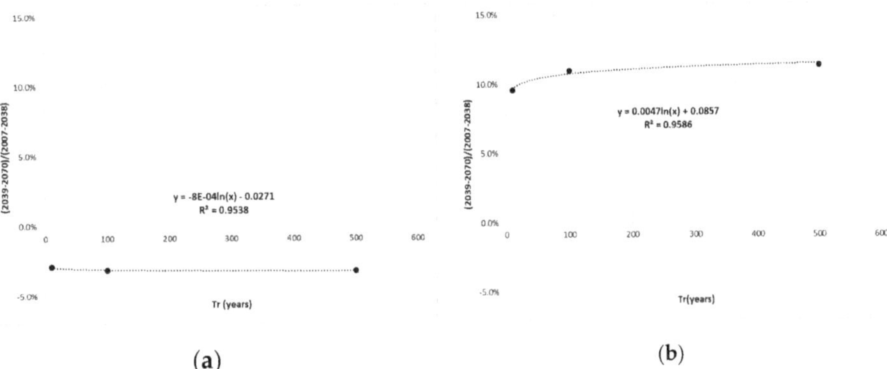

Figure 12. Second Stage (2039–2070)/1st Stage (2007–2038) flood surface ratio trend line in terms of return periods. (**a**) ANA—BCC-CSM1-1-m; (**b**) ANA—MRI-CGCM3.

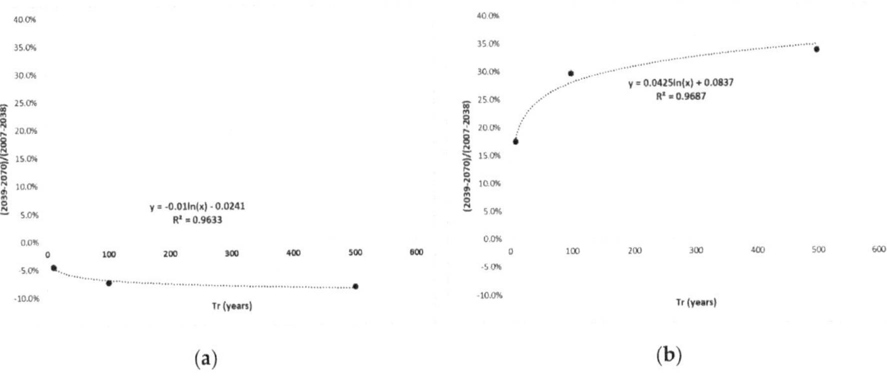

Figure 13. Second Stage (2039–2070)/1st Stage (2007–2038) economic damage ratio trend line in terms of return periods. (**a**) ANA—BCC-CSM1-1-m; (**b**) ANA—MRI-CGCM3.

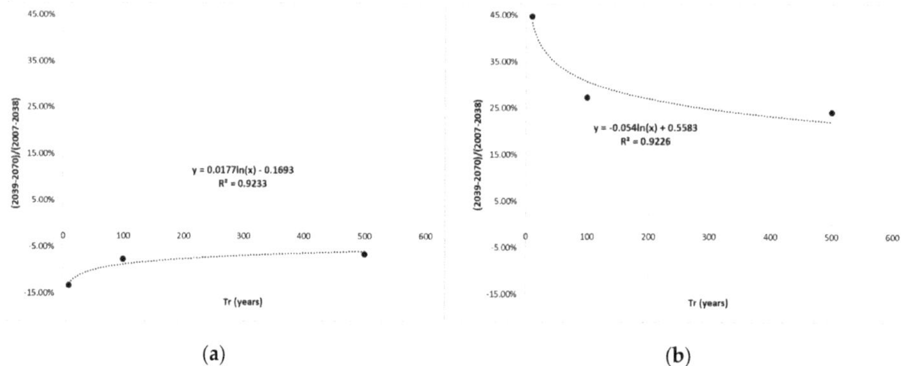

Figure 14. Second Stage (2039–2070)/1st Stage (2007–2038) inhabitants potentially affected ratio trend line in terms of return periods. (**a**) ANA—BCC-CSM1-1-m; (**b**) ANA—MRI-CGCM3.

Flooded surface, inhabitants potentially affected and flood economic damage value can be calculated in absolute terms and are shown in Table 4 for ES091_HIJ_01_02_04_05_06.

Table 4. Absolute terms for the flow rate, flood surface, inhabitants potentially affected and flood damage's economic value for selected climate change projections [1].

ANA—BCC-CSM1-1-m				
Flow Rate (m^3/s)	Return Period (years)	Surface (km^2)	Inhabitants	Economic Value (€)
132.4	10	2.703	300	27,887,407
205.1	100	3.722	907	52,640,944
261.5	500	4.472	1386	83,327,840
ANA—MRI-CGCM3				
Flow Rate (m^3/s)	Return Period (years)	Surface (km^2)	Inhabitants	Economic Value (€)
156.7	10	3.049	501	34,286,399
245.6	100	4.263	1250	73,649,691
314.7	500	5.145	1843	121,426,897

[1] Obtained for overall ES091_HIJ_01_02_04_05_06 potential significant flood risk area, including the Spanish Public Water Domain (RDPH).

4. Discussion

4.1. Climate Change Effect

Comparison between observed meteorological data and projections of climate scenarios, for a common time interval, has revealed a significant difference between the series. This means that the results must in any case be interpreted in relative terms, evaluating trends and not absolute values.

Results for both selected climate models show important differences between each other in absolute and relative terms, indicating that there is significant uncertainty in the projections, especially due to rainfall; and they suggest that for local impact studies, further analysis to regionalize outputs from global climate models still needs to be done. For the results to be reliable, climate models must adequately represent extreme phenomena, as most of them are suitable for evaluating water resource evolution.

There is a large uncertainty in climate projections that influences the estimation of future flood risk. The crucial action in the case of adaptation to floods is to carry out a detailed analysis to choose the best climate model and use results in relative terms between the short term (1st Stage) and long term (2nd Stage) of the projection period [4].

The proposed methodology allows one to determine in terms of hazard and flood risk, the effect that climate change has on a given territory, considering both variable rainfall and temperature, and enables a fast evaluation of various climate models and a comparison of their differences.

This evaluation enables one to define actions in terms of a cost–benefit analysis and prioritize the ones that should be included in flood risk management plans due to climate change, as these management plans must contain effective actions in the short, medium and long term and cannot ignore the effect of climate change.

In the specific case of Híjar River Basin, and Reinosa, MRI-CGCM3 model trend analysis shows that climate change will increase damage to society, both economic and personal, while BCC-CSM1-1-m shows a small decrease. As the MRI-CGCM3 model fits better with annual maximum real daily flow rates than BCC-CSM1-1-m, it can be concluded that climate change will cause a significant increase of potential affected inhabitants and economic damage due to flood risk that will require mitigation actions.

This assumption might match with previous years' observations, especially given the high number of low probability events recorded in Spain. This is the specific case of the Hijar River that in 2019 registered two floods, the one on December 2019 being the largest ever registered in Reinosa Gauging Station and related with a low probability event (300 years return period). This conclusion is consistent with the Fifth Assessment Report (AR5) of the IPCC (2013–14), which points out that it is likely that the frequency or intensity of intense rainfall has increased in Europe.

4.2. Future Research

As already mentioned, this methodology could be applied in every potential significant flood risk area through a water authority flood risk management plan in order to prioritize actions.

Firstly, all available regional climate models could be selected and evaluated. Finally, a sensitive RCP analysis with RCP 4.5 scenario could be carried out.

Supplementary Materials: The following are available online at http://www.mdpi.com/2073-4441/12/4/1114/s1. Spreadsheet1: Climate_Models_Input_Data.xlsm.

Author Contributions: Conceptualization, E.L. and G.C..; methodology, E.L. and G.C.; project administration, G.C.; software, E.L.; validation, G.C.; formal analysis, E.L.; writing—original draft preparation, E.L.; writing—review and editing, E.L. and G.C.; visualization, F.J.T.; supervision, F.J.T. All authors have read and agreed to the published version of the manuscript.

Funding: This research received no external funding.

Acknowledgments: The authors acknowledge ASTER model developer J. A. Collado (SPESA Ingenieria); F. J. Sanchez and M. Aparicio (Spanish Ministry for Ecological Transition and the Demographic Challenge); M. L. Moreno and L. Polanco (Ebro Water Authority); and the Ebro Water Authority.

Conflicts of Interest: The authors declare no conflict of interest.

References

1. EEA. *Mapping the Impacts of Natural Hazards and Technological Accidents in Europe*; EA Technical Report No 13/2010; European Environment Agency: Copenhagen, Denmark, 2010; Available online: http://www.eea.europa.eu/publications/mapping-the-impacts-of-natural (accessed on 10 January 2020).

2. Planes de Gestión del Riesgo de Inundación. Ministerio para la Transición Ecológica y el Reto Demográfico. Available online: https://www.miteco.gob.es/es/agua/temas/gestion-de-los-riesgos-de-inundacion/planes-gestion-riesgos-inundacion/ (accessed on 15 December 2019).

3. Yuanyuan, L.; Sayers, P.; Yuanyuan, L.; Galloway, G.; Penning-Rowsell, E.; Fuxin, S.; Kang, W.; Yiwei, C.; Quesne, T.L.; Asian Development Bank; et al. Flood Risk Management: A Strategic Approach. Available online: https://unesdoc.unesco.org/ark:/48223/pf0000220870 (accessed on 16 January 2020).

4. Doroszkiewicz, J.; Romanowicz, R.J.; Kiczko, A. The Influence of Flow Projection Errors on Flood Hazard Estimates in Future Climate Conditions. *Water* **2019**, *11*, 49. [CrossRef]

5. Zhu, T.; Lund, J.R.; Jenkins, M.W.; Marques, G.F.; Ritzema, R.S. Climate change, urbanization, and optimal long-term floodplain protection. *Water Resour. Res.* **2007**, *43*, 122–127. [CrossRef]

6. Nyaupane, N.; Thakur, B.; Kalra, A.; Ahmad, S. Evaluating Future Flood Scenarios Using CMIP5 Climate Projections. *Water* **2018**, *10*, 1866. [CrossRef]

7. Intergovernmental Panel on Climate Change. Available online: https://archive.ipcc.ch/ (accessed on 20 December 2019).

8. European Flood Awareness System (EFAS). Available online: https://www.efas.eu/en/news/summary-efas-notifications-2019 (accessed on 3 January 2020).

9. European Commission. Directive 2007/60/Ec of the European Parliament and of the Council of 23 October 2007 on the assessment and management of flood risks. *Off. J. L 288/27* **2007**, *8*, 27–34.

10. Garijo, C.; Mediero, L. Influence of climate change on flood magnitude and seasonality in the Arga River catchment in Spain. *Acta Geophys.* **2018**, *66*, 769–790. [CrossRef]

11. Garijo, C.; Mediero, L.; Garrote, L. Usefulness of AEMET generated climate projections for climate change impact studies on floods at national-scale (Spain). *Ingeniería del Agua* **2018**, *22*, 153–166. [CrossRef]

12. Cantarino, I. *Modelo *Aster. Fundamentos y Aplicación*; Ministerio de Medio Ambiente: Madrid, Spain, 1998.

13. Cobos, G.; Collado, J.A. ASTER. Modelo Hidrológico De Simulación Y Previsión Aplicado A Cuencas Donde El Fenómeno Nival Es Relevante. Manual De Usuario. Available online: http://www.spesa.es/paginas/basededatos/ASTER_Manual_Usuario.pdf (accessed on 27 December 2019).

14. Jiménez, A.; Montañés, C.; Mediero, L.; Incio, L.; Garrote, J. El Mapa De Caudales Máximos De Las Cuencas Intercomunitarias. *Revista De Obras Públicas* **2012**, *3533*, 7–32. Available online: https://www.miteco.gob.es/es/agua/temas/gestion-de-los-riesgos-de-inundacion/snczi/Mapa-de-caudales-maximos/ (accessed on 28 December 2019).

15. Propuesta de mínimos para la realización de los mapas de riesgo de inundación. *Directiva de inundaciones – 2° ciclo*; Ministerio para la Transición Ecológica: Madrid, Spain, 2019. Available online: https://www.miteco.gob.es/en/agua/temas/gestion-de-los-riesgos-de-inundacion/Metodologia%20mapas%20de%20riesgo%20Dir%20Inundaciones%20JULIO%202013_tcm38-98530.pdf (accessed on 29 December 2019).

16. Policy-Relevant Assessment of Socio-Economic Effects of Droughts and Floods, To Establish a Damage-Water Depth Relationship. Available online: http://www.feem-project.net/preempt/ (accessed on 29 December 2019).

17. Cobos, G. Cuantificación De Las Reservas Hídricas En Forma De Nieve Y Previsión En Tiempo Real De Los Caudales Fluyentes De La Fusión. Aplicación Al Pirineo Español: Cuenca Alta Del Río Aragón. Ph.D. Thesis, Universitat Politècnica de València, Valencia, Spain, 2004.

18. Cobos, G.; Frances, M.; Arenillas. The ERHIN Programme. Hydrological-Nival Modelling for the Management of Water Resources in the Ebro Basin. *La Houille Blanche* **2010**, *3*, 58–64. [CrossRef]

19. Anderson, E.A. Development and Testing of Snow Pack Energy Balance Equations. *Water Resour. Res.* **1968**, *4*, 19–37.

20. Anderson, E.A. *National Weather Service River Forecast System. Snow Accumulation and Ablation Model*; US Department of Commerce, National Oceanic and Atmospheric Administration, National Weather Service: Silver Spring, MA, USA, 1973.

21. Anderson, E.A. *A Point Energy and Mass Balance Model of a Snow Cover*; US Department of Commerce, National Oceanic and Atmospheric Administration, National Weather Service, Office of Hydrology: Silver Spring, MA, USA, 1976.

22. Anderson, E.A. *Snow Accumulation and Ablation Model—SNOW 17*; US Department of Commerce, National Oceanic and Atmospheric Administration, National Weather Service, National Weather Service River Forecast System: Newburyport, MA, USA, 2006; p. 61.

Article

On the Influence of the Main Floor Layout of Buildings in Economic Flood Risk Assessment: Results from Central Spain

Julio Garrote * and Nestor Bernal

Department of Geodynamics, Stratigraphy and Paleontology, Complutense University of Madrid, E-28040 Madrid, Spain; nbernal@ucm.es
* Correspondence: juliog@ucm.es; Tel.: +34-91-394-4850

Received: 30 December 2019; Accepted: 26 February 2020; Published: 1 March 2020

Abstract: Multiple studies have been carried out on the correct estimation of the damages (direct tangible losses) associated with floods. However, the complex analysis and the multitude of variables conditioning the damage estimation, as well as the uncertainty in their estimation, make it difficult, even today, to reach one single, complete solution to this problem. In no case has the influence that the topographic relationship between the main floor of a residential building and the surrounding land have in the estimation of flood economic damage been analysed. To carry out this analysis, up to a total of 28 magnitude–damage functions (with different characteristics and application scales) were selected on which the effect of over-elevation and under-elevation of the main floor of the houses was simulated (at intervals of 20 cm, between −0.6 and +1 m). According to each of the two trends, an overestimation or underestimation of flood damage was observed. This pattern was conditioned by the specific characteristics of each magnitude–damage function, meaning that the percentage of damage became asymptotic from a certain flow depth value. In a real scenario, the consideration of this variable (as opposed to its non-consideration) causes an average variation in the damage estimation around 30%. Based on these results, the analysed variable can be considered as (1) another main source of uncertainty in the correct estimation of flood damage, and (2) an essential variable to take into account in a flood damage analysis for the correct estimation of loss.

Keywords: floods; economic damage; building first floor; magnitude–damage models; Navaluenga

---

1. Introduction

Floods are probably the most frequently recurring natural process affecting society in terms of time and space, regardless of their geographical location, as shown by the data collected by the International Disasters Database for the period 1900–2018 [1]. This is the main reason why flood risk management has become an essential tool from both a social and an economic perspective, with the objective of reducing losses associated with both sets of factors (features that constitute tangible and intangible flood damage). Assessing direct tangible damages (those direct damages resulting from the physical contact of floodwater with property and its contents), the economic flood losses have been increasing throughout the past half-century. In the last decade (2008–2018), the economic losses associated with floods exceeded 35 billion dollars [1] and, within this period, the flood losses exceeded 19 billion dollars (USD) in 2012 alone [2–4].

In this flood risk survey, the extension of the study area conditioned the analytical approach, and, as proposed by de Moel et al. [5], each work scale (supra-national, national or macro-scale, regional or meso-scale, and local or micro-scale) required the use of different methods of analysis, with their results needing to satisfy different purposes. A state-of-the-art review of study-related approaches, analysis techniques, results, uncertainties and validation processes of the results associated with each of these four work scales is collected in Reference [5]. Regarding this last point, the validation and calibration of the results involved fundamental elements that will require further study in the future (regardless of the work scale in the analysis), since they condition the usefulness of these studies for end users [6].

Flood risk study methodology changes significantly depending on the study area, so the work scales associated with supra-national or national assess are either generally based on multi-criterion analysis techniques (e.g., References [7–9]) or on scaled models (such as those proposed by References [3,10]). These approaches change when studies have a regional character, a situation where the combination of hydraulic models predominates (mainly those that combine 1D and 2D analysis, coupled 1D/2D models; e.g., References [11,12]) together with magnitude–damage functions associated with different land uses and obtained from the aggregation of damage models (e.g., References [13,14]) developed on a micro-scale level.

Finally, regarding micro-scale studies (local scale), the analysis techniques are similar to those used in regional studies, with an increase in the detail and specificity of the magnitude–damage functions used (e.g., References [15,16]) and in the precision of the hydraulic models, which in many cases directly become 2D models [17]. For the magnitude–damage functions, the value of the flow depth is normally used in the description of the flood magnitude [18], but there are still other factors (speed, duration, presence of pollutants, sediment concentration, etc.) that also play an important role in flood damage [19]. However, as described by Soetano and Proverbs [20], the flow depth variable is considered to be the main factor in the estimation of direct damage to buildings due to flooding, conditioning the spatial extent of the flood in X, Y, and Z. In hydraulic models, the main difference with regard to analyses carried out at other scales is their precision. This precision depends on the topography used, and although Moel et al. [5] determined a range between 1 and 25 m for the pixel size of the DTMs (digital elevation models) used, that value is currently set in the 1–5 m range (e.g., Reference [21]).

Nevertheless, not only the spatial resolution of the DTMs used is important for correct damage estimation [22], but also the accuracy of their altitude. This aspect is directly related to proper hydraulic modelling and the definition of flood areas, which have been studied since the end of the last century (e.g., References [23–25]). The effect of topographic precision is evident mainly in low relief areas [26], where the delimitation of flood zones may cause significant differences. More recently, the incorporation of LiDAR (light detection and ranging) data for the generation of DTMs has led to the increase in their accuracy and spatial resolution, which is a clear indication of progress made in the improvement of hydraulic models associated with complex topographic configurations such as urban areas. A clear example is the increase in the capacity to include buildings as topographic elements within 2D hydraulic models, which significantly improves damage estimation in areas where the availability of this type of information is limited [27]. Even so, these LiDAR data requires post–processing aimed at the correct classification of the points (land, vegetation, buildings, etc.) and their filtering, in order to enhance the quality and precision of 3D–derived surfaces. In this sense, Bodoque et al. [28] showed the need to post–process LiDAR data for their subsequent application in the estimation of flood damage, and how the accuracy of these data is influenced by the different types of terrain. This estimated flood damage can vary significantly in urban areas depending on how the initial topographic information (LiDAR data) is used [29], so that the control and validation of LiDAR data in these areas/built–up areas with tall buildings seems indispensable. On the other hand, since flow depth is the main variable from which the damages are estimated, the variable method assigned to each of the exposed elements is another factor in the uncertainty of correct damage estimation in urban areas [30].

From the above, increasing interest has been shown in the resolution and precision of the DTMs used as the basis for hydrodynamic (2D) flow modelling in the analysis of flood damage. These hydrodynamic models will provide us with information (mainly) on flow depth or flow velocity on which the quantification of flooding magnitude can be based. This magnitude can be subsequently correlated with the damage function used to estimate direct economic losses due to floods. However, throughout this flow analysis, an important component for correct damage estimation seems to have been left out: the relationship between the height of the main floor of the buildings and the elevation of the adjacent land. It is usually assumed that the topographic elevation of the land around a given building (streets, sidewalks, patios, unbuilt lots) is the same as that of the main floor of the adjacent house, serving as a reference for the start of the magnitude (depth)–damage functions. In this sense, Garrote–Revilla et al. [31], in a preliminary study carried out in the centre of the Iberian Peninsula, based on the use of the magnitude–damage functions proposed by the USACE (United States Army Corps of Engineers) [32,33], suggested that the influence of this variable can significantly modify (variations greater than 20%) the damage estimation associated with flood events.

This work constituted the first detailed research on the influence of the real topographic relationship between the main building floor (height of the first floor that directly leads to the doorway) and the correct estimation of the direct tangible damage associated with floods. For this reason, its results can be used as a reference for the suitability or need to consider this variable in risk studies on a micro–scale level. Application to wide areas may be more difficult due to the presence of heterogeneous building stocks. From the last point of view, more micro–scale analysis may be useful for establishing regional or general trends. The need for or suitability of the use of the variable should based on the differences found in the estimation of damages, and the significance level of those differences depending on the damage models (magnitude–damage functions) considered. Although the uncertainty (which may also be exposure–related, caused by the lack of knowledge about building characteristics) in damage estimation in houses caused by floods has been widely discussed, the relationship between the variable considered in the present paper and its uncertainty has not yet been analysed. The other two variables that are used in damage estimation must not be neglected: the exposure (depending on the elements that are present at the location involved and on the spatial distribution of the flood spot) and the vulnerability (defined by the lack of resistance of exposed elements to the flooding), which were not taken into account as independent variables in this survey, as vulnerability models were taken from previously proposed models. The results obtained showed a scenario in which the variations in the estimation of economic damages in residential buildings reach percentages and magnitudes similar to those obtained in previous studies focusing on other variables. The topographic difference between the main floor of a house and the surrounding land should be considered in damage mitigation or flood management as a basic factor for the correct estimation of damages and, consequently, as another possible source of uncertainty.

2. Materials and Methods

As mentioned in the previous section, the vulnerability of exposed elements is usually assessed based on damage models according to magnitude–damage functions. The present study assembled different magnitude–damage functions from various countries, extracted from international scientific and technical literature, to assess the direct and tangible damages in residential buildings. In addition to different geographical areas, these functions covered a wide spectrum of characteristics. They took into account whether there were basements in buildings or not, considering or not each separate building and its content and damages either as a monetary value or as a percentage of damage of the set of buildings and their content.

The ensemble of magnitude–damage functions used [21,32–36] and their characteristics (for USACE functions, Type I = one storey with basement, Type II = one storey without basement, Type III = two storeys with basement and Type IV = two storeys without basement) are shown in Table 1. There were 11 damage estimation models (proposed for different countries or regions), consisting of 28 magnitude–damage functions (Figure 1).

Table 1. List of magnitude–damage models and their main characteristics.

Magnitude–Damage Function	Proposed Working Scale	Considered Elements				Potential Damage		Curve Shape	Maximum Flow Depth	Maximum Damage (%)	Flooding Type
		Content	Structure	Combined	Basement	€	%				
HKV Consultants–Structure	Meso–scale		x				x	asymptotic at 5 m	6 m	100	River flooding
HKV Consultants–Content	Meso–scale	x					x	asymptotic at 5 m	6 m	100	River flooding
Australia–Structure	Macro–scale		x				x	no asymptotic	3 m	41	River flooding
Australia–Content	Macro–scale	x					x	asymptotic at 1.8 m	3 m	89	River flooding
Spain (Pajares Pedraza)–V1P	Micro–scale			x		x		no asymptotic	2 m	100	Flash Flood
Spain (Pajares Pedraza)–V1O	Micro–scale			x		x		no asymptotic	2 m	100	Flash Flood
Spain (Pajares Pedraza)–V2P	Micro–scale			x		x		no asymptotic	2 m	100	Flash Flood
Spain (Pajares Pedraza)–V2O	Micro–scale			x		x		no asymptotic	2 m	100	Flash Flood
USACE–Type I–Content	Macro–scale	x			x		x	asymptotic at 3.4 m	5 m	39.1	River flooding
USACE–Type I–Structure	Macro–scale		x		x		x	asymptotic at 3.4 m	5 m	81.1	River flooding
USACE–Type II–Content	Macro–scale	x					x	asymptotic at 4 m	5 m	40	River flooding
USACE–Type II–Structure	Macro–scale		x				x	no asymptotic	5 m	80.7	River flooding
USACE–Type III–Content	Macro–scale	x			x		x	no asymptotic	5 m	52.6	River flooding
USACE–Type III–Structure	Macro–scale		x		x		x	asymptotic at 4.6 m	5 m	76.4	River flooding
USACE–Type IV–Content	Macro–scale	x					x	no asymptotic	5 m	37.2	River flooding
USACE–Type IV–Structure	Macro–scale		x				x	no asymptotic	5 m	69.2	River flooding
Europe–Content	Macro–scale	x					x	no asymptotic	6 m	100	River flooding
Switzerland–Content	Meso–scale	x					x	asymptotic at 3 m	6 m	64	River flooding
Czech Republic–Content	Meso–scale	x					x	no asymptotic	6 m	42	River flooding
Netherlands–Content	Macro–scale	x					x	asymptotic at 5 m	6 m	58	River flooding
Germany–Content	Macro–scale	x					x	no asymptotic	6 m	84	River flooding
United Kingdom–Content	Macro–scale	x					x	asymptotic at 3 m	6 m	100	River flooding
Spain (Poyo Ravine)–Single–family, no garage (Type I)	Micro–scale			x		x	x	asymptotic at 1.3 m	2 m	100	Flash Flood
Spain (Poyo Ravine)–Single–family, garage ground floor (Type II)	Micro–scale			x		x	x	asymptotic at 1.3 m	2 m	100	Flash Flood
Spain (Poyo Ravine)–Single–family, garage basement (Type III)	Micro–scale			x	x	x	x	asymptotic at 1.2 m	2 m	100	Flash Flood
Spain (Poyo Ravine)–Multi–family, no garage (Type IV)	Micro–scale			x		x	x	asymptotic at 1.2 m	2 m	100	Flash Flood
Spain (Poyo Ravine)–Multi–family, garage ground floor (Type V)	Micro–scale			x		x	x	asymptotic at 1.4 m	2 m	100	Flash Flood
Spain (Poyo Ravine)–Multi–family, garage basement (Type VI)	Micro–scale			x	x	x	x	asymptotic at 1.4 m	2 m	100	Flash Flood

The estimation of the variable magnitude within magnitude–damage functions is generally carried out directly from intersection of the water depth data and the entities (generally polygons representing the boundaries of the housing) that indicate the exposed elements or through neighbour buffers of those polygons. This second option is widely used when the exposed elements (mainly buildings) are represented by cell height elevation (the most common method of representation; e.g., References [37,38]) up to the height of the roof of buildings in the digital surface model (DSM) used for hydraulic modelling. In these situations, water is forced to flow through the streets and cannot enter the buildings. Therefore, if no influence area of the building were used, the intersection between flow depth data and the exposed elements would not yield results. The flow depth value is generally defined by either the maximum value or the average value. The second option limits the possibilities of obtaining erroneous values linked to deficiencies in the topographic representation of the land in the DSM. However, as stated by Bermúdez and Zischg [30], the use of the maximum value makes the estimation of the flow depth more independent of the 3D mesh–building representation method of hydrodynamic models.

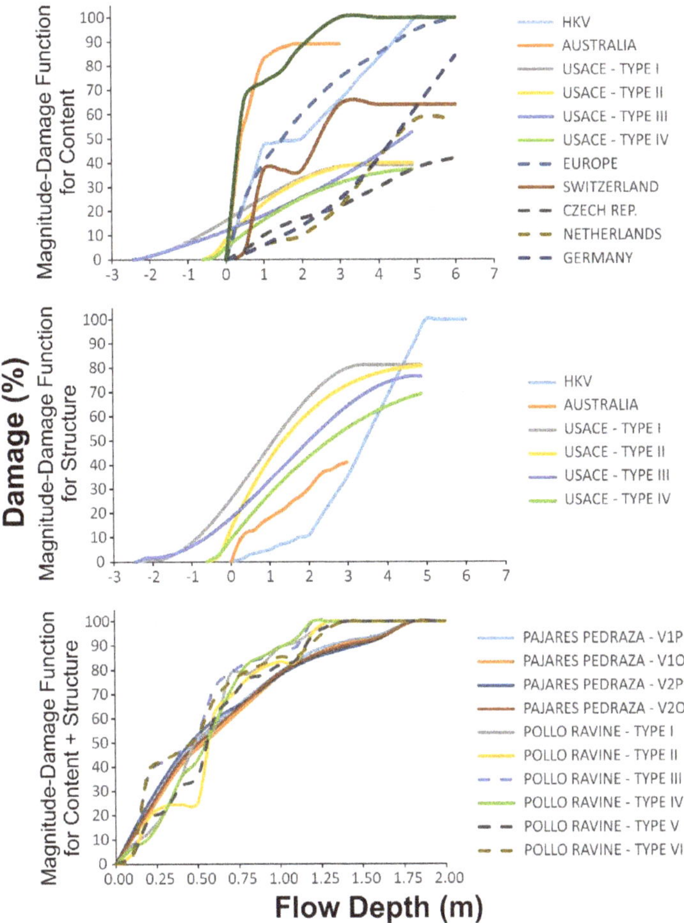

Figure 1. Magnitude–damage models used. The models are classified in relation to their consideration of content, structure, or content + structure.

Therefore, the estimation of the water height (flood magnitude) will be made with respect to the flow depth value in the land surrounding the buildings. Thus, these situations do not consider the possibility of the building being over– or under–elevated compared to the surrounding land (Figure 2). For this reason, for each of the magnitude–damage functions used, different topographic scenarios were considered in which the effect of the possible over–and under–elevation of the buildings compared to the surrounding territory were analysed.

Figure 2. Examples of houses with topographic over– and under–elevation with respect to the surrounding land.

For each of these topographic scenarios, the variation in the damages with respect to the initial scenario was estimated, wherein no type of variation between the elevation of the main building floor and the surrounding land was taken into account (Figure 3). Depending on the characteristics of each magnitude–damage function used, the existence of basements was considered in the variations obtained in the damage estimation.

Hence, for every function analysed, successive scenarios were considered in which a topographic variation of 20 cm between the height of the main floor of the house and the adjacent land was incorporated. The range of variation between the level of the main floor and the surrounding land was varied between –60 cm and +100 cm, resulting in a series of nine analysis scenarios (including the initial one) of direct tangible damage due to flooding. The application of these over– and under–elevation thresholds to the main floor was based on the field work carried out by Garrote–Revilla et al. [31].

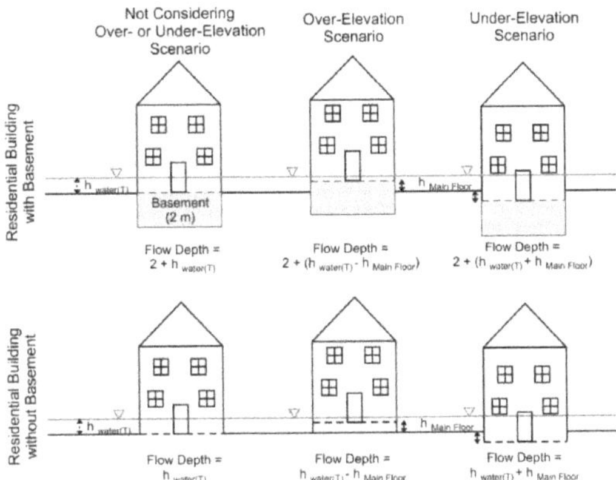

Figure 3. Flow depth estimation in the houses considering topographic over– or under–elevation with respect to the surrounding land, in the case of residential buildings both with and without a basement.

From the set of results obtained for each of the topographic scenarios considered, descriptive statistics and trend adjustment methods (mainly by estimating the changes in the damage to homes rates, and the graphic representation of these rates linked to the water depth) were applied to analyse the influence of the topographic relationship between the height of the main floor in the housing and the adjacent land. Thus, the influence of this variable was observed in the damage estimation, both in general (all the damage models used) and in a segregated manner in subsets defined according to the characteristics of the magnitude–damage functions used.

Finally, the results obtained in a real scenario are detailed in a later section, in which the estimation of the direct economic losses associated with floods in a small village in the centre of the Iberian Peninsula was carried out. In this specific case, residential dwellings were divided into the categories considered in each of the magnitude–damage functions proposed for use in this analysis. Subsequently, the average under– or over–elevation of the main floor in each of these categories was estimated with respect to the surrounding roads (streets). Lastly, the variation in the EAD value (expected annual damage) was determined based on the consideration or not of the variable "topographic relationship between main floor housing and roads" in the estimation of economic losses. In this last analysis (and due to the lack of real–time data) the economic valuation of the content and structure of residential buildings was considered as a constant value for each type of building.

3. Results

3.1. General Results for Magnitude–Damage Models

The damage models or magnitude–damage functions used were originally derived from a set of tables in which the percentage of damage associated with a series of flow depth values is related, instead of a mathematical formulation that determines a continuous function. For this reason, the first task required was to obtain a mathematical equation that defined the relationship between the "percentage of damage" variable with regard to the "flow depth" value. To attain this objective, the original data of a third– or fourth–degree polynomial function was adjusted, analysing the degree of adjustment of this polynomial function based on the R^2 value. In general, R^2 reached a value greater than 0.99. However, in the case of the magnitude–damage function used in Switzerland, a lower value was indicated (0.93). In this specific case, the magnitude–damage function showed a decrease in R^2 due to the stepped form (Figure 4) presented by the original data.

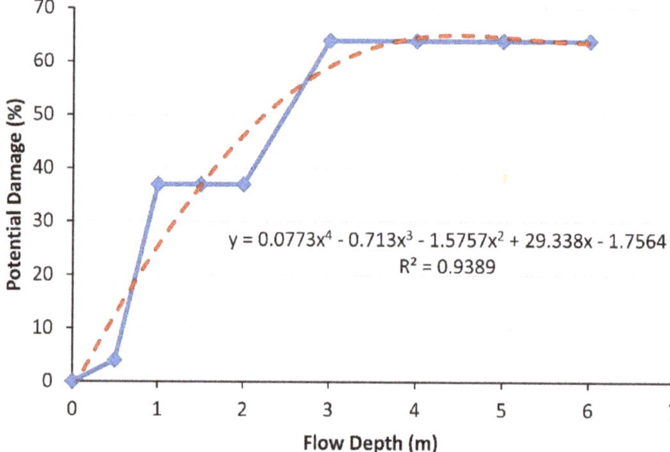

Figure 4. Switzerland magnitude–damage model for river flooding, showing its characteristic staggered shape.

The results obtained indicated how the flood damage estimation positively correlates with the topographic difference between the main floor of houses and the public roads or land surrounding these buildings. Thus, when buildings are over–elevated with regard to the adjacent land, there is a clear decrease in the percentage of estimated economic losses, and this effect is shown inversely in situations of under–elevation. In contrast, it can be seen that the sensitivity of the set of magnitude–damage functions or "damage models" used is varied (either between functions, or within the same function).

In general, the variation in the estimate of economic losses (damage rates) associated with flow depth values does not exceed a maximum of 100%. This situation occurs when the new topographic scenario considered eliminates the damages associated with a flow depth, so that the new percentage of damage is 0%. This pattern (Figure 5) was repeated in damage models with over a 1 m depth range (the maximum value considered for over–elevation in buildings with respect to the surrounding land), regardless of the damage model applied. It also considered either the "content" or the "structure", or that these models have been proposed with a general scope (at the country level) or detail (at the locality or population level).

In the case of some specific functions such as Switzerland (Figure 4) or the Netherlands, both proposed in the JRC Report, or the equation of Australia (proposed by HKV Consultants) the regression lines did not reach 100% as the maximum value, as seen in Figure 4. This trend was due to the fact that these functions, which originated from statistical calculations, were not designed for extreme flood events such as those resulting from 4.5 m flow depth floods (Figure 4).

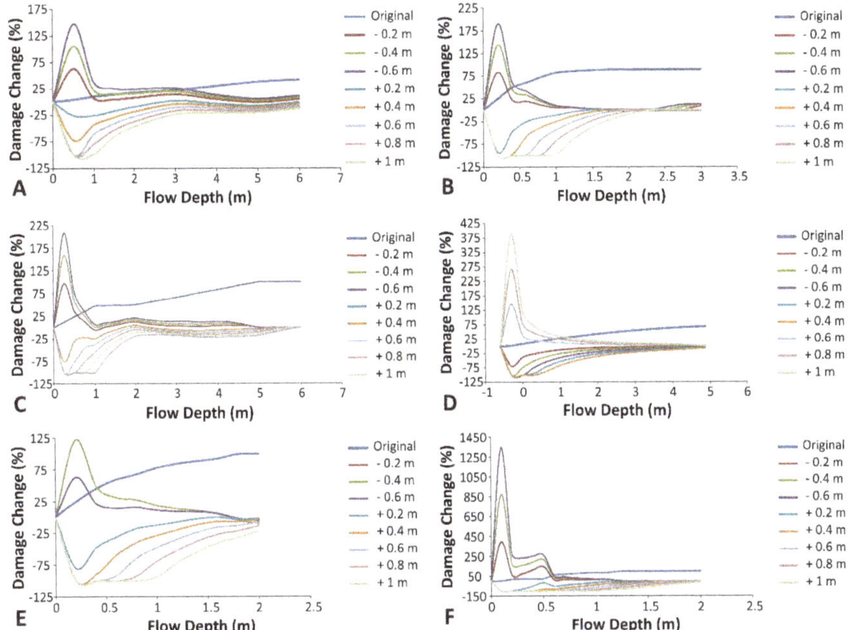

Figure 5. River flooding economic damage change (%) by considering topographic over– and under–elevation of buildings from magnitude–damage models of the Czech Republic (**A**), Australia–Content (**B**), HKV–Content (**C**), USA–Type IV–Content (**D**), Pajares de Pedraza–V2P (**E**) and Pollo Ravine–Type II (**F**).

When the topographic effect analysed was the under–elevation of the house, the difference in the results (not considering or considering height difference between the main floor and surrounding land) was greater than in the over–elevation effect. This fact was not due to the trend of these results, since this always showed an increase in the percentage of economic damage, but in terms of the magnitude of these damage percentage variations. The increases in the percentage of damages (according to the results obtained) seemed greater in those magnitude–damage functions proposed for detailed analysis, than in the cases of those proposed for regional analysis, although it was not possible to affirm that the damage increase value ranges were totally independent. In scenarios in which an under–elevation of the building was simulated with respect to the surrounding land, there were more abrupt changes (than in over–elevation situations) in the variations in the percentage of economic damage, with a rapid rise and subsequent decrease in these variations.

In any of the situations (over– or under–elevation), as we moved towards greater water depths there was a decrease in the differences of damage rates, so the effect of the main floor elevation from the surrounding land became negligible. When we moved towards the maximum flow depth values analysed in each magnitude–damage function, the damage percentage variations (associated with the different topographic scenarios proposed) were generally below 10%. This situation can be seen clearly in the damage change–flow depth graphs on the right of the Figure 6, related to magnitude–damage models of Switzerland (A), United Kingdom (B) and Germany (C). In the three examples, the damage change percentage increased with respect to the original damage values when the height of the main floor was under the street level. In contrast, when raising the elevation, the damage change percentage decreased, reaching negative values. Both trends stabilized when flow depth reached between 2 and 3 m, because at this point, the over– or under–elevation of the main floor level dis not influence the final damage.

The characteristics of the magnitude–damage function used had a clear influence on the damage percentage variation, both in the maximum flow depth considered (ranging from 2 m to more than 5 m), and in the maximum percentage of damage associated with these depths (which varied from percentages slightly higher than 40% to percentages of 100%). In the case of functions where the percentage of damage became asymptotic around a given flow depth value, the variations in that percentage associated with the topographic scenarios proposed were minor. An example of the above (Figure 6) would be the magnitude–damage functions related to countries such as the United Kingdom, Switzerland and the USA, or those proposed by Salazar Galán [35].

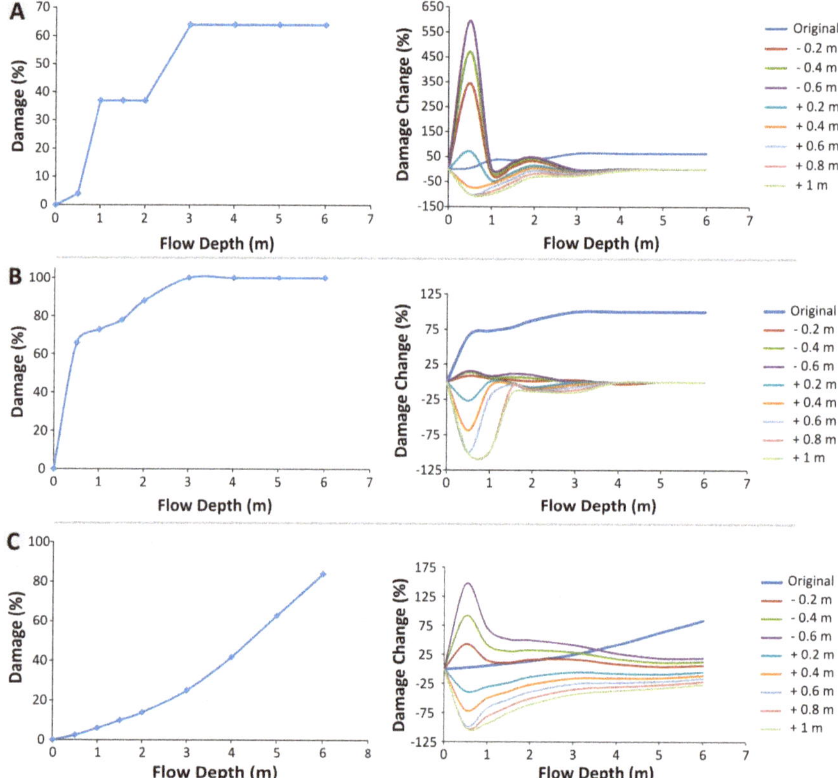

Figure 6. Examples of the effect of the magnitude–damage model graphic shape in the economic damage change evolution when considering the topographic over– or under–elevation of building main floor. The examples are related to magnitude–damage models of Switzerland (**A**), United Kingdom (**B**) and Germany (**C**).

3.2. Local Results at Navaluenga (Central Spain)

As a practical example of the analyses detailed above, the town of Navaluenga (Ávila, Central Spain) was used in order to determine the influence that the topographic relationship between buildings and surrounding lands had on the estimation of flood damages. The magnitude–damage functions applied were those proposed by the USACE and the buildings were classified into four groups. Depending on the number of dwellings included in each group, the topographic relationship (Z axis) between the main floor of the building and the surrounding land was estimated either for the total class of buildings or for a subset (statistically representative) of the buildings of the corresponding class.

In the studied town, the total number of dwellings amounts to 2256, which are distributed as follows (depending on the categories covered by USACE): 109 Type I housing, 621 Type II housing, 64 Type III housing and 1462 Type IV housing.

The results obtained (Table 2) showed systematic variation in the percentage of economic losses due to floods for the return periods considered (T50, T100 and T500, defined according to the Basic Guidelines of Civil Protection applied to flood risk planning). The reduction in economic losses was 32%, if neither the type of housing nor the period of return of the flood were taken into account. However, a certain degree of variability in the results was observed depending on the building typology, which reproduced an effect that had already been observed in the results described in the previous section.

Table 2. Navaluenga flood economic damage reduction considering topographic over–or under–elevation of houses.

Housing Type [1]	Economic Damage Reduction (%)					
	T_{50}		T_{100}		T_{500}	
	Content	Structure	Content	Structure	Content	Structure
Type I	61.34	67.81	54.82	59.17	50.20	44.73
Type II	17.55	18.22	14.57	15.44	12.46	13.19
Type III	34.26	42.78	33.77	41.26	30.47	34.55
Type IV	24.36	23.48	22.12	21.48	18.64	18.25

[1] Type I: one storey with basement; Type II: one storey without basement; Type III: two or more storeys with basement; Type IV: two or more storeys without basement.

Thus, in one–storey houses (Types I and II) with basements, the effect of the variable analysed caused a reduction in the estimate of economic losses (damage rate) of over 50% (average value for the return periods of the considered flood). This average value was quite homogeneous when assessing the losses (damage rate) associated with the content and the structure of the building (55% and 57%, respectively). In the case of "two or more storey houses" (Types III and IV), the reduction in the estimation of economic losses (damage rate) was below the damage percentages obtained for one storey houses, with a reduction of 33% for content and 40% for the housing structure.

In houses without basements, the reduction of economic losses (damage rate) was, in general, less than for houses with basements (Figure 7). When analysing the one–storey houses, this was around 15%, regardless of whether the content or structure of the houses was considered. Finally, in the case of houses with more than one storey, the reduction of economic losses was also very homogeneous in accordance with the content or the structure, placing this reduction at 22% and 21%, respectively.

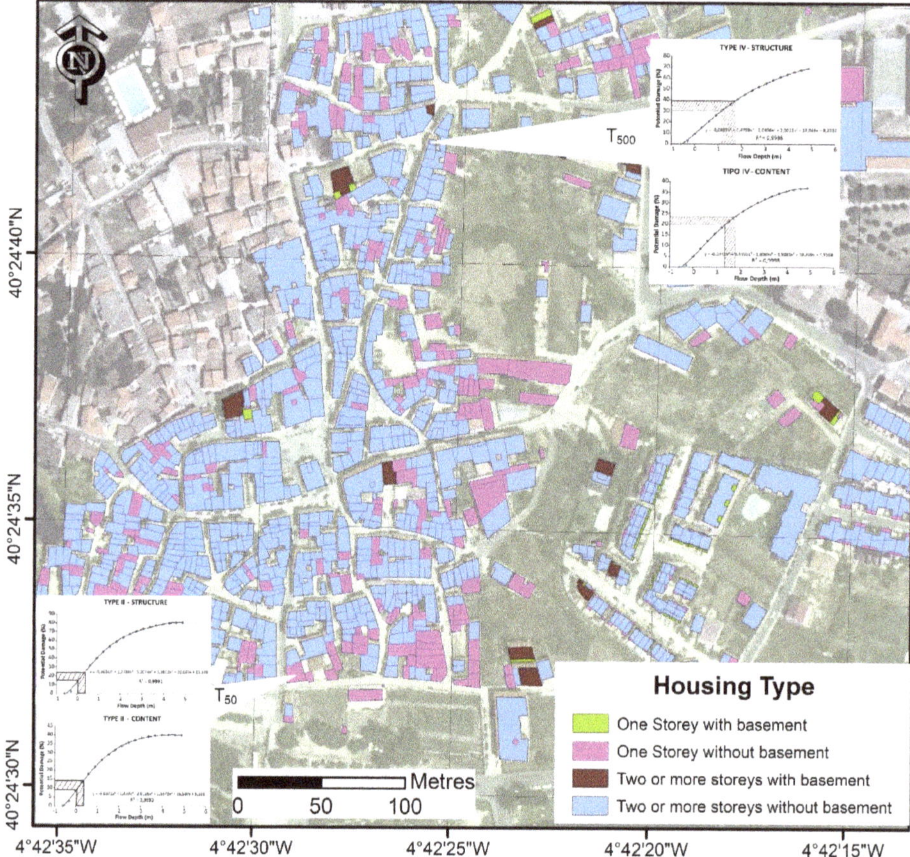

Figure 7. Navaluenga building classification according to USACE building types; and examples of flood economic damage change by considering the topographic over– and under–elevation of houses against its surrounding land.

Finally, regarding the influence of the topographic relationship between the main floor of the houses and their surrounding land in the EAD estimation, results showed a clear reduction in the EAD value (Figure 8) regardless of the content, the structure or the combination of both. Thus, taking into account the damages associated with floods for a 50, 100, and 500 year return period, the EAD values were reduced by a percentage of over 22%. Specifically, for the whole structure and content of the dwellings, the EAD value for Navaluenga town was reduced from €132,000 to €102,500.

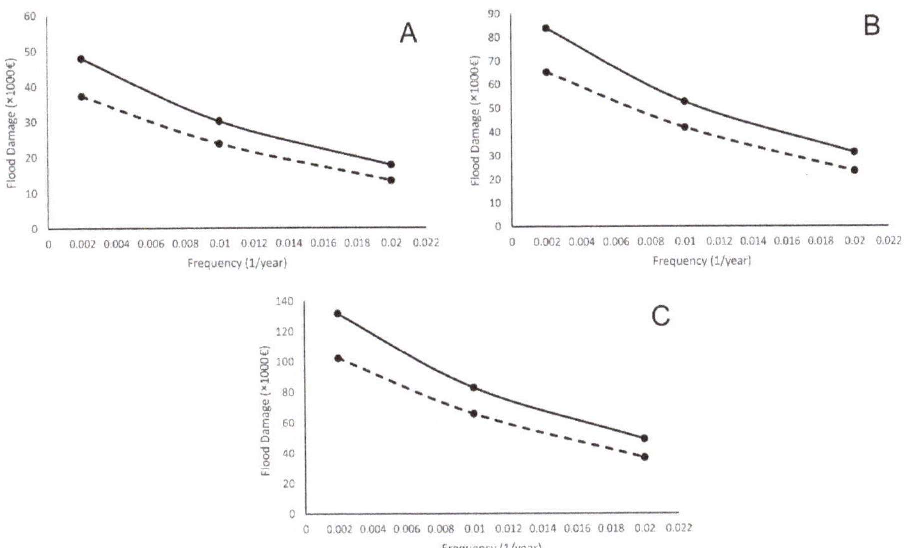

Figure 8. Damage–frequency curves for Navaluenga town considering content (**A**), structure (**B**) or content + structure (**C**). The solid line represents the original results, and the dashed line represents the results linked to the consideration of the topographic difference between the main floor of housing and surrounding land.

4. Discussion

Uncertainty in the estimation of damage associated with floods is a recurring subject addressed in this type of analysis [18,38]. This is due to the important variations observed and shown by different studies (e.g., References [39–41]) in terms of estimating damages. However, the complete elimination of uncertainty in flood damage estimation seems to be an impossible task, and this is partly due to the multiple sources of uncertainty linked to these analyses.

With regard to the damage estimation and making the issue more complex, there is no unanimity in the attribution of weights or influences on the uncertainty of the different sources linked to estimation, but it may be considered that there is a set of variables that constitute the greatest source of uncertainty. This set of variables can include the correct estimate of the flow frequency distribution function (which determines the magnitude and relative frequency of flooding in the territory), the flow depth, the damage models and the curves or magnitude–damage functions.

The first of the aforementioned variables, the correct estimation of the flow frequency distribution function, is recognized as one of the main sources of uncertainty in the damage estimation associated with floods (e.g., Reference [16,42]). Ballesteros–Cánovas et al. [41], in the town of Navaluenga (taken as a practical case study in the present study) affirmed that the percentage of uncertainty in flood damage estimation (associated with a correct estimate of the flow frequency law) in this town was above 40%. In this study, the flow frequency distribution function was first estimated using the data series of the gauging station, and this frequency law was later improved by the use of non–systematic information linked to dendro–geomorphological analyses. The uncertainty associated with this variable was much higher than other sources of uncertainty considered, such as the calibration of the hydrodynamic model, or the damage model used [41].

However, correct calibration of the hydrodynamic model should control and improve the estimation of flow depths, and this variable is considered essential in the uncertainty analysis in de Moel and Aerts' [39] research. In their study, these authors associated variations of 10 cm in the flow depth with close to 6% variations in the flood damage estimation. This percentage was not negligible if multiple factors affecting the correct estimate of flow depths (quality and spatial resolution of the digital elevation model and land use model, for instance) were taken into account. The influence of the damage model used has also been considered a main source of uncertainty in flood damage estimation. Thus, Wagenaar et al. [43] calculated that the uncertainty associated with the use of different damage models (HIS–SSM for the Netherlands; HAZUS–MH in USA; FLEMO in Germany; Multi-Coloured Manual in UK; Rhine Atlas in Germany too; and others), all of them based on the concept of unit lost method, can be assigned an order of magnitude from 2 to 5. These authors identified this high uncertainty in the use of different damage models with the fact that these models were based on data related to a limited number of flood events. According to Thieken et al. [19], flood damage data are rarely collected systematically. In such a way, this limited number of events does not represent the multiple causative factors and particular characteristics of other floods or territories. This consideration is very important, and will affect the entire set of damage models or magnitude–damage functions, due to the near impossibility of such models or functions being used to replicate the assessment of flood damage in other locations, where both the features of the floods (flow depth, water velocities, solid load, response time, etc.) and the socioeconomic characteristics of the population can vary significantly. Additionally, linked to the damage model, Chinh et al. [44] estimated the uncertainty associated with this variable (in a study carried out in Vietnam) to be between 10% and 20% for damage to the structures of houses and between 8% and 15% for the content. However, these authors concluded that the uncertainty linked to hazard estimation may have been double that linked to the damage model.

On the other hand, Saint–Geours et al. [41] learned from their research that the main component of the uncertainty in estimating flood damage in houses is linked to the damage model used (a magnitude–damage curve in this case). In this study, the sensitivity index of the damage estimation model associated with the magnitude–damage function had a value of 0.78, compared to 0.22 linked to the flow frequency law. These results differed from those obtained for economic activities, where the main source of uncertainty was associated with the correct application of the flow frequency law.

All of the above show the variability of results related to the identification of the main uncertainty component in the flood damage estimation. The topographic variable (the height of the main floor and surrounding land) analysed in the present study was not considered in any of the cases, at least independently; Saint–Geours et al. [41] did seem to take it into account in their flood exposure model. This model particularly included land uses and their relation to economic activities, showing that their sensitivity index (0.03) was really insignificant and within the total estimation model of economic losses for houses compared to those mentioned above.

However, in no case was the direct influence of the topographic variable (difference in height between the main floor of the house and surrounding land) analysed in the estimation of flood damages in housing. As was shown in the present study, this variable determines the percentage of losses assigned to each dwelling, and the uncertainty introduced in the results reached similar levels to those pointed out by other authors for different variables. In addition, it is important to note that this variable influences the behaviour of other variables commonly used in flood damage estimation: flow depth and hazard models, flood exposure models, etc.

The assessment of economic damage due to flooding, or the variation in damage rate when considering the topographic relationship between the main floor of a house and the surrounding land, also implies uncertainty. This uncertainty is mainly associated with the damage model (magnitude–damage function) used. As we have seen, different models associate different damage rates to the same water depth value. The use of all considered magnitude–damage functions allows the calculation of an average economic damage value (%) associated with each flood return period, as well as the estimation of the confidence intervals (90%, 95% and 99%) related to those average damage values. The results of this analysis are shown in Figure 9, which shows the general reduction of damage rates when considering the topographic relationship between the main floor of a house and the surrounding land.

The uncertainty in average damage value was very similar regardless of the consideration of the topographic relationship between the main floor of the house and the surrounding land. In both cases, the uncertainty was slightly higher for the 50 year return period. When we considered the 90% confidence interval, the uncertainty in the estimation of the average value of flood damage ranged from 20% to 30%. This uncertainty reached a value of up to 45% (damage rate = 27.64 ± 12.58) when we considered the 99% confidence interval and a 50 year return period. These uncertainty values indicated a significant variability in the results depending on the model (magnitude–damage function) used, and reaffirmed the need for the development of detailed magnitude–damage functions at a local scale (as opposed to regional) for a correct estimate of economic damage due to flooding [21].

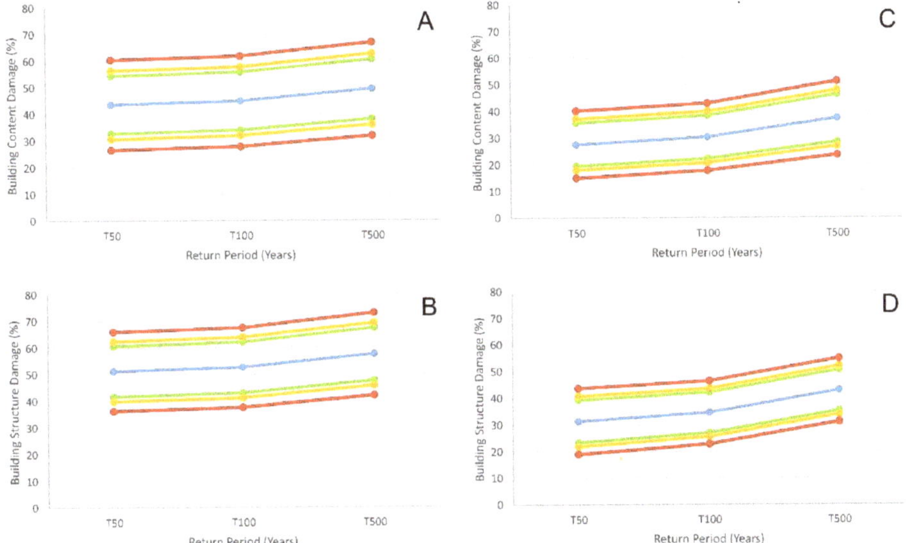

Figure 9. Average values and confidence intervals (90%: green line; 95%: yellow line; 99%: red line) for flood economic damage (%) at Navaluenga (Spain) from all magnitude–damage functions considered. Results not taking into account the topographic relationship between the main floor of the house and the surrounding land (**A**,**B**), and taking the topographic relationship into account (**C**,**D**).

Furthermore, as the difference in height between the main floor of houses and the surrounding land is a variable that is independent of the damage model used (magnitude–damage function), the errors associated with this variable are independent as well. For this reason, the benefits of using a specific magnitude–damage function of a location (which takes into account the socio–economic conditions of the area) can disappear against the error introduced by this variable.

It should be noted that the influence of the variable that relates the height of the main floor of a house to the elevation in the surrounding land is not constant and independent of the value thereof. This influence is maximised in average values of flow depths (in the order of 0.2–0.6 m), which could be more frequently related to housing. In major flood depths, the influence of the analysed variable is also important; in flow depths above 1.5–2 m (the least common in urban areas where housing is located) the "height of main floor–surrounding land" relationship begins to have a lower impact. This is because the expected damages associated with these flow depths will be so serious (including the risk of collapse of structures) that they are usually associated with poorly developed urban areas.

The effect of the variable that relates the height of the main floor of the house to the surrounding land with respect to flood management and its mitigation measures can be raised. According to the field work carried out in the present study, the most commonly observed situation is one in which there is an over–elevation of the main floor. Therefore, although this may be a local effect not easily extrapolated to other locations, the most common result was the over–estimation of the economic damages linked to flood events when applying the classical methods. As a consequence, this over–estimation of the economic losses will affect the cost–benefit analyses linked to projects for the implementation of flood risk mitigation measures, extending to flood management by the persons and authorities responsible.

The dominant model of over–elevation has been considered previously, as a value of 0.3 m [45] was proposed to reduce the uncertainty in flood damage estimation analysis at regional or national scales. However, that general value may be not appropriate for local assessments, as our results pointed out. For Navaluenga, average over–elevation values ranged from 0.26 m for USACE Type II housing to 1.51 m for USACE Type I housing, with a general average value (independent from housing type) of 0.75 m. In general, in houses with basements, the over–elevation was higher. These differences highlight the variability associated with local construction models and the necessary estimation of the topographic relationship between the first floor and the surrounding land where local assessment will be carried out.

Finally, it must be noted that the results cannot be considered homogeneous, and they show that the construction model of houses will have a decisive influence on the economic losses linked to the different scenarios considered. Type of housing and building style (existence of a significant number of over– or under–elevated houses) can lead to the results regarding the percentage of flood damage differing greatly from those shown here. Thus, the applicability of the results obtained in the present assessment to other regions may be difficult, and more analysis of the relationship of flood damage and the topographic relationship between the height of the main floor and the surrounding land is necessary. However, even taking this into account, the results shown in the present study highlight the need to consider this variable in order to obtain estimates of economic damage that are closer to reality.

5. Conclusions

The correct estimation of damages (direct tangible losses) associated with floods is an essential aspect of efficient flood risk management. Moreover, the influence of DEM quality, the correct and precise inclusion of the buildings in the DSM used as the basis of hydrodynamic models, improvement in the estimation of the flow frequency law, calibration of hydrodynamic models for a correct flow depth estimation, implementation of different damage models, the development of specific magnitude–damage functions that take into account the socio–economic characteristics of the study area, etc. are all variables that have an influence on the estimation of flood damages, with the uncertainty associated with them being subject to study. However, no studies previously carried out have analysed the influence of the topographic relationship between the height of the main floor in dwellings and the surrounding land on flood damage estimation.

To analyse the behaviour and influence of this variable in flood damage estimation, 28 magnitude–damage functions of different types were considered. As expected, in all cases, a correlation was observed between the damage estimation trend (compared to non–consideration of this variable) and the topographic scenario proposed. It was alos noted that the influence of the variable was not homogeneous for the entire range of flow depth values analysed in the magnitude–damage functions. This influence on the flow depth range was maximal between 0.2 and 0.6 m and progressively decreased to below 10% in relation to flow depths of around 2 m or more.

For a real scenario, and using the magnitude–damage functions proposed by the USACE in the town of Navaluenga (Spain), an average over–estimation of more than 30% was obtained of the economic damages associated with floods with return periods of 50, 100 and 500 years. These percentages were not homogeneous and varied depending on the type of residential building and on the structure or content of the dwelling. Thus, in the case of one–storey houses with basements, the reduction in flood losses was maximal, reaching values between 55% and 60% for the structure and content of the house. By incorporating these variations in the economic damage estimation commonly used in the cost–benefit analysis of risk mitigation measures (EAD value of the town of Navaluenga), the over–elevation of the topographic height of the main floor in houses with regard to the surrounding land represented a 22.5% reduction in the EAD estimation.

All the data shown above helped us to determine that the variable "topographic relationship between the main floor of houses and the surrounding land" is one of the main sources of uncertainty in the estimation of flood damage in houses, similarly to others such as the correct determination of the flow frequency law, flow depth or damage model. It seems, therefore, that its consideration is essential for a proper estimation of economic losses due to floods, in the design and implementation of flood risk mitigation measure projects and for the general purpose of comprehensive flood risk management.

Author Contributions: All authors have a significant contribution to final version of the paper. Conceptualization, J.G.; Flood damage models analysis, N.B.; Writing—original draft, J.G.; Writing—review & editing, all authors. All authors have read and agreed to the published version of the manuscript.

Funding: This research was funded by the project DRAINAGE, CGL2017–83546–C3–R (MINEICO/AEI/FEDER, UE), and it is specifically part of the assessments carried out under the coverage of the sub–project DRAINAGE–3–R (CGL2017–83546–C3–3–R).

Acknowledgments: The authors would like to thank to Andrés Díez–Herrero (Geological Survey of Spain, IGME) for his constructive comments and help in conceptualizing the manuscript; and to Jose M. Bodoque and Estefanía Aroca (University of Castilla–La Mancha, Spain) for their flow depth data at Navaluenga town.

Conflicts of Interest: The authors declare no conflict of interest.

References

1. Centre for Research on the Epidemiology of Disasters. The International Disaster Database. Available online: http://emdat.be/emdat_db/ (accessed on 15 July 2019).
2. Munich RE. NatCatSERVICE Database. Available online: https://natcatservice.munichre.com/ (accessed on 17 July 2019).
3. Ward, P.J.; Jongman, B.; Sperna Weiland, F.C.; Bouwman, A.; Van Beek, R.; Bierkens, M.; Ligtvoet, W.; Winsemius, H.C. Assessing flood risk at the global scale: Model setup, results, and sensitivity. *Environ. Res. Lett.* **2013**, *8*, 044019. [CrossRef]
4. Visser, H.; Bouwman, A.; Petersen, A.; Ligtvoet, W. *A Statistical Study of Weather–Related Disasters: Past, Present and Future*; PBL Netherlands Environmental Assessment Agency: The Hague, The Netherlands, 2012; p. 87.
5. De Moel, H.; Jongman, B.; Kreibich, H.; Merz, B.; Penning-Rowsell, E.; Ward, P.J. Flood risk assessments at different spatial scales. *Mitig. Adapt. Strateg. Glob. Chang.* **2015**, *20*, 865–890. [CrossRef] [PubMed]
6. Sayers, P.; Lamb, R.; Panzeri, M.; Bowman, H.; Hall, J.; Horritt, M.; Penning-Rowsell, E. Believe it or not? The challenge of validating large scale probabilistic risk models. In Proceedings of the Floodrisk2016, Lyon, France, 17–21 October 2016; p. 11004. [CrossRef]
7. Kourgialas, N.N.; Karatzas, G.P. Flood management and a GIS modelling method to assess flood–hazard areas—A case study. *Hydrol. Sci. J.* **2011**, *56*, 212–225. [CrossRef]
8. Zhao, G.; Pang, B.; Xu, Z.; Yue, J.; Tu, T. Mapping flood susceptibility in mountainous areas on a national scale in China. *Sci. Total Environ.* **2018**, *615*, 1133–1142. [CrossRef] [PubMed]
9. Pinto Santos, P.; Reis, E.; Pereira, S.; Santos, M. A flood susceptibility model at the national scale based on multicriteria analysis. *Sci. Total Environ.* **2019**, *667*, 325–337. [CrossRef]
10. Winsemius, H.C.; Van Beek, R.; Jongman, B.; Ward, P.J.; Bouwman, A. A framework for global river flood risk assessments. *Hydrol. Earth Syst. Sci.* **2013**, *17*, 1871–1892. [CrossRef]
11. Wilson, M.; Bates, P.; Alsdorf, D.; Forsbert, B.; Horritt, M.; Melack, J.; Frappart, F.; Famiglietti, J. Modeling largescale inundation of Amazonian seasonally flooded wetlands. *Geophys. Res. Lett.* **2007**, *34*, L15404. [CrossRef]
12. Biancamaria, S.; Bates, P.D.; Boone, A.; Mognard, N.M. Large–scale coupled hydrologic and hydraulic modelling of the Ob River in Siberia. *J. Hydrol.* **2009**, *379*, 136–150. [CrossRef]
13. Tang, J.C.S.; Vongvisessomjai, S.; Sahasakmontri, K. Estimation of flood damage cost for Bangkok. *Water Resour. Manag.* **1992**, *6*, 47–56. [CrossRef]
14. Kreibich, H.; Seifert, I.; Merz, B.; Thieken, A.H. Development of flemocs—A new model for the estimation of flood losses in the commercial sector. *Hydrol. Sci. J.* **2010**, *55*, 1302–1314. [CrossRef]

15. Merz, B.; Kreibich, H.; Schwarze, R.; Thieken, A. Review article 'Assessment of economic flood damage'. *Nat. Hazards Earth Syst. Sci.* **2010**, *10*, 1697–1724. [CrossRef]

16. Merz, B.; Hall, J.; Disse, M.; Schumann, A. Fluvial flood risk management in a changing world. *Nat. Hazards Earth Syst. Sci.* **2010**, *10*, 509–527. [CrossRef]

17. Ernst, J.; Dewals, B.J.; Detrembleur, S.; Archambeau, P.; Erpicum, S.; Pirotton, M. Micro–scale flood risk analysis based on detailed 2D hydraulic modelling and high resolution geographic data. *Nat. Hazards* **2010**, *55*, 181–209. [CrossRef]

18. Merz, B.; Kreibich, H.; Thieken, A.; Schmidtke, R. Estimation uncertainty of direct monetary flood damage to buildings. *Nat. Hazards Earth Syst. Sci.* **2004**, *4*, 153–163. [CrossRef]

19. Thieken, A.H.; Müller, M.; Kreibich, H.; Merz, B. Flood damage and influencing factors: New insights from the August 2002 flood in Germany. *Water Resour. Res.* **2005**, *41*, W12430. [CrossRef]

20. Soetanto, R.; Proverbs, D.G. Impact of flood characteristics on damage caused to UK domestic properties: The perceptions of building surveyors. *Struct. Surv.* **2004**, *22*, 95–104. [CrossRef]

21. Garrote, J.; Alvarenga, F.M.; Díez-Herrero, A. Quantification of flash flood economic risk using ultra–detailed stage–damage functions and 2–D hydraulic models. *J. Hydrol.* **2016**, *541*, 611–625. [CrossRef]

22. Horritt, M.S.; Bates, P.D. Effects of spatial resolution on a raster based model of flood flow. *J. Hydrol.* **2001**, *253*, 239–249. [CrossRef]

23. Nicholas, A.P.; Walling, D.E. Modelling flood hydraulics and overbank deposition on river floodplains. *Earth Surf. Proc. Land.* **1997**, *22*, 59–77. [CrossRef]

24. Horritt, M.S.; Bates, P.D. Predicting floodplain inundation: Rasted–based modelling versus the finite–element approach. *Hydrol. Process.* **2001**, *15*, 825–842. [CrossRef]

25. Casas, M.A.; Benito, G.; Thorndycraft, V.R.; Rico, M.T. The topographic data source of digital terrain models as a key element in the accuracy of hydraulic flood modelling. *Earth Surf. Proc. Land.* **2006**, *31*, 444–456. [CrossRef]

26. Bates, P.D.; De Roo, A.P.J. A simple raster–based model for flood inundation simulation. *J. Hydrol.* **2000**, *236*, 54–77. [CrossRef]

27. Shen, D.; Qian, T.; Chen, W.; Chi, Y.; Wang, J. A Quantitative Flood–Related Building Damage Evaluation Method Using Airborne LiDAR Data and 2–D Hydraulic Model. *Water* **2019**, *11*, 987. [CrossRef]

28. Bodoque, J.M.; Guardiola–Albert, C.; Aroca–Jimenez, E.; Eguibar, M.A.; Martinez–Chenoll, M.L. Flood damage analysis: First floor elevation uncertainty resulting from LiDAR derived digital surface models. *Remote Sens.* **2016**, *8*, 604. [CrossRef]

29. Arrighi, C.; Campo, L. Effects of digital terrain model uncertainties on high resolution urban flood damage assessment. *J. Flood Risk Manag.* **2019**, e12530. [CrossRef]

30. Bermúdez, M.; Zischg, A.P. Sensitivity of flood loss estimates to building representation and flow depth attribution methods in micro–scale flood modelling. *Nat. Hazards* **2018**, *92*, 1633. [CrossRef]

31. Garrote-Revilla, J.; Díez-Herrero, A.; Aroca-Jiménez, E.; Bodoque, J.M.; Bernal, N.; Martins, L.R. The relationship between the variables "main floor of houses with respect to their urban surroundings" and models for the estimation of economic losses owing to floods. In *Interdisciplinarity in Practice and in Research on Society and the Environment: Joint Paths towards Risk Analysis, Proceedings of the SRA–E–Iberian Chapter (SRA–E–I) Conference, Toledo, Spain, 6–7 September 2018*; Amérigo, M., García, J.A., Gaspar, R., Luís, S., Eds.; Ediciones de la Universidad de Castilla–La Mancha: Toledo, Spain, 2018; Volume 27811, pp. 111–112.

32. USACE. *Economic Guidance Memorandum (EGM) 01–03, Generic Depth–Damage Relationships*; USACE: Washington, DC, USA, 2000; p. 11.

33. USACE. *Economic Guidance Memorandum (EGM) 04–01, Generic Depth–Damage Relationships for Residential Structures with Basements*; USACE: Washington, DC, USA, 2003; p. 17.

34. Van Der Sande, C. *River Flood Damage Assessment Using IKONOS Imagery*; European Commission, Joint Research Centre, Space Applications Institute (SAI): Ispra, Italy, 2001; p. 86.

35. Salazar Galán, S.A. Metodología Para el Análisis y la Reducción del Riesgo de Inundaciones: Aplicación en la Rambla del Poyo (Valencia) Usando Medidas de "Retención de Agua en el Territorio". Ph.D. Thesis, Polytechnic University of Valencia, Valencia, Spain, 2013.

36. Huizinga, J.; De Moel, H.; Szewczyk, W. *Global Flood Depth–Damage Functions: Methodology and the Database with Guidelines*; Joint Research Centre: Luxembourg, 2017; p. 114. [CrossRef]

37. Bellos, B.; Tsakiris, G. Comparing Various Methods of Building Representation for 2D Flood Modelling In Built–Up Areas. *Water Resour. Manag.* **2015**, *29*, 379–397. [CrossRef]

38. Merz, B.; Thieken, A. Flood risk curves and uncertainty bounds. *Nat. Hazards* **2009**, *51*, 437–458. [CrossRef]

39. De Moel, H.; Aerts, J.C.J.H. Effect of uncertainty in land use, damage models and inundation depth on flood damage estimates. *Nat. Hazards* **2011**, *58*, 407–425. [CrossRef]

40. Ballesteros–Cánovas, J.A.; Sánchez–Silva, M.; Bodoque, J.M.; Díez–Herrero, A. An Integrated Approach to Flood Risk Management: A Case Study of Navaluenga (Central Spain). *Water Resour. Manag.* **2013**, *27*, 3051–3069. [CrossRef]

41. Saint–Geours, N.; Grelot, F.; Bailly, J.S.; Lavergne, C. Ranking sources of uncertainty in flood damage modelling: A case study on the cost–benefit analysis of a flood mitigation project in the Orb Delta, France. *J. Flood Risk Manag.* **2015**, *8*, 161–176. [CrossRef]

42. Apel, H.; Thieken, A.H.; Merz, B.; Blöschl, G. Flood risk assessment and associated uncertainty. *Nat. Hazards Earth Syst. Sci.* **2004**, *4*, 295–308. [CrossRef]

43. Wagenaar, D.J.; de Bruijn, K.M.; Bouwer, L.M.; de Moel, H. Uncertainty in flood damage estimates and its potential effect on investment decisions. *Nat. Hazards Earth Syst. Sci.* **2016**, *16*, 1–14. [CrossRef]

44. Chinh, D.T.; Dung, N.V.; Gain, A.K.; Kreibich, H. Flood Loss Models and Risk Analysis for Private Households in Can Tho City, Vietnam. *Water* **2017**, *9*, 313. [CrossRef]

45. Maqsood, T.; Wehner, M.; Ryu, H.; Edwards, M.; Dale, K.; Miller, V. *GAR15 Vulnerability Functions: Reporting on the UNISDR/GA SE Asian Regional Workshop on Structural Vulnerability Models for the GAR Global Risk Assessment*; Geoscience Australia: Canberra, Australia, 2014; p. 74. [CrossRef]

Review

Flood Risk Analysis and Assessment, Applications and Uncertainties: A Bibliometric Review

Andrés Díez-Herrero [1,†] and Julio Garrote [2,*]

[1] Geological Hazards Division, Geological Survey of Spain (IGME), 28003 Madrid, Spain; andres.diez@igme.es
[2] Department of Geodynamics, Stratigraphy and Paleontology, Complutense University of Madrid, 28040 Madrid, Spain
* Correspondence: juliog@ucm.es; Tel.: +34-91-394-4850
† Associated Researcher to IMDEA-Water.

Received: 22 May 2020; Accepted: 16 July 2020; Published: 18 July 2020

Abstract: Studies looking at flood risk analysis and assessment (FRA) reviews are not customary, and they usually approach to methodological and spatial scale issues, uncertainty, mapping or economic damage topics. However, most of these reviews provide a snapshot of the scientific state of the art of FRA that shows only a partial view, focused on a limited number of selected methods and approaches. In this paper, we apply a bibliometric analysis using the Web of Science (WoS) database to assess the historic evolution and future prospects (emerging fields of application) of FRA. The scientific production of FRA has increased considerably in the past decade. At the beginning, US researchers dominated the field, but now they have been overtaken by the Chinese. The Netherlands and Germany may be highlighted for their more complete analyses and assessments (e.g., including an uncertainty analysis of FRA results), and this can be related to the presence of competitive research groups focused on FRA. Regarding FRA fields of application, resilience analysis shows some growth in recent years while land planning, risk perception and risk warning show a slight decrease in the number of papers published. Global warming appears to dominate part of future FRA production, which affects both fluvial and coastal floods. This, together with the improvement of economic evaluation and psycho-social analysis, appear to be the main trends for the future evolution of FRA. Finally, we cannot ignore the increase in analysis using big data analysis, machine learning techniques, and remote sensing data (particularly in the case of UAV sensors data).

Keywords: flood risk analysis; flood risk assessment; flood damage uncertainty; flood hazard; flood vulnerability

---

1. Introduction, Materials and Methods

Flood risk analysis and assessment (FRA) is a scientific subdiscipline or a set of techniques regarding the quantitative analysis and evaluation of flood risk, its components, variables and parameters ([1], in this Special Issue). The modern concept of FRA is based on a combination of flood hazard, probability and potential negative consequences of floods for human health, economic activities, the environment and cultural heritage [2].

There are several published papers and chapters of books that present overall reviews of FRA [2–7]. Thematic reviews have also been published of different FRA approaches, regarding: methodological and spatial scale issues [8–10]; uncertainty [11,12]; mapping [13]; global scale risk analyses and assessments [14] and economic damage [15], among others.

Nevertheless, most of these thematic reviews provide a snapshot or a partial point of view of FRA. They focus only on a small number of selected methods and approaches (usually developed by the authors of the review). Additionally, they do not allow us to understand the temporal evolution of this subdiscipline or interpret the emerging fields of applications and the main sources of uncertainty. Only a bibliometric analysis of scientific databases that includes all of the technical literature can show us the range of worldwide flood risk analysis and assessment work over time.

Bibliometric studies have become widespread in many fields since 2006. In medical science, this technique has allowed us to analyze the emerging trends, and in economics and finance it has allowed us to discover useful information that is not visible using other bibliographical approaches [16]. In the scientific field of FRA, the relevant literature has been abundant (over 7000 references; see the following sections), but there are not many bibliometric systematic analyses (less than 20 references). Only partial bibliometric analyses are available: urban flood risk in China [17]; urban flood vulnerability [18]; coastal flooding [19]; group decision-making techniques for risk analysis and assessment [20]; environmental risks and impacts [21]; multi-criteria decision-making methods in risk analysis and assessment [22]; availability of global-scale datasets to support the study of water-related disasters [23] and among other specific topics and regional reviews. The objective of this global review is thus to carry out a bibliometric analysis of all FRA literature using a well-known worldwide database: Web of Science (WoS).

Web of Science (previously known as 'Web of Knowledge', WoK) is a website that provides subscription-based access to six large interconnected databases ('Science Citation Index Expanded', 'Social Sciences Citation Index', 'Arts and Humanities Citation Index', 'Emerging Sources Citation Index', 'Book Citation Index' and 'Conference Proceedings Citation Index') and several regional databases (Latin America, China, Korea and Russia) that provide comprehensive citation data for many different academic disciplines worldwide. It was originally produced by the Institute for Scientific Information (ISI) and is currently maintained by Clarivate Analytics (previously the Intellectual Property and Science business of Thomson Reuters; [24]). Web of Science Core Collection contains over 1.7 billion cited references from over 155 million records, covering over 34,000 journals. WoS has been chosen as the database to analyze over other sources such as Scopus (Elsevier) or Google Scholar (Google), because it is indisputably the largest citation database available. Moreover, it is more selective in including scientific journal titles in its 'Journal Citation Reports' lists; and Web of Science analytics are the most accurate and trusted source of research evaluation [25]. Of course, the WoS databases are focused on academic based type publications (universities and research centers), their records are skewed to English language science journals from the major academic publishers (i.e., Taylor and Francis, Elsevier, Nature-Springer) and ignore the huge grey literature coming from technical application of this research to management (public administrations) and practice (companies). However, they are the more homogeneous reviewed databases and avoid the need to eliminate redundancies and duplicate records. At the same time, WoS give us homogeneity about published papers quality, as all of them have been revised in a peer-review process (this key point is not always satisficed by grey literature).

The 'Core Collection' of WoS was consulted and several thousand records were extracted using searches to perform a classical bibliometric analysis, using the available tools on the 'Results Analysis' menu. These include the fields (Figure 1): research areas, publication years, databases, document types, authors (with and without Chinese), countries/regions (±Chinese), source titles (±Chinese), meeting titles, group/corporate authors, languages, institutions (±Chinese) and general categories. For each field of results analysis, a visualization treemap or bar graph can be shown; and tables sorted by 'Record count' or by a 'Selected field'; can both be downloaded as image files (.jpg) or tab-delimited text files (.txt), respectively. All these data have been analyzed using a common spreadsheet and other office automation applications.

All of these bibliometric results and statistics were interpreted and discussed by consulting and comparing the scientific literature and selected webpage content (see the References section). Finally, the main conclusions regarding FRA trends are highlighted, as well as the current main topic fields, and those that are currently underdeveloped.

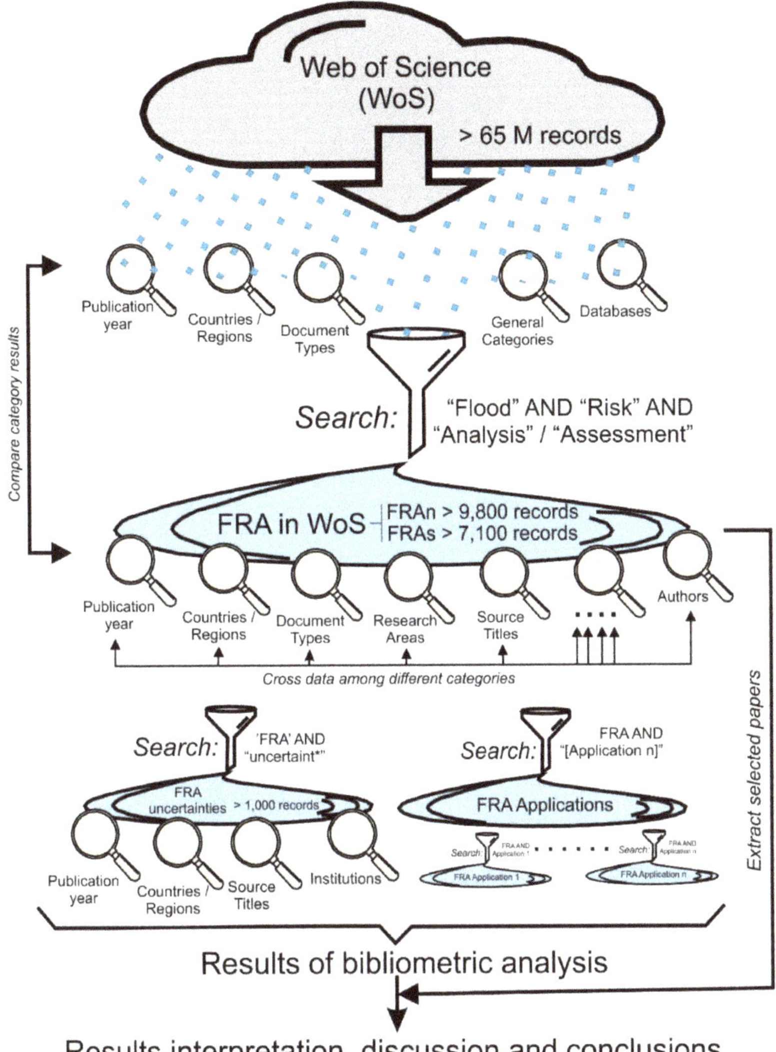

Figure 1. General methodological scheme from the data sources (Web of Science (WoS) databases) to final conclusions, throughout the different bibliometric analyses (magnifying glasses), actions (in cursive letters), search criteria (funnels) and record sets (blue pools).

2. Recent Evolution of FRA in the Scientific Literature

Flood risk analysis and assessment is a recurrent theme in scientific and technical literature, at least in the past decades. Some of the logical queries using different combinations of FRA terms in WoS databases are shown in Table 1. The more statistically representative combinations of query terms related to all FRA words are: the sum of terms "Flood" AND "Risk" AND "Analysis" (FRAn, 9840 records, updated at 22 June 2020); and the sum of terms "Flood" AND "Risk" AND "Assessment" (FRAs, 7166 records; updated at 19 January 2020). So, both record sets (FRAn and FRAs) will be analyzed in detail in the next sections.

Table 1. List of different logical queries using the flood risk analysis and assessment (FRA) terms in the WoS databases and the obtained records (updated at 22 June 2020; * updated at 19 January 2020; ** updated at 4 July 2020).

Logical Query of FRA Terms in WoS Databases	Records
"Flood" AND "Risk" AND "Analysis" (FRAn)	9840
"Flood" AND "Risk" AND "Assessment" (FRAs)	7166 *
"Flood" AND "Risk" AND "Analysis" AND "Assessment"	3969
"Flood" AND "Risk" AND "Analysis" NOT "Assessment"	5871
"Flood" AND "Risk" NOT "Analysis" NOT "Assessment"	9439
"Flood Risk Analysis" AND "Assessment"	156
"Flood Risk Analysis" NOT "Assessment"	167
"Flood Risk Assessment" AND "Analysis"	493
"Flood Risk Assessment" NOT "Analysis"	405
"Flood Risk Analysis" NOT "Flood Risk Assessment"	279
"Flood Risk Assessment" AND "Flood Risk Analysis"	44
"Flood Risk Assessment" NOT "Flood Risk Analysis"	854
"Flood" AND "Risk" AND "Analysis" AND "Uncertain *"	1444
"Flood" AND "Risk" AND "Assessment" AND "Uncertain *"	1093 *
"Flood" AND "Risk" AND "Forecasting"	1551 **
"Flood" AND "Forecasting"	7841 **
"Flood" AND "Risk" AND "Manage *" (FRMg)	11,489
"Flood" AND "Risk"	22,934

The search in the WoS webpage using key words in the 'Topic' field, only guarantees the search in the publication title, abstract and author keywords (and Keywords Plus by WoS), but not in the full text of the publication. The statistical distribution of these records according to different criteria allows us to interpret the worldwide thematic, temporal and geographical evolution of FRA.

2.1. Document Types, Databases and Year of Publication

Most of the FRA publications in the WoS databases are 'Articles' or journal papers (86% FRAn and 84.5% FRAs), followed by communications and presentations at 'Meetings' and conferences (17.5% FRAn and 19.8% FRAs), 'Reviews' (2.8% FRAn and 3.8% FRAs) and 'Books' (2.0% FRAn and 3.1% FRAs). Other document types, like editorials, early access, abstracts and news have percentages of less than 1.5%. This is a logical distribution, taking into account that, in the global database, articles also accounted for 66% of the records. What is striking is the very low percentage of patents (0.1%) compared to their presence in the entire WoS database on any subject (22%). Both figures denote a more scientific approach to FRA (with a higher percentage of articles) than a technological or industrial approach (with more patents).

Most FRA publications are in the 'Web of Science' nuclear database (WoS; 9377 records, 95.3% FRAn; 6914 records, 96.5% FRAs), followed by 'Current Contents Connect' (CCC; 6371 records, 64.7% FRAn; 4775 records, 66.6% FRAs) and finally by Medline (1577 records, 16% FRAn; 1326 records, 18.5% FRAs). Articles from the rest of the WoS databases like KJD, SCIELO, RSCI and DIIDW (Clarivate Analytics, 2018) are almost testimonial, less than 2.5%. This distribution is very similar to the global distribution of the database, although with a greater presence of the WoS nuclear database (95.3–96.5% vs. 70%), double the percentage in CCC (64.7–66.6% vs. 34.2%) and half of the percentage in Medline (16–18.5% vs. 32.7%). Once again, we see a more scientific vision of the FRA and its technical application (CCC), compared to more research related to biomedical sciences (Medline).

In the past 25 years (1996–2020), there has been a significant increase in publications on FRA, greater than one order of magnitude, and this growth has continued over time, without any significant decreases. FRA is thus an emerging discipline that has not reached its peak. Curiously, the series shows that when there is a large annual increase in publications, the following year is often similar (or even decreases slightly), forming pairs of characteristic years of growth stabilization, such as

2006–2007, 2013–2014, 2016–2017 and 2018–2019. The most significant increases occurred between 2014 and 2015, and between 2017 and 2018 (Figure 2). These could be related to international provisions or programs that provide impetus for the scientific-technical community, such as the application of the European Flood Directive (EC 60/2007) and its successive cycles, although perhaps with a time lag due to the editing and publication times of the manuscripts. This progressive increase in records over time follows a parallel evolution with the WoS database as a whole, although there is a sharp drop in WoS records between 2009 and 2010 that is not reflected in the temporal evolution of FRA records. Quite the contrary, FRA publications show an increase in the number of registrations per million (Figure 2), which may be related to the effects of the aforementioned European Directive, with a delay of three years.

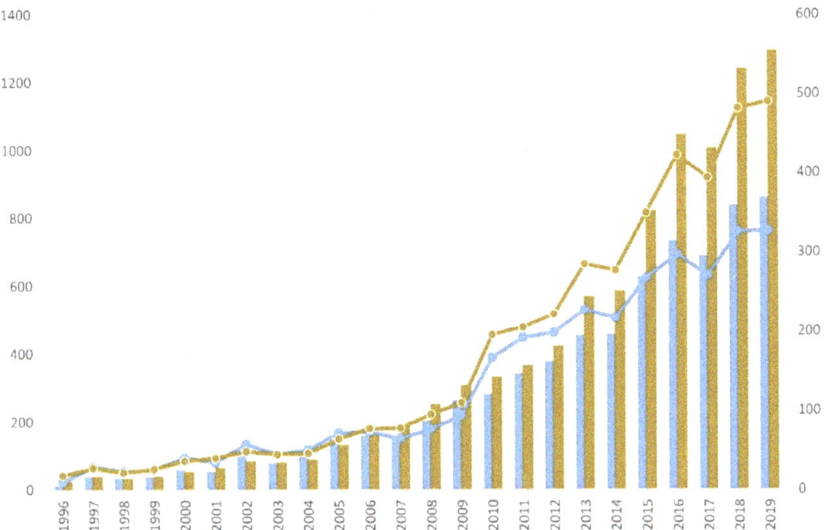

Figure 2. Bar chart showing the number of FRA references (brown bars, FRAn; blue bars, FRAs; records in left axis) in WoS databases classified by 'Publication years', for the last quarter century (1996–2019); dots and lines correspond to the rate per million of FRA records with respect to the total from the WoS database (brown ones, FRAn; blue bars, FRAs; right axis). The sharp jump between 2009 and 2010 must correspond to a change in the WoS database registration system.

2.2. General Categories, Research Areas and Source Titles

In the WoS database there are several general categories, some of which stand out for the number of records in the scientific-technical categories, in this order: science and technology (95.2% FRAn and 97.3% FRAs), physical sciences (73.7% FRAn and 78.9% FRAs), life sciences and biomedicine (63.3% FRAn and 66.9% FRAs), social sciences (44.7% FRAn and 40.3% FRAs) and technology (43% FRAn and 43.2% FRAs). Not so evident are arts and humanities and other categories, which have less than 200 records (<3%). This distribution is logical and correlates with that of the WoS database as a whole, although for FRA the scientific and technological values are slightly higher (FRA 95.2–97.3% vs. WoS 93.3%), and more than double those related to social sciences (FRA 44.7–40.3% vs. WoS 15.8%). This is due to the duality of FRA: a scientific bias for hazard analysis and a social bias for the analysis of exposure and vulnerability.

Regarding 'Research Areas', publications from the 'Environmental Sciences' area predominated (57.6% FRAn; 63% FRAs), followed by 'Water Resources' (45.7% FRAn; 51% FRAs) and 'Meteorology and Atmosphere' (36.7% FRAn; 39% FRAs); and to a lesser extent geology, engineering, mathematics, etc.

(Figure 3). This distribution is logical since FRA often requires multi and interdisciplinary evaluations of flood hazard, exposure and vulnerability, such as those usually developed by environmental sciences. In second place, as expected, was the area that manages flood risks and applies prevention measures (water resources), and following that, the research area that studies one of its main causes and early warning and forecasting (meteorology and atmosphere).

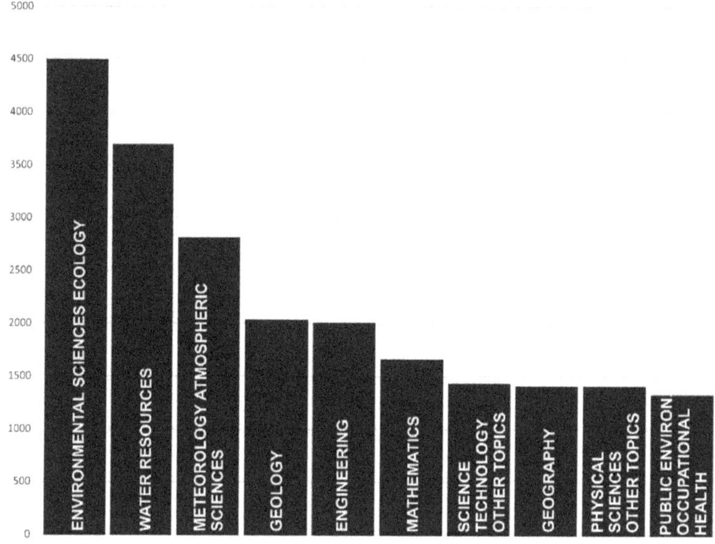

Figure 3. Bar chart showing the number of FRAs references (records) in WoS databases classified by 'Research Areas', indicating the ten main areas.

Among the journals that publish FRA papers, there are three sets of source titles: (i) a high level journal (>350 records), including only 'Natural Hazards' (NatHaz); (ii) middle level journals (between 140 and 250 records for each title), including 'Natural Hazards and Earth System Sciences' (NHESS), 'The Science of the Total Environment' (STOTEN), 'Journal of Hydrology' (JoH), 'Water' and 'Journal of Flood Risk Management' (JFRM) and (iii) low level journals (<125 records), including 'International Journal of Disaster Risk Reduction' (IJDRR), 'Hydrological Processes (HP), 'Water Resources Research' (WRR), 'Water Science and Technology' (WST), 'Risk Analysis' (RA), among others. There are two possible explanations for this polymodal distribution of the source titles: either some of these journals publish a greater number of volumes and issues, and they therefore have a higher number of records (e.g., NatHaz reached a record production of 648 papers in 2015); or the scientific-technical community publish their works in a low number of journals due to their prestige (e.g., NatHaz had an impact factor of 2.319 in 2018) or due to their short editorial process times (e.g., NatHaz has a time span of 277 days of submission to acceptance); or a combination of both. In fact, most referenced sources are categorized as Q1 or Q2 in journal rankings (such as 'Journal Citation Reports' and 'Scopus') in recent times. Another factor that influences this distribution of records is the journals' age, since titles such as 'Water' and JFRM (published since 2009 and 2008, respectively) have monopolized a large part of these works on FRA; compared to older ones such as JoH (1963), WRR (1965), STOTEN (1972), RA (1981), WST (<1982), HP (1986) or NatHaz (1988). NHESS (2001) could possibly be an intermediate case because it appeared at the time when FRA studies began to increase and it monopolized a large part of them. Another case is IJDRR (2012), the most recent, which is a journal that also seems directly focused on that kind of works. Indeed, if these journals are organized taking into account the percentage of articles on FRA they represent with respect to the total scientific production of the journal, three groups can be seen:

(i) a journal with a high proportion of FRAs works, JFRM (24.5%); (ii) journals with an intermediate proportion (6–7%), such as NH, NHESS and IJDRR and (iii) journals with a percentage lower than 2%, including 'Water', JoH, HP, RA, STOTEN, WRR and WST. Furthermore, the evolution of these percentages over time was increasing for the first two groups of journals (Figure 4), but especially in the case of the JFRM, which is on the way to becoming the journal of choice for the scientific-technical community for works on FRA.

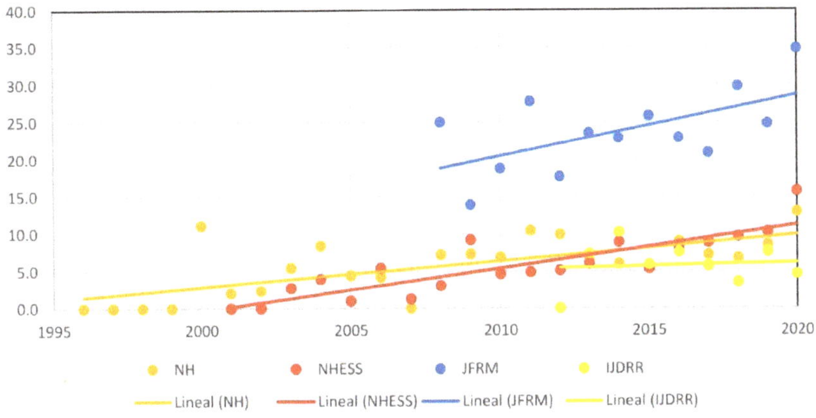

Figure 4. Percentages showing FRAs papers published in each journal versus the total papers published each year, and linear data adjustments.

Regarding the 'Meeting Titles', there is one congress ('3rd European Conference on Flood Risk Management' FLOODrisk 2016, held in Lyon, France, and supported by the Flood Risk Management Research Consortium) that has more than double the number of records (66 FRAn and 61 FRAs) than the next biggest contributors, which are 35th and 36th IAHR World Conferences, held in The Hague, The Netherlands, and organized by the International Association for Hydro-Environment Engineering and Research; (28 FRAn and 23 FRAs records). The rest of the meetings (FRIAR, IEEE IGARSS, 12th International IAEG Congress) all contribute between 10 and 20 publications.

Related to the main source titles and meeting proceedings, published exclusively in English, FRA papers contained in the WoS were mostly written in English (94.2% FRAn and 95.8% FRAs), as were most of the publications in this database (93%). FRA contributions in other languages (Table 2), such as Korean, French, German, Spanish, Chinese and Russian, were almost incidental in WoS (less than 2.5% of total FRA records).

Table 2. Number of records of FRA terms in WoS databases translated to the other main languages (updated at 22 June 2020).

Query Words Translated to Other Languages	Language			
	Korean	Spanish	French	German
"Flood" AND "Risk"	111	93	7	6
"Flood" AND "Risk" AND "Analysis"	66	35	4	1
"Flood" AND "Risk" AND "Evaluation"	60	10	2	0
"Flood" AND "Risk" AND "Assessment"	60	0	0	1
"Flood" AND "Risk" AND "Management"	1	23	2	5
"Flood" AND "Risk" AND "Estimation"	0	5	0	0

2.3. Countries, Institutions, Corporate/Groups and Authors

The geographical distribution of the countries and regions of the FRA articles, based on the study area and author affiliation, was heterogeneous. The records associated with the United States of America (19.2% FRAn and 18.1% FRAs) stand out, followed at a distance by those from the People's Republic of China (14% FRAn and 12.7% FRAs) and England (10.5% FRAn and 12% FRAs). A succession of European countries (Germany, Italy, The Netherlands) held a smaller percentage (6.5–9%), in addition to other developed countries of Europe, Oceania, North America and Asia (Figure 5). It is a reflection of North American leadership in science and technology in the late 20th century-early 21st century and the Anglo-Saxon dominance in scientific journals, especially in the source titles contained in the WoS. However, the emerging scientific production of FRA in China has been gaining on the US in the past years (and surpassed it in 2019; Figure 6). This is due to the huge population of China and its growing university community, the fact that flood risk is one of the main problems of this country, and the incisive policies of scientific and technological production that have led to the opening of specific sections for Chinese authors in bibliographic databases such as the WoS.

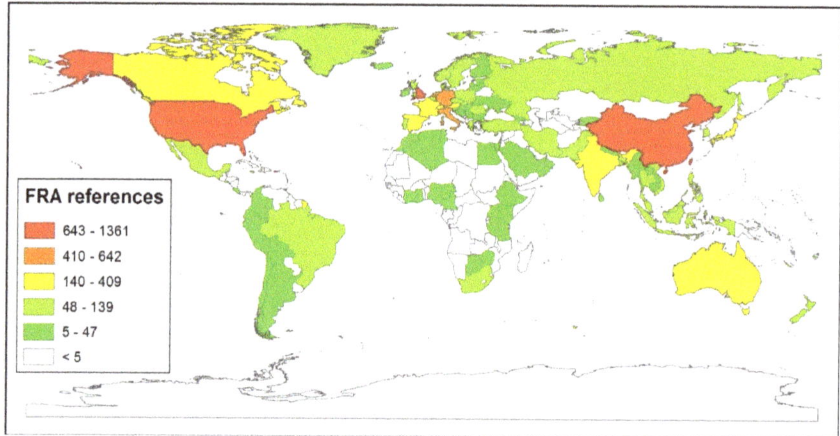

Figure 5. World map showing the number of FRAs references (records) in WoS databases classified by 'Countries /Regions', indicating the top 100 countries.

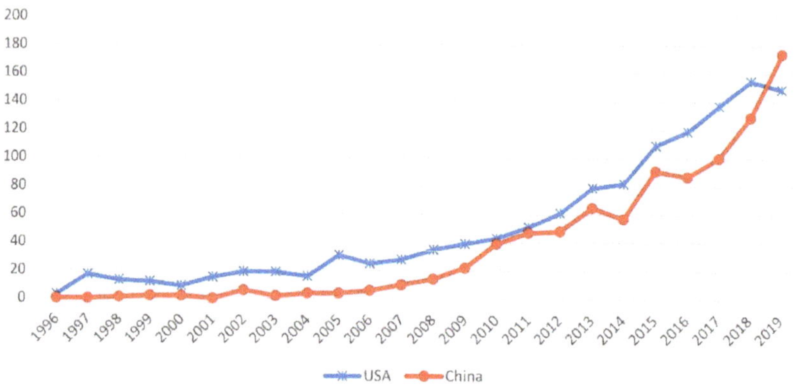

Figure 6. Temporal evolution (1996–2019) of FRAs references (number of records) in WoS databases classified by 'Countries /Regions', comparing production by the US and China.

Related to those emerging countries in these techniques and their scientific policy strategies, the two institutions leading the production of FRA papers are the Helmholtz Association and the Chinese Academy of Sciences. This is followed by a uniform distribution of publications among other many European and North American institutions (Figure 7), where the production of manuscripts by institutions from a relatively small country, like The Netherlands, must be highlighted due to the third and fourth position of the universities of Amsterdam and Delft.

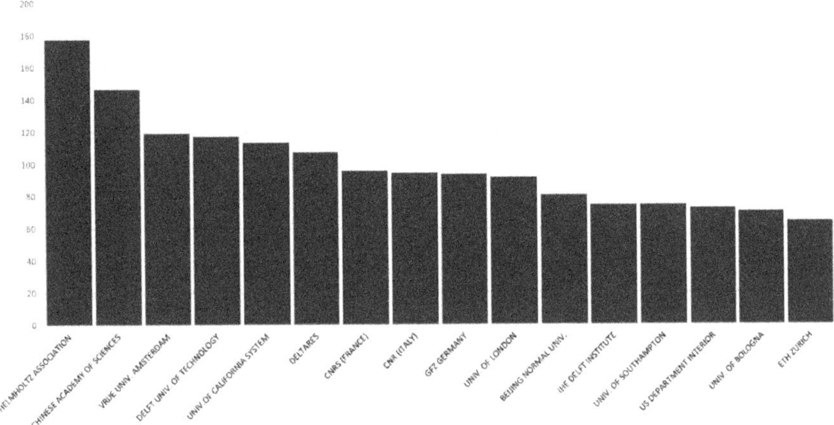

Figure 7. Bar chart showing the number of FRAs references (records) in WoS databases related to the main international R&D Institutions.

The Helmholtz Association of German Research Centers (in German, Helmholtz-Gemeinschaft Deutscher Forschungszentren) was created in 1995 to formalize existing relationships between several globally renowned independent research centers [26]. The Helmholtz Association contributes to solving important challenges facing society, science and industry by conducting top-level research in the fields of aeronautics, space and transport; Earth and environment; energy; health; matter and key technologies. It distributes core funding from the German Federal Ministry of Education and Research (BMBF) to its, now, 19 autonomous research centers and evaluates their effectiveness against the highest international standards. Among them, FRA production comes mainly from centers like the German Research Center for Geosciences (GFZ), at Potsdam. GFZ currently has a staff of around 1180 people and is active in all fields of Earth science. Its hydrology section has internationally recognized experience in research in the entire flood risk chain and in risk analysis and assessment methods. They develop simulation models of the various processes and calculate probabilities of occurrence and intensity for extreme hydrological events.

The Chinese Academy of Sciences (CAS) is the cornerstone of China's ambition to explore and harness high technology and natural sciences for the benefit of both this country and the rest of the world. CAS was established on 1 November 1949, in Beijing, where it is headquartered. It was created by bringing together several existing scientific institutes and soon welcomed over 200 returning scientists who provided CAS with the high-level expertise they had acquired abroad [27]. Among the most important CAS branches and institutes for FRA production are the Institute of Atmospheric Physics (IAP), the Center for Earth Observation and Digital Earth or the Institute of Hydroecology, Ministry of Water Resources.

Finally, the Department of Water and Climate Risk (Vrije Universiteit Amsterdam) is a dynamic group of 45 international researchers, faculty members and PhD students in the Institute for Environmental Studies (IVM). The department is led by Prof. Jeroen Aerts (Head) and Mr. Philip Ward (Deputy Head).

For the United States of America, despite being the country with the highest scientific production in FRA, the records are distributed among multiple universities (i.e., University of California system; included as a whole in the WoS databases) and isolated research centers, so that none stands out among the top positions. Conversely, the Chinese Academy of Sciences stands out by far for Chinese scientific production, providing most of the records from that country. Institutions in European countries and other centers in North America, Asia and Oceania are in an intermediate situation, except the German institutions, mainly concentrated in the Helmholtz Association.

Contrary to what happens with research institutions, there is no established tradition for using the names of research groups or corporate authors in scientific-technical publications. For this reason, in the list of research groups that undertake FRA studies, the records are few and widely scattered. Only one corporate author, the Institute of Electrical and Electronics Engineers (IEEE), accounts for most of the authors of publications (137 FRAn and 85 FRAs). It is followed by the Institute of Physics (IOP; 43 FRAn and 21 FRAs) and Surveying Geology and Mining Ecology Management-SGEM (42 FRAn and 32 FRAs), with about a third of the IEEE's production; and other minor group/corporate authors (The American Society of Mechanical Engineers, ASME; 32 FRAn and 23 FRAs). The other groups (Geobalcanica Soc., United States Soil Department-USSD, Centers for Disease Control and Prevention-CDC, United States Committee on Large Dams-USCOLD, etc.) have less than 10 FRA records. IEEE, more than a group of corporate authors, is actually a global community. Its aim is to innovate for a better tomorrow through its 419,000+ members in over 160 countries by way of its highly cited publications, conferences, technology standards and professional and educational activities in engineering, computing and technology information. Similarly, SGEM is a group of scientists and technicians that organizes the International Multidisciplinary Scientific GeoConference. Their aim is to bring together leading academic scientists, researchers and research scholars to exchange and share their experiences and research results in the SGEM conferences on all aspects of Earth and planetary sciences. It also provides a premier interdisciplinary platform for researchers, practitioners and educators to present and discuss the most recent innovations, trends and concerns, as well as the practical challenges encountered and the solutions adopted in the fields of Earth and planetary sciences.

Nevertheless, several featured authors account for most of the group's scientific production on FRA. Only a dozen authors exceed 30 records FRAn or 20 records FRAs (nFRAn-nFRAs) in the WoS database: Kreibich, H. (49–48); Merz, B. (43–45); Aerts, J.C.J.H. (41–35); Nicholls, R.J. (30–33); Hall, J.W. (43–30); Thieken, A.H. (27–30); Ward, P.J. (25–27); Singh, V.P. (32–8); Jonkman, S.N. (29–24); Fuchs, S. (18–23); Wang, J. (36–23); Zelenakova, M. (17–23); Zhang, Y. (32–23); Bouwer, L.M. (14–22); Li, Y. (32–22) and Chen, C.W. (24–21). All of them are prestigious researchers with high impact parameters, with a Publons h-index and ResearchGate Score-RG usually over 30. They lead groups with dynamic scientists and technicians that are attached to powerful institutions. This leads us to suppose that, at least in the FRA research field, it is more productive (in terms of publications) to have teams that provide a critical mass than the individual works of isolated researchers. An overview of some of their research profiles (FRA records > 25) is:

- Heidi Kreibich (h = 31; RG = 37.51) is head of the working group "Flood risk and climate adaptation" (8 members) at the Hydrology Section, German Research Center for Geosciences (GFZ). Her research is focused on flood risk assessment and mitigation with a background in Environmental engineering, Hydrology and Geography.

- Bruno Merz (h and RG unavailable) is head of the Hydrology Section in the German Research Center for Geosciences (GFZ), where he led one of the most influential research groups on FRA studies and on European flood prevention policies. His research interests are: flood risk, hydrological extremes, multi-hazards, monitoring and simulation of hydrological/hydraulic processes, risk analysis and non-linear statistical methods.

- Jeroen C.J.H. Aerts (h = 46; RG = 29.48), holds a Ph.D. in Hydrology and Risk Assessment and is director of the Institute for Environmental Studies (IVM), as well as head of the 'Water and Climate

Risk' department, and currently supervises a research group, with 6 academic staff, 20 PhD students and 5 post-docs addressing various aspects of water and climate risk management.

- Robert J. Nicholls (h = 53; RG= -) is Professor of Coastal Engineering at the Engineering and Physical Science Department at the University of Southampton (United Kingdom). His research is mainly focused on long-term coastal engineering and management, especially the issues of coastal impacts and adaptation to climate change, with emphasis on sea-level rise.
- Jim W. Hall (h = 18; RG = 42.38) is Professor of Climate and Environmental Risks at the University of Oxford (United Kingdom) and Director of Research in the School of Geography and the Environment. He is internationally recognized for his research on risk analysis and decision making under uncertainty for water resource systems, flood and coastal risk management, infrastructure systems and adaptation to climate change.
- Annegret H. Thieken (h = 37; RG = 36.44) currently works at the Institute of Environmental Science and Geography (part of the former Institute of Earth and Environmental Science), Universität Potsdam, Germany, where she is head of the working group "Geography and Disaster Risk Research" (12 members). Her team investigates natural hazards and risks, particularly flooding. One ongoing project is the research training group NatRiskChange—Natural Hazards and Risks in a Changing World.
- Philip J. Ward (h = 37; RG = 39.04), holds a Ph.D. in Earth and Life Sciences and is senior researcher at the Institute for Environmental Studies (IVM) at the Vrije Universiteit Amsterdam (Amsterdam, The Netherlands), where he is Lab head supervising five members.

As can be seen, several of the top authors work in the same research centers (GFZ: Kreibich, Merz and Thieken; IVM: Aerts and Ward), which again demonstrate the importance of having a critical mass of researchers with interrelated teams. On the other hand, it is curious that some famous and prestigious authors in the field of flood risk, such as Günter Blöschl (Vienna University of Technology, Vienna, Austria), Marco Borga (University of Padova, Padova, Italy), Bruno Mazzorana (Austral University of Chile) or Lorenzo Marchi (Italian National Research Council), among others, do not lead the ranking in number of FRA publications in the WoS database. Perhaps many of their contributions are categorized as flood hazard analysis or simply as basic studies of the flooding phenomenon.

2.4. Research Spotlight on FRAs

Regarding the research spotlight on FRAs, a citation report has been created for the 7421 FRAs records in the WoS (20 April 2020). The average number of citations per item was 16.91, and per year 2852. The sum of times cited was 125,478 (106,991 without self-citations). Citing articles were 77,718 (73,738 without self-citations). The oldest record was published in 1976 and the newest one this year (2020).

Sorting all records by number of times cited, there were: 80 items with citations over ten times the average (170); 16 items over twenty times (339) and only one item over thirty times or more. Of the top 20 items by times cited, there were at least 14 records directly related to FRAs as the main theme of papers (Table 3). Note that the main themes of the top-cited papers were evaluated subjectively based primarily on title and abstract.

These 14 items share a number of common characteristics. Most of them were published recently (2005–2015), and despite this, they had a high number of citations in a short time. All of them had two or more authors (2–17), with a mean of 5.2 authors per paper. This is because some of the selected items were reviews and syntheses made by huge groups of collaborators and international consortia. Global or continental studies predominate (17/20) over regional or national papers (3/20). The most frequently repeated themes were related to climate change (global change, greenhouse effect and sea-level rise) and their impacts on flood risk (9/20). Other common themes were flood risk perception and communication (4/20), coastal flood risk (4/20) and relation to other natural hazards, such as landslides (2/20) or glacial hazards. Finally, most research spotlights were papers published in some of the most prestigious scientific journals (Q1), related to climate change and natural risks, like Nature Climate Change (2), Risk Analysis (2), Nature, PNAS, PloS One, Climate Change, Global and Planetary Change, NHESS, etc.

Table 3. List showing the 20 most cited FRAs records (research spotlights) in WoS databases, including: times cited, publication year, number of authors (No. A.), journal title abbreviations and the main theme of papers. *, directly related to FRA; and the mean values and modal words (M).

No.	Cites	Year	No. A.	*Journal Abbreviation*	Paper Main Theme
1	1299	1999	4	*Geomorphology*	Landslide hazard Italy
2	776	2005	2	*Global Environ Chang*	Adaptation to climate change *
3	670	2013	4	*Nat Clim Change*	Flood losses in coastal cities *
4	651	2013	8	*Nat Clim Change*	Flood risk under climate change *
5	559	2010	3	*Annu Rev Plant Biol*	Arsenic as food contaminant
6	508	2008	3	*Psychol Med*	Psychology of disasters Review *
7	487	2011	8	*Nature*	Greenhouse and Flood Risk England *
8	463	2006	5	*Clim Change*	Global Change, Floods and Droughts *
9	455	2010	4	*Nat Hazard Earth Sys*	Economic flood risk assessment *
10	422	2000	2	*Quatern Int*	Glacial hazards Himalayas
11	400	2015	4	*Plos One*	Exposure and coastal flooding *
12	395	2013	4	*Risk Anal*	Risk perception and communication *
13	381	2006	5	*Global Planet Change*	Sea-level rise and deltas flooding *
14	368	2014	10	*P Natl Acad Sci USA*	Coastal flood damages *
15	361	2009	5	*Environ Int*	Climate Change and water quality
16	344	2010	3	*J Geophys Res-Atmos*	IPCC scenarios and precipitation *
17	329	2010	7	*Philos T Roy Soc B*	C.C. and agricultural productivity
18	316	2014	17	*Hydrolog Sci J*	Flood risk and Climate Change *
19	292	2012	3	*Risk Anal*	Flood risk perception *
20	290	2005	3	*Remote Sens Environ*	Remote sensing and landslides
M	488.3	2009	5.2	*'Change'*	Flood and Climate Change

Among these 14 selected items, five papers can be highlighted for their global contributions to FRA research, especially regarding the consequences of global change for flood risk. Two of them were published in the same issue and volume of Nature Climate Change. "Future flood losses in major coastal cities", Hallegatte et al. [28], provided a quantification of present flood losses in the 136 largest coastal cities (average of 6 billion USD per year). It identified the cities that seem most vulnerable to climate change trends, that is, where the largest increase in losses can be expected in the future (up to 50%). "Global flood risk under climate change", Hirabayashi et al. [29], proposed a global river routing model with an inundation scheme to compute river discharge and inundation area. It showed that global exposure to floods would increase in certain areas of the World depending on the degree of warming. Another review article on "Assessment of economic flood damage" (Merz et al., [15]) reviewed the state of the art and identified research directions of economic flood damage assessment, which required much greater efforts for empirical and synthetic data collection and for providing consistent data to scientists and practitioners. The paper "Flood risk and climate change: global and regional perspectives" (Kundzewicz et al. [30]) provided a holistic perspective on changing rainfall-driven flood risk for the late 20th and early 21st centuries, assessing the literature included in the IPCC SREX report and new literature published since then. Finally, the paper "A Review of Risk Perceptions and Other Factors that Influence Flood Mitigation Behavior" (Bubeck et al., [31]) provided a review of factors that motivate precautionary behavior other than risk perceptions. It concluded that the current focus on risk perceptions as a means to explain and promote private flood mitigation behavior is not supported on either theoretical or empirical grounds.

2.5. Including "Uncertainty" as a Theme

The analysis of uncertainties is increasingly important in FRA, not only because the analysis and assessment involve many different data sources and methodologies, each one with their associated uncertainties, but because of their spread or propagation throughout the analysis and assessment process. Two different sources of uncertainty, aleatory uncertainty (due to natural and anthropogenic variability) and epistemic uncertainty (due to incomplete knowledge of the system), must be analyzed [11].

If a new term related to uncertainty ("uncertaint*") was entered in the search string (Figure 1), the number of search results dropped significantly, going from the previous 9840 FRAn and 7166 FRAs to only 1444 and 1093 records, respectively. Furthermore, the number of FRA records related to uncertainties increased exponentially over the last quarter century (Figure 8). However, this effect is misleading, because the percentage of these items with respect to the total number of FRA publications (which increased in the same proportion) remained between 10 and 20%, without a clear trend. This shows that although all definitions of risk, risk analysis and assessment include uncertainties (see Díez-Herrero and Garrote [1]; in the Editorial of this Special Issue), the reality is that only a sixth of the FRA publications included the concept in their key topics. The reason for this may be the difficulty of incorporating quantitative uncertainty analysis, and especially its dissemination through the risk analysis and assessment process.

Among the search results, including uncertainties and their distribution according to different categorization criteria, only the following items show remarkable changes with respect to previous analyses (see Sections 2.1–2.3):

- Research areas: the publications in the environmental sciences and water resources areas were more equally represented (65.8% FRAn and 65.6% FRAs; and 63.5% FRAn and 64.7% FRAs; respectively), because it was precisely in the latter that a greater number of uncertainty analyses were carried out on parameters of frequency and magnitude of floods for hazard analysis.

- Publication years: although the temporal evolution shows an upward trend, the growth in publications over time was not as continuous and progressive, but rather shows successive decreases and peaks in three-year cycles (compare bar charts in Figures 2 and 8). There were also significant leaps in 2009 first, and then in 2013, possibly due to improvements in uncertainty analysis procedures.

- Countries/regions: the number of publications from England and the US became similar; the People's Republic of China moved from second (see Figure 5) to third position (FRAn) or to the sixth position (FRAs) and the rest of the European countries appeared with lower percentages in differences, especially Germany, Italy and The Netherlands. Perhaps English-speaking countries have a longer tradition of analyzing uncertainties than the Chinese, who conduct research projects that are not as conceptual and more pragmatic. This could indicate that those countries with a longer tradition (such as the US and the United Kingdom) are also ahead in terms of including uncertainty, while China, which is a newer player, mostly uses methodologies that were standardized long ago and that do not consider uncertainty.

- Source titles: Journal of Hydrology (FRAn) and NHESS journal (FRAs) surpassed NatHaz. Additionally, the percentages of the main sources were more equal than in previous analyses, especially those that were more quantitative and that promoted more scientific publications, compared to the more descriptive and applied ones.

- Institutions: Delft University of Technology surpassed Helmholtz for FRAn; although the first position in the ranking of publications for FRAs was still occupied by Helmholtz. The following places were now occupied by prestigious European technical institutes with a strong focus on river engineering (Bristol, Deltares, Oxford) and not so much by scientific institutions. This could indicate that the institutions that the use of more of an engineering approach may be better prepared to cope with the spread of uncertainty throughout the process, unlike those with a different focus that may have less statistical knowledge.

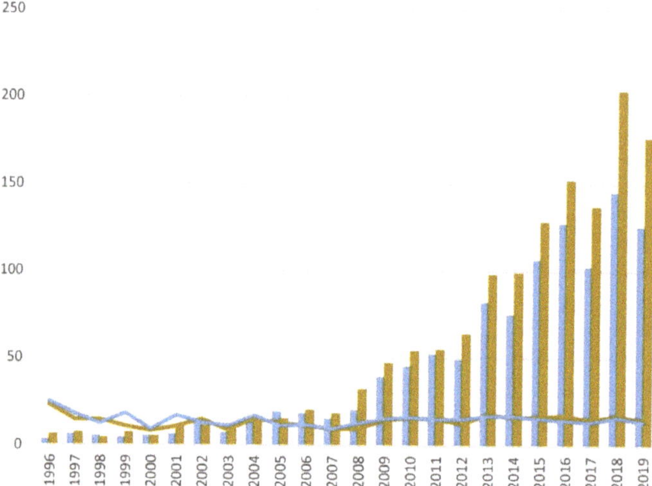

Figure 8. Bar chart showing the number of FRA references (records) in WoS databases including uncertainties, for the last quarter century (1996–2019); brown color, FRAn; blue color, FRAs. The lines represent the percentages of the uncertainty-related records versus the total FRA records over time.

2.6. Regarding the Most Used Approaches and Methodologies

Regarding the most frequent methodological approaches for flood risk analysis and assessment (frequency estimation, use of magnitude-damage functions, etc.), the specialized bibliography, following some of the items listed by Merz et al. [15], offers the results shown in Table 4.

The number of records in the WoS database shows that flood risk analyses and assessments mainly used hazard analysis, to a lesser extent vulnerability, and even much less exposure. Within flood hazard analysis, geological methods may seem to be predominant, but this misleading result is due to references to the geological context of the studies, and not so much to the use of geological analysis itself. It is evident that hydrological modeling and the flood frequency analysis were used the most, compared to other methods and data sources, such as historical-documentary and palaeohydrological, which were almost incidental. Within the hydrologic–hydraulic methodologies, the recent trend shows an increase in hydrologic and hydraulic modeling (probably related to an increase in the study of small ungauged basins), and a decrease in flood frequency analysis (Figure 9).

Regarding flood exposure, published studies mainly deal with population and land use data, compared to a limited use of the cadastre and the CORINE-Land Cover inventory. Regarding flood vulnerability, although frequently cited, quantitative analyses are performed only occasionally, and in those cases they mainly use the damage functions, specifically those that look at depth-damage. Finally, the widespread use of geographic information systems (GIS) and maps in most of the published works is striking. Common methodologies and techniques, like cost–benefit analysis (CBA) and Monte Carlo, had only been cited in 7% and 2.5% publications, respectively. The three main components of flood risk (hazard, exposure and vulnerability), as well as the use of GIS and flood mapping, had evolved over time (1996–2019) with a sustained upward and parallel trend, with no significant variations between them.

Table 4. Number of records in the Web of Science (WoS) database regarding different methods, approaches and techniques of "flood risk analysis and assessment". The asterisk (*) is a commonly used wildcard symbol that broadens a search by finding words that start with the same letters. Searched on 17 March 2020.

Risk Factor	Method/Approach	Technique	Search Word(s)	WoS Records	
				FRAn	FRAs
HAZARD	Hydrologic-Hydraulic	Hydrologic Modeling	"hydrologic * model *"	416	317
		Rainfall-runoff transf.	"rainfall-runoff"	197	147
		Frequency Analysis	"frequency analysis"	610	234
		Hydraulic Modeling	"hydraulic * model *"	303	222
	Geosciences	Geological	Geologic *	455	423
		Geomorphological	Geomorphologic *	196	126
	Historical Paleo-hydrol.	Documentary	documentary	65	35
		Paleo-hydrological	pal(a)eo-hydrologic *	7 + 7	4 + 7
	Hazard analysis		"hazard anal *"	96	58
	Hazard		hazard	2775	2540
EXPOSURE	Social	Census	census	105	77
		Population	population	1464	1133
	Economic	Land uses	"land use *"	947	820
		Cadastre	cadastre	10	8
		CORINE	corine	7	5
	Exposure analysis		"exposure anal *"	20	11
	Exposure		exposure	895	868
VULNERABILITY	Social Vulnerability		"social vulnerability *"	214	189
	Economic Vulnerability		"economic * vulnerability *"	30	33
	Damage functions	Damage functions	"damage function*"	79	74
		Magnitude-damage	"magnitude* damage *"	3	1
		Magnitude-frequency	"magnitude* frequenc *"	15	13
		Depth-damage	"depth* damage *"	39	51
		Depth-frequency	"depth* frequenc *"	1	0
	Vulnerability analysis		"vulnerability anal *"	109	56
	Vulnerability		vulnerability	1842	1689
OTHER	Susceptibility analysis		"susceptibility anal *"	10	9
	Floodprone areas		floodprone	6	4
	Common methodologies	Cost-benefit analysis	Benefit *	812	507
			cba	23	10
		Monte Carlo	"monte * carlo *"	310	183
	Representation techniques	GIS	gis	1020	889
		Mapping	mapping	1646	1401

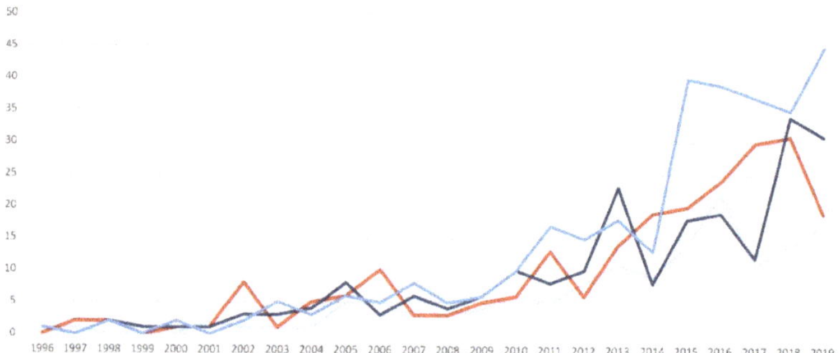

Figure 9. Temporal evolution (1996–2019) of FRAs references (number of records) in WoS databases classified by hydrologic–hydraulic methodologies: red line, flood frequency analysis; clear blue, rainfall–runoff transformation; middle blue, hydrologic modeling; deep blue, hydraulic modeling.

2.7. Fields of Application

Another interesting exercise is to organize scientific-technical production according to the main fields of application of flood risk analysis and assessment [13], which allows us to have a perspective of their past and future usage. The results of organizing the fields of application into general aspects and predictive, preventive and corrective applications are presented in Table 5.

Preventive applications that focus on risk management predominated, as opposed to corrective ones in civil protection, which were a minority. This distribution is logical, bearing in mind that the scientific-technical community is much more dedicated to preventive studies than to acting in emergencies.

Non-structural preventive applications predominated and, among them, land planning applications. This is more common in natural areas than in those related to urban planning. The next group would be applications for improving resilience, which were much more common than other non-structural applications of a psycho-social type, such as increasing perception, communication and dissemination of flood risk or the establishment of insurance and reinsurance systems. Among the structural measures employed, civil works covered most of the references and, although they were gaining more and more prominence (Figure 10), there were still few references to natural water retention measures (NWRM) or nature-based solutions (NbS). Predictive applications were dominated by early warning systems associated with weather models; and among the corrective applications, those aimed at emergency management, with fewer references to civil protection.

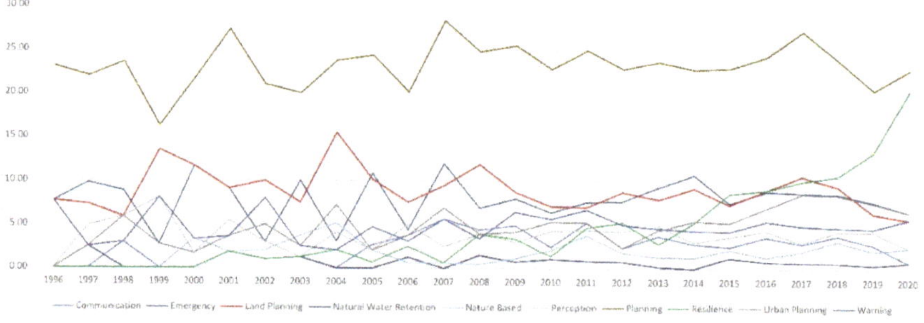

Figure 10. Temporal evolution (1996–2020) of the percentage of FRAs references in WoS databases related to each field of application versus the total of FRAs publications in each year.

Table 5. Number of records in the Web of Science (WoS) database regarding different fields of applications of "flood risk assessment". The asterisk (*) is a commonly used wildcard symbol that broadens a search by finding words that start with the same letters. Searched on 17 March 2020.

Mitigation Measure	Field of Application	Subfield of Application	Search Word(s)	WoS Records	
				FRAn	FRAs
General	Management	-	management	4934	4107
	Application	-	application	1592	1301
Predictive	Warning	General Warning	warning	450	357
			early warning	229	192
	Meteorology	General	Meteorol *	1964	1490
		Modeling	meteorol * model *	1127	885
Preventive	Structural measures	Infrastructures	infrastructure	1031	842
			critical infrastructure	186	160
		Works	works	1157	825
			civil works	61	42
			natural water retention	58	51
			nature based	197	147
	Non-structural measures	Land and Urban Planning	planning	2129	1715
			urban planning	583	439
			land * planning	748	630
		Insurance	insurance	443	278
		Psycho-social measures	education	178	96
			communication	364	213
			dissemination	311	236
			resilience	710	522
			perception	547	282
			awareness	294	206
Corrective (post-disaster)	Emergency	Civil protection Emergency	civil protection	106	72
			emergency	780	607
	Financial	Compensation	compensation	57	31

Another interesting aspect is the evolution over time of civil works referring to various groups of these applications, which were on the rise in almost all cases and in scientific production overall. However, as the total production of publications had also undergone sustained growth, it was necessary to normalize the total number of FRA works published each year, which allowed us to see more clearly the changes in priorities in mitigation policies and their reflection in international scientific-technical production (Figure 10).

As can be seen in Figure 10, the different fields of application analyzed for FRAs maintained similar percentages of scientific production over the past 25 years, with the exception of resilience, which went from zero between 1996 and 2020, to around 20% in 2020, with sustained growth over time. Although the rest of the applications remained in a band of percentages, between 1996 and 2010 they registered an enormous year-on-year fluctuation, with peaks and valleys in the percentages, compared to a period of greater stability between 2011 and 2020. This can be interpreted as a stabilization of the scientific production of these specialties. Only land planning ("land* planning") shows a decrease in production percentage with 15% at the beginning of the century, and barely 5% at present. This may be associated with the increased precision of spatial data. Previously, studies could be less accurate and used a territorial scope to quantify flood damage (economic valuation, €/m^2, occupation and land-use units). Whereas now, as the precision of the starting data increased, we tended to perform more local (urban) analyses and the concept of land use was slightly lost. However, applications that focus on planning had generally been maintained over time, with fluctuations between 20 and 25% of the scientific-technical production.

3. Discussion

This bibliometric review of FRA scientific-technical production presents limitations derived both from the data sources and from the analysis methodology used. The sole use of the WoS database, compared to other bibliometric studies that combine several databases, has both disadvantages and advantages. Hudson and Botzen [32] used three databases (Web of Science, Google Scholar and Google Search) because the use of multiple sources reduces the likelihood of a biased sample [33]. However, the use of a unique and high-quality database (like WoS; Cho and Chang [18] and this work) means that the sample is homogenous and avoids the need to eliminate redundancies and duplicate records. Furthermore, the use of Google's bibliographic databases and its criteria for record selection, classification and keyword searches have been widely criticized for over a decade [34].

However, authors are aware that the bibliometric analysis is not complete when the FRA grey literature, other databases and articles in non-English languages are excluded. The implications of excluding the grey literature in particular, which is more focused the actual application of this research to management and practice, will make a difference. Moreover, perhaps the type of publication included in WoS also biases our search towards a particular conception of conceptual and core research. It is true that technology has been increasingly science-based; but science has been increasingly technology-based—with the choice of problems and the conduct of research often inspired by societal needs [35]. Traditionally, it has been imagined that 'basic' research tends to happen in universities and then later gets applied in practice. However, increasing this division of labor is no longer so clear, particularly in the technological sector, where most of the important innovations take place in the private sector. It has even been proposed to strengthen the public sector, harnessing science and securing societal impacts from publicly funded research is needed [36]. Even in flooding, many of the most important innovations in flood forecasting and flood risk mapping, for example, are getting rolled out in government laboratories (Hydrologic Engineering Center, US) or by private companies, like JBA or Fathom in the UK, or the First Street Foundation in the US. Our search methods could tend to miss out on the work they are doing. So, the implications of this paper focused on academic WoS outputs that will tend to: (i) overlook practitioner grey literature research putting things into practice and (ii) give less emphasis to private sector research, which is increasingly where innovation in the Tech world, and even in FRM, is now coming. However, the complete incorporation of all the grey literature from the technical and governmental worlds is an endless task and, above all, impossible to carry out objectively and homogeneously. Additionally, as we point out before, their inclusion break with the homogeneity linked to the use of WoS database about quality, accessibility or field expert peer-review process of manuscripts.

Regarding the analysis methodology, the search tools provided by WoS do not ensure that the results obtained are complete. As we explained in the Section 2, the search in the WoS application using key words in the 'Topic' field, only guarantees the search in publication title, abstract and author keywords (and keywords Plus by WoS), but not in the full text of the publication. Hudson and Botzen [32] established that not all abstracts and titles were sufficiently informative to judge whether the inclusion criteria are met; so they read the full document to confirm that it met all required criteria. However, this procedure is only possible for sets of tens or hundreds of records. For data sets with over 7000 records (such as the present study), detailed reading of the complete texts is impossible. Moreover, no additional programs to perform a bibliometric analysis, like Sci2 and Gephi, as have been used [18] to generate representative results (e.g., network maps). As the used search terms (words) define the field tends to focus on time-independent assessments of flood risk, it excludes an important body of time-dependent dynamic flood forecasting that sits at the interface of hydrology and meteorology.

However, regarding flood forecasting, it could be noticed that even though there were more than 7841 records in WoS databases, only a fifth of them (1551 records) seemed to include a complete flood risk study (see Table 1). This scenario point out to a major relationship between flood forecasting and flood hazard, more than flood forecasting and flood risk.

Nevertheless, this bibliographical review presents some advantages and significant contributions in comparison with previous partial reviews. Firstly, this review covered all FRA topics and all possible approaches (basis, methodologies and applications), not just a specific subset (e.g., cost–benefit analysis [32] or urban flood vulnerability [18]). Secondly, this review explored all possible search criteria, including their evolution over time by analyzing publications by year (not included in previous reviews) and not only a short period [18]. Finally, all results were interpreted in terms of future prospects and practical conclusions, useful to guide future research.

From the results, a continuous increase in the number of bibliographic references associated with both search terms ("analysis" and "assessment") could be observed, which have different meaning nuances. Additionally, therefore, it cannot be argued on the basis of these results that there has been a tendency to temporarily replace the term "analysis" by that of "assessment". For the Society for Risk Analysis, the main difference between terms 'assessment' and 'analysis' is that the former implies an evaluation and quantification, not only a qualitative decomposition of a complex element into its components or factors. In that way, the productions of more complete studies (as shown in Figure 2) about flood risk were increased with time.

The results show several future trends in FRA research and development. There will be a predominance of the papers or journal articles compared to other types of publication. FRA publications will increase significantly over time (which could be related to the general increase over time of scientific manuscript production), but including pairs of characteristic years of growth stabilization. There may be increases in scientific production in 6-year cycles, due to regulatory requirements (e.g., European Floods Directive). There will be a predominance of multi- and interdisciplinary studies covering several research categories (science and technology, physical sciences and social sciences) and comprising of various research areas (environmental sciences, water resources and meteorology). A small group of source titles (and mainly recently published journals) will retain most of the publications. Although the US and the European countries will continue to lead in publications, Asian and South American countries will grow in number and quality. Research groups with a large critical mass of scientists, or groups of centers and institutes in national federated institutions (such as Helmholtz or CAS) or international associations (such as IEEE) will continue to have priority in high impact publications. Finally, only a reduced group of twenty researchers will maintain leadership. The results show the importance of the consolidated research groups or centers dedicated to these analyses, in the advancement of FRA knowledge and techniques, as opposed to the dissemination of production in different small centers.

At this point, there is a large gap in FRA in the African continent, which can be clearly seen in Figure 5 where the largest region of very low production of papers about FRA is located just in the African continent. This is not related to the absence of flood events, but to the limited number of scientific institutions and data sources (only 261 of the 7454 FRA records in the WoS databases). The lack of appropriate infrastructures and data make development of FRA difficult in African countries, beyond the inclusion of this continent in global risk studies, such as those cited in Ward et al. [14]. The same causes are behind the absence of FRA manuscripts in high impact journals. For example, the first African country in the FRA production ranking is South Africa (50 records, 44th place); followed by Egypt (47, 45th), Nigeria (28, 58th), Ghana (15, 63rd), Morocco (13, 66th), Uganda (13, 67th) and Ethiopia (12, 70th). In most African countries, the best option is the 30 m spatial resolution digital elevation model (DEM) from the SRTM project or the ASTER project, or the SRTM derived 'Bare-Earth' DEM and Multi-Error-Removed Improved-Terrain (MERIT) DEM. The lack of detailed DEMs can be considered as one but not the only key factor of the limited FRA in Africa. Apart from the lack of DEMs for flood hazard analysis, the poor quality or limited availability of flow data must be kept in mind. Furthermore, the analysis of the risk exposure and vulnerability components, which are closely linked to the availability of social data, is also highly compromised by the absence of systematic and quality social data.

Another unique case is that of the Australian continent, where Australia has a production of 401 records (8th position) and New Zealand (75 records, 31st position). However, there is a significant absence of research groups, institutions or individual researchers among those that stand out for their FRA manuscript production.

By examining the main articles that represent the research spotlight on FRA (see Section 2.4), four fields of emerging research and development could be recognized: (i) the impact of global change and its components (greenhouse effect, climate change, global warming and sea-level rise) in the frequency and magnitude of floods and their associated damages, as highlighted by other papers [29,30]; (ii) coastal floods and their economic consequences, especially in the main coastal urban centers, as reported in previous papers [15,28]; (iii) multi-risk studies and interactions between the risk of flooding with other types of natural hazards, such as landslides, volcanic eruptions, seismic phenomena, etc., and (iv) incorporation of the psycho-social aspects of flood risk, with an analysis of the perception and importance of risk communication (e.g., Bubeck et al. [31]).

Apart from the four emerging fields mentioned above, the studies carried out on a global scale (specifically reviewed by Ward et al. [14]) deserve individual mention. These analyses focus mainly on the past decade, sharing some similarities with the majority of meso- or micro-scale studies (an increase in the spatial resolution of the results obtained in recent studies; predominance of studies that focus on flood hazard analysis; prevalence of studies evaluating direct flood damages; importance and number of studies focused on coastal floods; etc.). They also show some differentiating characteristics, among which the conception or use of a time horizon for calculations could be highlighted, which generally cover the entire 21st century and include climatic change scenarios.

As a check and validation of future FRA trends and the fields and topics that seem to have the greatest future projection, the publications of the authors mentioned in Section 2.3 have been analyzed for the last two years (2019 and 2020). The result is illuminating, because the 28 records in the WoS databases published by these seven authors in this last half year (January–June 2020) on FRA, correspond to: psycho-social aspects (six records; e.g., [37,38]), the impact of global change (five records; e.g., [39]), coastal floods (five records; e.g., [40]) and global scale studies (four records; e.g., [41]); or combination of two or more (e.g., [42]). Analyzing the 112 records on FRA of these 7 authors in the years 2019 and 2020, all the emerging topics are also represented, since all but one of the records (111) include references to the global change.

Other possible emerging studies fields that were analyzed in the total production on FRA of the years 2019 and 2020 (total of 1257 records) are: machine learning (33 records), big data (10), location intelligence (1), decision support systems (DSS; 1), remote sensing (74) and socio-environmental systems (1). As can be seen, its representation in the last works carried out on FRA is still minimal compared to the total number of works published, except for the machine learning field and especially for the remote sensing field, which is very broad and used for decades, so it cannot strictly be considered an emerging field (although possibly the use of data provided by UAVs can be considered). So, it is not justified to call location intelligence, DSS and socio-environmental systems "emerging fields" on the basis of a single record in 2019–2020.

Although the percentage of papers (per year) that deal with uncertainty has remained more or less constant, the number of publications has increased along with the total number of papers. This increase should continue in the future, in such a way that the percentage of manuscripts that analyze the uncertainty of their results with respect to the total number of jobs produced in the field of FRA will increase significantly. It is also necessary to identify the contributions of individual sources of uncertainty to overall uncertainty [11]. Moreover, in the coming years, due to factors such as population growth, increasing urbanization, infrastructure decay and the potential impact of climate change, FRA studies must incorporate the uncertainties associated with the spatio-temporal dimension of these factors in flood risk modeling [7].

Within the methodologies for flood hazard analysis, in urban areas with mild terrain and complicated topography, 2-D fully dynamic flood modeling is usually employed [6]. Regarding the methodologies used to assess economic flood damage, considerable research has been carried out and progress made on damage data collection, data analysis and model development in recent years. However, there still seems to be a mismatch between the relevance of damage analyses and assessments and the quality of the available models and datasets [15]. Future FRA studies must be better balanced, paying the same attention to both damage and hazard. Cho and Chang [18] concluded that the continuously increasing number of comprehensive approaches to urban flood risk assessment shows evidence of a new paradigm shift toward a more inclusive way to understand multi-dimensional aspects of urban flood vulnerability across disciplines and different knowledge systems.

The complexity of the flooding process and its analysis and assessment are another obstacle that we must try to overcome, with the aim of transferring analysis methodologies from the scientific research domain to that of private companies. In this sense, the objective would be to try to transfer robust but simple analysis methodologies, in which most of the aspects linked to the uncertainty of the results were easy to implement or could already be tabulated based on the experience and results obtained by scientists.

Ologunorisa and Abawua [3] concluded that the GIS technique appears to be the most promising as it is capable of integrating all the other techniques of flood risk analysis and assessment. Yet, the recent evolution of publications using this tool shows a certain stagnation since it is a complementary technique and not the main tool of FRA studies. This stagnation may be due to the fact that GIS are used as a receptacle for the results of different analyses and as a map generator, but not as a real analytical tool. There are at least two reasons for this. Firstly, information sources do not have standardized formats, which makes it difficult to systematically use GIS as a standard analysis tool. The information must be "formatted" so that the GIS can understand and analyze it. Secondly, there has not been a sufficiently flexible or adaptable development of GIS packages for flood risk analysis, probably due to the aforementioned reason. Only watertight applications have been developed (which do not allow modification or application in other parts of the planet other than those where they were implemented). Many of these have had a macroscale approach (applying land uses in the vulnerability of the territory), like the FRA compute option of HEC-WAT software [43] or the applicability of HEC-FDA. FRA models that do not focus on flood damage analysis and assessment (such as HIS–SSM for The Netherlands, Flood Loss Estimation Model (FLEMO) in Germany, the Multi-Coloured Manual in UK and Rhine Atlas in Germany) are very scarce. Examples that could be mentioned include Hazards U.S. Multi-Hazard (HAZUS-MH) [44,45] in the US, HAZUS-MH 2.1 in UK or the Flood Risk Assessment Model for Companies (FRAC, [46]).

In relation to mitigation measures, applications of NWRM and NbS will increase to the detriment of traditional civil works. All these measures for flood risk mitigation will need a cost–benefit analysis. To this end, Hudson and Botzen [32] proposed a multistep process: (i) determine the spatial extent of the zoning policy and how interconnected the zoned area is to other locations; (ii) conduct a CBA using estimated monetary costs and benefits from an integrated hydro-economic model to investigate whether total benefits exceed total costs; (iii) conduct a sensitivity analysis regarding the main assumptions and (iv) conduct a multicriteria analysis (MCA) of the normative outcomes of a zoning policy.

4. Conclusions

The main characteristics that we can expect from future FRA studies can be inferred from the results of this bibliographic review. We can expect an increase in these studies over time and a more heterogeneous distribution around the world.

By interpreting the past evolution and future prospects and trends in FRA research, we can conclude that countries such as the United States, China and some European countries (Germany and

The Netherlands) will continue to lead. However, certain aspects such as uncertainty analysis in FRA, highlighting production in the United Kingdom and in other Asian countries, will be incorporated in the future. However, it remains unclear how the African countries will join the development and production of FRA.

The success of the concentration of research efforts in large centers or consortia of universities and research institutes (Helmholtz, Chinese Academy of Sciences) is evident. They not only produce a greater number of publications but also those of higher quality and worldwide impact. This situation can have advantages and disadvantages. On the one hand, you have FRA development accompanied by the seal of quality provided by the work done by competent and consolidated research groups. On the other, there may be a loss of "small big-ideas" proposed by isolated researchers, which do not achieve the necessary scientific repercussion. Other countries or continents (such as South America or Africa) will copy and adapt this model to achieve greater weight in world science, trying to overcome the drawbacks mentioned.

Faced with the enormous increase in scientific production, a select few very specialized journals, (e.g., JoH, JFRM, IJDRR and NHESS) will concentrate the majority of scientific production on FRA, becoming the preferred communication mechanisms of the scientific-technical community. Of course, there will be an increasing acceptance of open access initiatives for scientific production, supported by public policies of funding agencies (e.g., EU Horizon, South American CONICYT agencies and US-National Academy of Science).

There are four themes within FRA research that mark future trends: global change, coastal floods, economic evaluation and psycho-social analysis. This last theme, including the analysis of the perception and communication of flood risk, is still undervalued and not sufficiently developed. The most widely used methodologies place a lot of emphasis on the analysis of flood hazard, but very little on the analysis of exposure and vulnerability, and their integration into risk. In the future, FRA studies should focus on the integration of the three factors by giving them equal weight and taking into account the uncertainties associated with the sources, methodologies and their propagation during the analysis. They should also consider how we transfer all this complex methodological analysis to private engineering or environmental companies that, in fact, will conduct most of the FRA in the current "urban-growth" scenario.

In short, FRA research has a relevant past, a prominent present and an innovative and hopeful future to provide solutions to humanity's problems in relation to the reduction of natural disasters.

Author Contributions: All authors have a significant contribution to final version of the paper. Conceptualization, A.D.-H. and J.G.; formal analysis, A.D.-H.; Writing—original draft, A.D.-H. and J.G.; Writing—review and editing, A.D.-H. and J.G. All authors have read and agreed to the published version of the manuscript.

Funding: This research was funded by the project "Design of a methodology to increase flood resilience compatible with improved status of water bodies and sustainable management of water resources (DRAINAGE)", CGL2017–83546–C3–R (MINEICO/AEI/FEDER, UE).

Acknowledgments: Authors gratefully acknowledge the comments of the three anonymous reviewers of the manuscript and the Editor suggestions and comments that have been included in the original manuscript improving its quality.

Conflicts of Interest: The authors declare no conflict of interest.

References

1. Díez-Herrero, A.; Garrote, J. Flood Risk Assessments: Applications and Uncertainties. *Water* **2020**.
2. Solin, L.; Skubincan, P. Flood risk assessment and management: Review of concepts, definitions and methods. *Geogr. J.* **2013**, *65*, 23–44.
3. Ologunorisa, T.E.; Abawua, M.J. Flood risk assessment: A review. *JASEM* **2005**, *9*, 57–63.
4. Cheng, W.; Chen, J.; Liu, D. Review on flood risk assessment. *J. Yangtze River Sci. Res. Inst.* **2010**, *20*, 44–46.
5. Winsemius, H.H.C.; Van Beek, L.P.H.; Jongman, B.; Ward, P.J.; Bouwman, A. A framework for global river flood risk assessments. *Hydrol. Earth Syst. Sci.* **2013**, *17*, 1871–1892. [CrossRef]

6. Tsakiris, G. Flood risk assessment: Concepts, modelling, applications. *Nat. Hazards Earth Syst. Sci.* **2014**, *14*, 1361–1369. [CrossRef]

7. Salman, A.M.; Li, Y. Flood Risk Assessment, Future Trend Modeling, and Risk Communication: A Review of Ongoing Research. *Nat. Hazards Rev.* **2018**, *19*, 04018011. [CrossRef]

8. Hall, J.W.; Deakin, R.; Rosu, C.; Chatterton, J.B.; Sayers, P.B.; Dawson, R.J. A methodology for national-scale flood risk assessment. *Proc. Inst. Civ. Eng. Water Marit. Eng.* **2003**, *156*, 235–247. [CrossRef]

9. Gouldby, B.; Sayers, P.; Mulet-Marti, J.; Hassan, M.A.A.M.; Benwell, D. A methodology for regional-scale flood risk assessment. *Proc. Inst. Civ. Eng. Water Manag.* **2008**, *161*, 169–182. [CrossRef]

10. De Moel, H.; Jongman, B.; Kreibich, H.; Merz, B.; Penning-Rowsell, E.; Ward, P.J. Flood risk assessments at different spatial scales. *Mitig. Adapt. Strat. Glob. Chang.* **2015**, *20*, 865–890. [CrossRef]

11. Apel, H.; Thieken, A.; Merz, B.; Blöschl, G. Flood risk assessment and associated uncertainty. *Nat. Hazards Earth Syst. Sci.* **2004**, *4*, 295–308. [CrossRef]

12. Apel, H.; Merz, B.; Thieken, A. Quantification of uncertainties in flood risk assessments. *Int. J. River Basin Manag.* **2008**, *6*, 149–162. [CrossRef]

13. Díez-Herrero, A.; Lain-Huerta, L.; Llorente-Isidro, M. *A Handbook on Flood Hazard Mapping Methodologies*; Publications of the Geological Survey of Spain: Madrid, Spain, 2009; 190p.

14. Ward, P.J.; Blauhut, V.; Bloemendaal, N.; Daniell, J.E.; De Ruiter, M.C.; Duncan, M.J.; Emberson, R.; Jenkins, S.F.; Kirschbaum, D.; Kunz, M.; et al. Review article: Natural hazard risk assessments at the global scale. *Nat. Hazards Earth Syst. Sci.* **2020**, *20*, 1069–1096. [CrossRef]

15. Merz, B.; Kreibich, H.; Schwarze, R.; Thieken, A. Review article Assessment of economic flood damage. *Nat. Hazards Earth Syst. Sci.* **2010**, *10*, 1697–1724. [CrossRef]

16. Ye, P.; Li, Y.; Zhang, H.; Shen, H. Bibliometric analysis on the research of offshore wind power based on web of science. *Econ. Res. Ekon. Istraživanja* **2020**, *33*, 887–903. [CrossRef]

17. Yin, J.; Ye, M.; Yin, Z.; Xu, S. A review of advances in urban flood risk analysis over China. *Stoch. Environ. Res. Risk Assess.* **2014**, *29*, 1063–1070. [CrossRef]

18. Cho, S.Y.; Chang, H. Recent research approaches to urban flood vulnerability, 2006–2016. *Nat. Hazards* **2017**, *88*, 633–649. [CrossRef]

19. Gao, C.; Ruan, T. Bibliometric Analysis of Global Research Progress on Coastal Flooding 1995–2016. *Chin. Geogr. Sci.* **2018**, *28*, 998–1008. [CrossRef]

20. Munhoz, R.F.; Ribeiro, L.C. Group Decision Making Techniques for Risk Assessment: A Literature Review and Research Directions. In Proceedings of the 2019 IEEE International Conference on Fuzzy Systems (FUZZ-IEEE), New Orleans, LA, USA, 18–21 June 2019; pp. 1–6.

21. Hu, W.; Li, C.-H.; Ye, C.; Wang, J.; Wei, W.-W.; Deng, Y. Research progress on ecological models in the field of water eutrophication: CiteSpace analysis based on data from the ISI web of science database. *Ecol. Model.* **2019**, *410*, 108779. [CrossRef]

22. Liu, H.-C.; Chen, X.-Q.; Duan, C.; Wang, Y.-M. Failure mode and effect analysis using multi-criteria decision making methods: A systematic literature review. *Comput. Ind. Eng.* **2019**, *135*, 881–897. [CrossRef]

23. Lindersson, S.; Brandimarte, L.; Mård, J.; Di Baldassarre, G. A review of freely accessible global datasets for the study of floods, droughts and their interactions with human societies. *Wires Water* **2020**, *7*, e1424. [CrossRef]

24. Clarivate Analytics. Available online: https://clarivate.com/news/acquisition-thomson-reuters-intellectual-property-science-business-onex-baring-asia-completed/ (accessed on 21 March 2020).

25. Clarivate Analytics. Available online: https://clarivate.com/webofsciencegroup/wp-content/uploads/sites/2/2019/08/d6b7faae-3cc2-4186-8985-a6ecc8cce1ee_Crv_WoS_Upsell_Factbook_A4_FA_LR_edits.pdf (accessed on 20 March 2020).

26. Helmholtz Association. Available online: https://www.helmholtz.de/en/about_us/the_association/ (accessed on 13 April 2020).

27. Chinese Academy of Sciences. Available online: http://english.cas.cn/about_us/introduction/201501/t20150114_135284.shtml (accessed on 14 April 2020).

28. Hallegatte, S.; Green, C.; Nicholls, R.J.; Corfee-Morlot, J. Future flood losses in major coastal cities. *Nat. Clim. Chang.* **2013**, *3*, 802–806. [CrossRef]

29. Hirabayashi, Y.; Mahendran, R.; Koirala, S.; Konoshima, L.; Yamazaki, D.; Watanabe, S.; Kim, H.; Kanae, S. Global flood risk under climate change. *Nat. Clim. Chang.* **2013**, *3*, 816–821. [CrossRef]

30. Kundzewicz, Z.W.; Kanae, S.; Seneviratne, S.I.; Handmer, J.; Nicholls, N.; Peduzzi, P.; Mechler, R.; Bouwer, L.M.; Arnell, N.; Mach, K.; et al. Flood risk and climate change: Global and regional perspectives. *Hydrol. Sci. J.* **2014**, *59*, 1–28. [CrossRef]

31. Bubeck, P.; Botzen, W.; Aerts, J.C. A Review of Risk Perceptions and Other Factors that Influence Flood Mitigation Behavior. *Risk Anal.* **2012**, *32*, 1481–1495. [CrossRef] [PubMed]

32. Hudson, P.; Botzen, W.J.W. Cost–benefit analysis of flood-zoning policies: A review of current practice. *Wires Water* **2019**, *6*, 1387. [CrossRef]

33. Paez, A. Gray literature: An important resource in systematic reviews. *J. Evid. Based Med.* **2017**, *10*, 233–240. [CrossRef] [PubMed]

34. The Chronicle of Higher Education. Available online: https://www.chronicle.com/article/Googles-Book-Search-A/48245/ (accessed on 3 March 2020).

35. Stokes, D.E. *Pasteur's Quadrant: Basic Science and Technological Innovation*; Brookings Institution Press: Washington, DC, USA, 1997; p. 196.

36. DeMeritt, D. Harnessing Science and Securing Societal Impacts from Publicly Funded Research: Reflections on UK Science Policy. *Environ. Plan. A Econ. Space* **2010**, *42*, 515–523. [CrossRef]

37. Kind, J.M.; Botzen, W.J.W.; Aerts, J.C.J.H. Social vulnerability in cost-benefit analysis for flood risk management. *Environ. Dev. Econ.* **2020**, *25*, 115–134. [CrossRef]

38. Kuhlicke, C.; Masson, T.; Kienzler, S.; Sieg, T.; Thieken, A.; Kreibich, H. Multiple Flood Experiences and Social Resilience: Findings from Three Surveys on Households and Companies Exposed to the 2013 Flood in Germany. *Weather Clim. Soc.* **2020**, *12*, 63–88. [CrossRef]

39. Haer, T.; Husby, T.G.; Botzen, W.W.; Aerts, J.C. The safe development paradox: An agent-based model for flood risk under climate change in the European Union. *Glob. Environ. Chang.* **2020**, *60*, 102009. [CrossRef]

40. Fang, J.Y.; Lincke, D.; Brown, S.; Nicholls, R.J.; Wolff, C.; Merkens, J.-L.; Hinkel, J.; Vafeidis, A.T.; Shi, P.; Liu, M. Coastal flood risks in China through the 21st century—An application of DIVA. *Sci. Total Environ.* **2020**, *704*, 135311. [CrossRef] [PubMed]

41. Bloemendaal, N.; Haigh, I.D.; De Moel, H.; Muis, S.; Haarsma, R.J.; Aerts, J.C.J.H. Generation of a global synthetic tropical cyclone hazard dataset using STORM. *Sci. Data* **2020**, *7*, 1–12. [CrossRef]

42. Toimil, A.; Losada, I.J.; Nicholls, R.J.; Dalrymple, R.A.; Stive, M.J. Addressing the challenges of climate change risks and adaptation in coastal areas: A review. *Coast. Eng.* **2020**, *156*, 103611. [CrossRef]

43. Dunn, C.; Baker, P.; Fleming, M. Flood risk management with HEC-WAT and the FRA compute option. In Proceedings of the FLOODrisk 2016—3rd European Conference on Flood Risk Management, Lyon, France, 17–21 October 2016.

44. Scawthorn, C.; Blais, N.; Seligson, H.; Tate, E.; Mifflin, E.; Thomas, W.; Murphy, J.; Jones, C. HAZUS-MH Flood Loss Estimation Methodology. I: Overview and Flood Hazard Characterization. *Nat. Hazards Rev.* **2006**, *7*, 60–71. [CrossRef]

45. Scawthorn, C.; Flores, P.; Blais, N.; Seligson, H.; Tate, E.; Chang, S.; Mifflin, E.; Thomas, W.; Murphy, J.; Jones, C.; et al. HAZUS-MH flood loss estimation methodology, II. Damage and loss assessment. *Nat. Hazards Rev.* **2006**, *7*, 72–81. [CrossRef]

46. Ryu, J.; Yoon, E.J.; Park, C.; Lee, D.K.; Jeon, S.W. A Flood Risk Assessment Model for Companies and Criteria for Governmental Decision-Making to Minimize Hazards. *Sustainability* **2017**, *9*, 2005. [CrossRef]

MDPI

St. Alban-Anlage 66

4052 Basel

Switzerland

Tel. +41 61 683 77 34

Fax +41 61 302 89 18

www.mdpi.com

Water Editorial Office

E-mail: water@mdpi.com

www.mdpi.com/journal/water